# STELLAR ATMOSPHERES

A Series of Books in Astronomy and Astrophysics

EDITORS: *Geoffrey Burbidge and Margaret Burbidge*

Jacques E. SCHOUTENS
June 24, 02.

# STELLAR ATMOSPHERES

Dimitri Mihalas

Yerkes Observatory
The University of Chicago

W. H. FREEMAN AND COMPANY
San Francisco

Printed in the United States of America

Library of Congress Catalog Card Number 77-116897

International Standard Book Number 0-7167-0333-5

1 2 3 4 5 6 7 8 9

*To Joelen*

# Preface

The study of stellar atmospheres is in many ways one of the most interesting and rewarding areas of modern astrophysics. It is not an exaggeration to state that most of what we know about stars and systems of stars is derived from an analysis of their radiation and that this knowledge will be secure only as long as the analytical technique is physically reliable. It is therefore important to have a sound theoretical framework upon which our inferences can be based with confidence.

The field of stellar atmospheres enjoyed a period of rapid growth during the last decade. On the one hand, great improvements were made in the quantity and quality of the observational material. Not only did ground-based observations provide a continuing flow of data, but, in addition, observations from balloons, rockets, and satellites opened broad horizons hitherto completely hidden from us. On the other hand, enormous progress was made in the development of theory. A concerted effort by astronomers and physicists filled in many of the gaps in our understanding of the basic physical processes occurring in stellar atmospheres. The availability of large-capacity, high-speed computing machines stimulated the development of powerful new mathematical techniques and allowed their application to a wide range of cases. Thus, tremendous gains were made in enlarging and clarifying the formal and conceptual basis of the subject.

One of the unfortunate side effects of this period of growth is that practically all of the available textbooks in this field of astronomy are badly out of date. Students and instructors alike must now attempt to digest a large body of scattered literature in order to learn of recent developments. There is, in my mind, a definite need for a new text on the subject, and the present book is an attempt to provide such a text on an introductory graduate level. It is based upon courses I have given to first- and second-year graduate students at Princeton University, the University of Colorado, and the University of Chicago. It represents what I feel is a minimum background for a student who wishes to understand the literature and to do research in the field. Naturally, it has been necessary to be selective in the material presented. In writing this book, I had in mind the goal of providing a basic synopsis of the theory that can be covered in two quarters, with the hope that the content of the third quarter of the normal academic year will be drawn by the instructor (and the students) from the current literature on topics of special interest to them. Although emphasis is given to the more modern approaches, I have also attempted to give a coherent review of the older methods and results. I feel it is important for students to be familiar with these classical approaches so that they will be aware of the limitations of such approaches and the conclusions based upon them.

It has been tempting to include a wider range of subjects, but I have avoided doing so in the belief that it is more worthwhile for the student to consider a smaller number of topics in depth than attempt to survey the entire field superficially. In this vein, I have purposely limited the comparison of theory with observation to a few of the more crucial and illustrative examples. Moreover, I have restricted most of the theoretical discussion to what may be called the *classical stellar atmospheres* problem—i.e., atmospheres in hydrostatic, radiative, and steady-state statistical equilibrium. This is ample material for a two-quarter course and is understood well enough to require little speculation. Even within this problem, I have limited the variety of techniques treated. For example, I personally favor using differential equations over using integral equations to solve transfer problems. Thus, although the latter method has enjoyed wide application and good success, particularly in the hands of the Harvard-Smithsonian Astrophysical Observatory group, there is little discussion of it in this book. This omission is not arbitrary, however, but is based upon the view that since the two methods are mathematically equivalent, discussion of one suffices and, in addition, that the one I have chosen seems to offer more promise in future applications—for example, to situations involving hydrodynamics (wherein lies the real frontier of the subject). On the other hand, in my experience, the physics background of astronomy students is often uneven; I have, therefore, not hesitated to develop those aspects of physical theory that are of special interest to the armospheres problem. In any case, I hope that users of this book will find it a helpful outline, which they can edit, alter, and enlarge upon as their needs dictate.

I wish to express my sincere thanks to a number of astronomers for helpful discussions about problems of stellar atmospheres and line formation, particularly to L. H. Auer, E. H. Avrett, J. L. Greenstein, D. G. Hummer, J. T. Jefferies, W. Kalkofen, G. Münch, J. B. Oke, G. Rybicki, the late J. C. Stewart, S. E. Strom, R. N. Thomas, and A. B. Underhill. I wish also to thank J. R. Cooper for comments on Chapter 9. Additional thanks are due to Hummer and Stewart for specific suggestions concerning the content and organization of this book and to L. G. Henyey for reading the manuscript and a number of helpful comments and corrections.

My greatest debt is to Lawrence H. Auer, with whom I have engaged in an enjoyable and fruitful collaboration over the past few years. The innumerable comments, discussions, and arguments he has offered have been of the greatest value.

Finally, I wish to thank my wife, Joelen, for her patience and skill in typing heavily edited intermediate drafts and for her encouragement.

*Williams Bay, Wisconsin* DIMITRI MIHALAS
*November 1969*

# Contents

# 1 | The Equation of Transfer

The basic goal of the theory of stellar atmospheres is to describe the flow of energy through the outermost layers of a star and to predict the observational characteristics of the emergent radiation. In the development of this theory, we may often treat the radiation field and its interaction with the material in the atmosphere from an essentially macroscopic point of view. We usually will not need to consider the detailed quantum or electromagnetic properties of radiation, unless these aspects happen to be of primary importance in understanding the physics of the situation under consideration. Thus, we will characterize the field in terms of quantities such as intensity, flux, and energy density—as a function of frequency.

In this chapter we shall introduce the basic definitions required in our analysis and formulate the fundamental equations describing radiation transport. We shall also state the restrictions and simplifications that define the problem which we consider the central theme of this book.

## 1-1. Basic Definitions

### THE SPECIFIC INTENSITY

We define the *specific intensity* $I_\nu(\mathbf{r}, \mathbf{n}, t)$ at position $\mathbf{r}$, traveling in direction $\mathbf{n}$ at time $t$, as the amount of energy, per unit frequency interval, passing through a unit area oriented normal to the beam, into a unit solid angle, in a unit time (see Figure 1-1). Thus if $\cos\theta$ denotes the angle between the normal $\hat{\mathbf{s}}$ of the reference surface $dA$ and the direction $\mathbf{n}$, then the energy passing through $dA$ is

$$dE = I_\nu(\mathbf{r}, \mathbf{n}, t)\, dA \cos\theta\, d\nu\, d\omega\, dt \tag{1-1}$$

For our present purposes, the specific intensity may be regarded as the entire description of the photon distribution function inasmuch as it contains complete information about the spatial time, angular, and frequency dependences of the radiation. Since we will consider in this book only time-independent problems in plane parallel geometry, we will drop the specification of $t$ and replace $\mathbf{r}$ and $\mathbf{n}$ by the simpler geometric description $(z, \theta, \varphi)$, where $z$ is the distance measured with respect to some prechosen reference level (e.g., the upper boundary of a stellar atmosphere), and $\theta$ and $\varphi$ are polar and azimuthal angles, respectively, measured relative to the normal (see Figure 1-2). It is

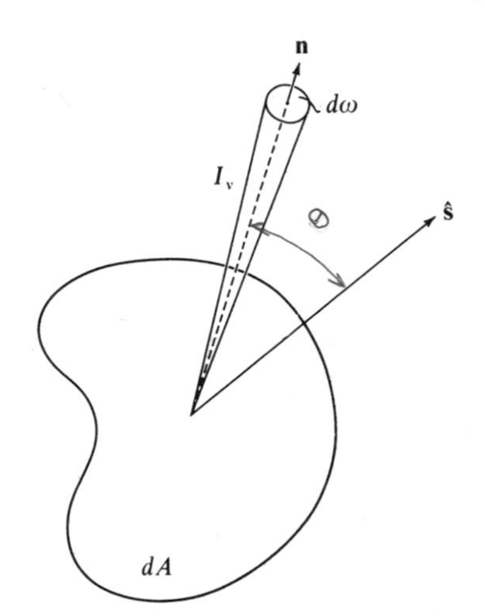


FIG. 1-1. Definition of specific intensity.

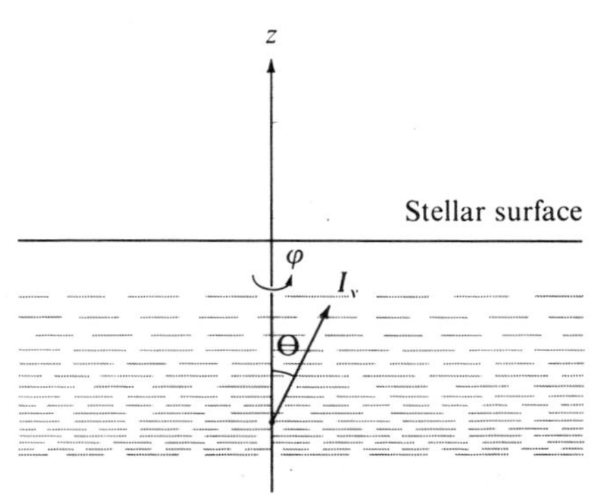


FIG. 1-2. Plane-parallel atmosphere.

customary to measure $z$ positive *upward* in the atmosphere (i.e., away from the center of the star); if the upper boundary is chosen as the reference level, $z$ will be positive outside the star and increasingly negative at deeper layers in the star.

### THE MEAN INTENSITY

The *mean intensity*, $J_\nu(z)$, is defined as the simple average of $I$ over all solid angles, i.e.,

$$J_\nu(z) = \frac{1}{4\pi} \oint I_\nu(z, \theta, \varphi)\, d\omega$$

$$= \frac{1}{4\pi} \int_0^{2\pi} d\varphi \int_0^{\pi} I_\nu(z, \theta, \varphi) \sin\theta\, d\theta \tag{1-2}$$

In all applications in this book, we will assume $I_\nu$ is independent of the azimuthal angle $\varphi$, and writing $\mu = \cos\theta$, we may simplify equation (1-2) to read

$$J_\nu(z) = \frac{1}{2} \int_{-1}^{1} I_\nu(z, \mu)\, d\mu \tag{1-3}$$

The assumption that there is no azimuthal dependence of the radiation field is in essence equivalent to assuming that the plane-parallel layers of the atmosphere are homogeneous. (We shall discuss the significance of this approximation later in this chapter.)

### THE FLUX

We define the *flux* as the net rate of energy flow across a unit area in the atmosphere. Quite generally the flux is a vector

$$\boldsymbol{\mathscr{F}}_\nu = \oint I_\nu(\mathbf{r}, \mathbf{n})\mathbf{n}\, d\omega \tag{1-4}$$

If we choose the coordinate system described above and ask for the flux passing through one of the plane surfaces, we may write

$$\mathscr{F}_\nu = \boldsymbol{\mathscr{F}}_\nu \cdot \hat{\mathbf{k}} = \oint I_\nu(z, \theta, \varphi) \cos\theta\, d\omega$$

$$= 2\pi \int_{-1}^{1} I_\nu(z, \mu)\mu\, d\mu \tag{1-5}$$

where in the last step we have again assumed no azimuthal variation of $I_\nu$. It is customary to absorb the factor $\pi$ into the definition of $\mathscr{F}_\nu$ and write the *astrophysical flux* $F_\nu$ as

$$F_\nu \equiv \frac{1}{\pi}\mathscr{F}_\nu = 2\int_{-1}^{1} I_\nu(\tau, \mu)\mu\, d\mu \tag{1-6}$$

The flux $F_\nu$ has an important connection with the energy received from a star by an observer. Consider the geometry shown in Figure 1-3. Assume the distance $D$ between star and observer is very much larger than the stellar

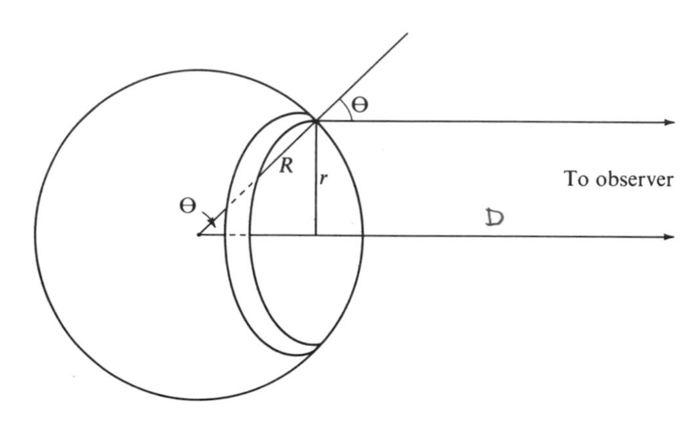


FIG. 1-3. Geometry of measurement of stellar flux. The annulus on the surface of the star has an area $2\pi r dr = 2\pi R^2 \sin\theta \cos\theta\, d\theta$ normal to the line of sight; this area subtends a solid angle $d\omega = 2\pi\,(R^2/D^2) \sin\theta \cos\theta\, d\theta$ as seen by the observer.

radius so that all rays emergent from the star and received by the observer are essentially parallel. Now the flux received by the observer is $df_\nu = I_\nu\, d\omega$, where $d\omega$ is the solid angle subtended by a differential area on the star's disk, as seen by the observer, and $I_\nu$ is the specific intensity emergent at the surface of the star. Noting that $r = R \sin\theta$, we can write the area of a differential annulus in the disk as $dA = 2\pi r\, dr = 2\pi R^2 \mu\, d\mu$ so that $d\omega = 2\pi(R/D)^2 \mu\, d\mu$. To the observer the radiation from this annulus emerges at an angle $\theta$ relative to the normal to the star's surface; thus the appropriate value of the specific intensity is $I_\nu(0, \mu)$. Integrating over the surface of the star, we then obtain

$$f_\nu = 2\left(\frac{R}{D}\right)^2 \pi \int_0^1 I_\nu(0, \mu)\mu\, d\mu = \left(\frac{R}{D}\right)^2 \mathscr{F}_\nu \tag{1-7}$$

In this calculation we have assumed there is no incident radiation upon the surface of the star, i.e., $I_\nu(0, -\mu) \equiv 0$. For stars, which we cannot resolve as disks, the quantity we observe is $f_\nu$. The sun, however, is close enough to be resolved, and we can therefore obtain $I_\nu(0, \mu)$ as well as $f_\nu$.

### MOMENTS OF THE RADIATION FIELD

We have seen above that we find certain angular integrals over the radiation field to be useful. We may define quite generally the $n$th *moment* over the radiation field as

$$M_\nu(z, n) \equiv \frac{1}{2}\int_{-1}^{1} I_\nu(z, \mu)\mu^n \, d\mu \tag{1-8}$$

Following Eddington, it is customary to use a special notation for the moment of order zero, the mean intensity,

$$J_\nu(z) = \frac{1}{2}\int_{-1}^{1} I_\nu(z, \mu) \, d\mu \tag{1-9}$$

the moment of order one, the *Eddington flux*,

$$H_\nu(z) = \frac{1}{2}\int_{-1}^{1} I_\nu(z, \mu)\mu \, d\mu \tag{1-10}$$

and the moment of order two, the so-called $K$-integral,

$$K_\nu(z) = \frac{1}{2}\int_{-1}^{1} I_\nu(z, \mu)\mu^2 \, d\mu \tag{1-11}$$

Moments of arbitrarily high order may be constructed, and certain theories of radiative transport employ them. In the developments described in this book, we will deal only with the moments $J$, $H$, and $K$.

### INVARIANCE OF THE SPECIFIC INTENSITY

One important property of the specific intensity is that it has been defined in such a way as to be independent of the distance between the source and the observer, in the absence of sources or sinks of energy along the line of sight. Thus, consider that pencil of rays which passes through both area $dA$ at $P$ and $dA'$ at $P'$ (see Figure 1-4). Then the amount of energy $dE_\nu$ passing through *both* areas can be written as

$$dE_\nu = I_\nu \, dA \cos\theta \, d\omega \, d\nu \, dt = dE_\nu' = I_\nu' \, dA' \cos\theta' \, d\omega' \, d\nu \, dt$$

where $d\omega$ is the solid angle subtended by $dA'$, as seen from $P$, and $d\omega'$ is the solid angle subtended by $dA$, as seen from $P'$. Now from Figure 1-3 we see that $d\omega = dA' \cos\theta'/r^2$ while $d\omega' = dA \cos\theta/r^2$ so that

$$dE_\nu = I_\nu \, dA \cos\theta \, \frac{dA' \cos\theta'}{r^2} \, d\nu \, dt = I_\nu' \, dA' \cos\theta' \, \frac{dA \cos\theta}{r^2} \, d\nu \, dt \tag{1-12}$$

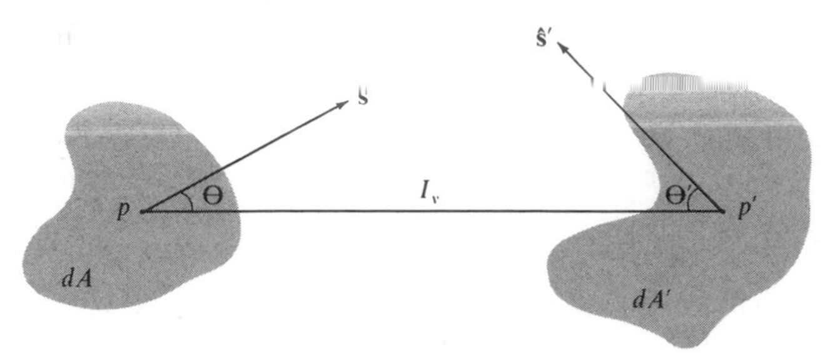


FIG. 1-4. Geometry used in proof of invariance of specific intensity. Area $dA$ subtends a solid angle $d\omega'$ at $P'$, and the area $dA'$ subtends a solid angle $d\omega$ at $P$.

from which it follows that $I_\nu = I_\nu'$. We also note that equation (1-12) shows that the energy per unit area falls as the inverse square of the separation of $P$ and $P'$, a fact also contained in equation (1-7).

THE ENERGY DENSITY

Consider an infinitesimal volume $V$ through which energy flows from all solid angles. The amount flowing from a specific solid angle $d\omega$ through an element of area $dA$ of the volume is

$$\delta E_\nu = I_\nu \, dA \cos\theta \, d\omega \, d\nu \, dt \tag{1-13}$$

Now consider only those photons in flight across $V$; if their path length while in $V$ is $l$, then the time they will be contained within $V$ is $dt = l/c$. Moreover, $l \, dA \cos\theta = dV$, the differential volume through which they sweep. Thus we may rewrite equation (1-13) in the form $\delta E_\nu = (1/c) I_\nu \, d\omega \, d\nu \, dV$, and by integrating over the entire volume, we have the total energy contained within it:

$$E_\nu \, d\nu = \frac{1}{c} \int_V dV \oint d\omega \, I_\nu \, d\nu \tag{1-14}$$

But if we pass to the limit as the size of the volume element becomes vanishingly small, we may assume that $I_\nu$ is independent of position in $V$; then the integrations can be carried out separately. Finally, writing the *energy density* $U_\nu$ as $U_\nu = E_\nu/V$, we have

$$U_\nu = \frac{1}{c} \oint I_\nu \, d\omega = \frac{4\pi}{c} J_\nu \tag{1-15}$$

PHOTON MOMENTUM TRANSPORT

Consider now the rate of momentum transport by the radiation field across a differential surface $dA$. The amount of momentum per photon is

$h\nu/c$. Thus, the amount of momentum to be associated with a pencil of energy $dE_\nu$ is $dE_\nu/c$. Taking now only the component perpendicular to the surface, we have per unit area and per unit time,

$$dp_R(\nu) = \frac{1}{dA}\frac{dE_\nu \cos\theta}{c} = \frac{(I_\nu \, dA \cos\theta \, d\omega)\cos\theta}{c \, dA} = \frac{I_\nu \cos^2\theta \, d\omega}{c} \tag{1-16}$$

Integrating over all solid angles, we obtain

$$p_R(\nu) = \frac{1}{c}\oint I_\nu \cos^2\theta \, d\omega = \frac{4\pi}{c} K_\nu \tag{1-17}$$

The quantity $p_R(\nu)$ is called the *radiation pressure*, but is not to be confused with the mechanical force that would be exerted by photons on an absorbing screen (we shall calculate this force in Section 1-3 below).

Note that for an *isotropic* radiation field, $I_\nu(\mu) \equiv I_\nu$; therefore we may rewrite equation (1-17) as

$$p_R(\nu) = \frac{4\pi}{c}\frac{I_\nu}{3}$$

while from equation (1-15) the energy density becomes

$$U_\nu = \frac{4\pi I_\nu}{c}$$

Thus, in the case of an isotropic radiation field $p_R(\nu) = \frac{1}{3}U_\nu$, and in addition $J_\nu = 3K_\nu$. We shall consider the significance of this result again at a later time.

## 1-2. Absorption and Emission Coefficients

### ABSORPTION VERSUS SCATTERING PROCESSES

For the present, we will consider the absorption and emission of radiation to be characterized by macroscopic coefficients and inquire into the atomic constituents of these processes in Chapter 4. In order to define these coefficients properly, we should make a distinction at the outset between the *true absorption* of radiation and the process of *scattering*. This distinction is important since as we shall see again and again as the theory develops, the effective interaction between the gas composing the atmosphere and the radiation field is quite different (in important physical ways) in these two cases. In the final analysis, it is not possible to separate these two processes in a strictly rigorous way, nor would it be important or useful to do so. We shall, therefore, merely give examples of typical occurrences of these two mechanisms with the intent of providing at least a conceptual comparison between the two.

We will identify *scattering* processes as those in which the photon interacts with the scatterer and emerges in a new direction with (perhaps) a slightly altered frequency. In this process the photon is not destroyed in the sense of being converted into kinetic energy of particles. By way of contrast, we will identify *absorption* processes as those in which the photon is destroyed and its energy is converted (at least partly) into kinetic energy of the particles composing the gas. The important point to recognize here is that scattering processes depend basically upon the *radiation field* at the point under consideration and have only a weak connection with other thermodynamic properties of the gas, such as local temperatures. Absorption processes, on the other hand, feed photon energy directly from the radiation field into the thermal kinetic energy of the gas and thus are more intimately connected with the local thermodynamic properties of the medium. Similarly, the inverse of absorption, *thermal emission*, directly couples the thermal state of the gas into the radiation field.

Examples of scattering processes are the following.

(a) Interaction with a photon, raising a bound electron of an atom from energy level $a$ to energy level $b$, followed by a direct radiative return of the electron from level $b$ to level $a$ with the emission of a photon. We should note in passing that since the atom is not really an isolated system but is imbedded in a plasma, levels $a$ and $b$ will normally have finite energy widths, and hence the emitted photon will, in general, have a slightly different frequency from the incident photon. We may thus often consider the bound levels to consist of a distribution of substates (or sublevels). When necessary, we will analyze transitions among substates, both between two levels and within a given level.

(b) Scattering of a photon by a free electron (Thomson scattering) or by a molecule (Rayleigh scattering).

The above selection is by no means exhaustive, but it contains the cases of principle interest to us.

Let us now list a few examples of absorption processes. Note that each process listed has a direct inverse, leading to the (thermal) emission of a photon.

(a) A photon is absorbed by an atom in a bound state, causing removal of the bound electron into the continuum. This process is called *photoionization* or *bound-free absorption*. The excess of the photon energy over the binding energy of the electron goes into the free electron's kinetic energy. The free electrons subsequently interact by collisions and establish a thermal velocity distribution. The inverse process of an electron passing from the continuum to a bound atomic state is referred to as direct *radiative recombination*.

(b) A photon is absorbed by a free electron moving in the field of an ion, resulting in an alteration of the electron's energy relative to the ion. The electron then, classically speaking, moves off on a different (hyperbolic) orbit around the ion. This process is known as *free-free absorption* since the electron is unbound both before and after absorbing the photon. The inverse process, leading to the emission of a photon, is referred to as *bremsstrahlung*.

(c) A photon is absorbed by an atom, leading to a transition of an electron from one bound state to another; this process is called *photoexcitation*. The excited atom is then de-excited by an inelastic collision with another particle. Energy is put into kinetic energy of the atom and collision partner and thereby ultimately ends up as part of the thermal pool. The photon is said to have been destroyed by a *collisional de-excitation*. The inverse process leads to the collisional creation of a photon at the expense of the thermal energy of the gas.

(d) Photoexcitation of an atom with subsequent *collisional ionization* of the excited electron into the continuum. Photon energy again contributes to the thermal energy of particles. The inverse process is referred to as *collisional recombination*.

The above lists are meant to be only illustrative not exhaustive. A variety of other types of processes may occur, e.g., two successive radiative processes. In many such cases a clear division into "scattering" versus "absorption" categories is no longer possible.

Bearing in mind the examples offered above, let us now define an *absorption coefficient* $\kappa_\nu$, per gram of stellar material, such that a differential element of material of cross-section $dA$ and length $ds$ absorbs from a beam of specific intensity $I_\nu$, incident normal to the ends of the element, an amount of energy

$$dE_\nu = \kappa_\nu I_\nu \, d\omega \, d\nu \, dt \, \rho \, dA \, ds \tag{1-18}$$

Similarly, we may define a *scattering coefficient* $\sigma_\nu$ such that the amount of energy scattered out of the beam under the above circumstances is

$$dE_\nu = \sigma_\nu I_\nu \, d\omega \, d\nu \, dt \, \rho \, dA \, ds \tag{1-19}$$

In the cases of interest to us in this book, both $\kappa_\nu$ and $\sigma_\nu$ may be assumed to have no dependence upon angle. The combined effects of both absorption and scattering processes in removing energy from the beam can be described by their sum $k_\nu = \kappa_\nu + \sigma_\nu$, which is referred to as the total *extinction coefficient*. When we know the values of $\kappa_\nu$ and $\sigma_\nu$ (the details of their calculation will be considered in Chapter 4), we have a complete macroscopic description of the rate at which energy may be removed from the beam of radiation.

## EMISSION PROCESSES

Let us now consider the emission of energy. We define the *emission coefficient*, $j_\nu$, such that the total energy returned to the beam into solid angle $d\omega$ is

$$dE_\nu = j_\nu \, d\omega \, d\nu \, dt \, \rho \, dA \, ds \tag{1-20}$$

In light of our previous discussion, we may from a phenomenological point of view regard the emission coefficient as consisting of a thermal term and a scattering term. Let us consider first the thermal emission. (Excellent further discussion may be found in the article by Milne in Ref. 24, pp. 93–98.)

If the material were in a cavity of uniform temperature $T$ with perfectly absorbing and emitting walls and if the material is assumed to be in equilibrium with the radiation field, then the requirement that as much energy be emitted as is absorbed demands that

$$j_\nu{}^t = \kappa_\nu B_\nu(T) \tag{1-21}$$

where $B_\nu(T)$ is the well-known Planck function which describes the photon distribution in the case of strict thermodynamic equilibrium ($TE$). Equation (1-21), the *Kirchhoff-Planck law*, is often adopted as an expression for the rate of thermal emission. Note that both thermal absorption and emission is independent of angle. However, the usual situation in a stellar atmosphere is not the one described above.

To begin with, we recognize that the simple fact that radiation emerges from the surface of the star into essentially empty space implies energy transport, which in turn (as we will see in detail later) implies the existence of a temperature gradient. Thus, we recognize immediately that we are not considering a closed system of uniform temperature but rather quite the opposite situation. In particular, at the very surface of the star we will encounter an extremely anisotropic radiation field. Moreover, as we shall see later, the atmospheric material may be much more transparent at some frequencies than at others so that the emergent radiation at different frequencies may effectively originate at very different physical depths in the atmosphere. If this emergent radiation in some sense reflects the temperature of the characteristic region from which it originates and there is, as mentioned above, a temperature gradient in the atmosphere, then it is evident that the radiation at the surface must necessarily be non-Planckian since it will be a superposition of contributions from points in the atmosphere that may have very different properties. In short, in the outermost layers of a star, none of the assumptions leading to equation (1-21) are valid, and we should not expect this relation to hold true in general.

Nevertheless, it is often *assumed* that the occupation numbers of the atoms, the opacity, the emission, and indeed all thermodynamic properties of a small volume of the material in the atmosphere are the same as their thermodynamic equilibrium values at the *local* values of the temperature $T$ and the electron density $N_e$. This assumption is known as the *local thermodynamic equilibrium* (LTE) approximation. Later we will discuss the conditions under which we expect LTE to be a valid assumption and then develop a general theory to replace this assumption when it fails. In the meantime, we will sometimes use LTE as a computational expedient since it so greatly simplifies the calculation of atmospheric properties and thus provides a convenient introduction into the basic techniques of stellar atmospheres theory. We must always bear in mind, however, that in general LTE *cannot* be an accurate approximation and that in the end we must carry through the general analysis which allows us to determine the coupled thermodynamic state of the gas and radiation field.

Let us now consider the radiation emitted by scattering. As described above, in scattering processes both the direction and the frequency of the photon may change. We may describe these changes by means of a *redistribution function*

$$R(\nu', \mathbf{n}'; \nu, \mathbf{n})\, d\nu'\, d\nu \frac{d\omega'}{4\pi} \frac{d\omega}{4\pi}$$

which gives the probability that a photon will be scattered from the solid angle $d\omega'$ and frequency range $(\nu', \nu' + d\nu')$ into the solid angle $d\omega$ and frequency range $(\nu, \nu + d\nu)$. We shall discuss the derivation and properties of redistribution functions in Chapter 10, but it is helpful to mention a few general properties of these functions at the present time. We assume R is normalized such that

$$\oint\oint \int_0^\infty \int_0^\infty R(\nu', \mathbf{n}'; \nu, \mathbf{n})\, d\nu'\, d\nu \frac{d\omega'}{4\pi} \frac{d\omega}{4\pi} = 1 \tag{1-22}$$

The redistribution function may be used to define a normalized scattering profile $\varphi(\nu')$ and a normalized emission profile $\psi(\nu)$ for the scattering process. Thus, from the physical meaning of the redistribution function, it is evident that if we integrate over all angles and emitted frequencies, we must obtain the scattering profile

$$\varphi(\nu')\, d\nu' = d\nu' \oint\oint \int_0^\infty R(\nu', \mathbf{n}'; \nu, \mathbf{n})\, d\nu \frac{d\omega}{4\pi} \frac{d\omega'}{4\pi}$$

which, by virtue of equation (1-22), is normalized such that

$$\int_0^\infty \varphi(\nu')\, d\nu' = 1$$

We may then write

$$\sigma_{\nu'} = \sigma\varphi(\nu')$$

where $\sigma$ represents a total scattering cross-section. Similarly, integration over all angles and absorbed frequencies must yield the emission profile

$$\psi(\nu)\, d\nu = d\nu \oint\oint \int_0^\infty R(\nu', \mathbf{n}'; \nu, \mathbf{n})\, d\nu' \frac{d\omega'}{4\pi} \frac{d\omega}{4\pi}$$

which again is normalized such that

$$\int_0^\infty \psi(\nu)\, d\nu = 1$$

The energy scattered by unit amount of material is then

$$dE_\nu = \sigma \, d\nu \, d\omega \oint \int_0^\infty I_{\nu'}(\mathbf{n}')R(\nu', \mathbf{n}'; \nu, \mathbf{n}) \, d\nu' \frac{d\omega'}{4\pi} = j_\nu{}^s \, d\nu \, d\omega \tag{1-23}$$

Normally the scattering of radiation is not treated in such generality. For example, if we are primarily interested in the redistribution of the radiation in frequency but not in angle, then we may define an angle-averaged redistribution function

$$R(\nu', \nu) \, d\nu' \, d\nu = d\nu' \, d\nu \oint\oint R(\nu', \mathbf{n}'; \nu, \mathbf{n}) \frac{d\omega'}{4\pi} \frac{d\omega}{4\pi} \tag{1-24}$$

Then the amount of energy scattered into $d\omega$ by unit amount of material is

$$\begin{aligned} j_\nu{}^s \, d\nu \, d\omega &= \sigma \, d\nu \, d\omega \int_0^\infty R(\nu', \nu) \, d\nu' \oint I_{\nu'}(\mathbf{n}') \frac{d\omega'}{4\pi} \\ &= \sigma \, d\nu \, d\omega \int_0^\infty R(\nu', \nu) J_{\nu'} \, d\nu' \end{aligned} \tag{1-25}$$

This approximation is a very useful one in line transfer problems. To give a bit more explicit example, suppose that there is no correlation between the frequency of the incoming and scattered photons but that their frequencies are independently distributed over the line profile $\varphi(\nu)$. This situation is referred to as *complete redistribution or complete noncoherence*. This would be a good approximation, for instance, when the scattering atoms are so strongly perturbed by collisions during the scattering process that the excited electrons are randomly redistributed over substates of the upper level. In this case, both the absorption and emission probabilities would be proportional to the number of substates at each frequency within the line, i.e., $\varphi(\nu)$ itself. Then we can write $R(\nu', \nu) = \varphi(\nu')\varphi(\nu)$, and the energy scattered by a unit amount of material is

$$\begin{aligned} j_\nu{}^s \, d\nu \, d\omega &= \sigma\varphi(\nu) \, d\nu \, d\omega \int_0^\infty \varphi(\nu') J_{\nu'} \, d\nu' \\ &= \sigma_\nu \, d\nu \, d\omega \int_0^\infty \varphi(\nu') J_{\nu'} \, d\nu' \end{aligned} \tag{1-26}$$

We note here that in this case *the emission profile is identical to the absorption profile.*

Another class of problems arises when we assume that the scattering is *coherent*, i.e., the incoming and outgoing photons have precisely the same frequency, though possibly redistribution in angle occurs. Then we might write

$$R(\nu', \mathbf{n}'; \nu, \mathbf{n}) = g(\mathbf{n}', \mathbf{n})\varphi(\nu')\delta(\nu - \nu') \tag{1-27}$$

where $\delta$ is the well-known Dirac function and $g$ is an angular phase function normalized such that

$$\oint g(\mathbf{n}', \mathbf{n}) \frac{d\omega'}{4\pi} = 1 \tag{1-28}$$

in Chandrasekhar p. 6

$\int p(\cos\Theta) \frac{d\omega'}{4\pi} = \varpi_0 \leq 1$

Two important phase functions are

$$g(\mathbf{n}', \mathbf{n}) \equiv 1 \tag{1-29}$$

which corresponds to *isotropic* scattering, and

$$g(\mathbf{n}', \mathbf{n}) = \tfrac{3}{4}(1 + \cos^2 \Theta) \tag{1-30}$$

where $\cos \Theta = \mathbf{n}' \cdot \mathbf{n}$, which is the phase function for dipole scattering. In this case equation (1-23) reduces to

$$j_\nu^s \, d\nu \, d\omega = \sigma_\nu \, d\nu \, d\omega \oint I_\nu(\mathbf{n}') g(\mathbf{n}', \mathbf{n}) \frac{d\omega'}{4\pi} \tag{1-31}$$

In the case of a spectrum line, coherent scattering would occur only if the lower level of the line were completely sharp and if the scattering atoms were at rest in the observer's frame. This is almost never the case, and, in fact, scattering in a line is much more accurately described by complete redistribution than by coherence. In the case of continuum scattering (e.g., by electrons), the assumption of coherence is not as bad since the frequency distribution of continuum radiation is smooth and essentially constant over typical frequency shifts caused by the scattering process. For this reason, continuum scattering processes are customarily treated as if they were coherent (though this may not be adequate if the profile of a spectral line is to be calculated). Moreover, as angular redistribution effects for scattering with a dipole phase function are usually small, it is customary to assume that continuum scattering is isotropic and to write the energy scattered as

$$j_\nu^s \, d\nu \, d\omega = \sigma_\nu J_\nu \, d\nu \, d\omega \tag{1-32}$$

It should be recognized that the discussion given in this section is purposely heuristic. A rigorous treatment of the form of absorption and emission terms requires a more detailed examination of the microscopic processes taking place and an appeal to the *equations* of *statistical equilibrium* (Chapter 5). We shall follow this more comprehensive approach in Chapters 7, 12, 13, and 14; the expressions given in this section will serve in the meantime to demonstrate the solution of typical transfer problems.

## 1-3. The Mechanical Force Exerted by Radiation

Before we consider the problem of transfer of radiation, let us discuss a neglected detail mentioned previously, namely, the force exerted by radiation upon absorbing material. Consider a plane slab of normal thickness $dz$, density $\rho$, and with absorption and scattering coefficients $\kappa_\nu$ and $\sigma_\nu$. If photons enter the slab at an angle $\theta$ to the normal (see Figure 1-5), an amount of energy

$$dE_\nu = I_\nu \, d\nu \, d\omega \, dt(dA \cos\theta)\rho(\kappa_\nu + \sigma_\nu) \, ds \tag{1-33}$$

where $ds = dz/\cos\theta$, will be removed from the beam. As in the derivation of equation (1-17), we multiply by $1/c$ to obtain the momentum associated with this energy and by a factor of $\cos\theta$ to obtain the normal component so that the momentum transferred (in unit time per unit area) from the photons to the material is

$$\frac{1}{c}(\kappa_\nu + \sigma_\nu)I_\nu \, d\nu \, d\omega \cos^2\theta \, \rho \, ds = \frac{1}{c}(\kappa_\nu + \sigma_\nu)I_\nu \, d\nu \, d\omega \cos\theta \, \rho \, dz \tag{1-34}$$

Integrating over all frequency and solid angles, we obtain the total rate of change of momentum per unit area on the material, which can be written as the differential pressure $dp_R$ exerted by the radiation upon the material. Thus, from Figure (1-5) and equation (1-34) we have

$$p_R(z) - \left[p_R(z) + \frac{dp_R}{dz}\,dz\right] = \frac{1}{c}\rho \, dz \int_0^\infty (\kappa_\nu + \sigma_\nu) \, d\nu \oint I_\nu \cos\theta \, d\omega$$

which, using equation (1-5), reduces to

$$\frac{dp_R}{dz} = -\frac{\rho\pi}{c}\int_0^\infty (\kappa_\nu + \sigma_\nu)F_\nu \, d\nu \tag{1-35}$$

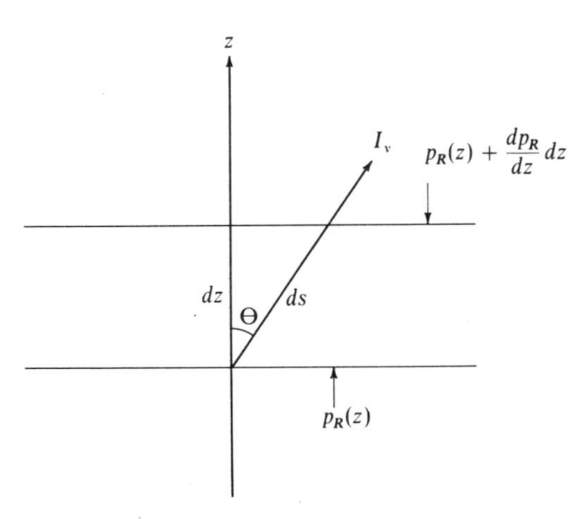


FIG. 1-5. Geometry used in calculation of mechanical force exerted by the radiation field.

The minus sign is due to our choice that $z$ increase outward in the atmosphere. Thus, as we proceed inward into the atmosphere, the mechanical force exerted by the radiation increases and acts in opposition to the gravitational force. Note here that we have assumed that the emitted radiation is isotropic so that there is no net recoil upon emission. We shall make use of equation (1-35) in our later discussions of the determination of the density structure of the atmosphere.

## 1-4. The Equation of Transfer

### THE SOURCE FUNCTION

Let us now examine the problem of radiative transport. Consider an infinitesimal volume element of length $ds$ and a normal cross-section $dA$ containing matter of density $\rho$ (see Figure 1-6). Then the energy change in the pencil of radiation passing through this element must equal the energy emitted by the element minus the energy absorbed. Thus

$$dI_\nu \, d\nu \, d\omega \, dA \, dt - j_\nu(\rho \, dA \, ds) \, d\nu \, d\omega \, dt - k_\nu I_\nu(\rho \, dA \, ds) \, d\nu \, d\omega \, dt \qquad (1\text{-}36)$$

or

$$\frac{\mu}{\rho}\frac{dI_\nu}{dz} = j_\nu - k_\nu I_\nu \qquad (1\text{-}37)$$

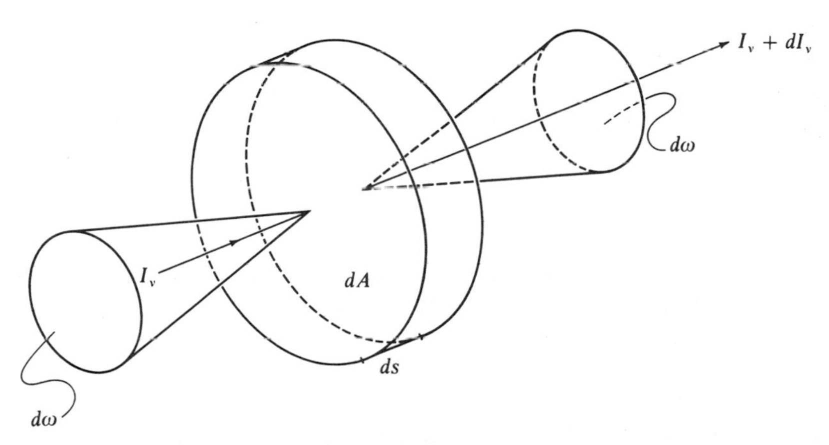


FIG. 1-6. Change in specific intensity across an element of material is the difference between energy emitted and energy absorbed.

where we have made use of the fact that $dz = \mu\, ds$ for a plane-parallel atmosphere. We may now define an *optical depth scale* $\tau_\nu$ by the relation

$$d\tau_\nu = -\mu k_\nu \, dz \tag{1-38}$$

(The minus appears because again we measure $z$ positively outward and want $\tau_\nu$ to run positive inward.) The optical depth scale is clearly of more fundamental importance in the description of the radiation field than the geometric depth because it is a direct measure of the absorptivity of the material along the pencil of radiation under consideration. In addition, we define the ratio of emissivity to opacity to be the *source function*

$$S_\nu \equiv \frac{j_\nu}{k_\nu} \tag{1-39}$$

The equation of transfer then assumes its standard form

$$\mu \frac{dI_\nu}{d\tau_\nu} = I_\nu - S_\nu \tag{1-40}$$

Note in passing that the $k_\nu$ in the above equations is understood to be the total extinction; i.e., if scattering and absorption are present, then we have $k_\nu = \kappa_\nu + \sigma_\nu$; if lines with opacity $l_\nu$ also contribute, then $k_\nu = \kappa_\nu + \sigma_\nu + l_\nu$. From the discussion in section 1-2 we may write prototype expressions for the source function. Thus, if we assume that absorption processes yield an LTE thermal emission term as given by equation (1-21) and a general scattering emission term as given by equation (1-23), the total emission coefficient can be written schematically as

$$j_\nu = j_\nu^t + j_\nu^s = \kappa_\nu \mathrm{B}_\nu + \sigma \oint \int_0^\infty I_{\nu'}(\mathbf{n}')R(\nu' \, \mathbf{n}'; \nu, \mathbf{n}) \, d\nu' \, \frac{d\omega'}{4\pi}$$

so that the source function is simply

$$S_\nu = \frac{\kappa_\nu}{\kappa_\nu + \sigma_\nu} \mathrm{B}_\nu + \frac{\sigma}{\kappa_\nu + \sigma_\nu} \oint \int_0^\infty I_{\nu'}(\mathbf{n}')R(\nu' \, \mathbf{n}'; \nu, \mathbf{n}) \, d\nu' \, \frac{d\omega'}{4\pi} \tag{1-41}$$

In strict LTE with no scattering, equation (1-41) simplifies to

$$S_\nu = B_\nu \tag{1-42}$$

For LTE thermal emission plus coherent isotropic scattering, the source function may be written, using equation (1-32), as

$$S_\nu = \frac{\kappa_\nu}{\kappa_\nu + \sigma_\nu} \mathrm{B}_\nu + \frac{\sigma_\nu}{\kappa_\nu + \sigma_\nu} J_\nu \tag{1-43}$$

For pure scattering (again coherent and isotropic), we obtain

$$S_\nu = J_\nu \tag{1-44}$$

Again we emphasize that the source functions given in equations (1-41) through (1-44) are based on essentially heuristic arguments and are meant to be only illustrative; we will return to the problem of a more rigorous formulation of the source function at a later time.

Equation (1-40) provides the basic framework for the calculation of radiation transport, and our goal will be to solve this equation subject to specified boundary conditions and other constraints (e.g., conservation of energy).

### BOUNDARY CONDITIONS

The solution of the transfer equation must be carried out subject to certain *boundary conditions.* Two problems of fundamental importance in astrophysics are those of a *finite slab* of material, or of a medium (e.g., a stellar atmosphere) that has a boundary on one side but is so thick that it can be imagined as extending to infinity on the other side—the semi-infinite atmosphere.

For the finite slab problem we may specify a total geometrical thickness $Z$ and a total optical thickness $T_\nu$. Following the convention explained earlier in this section, the optical depth is taken to run from 0 to $T_\nu$ away from the observer while the geometrical depth scale runs from 0 to $Z$ toward the observer (see Figure 1-7). To obtain a unique solution of the transfer equation, we must specify the incident radiation field on both faces of the slab. Measuring $\theta$ positive away from the direction to the observer, $\mu = \cos\theta$ will be greater than zero for pencils of radiation moving toward the observer and less than zero for pencils moving away. Thus we specify the boundary functions $f$ and $g$ such that

$$I_\nu(0, \mu) = f_\nu(\mu) \tag{1-45}$$

for $-1 \le \mu \le 0$ at the upper boundary, and for $0 \le \mu \le 1$,

$$I_\nu(T_\nu, \mu) = g_\nu(\mu) \tag{1-46}$$

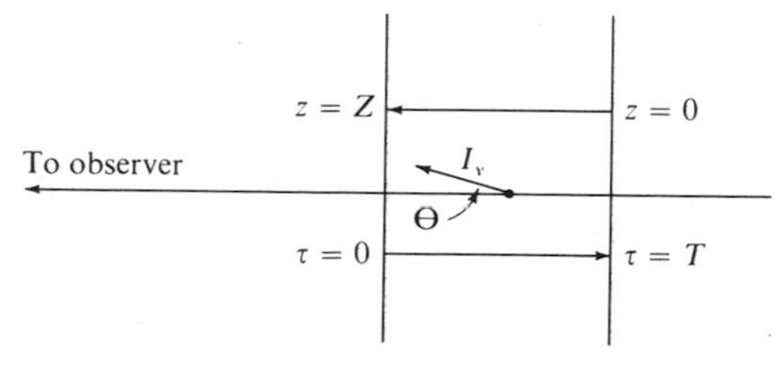


FIG. 1-7. Geometry of a finite slab. Optical depth increases from 0 to $T$ away from observer; geometrical thickness increases from 0 to $Z$ toward observer.

at the lower boundary. In equations (1-45) and (1-46) the first argument of the specific intensity is the optical depth. Given these boundary conditions and a complete specification of sources and sinks of energy within the slab, the radiation field follows directly from the solution of the transfer equation (1-40).

In the semi-infinite case we need to specify only the radiation field incident upon the upper boundary, i.e., equation (1-45). In the usual stellar atmospheres analysis we assume that the incident radiation field is negligible, and therefore we may set $f_\nu(\mu) \equiv 0$; this will not be a valid approximation if the star finds itself in close proximity to an intense source of radiation, e.g., a binary companion. Since there is no lower boundary in the semi-infinite case equation (1-46) is no longer applicable but is replaced by a *boundedness* requirement. Specifically, we demand that at great depth the radiation field shall satisfy the relation

$$\lim_{\tau\to\infty} I_\nu(\tau_\nu, \mu)e^{-\tau_\nu/\mu} = 0 \tag{1-47}$$

The reasons for this particular choice will become more evident in the discussion that follows.

## SIMPLE EXAMPLES

Before writing the formal solution of the equation of transfer, it is instructive to consider a few simple examples.

(a) Suppose no material is present. Then $k_\nu$ and $j_\nu$ are both zero; thus from equation (1-37) $\mu\, dI_\nu/dz = 0$ or $I_\nu =$ constant. This result is consistent with our earlier proof of the invariance of the specific intensity when no sources or sinks are present.

(b) Suppose that emitting material is present but there is no opacity at the frequency under consideration. Then

$$\frac{\mu}{\rho}\frac{dI_\nu}{dz} = j_\nu \tag{1-48}$$

and for a finite slab the emergent radiation is given by the expression

$$I_\nu(0, \mu) = \frac{1}{\mu}\int_0^Z \rho(z) j_\nu(z)\, dz + I_\nu(T, z) \tag{1-49}$$

This expression is of interest physically in the formation of optically forbidden lines in nebulae. In such lines atoms are excited to metastable levels by collisions, and subsequently some of these decay and emit a photon. The absorption probability for such a forbidden transition is negligible compared with other processes that are depopulating the lower state. Thus, in effect, photons are created at the expense of the thermal energy pool of the gas, but none are destroyed by adsorption. This situation, incidentally, is far from LTE.

(c) Suppose that there is absorption of radiation from the beam but no emission. Then

$$dI_\nu = -k_\nu \rho I_\nu \, dz/\mu \tag{1-50}$$

and again for a finite slab the emergent radiation is given by the expression

$$\begin{aligned} I_\nu(0, \mu) &= I_\nu(T_\nu, \mu)\exp\left(-\frac{1}{\mu}\int_0^Z k_\nu \rho \, dz\right) \\ &= I_\nu(T_\nu, \mu)\exp(-T_\nu/\mu) \end{aligned} \tag{1-51}$$

Equation (1-51) is of relevance physically when photons absorbed in the material (e.g., a filter) are converted into photons of another frequency before being reemitted or are destroyed and converted directly to the kinetic energy of the particles in the absorbing medium.

### FORMAL SOLUTION

Let us now obtain a formal solution to the equation of transfer. Mathematically, equation (1-40) (supressing the subscripts $\nu$ for convenience) is a linear first-order differential equation with constant coefficients,

$$\mu \frac{dI}{d\tau} = I - S$$

and must therefore have an integrating factor, namely, $\exp(-\tau/\mu)$. Thus

$$\frac{d}{d\tau}(Ie^{-\tau/\mu}) = -\frac{Se^{-\tau/\mu}}{\mu} \tag{1-52}$$

so that

$$\Big|_{\tau_1}^{\tau_2} Ie^{-\tau/\mu} = -\int_{\tau_1}^{\tau_2} S(t)e^{-t/\mu}\frac{dt}{\mu} \tag{1-53}$$

or

$$I(\tau_1, \mu) = I(\tau_2, \mu)e^{-(\tau_2-\tau_1)/\mu} + \int_{\tau_1}^{\tau_2} S(t)e^{-(t-\tau_1)/\mu}\frac{dt}{\mu} \tag{1-54}$$ *

For example, suppose we set $\tau_1 = 0$ and take the limit as $\tau_2 \to \infty$; thus we compute the emergent intensity of a semi-infinite atmosphere. Then by virtue of equation (1-47) we have

$$I(0, \mu) = \int_0^\infty S(t)e^{-t/\mu}\frac{dt}{\mu} \tag{1-55}$$

Physically, this merely states that the emergent intensity is given by a weighted mean over the source function—the weighting factor corresponds to the fraction of energy that penetrates to the surface from each element of optical depth. In particular, suppose $S(t)$ is a linear function of depth, $S = a + bt$; then

$$I(0, \mu) = a + b\mu$$

which simply shows that the weighting process is such that the radiation emerging is characteristic of optical depth unity along the line of sight. Mathematically, we see that the specific intensity is the Laplace transform of the source function—a property which may be used to determine the source function in certain problems.

As a second example, consider a finite slab of thickness $T$, within which $S$ is constant and upon which there is no incident radiation. Then the normally emergent radiation is

$$I(0, 1) = S(1 - e^{-T})$$

For $T \gg 1$, $I = S$. This is reasonable physically since the energy that emerges should consist of those photons emitted over the mean free path for escape. The rate of emission is $\rho j$, and the mean free path is $1/\kappa\rho$, so it is reasonable that the intensity saturates to $S = j/\kappa$. For $T \ll 1$, $e^{-T} \approx 1 - T$, so $I \approx ST$. Here again the answer is sensible physically because in the optically thin case we can see through the entire volume. Thus the energy emitted (per unit area) must be the emissivity $\rho j$ times the total path length $Z$ through the volume, so $I = \rho j Z = (j/\kappa)(\kappa\rho Z) = ST$. In this limit we have recovered equation (1-49).

Suppose now we consider an *arbitrary* interior point in a semi-infinite atmosphere and apply the usual boundary conditions, namely, equation (1-47) and equation (1-45) with $f_\nu \equiv 0$. Then considering first the case $\mu \geq 0$, i.e., outgoing radiation, we have from equation (1-54)

$$I(\tau, \mu) = \lim_{\tau_2 \to \infty} I(\tau_2, \mu) e^{-(\tau_2 - \tau)/\mu} + \int_\tau^\infty S(t) e^{-(t-\tau)/\mu} \frac{dt}{\mu}$$

or from equation (1-47),

$$I(\tau, \mu) = \int_\tau^\infty S(t) e^{-(t-\tau)/\mu} \frac{dt}{\mu}, \qquad (0 \leq \mu \leq 1) \tag{1-56}$$

Considering now $\mu \leq 0$, i.e., incoming radiation, we take $\tau_2 = 0$, so that

$$I(\tau, \mu) = I(0, \mu) e^{\tau/\mu} + \int_\tau^0 S(t) e^{-(t-\tau)/\mu} \frac{dt}{\mu}$$

or from equation (1-45)

$$I(\tau, \mu) = -\int_0^\tau S(t) e^{(\tau - t)/\mu} \frac{dt}{\mu}, \qquad (-1 \leq \mu \leq 0) \tag{1-57}$$

Equations (1-56) and (1-57) constitute a complete solution of the transfer equation if the source function $S(\tau)$ is given.

THE SCHWARZSCHILD-MILNE EQUATIONS

Having obtained a formal solution of the transfer equation, we may now perform integrations over angles to derive moments of the specific intensity and to write the solution in a concise and useful form. Consider first the mean intensity; making use of the definition in equation (1-3) and the solution in equations (1-56) and (1-57), we have

$$J(\tau) \equiv \frac{1}{2}\int_{-1}^{1} I(\tau, \mu)\, d\mu$$

$$= \frac{1}{2}\int_{0}^{1} d\mu \int_{\tau}^{\infty} S(t)e^{-(t-\tau)/\mu}\,\frac{dt}{\mu} - \frac{1}{2}\int_{-1}^{0} d\mu \int_{0}^{\tau} S(t)e^{(\tau-t)/\mu}\,\frac{dt}{\mu} \tag{1-58}$$

We now interchange the order of integration and in the first integral define $w = 1/\mu$ so that $-d\mu/\mu = dw/w$, while in the second we set $w = -1/\mu$ so that $-d\mu/\mu = dw/w$ again. Then

$$J(\tau) = \frac{1}{2}\int_{\tau}^{\infty} S(t)\, dt \int_{1}^{\infty} e^{-w(t+\tau)}\,\frac{dw}{w} + \frac{1}{2}\int_{0}^{\tau} S(t)\, dt \int_{1}^{\infty} e^{-w(\tau-t)}\,\frac{dw}{w} \tag{1-59}$$

The integrals against $w$ are of a well-known form and are called the first *exponential integral*. In general, for integer values of $n$, one defines the $n$th exponential integral by the expression

$$E_n(x) \equiv \int_{1}^{\infty} \frac{e^{-xt}\, dt}{t^n} = x^{n-1}\int_{x}^{\infty} \frac{e^{-t}\, dt}{t^n} \tag{1-60}$$

Thus in terms of $E_1(x)$, equation (1-59) may be rewritten as

$$J(\tau) = \frac{1}{2}\int_{\tau}^{\infty} S(t)E_1(t-\tau)\, dt + \frac{1}{2}\int_{0}^{\tau} S(t)E_1(\tau - t)\, dt \tag{1-61}$$

or, re-introducing the frequency subscripts,

$$J_\nu(\tau_\nu) = \frac{1}{2}\int_{0}^{\infty} S_\nu(t_\nu)E_1\,|t_\nu - \tau_\nu|\, dt_\nu \tag{1-62}$$

Equation (1-62) was first derived by K. Schwarzschild, and is named in his honor. Schwarzschild's analysis (see Ref. 26, p. 35) is one of the foundation stones of the theory of radiative transfer and merits careful reading.

Because the particular integral appearing in equation (1-62) often occurs in the theory of radiative transfer, it has been abbreviated to an operator notation by writing

$$\Lambda_\tau[f(t)] \equiv \frac{1}{2}\int_{0}^{\infty} f(t)E_1\,|t-\tau|\, dt \tag{1-63}$$

In a similar way, we may use the definition in equation (1-6) and the solution in equations (1-56) and (1-57) to derive a concise expression for the flux. We obtain

$$F_\nu(\tau_\nu) = 2\int_{\tau_\nu}^{\infty} S_\nu(t_\nu)E_2(t_\nu - \tau_\nu)\,dt_\nu - 2\int_0^{\tau_\nu} S_\nu(t_\nu)E_2(\tau_\nu - t_\nu)\,dt_\nu \tag{1-64}$$

an expression derived by Milne (see Ref. 26, p. 77). As before we define another operator such that

$$\Phi_\tau[f(t)] \equiv 2\int_\tau^{\infty} f(t)E_2(t-\tau)\,dt - 2\int_0^{\tau} f(t)E_2(\tau - t)\,dt \tag{1-65}$$

By a completely analogous argument for the $K$-integral, we may show that

$$K_\nu(\tau_\nu) = \frac{1}{2}\int_0^{\infty} S_\nu(t_\nu)E_3\,|t_\nu - \tau_\nu|\,dt_\nu \tag{1-66}$$

and we define

$$X_\tau[f(t)] \equiv 2\int_0^{\infty} f(t)E_3\,|t - \tau|\,dt \tag{1-67}$$

Thus, knowledge of $S_\nu(\tau_\nu)$ allows direct computation of $J_\nu$, $F_\nu$, and $K_\nu$.

We should emphasize here that the solution of the transfer equation, given either by equations (1-56) and (1-57) or by equations (1-62), (1-64), and (1-66), is only formal, and its apparent simplicity is in many ways illusory. This may readily be seen as follows. Suppose that equation (1-43) is the correct expression for the source function. Then clearly the source function itself depends upon the radiation field and hence upon the solution of the transfer equation. Thus the solution of the transfer equation will, in general, require more than a simple quadrature.

Complications of this kind are also introduced by other constraints upon the solution. For example, we shall show in Section 1-5 that the requirement of energy balance couples together the radiation field and the temperature structure of the atmosphere. Thus even if equation (1-42) were valid, we would not usually be able to prespecify the run of $T(\tau_\nu)$ and hence of $B_\nu[T(\tau_\nu)]$, but we would find that the temperature structure and hence the source function distribution gain depends upon the radiation field and hence upon the solution of the transfer equation. When the more general non-LTE situation is considered, the coupling becomes even more subtle and complex. The development of methods to treat these interactions will occupy a major portion of the work we shall undertake in this book.

### MATHEMATICAL PROPERTIES OF THE FUNCTIONS $E_n(x)$ AND THE OPERATORS $\Lambda$, $\Phi$, AND $X$

Because they determine many of the characteristic features of radiative transfer problems, we should discuss briefly here the mathematical properties of the functions $E_n(x)$ and the operators $\Lambda$, $\Phi$, and $X$.

Note first that we may write

$$E_n'(x) = \frac{\partial}{\partial x}\int_1^\infty \frac{e^{-xt}\,dt}{t^n} = -\int_1^\infty \frac{e^{-xt}\,dt}{t^{n-1}} = -E_{n-1}(x) \tag{1-68}$$

By integrating equation (1-60) by parts, one may easily show that

$$E_n(x) = \frac{1}{n-1}[e^{-x} - xE_{n-1}(x)], \qquad n > 1 \tag{1-69}$$

which is a useful recursion relation that allows for computation of higher order exponential integrals. An asymptotic expression for $E_1(x)$, valid for $x \gg 1$ is

$$E_1(x) = \frac{e^{-x}}{x}\left[1 - \frac{1}{x} + \frac{2!}{x^2} - \frac{3!}{x^3} + \cdots\right] \tag{1-70}$$

Equation (1-70) shows that the asymptotic behavior of the exponential integral is $E_1(x) \sim e^{-x}/x$. A series that is uniformly convergent for all $x$, though in practice useful mainly for small $x$, is

$$E_1(x) = -\gamma - \ln x + \sum_{k=1}^{\infty}(-1)^{k-1}\frac{x^k}{kk!}, \qquad (x > 0) \tag{1-71}$$

where $\gamma = 0.5572156\ldots$. Several useful numerical approximation formulae for $E_1(x)$ exist (see e.g., Ref. 1, pp. 228–37).

From equations (1-70) and (1-71) we may note two important properties of $E_1(x)$: It is singular at the origin, and it has a relatively rapid decay at large $x$. Only $E_1(x)$ is singular at the origin; the recursion formula shows that for $n > 1$, $E_n(0) = 1/(n-1)$. Note however that since $E_2'(0) = -E_1(0)$, $E_2(x)$ has a singularity in its derivative at the origin even though the function itself is finite. The physical implications of these facts will be commented upon in later discussions.

Because they are of use in many kinds of analysis, let us summarize the $\Lambda$, $\Phi$, and $X$ transforms of a few elementary functions.

These may be verified by direct integration:

$$\Lambda_\tau(1) = 1 - \tfrac{1}{2}E_2(\tau) \tag{1-72}$$

$$\Lambda_\tau(t) = \tau + \tfrac{1}{2}E_3(\tau) \tag{1-73}$$

$$\Lambda_\tau(t^p) = \frac{1}{2}p!\left[\sum_{k=0}^{p}\frac{\tau^k}{k!}\delta_\alpha + (-1)^{p+1}E_{p+2}(\tau)\right] \tag{1-74}$$

where $\delta_\alpha = 0$ if $\alpha \equiv p + 1 - k$ is even, and $\delta_\alpha = 2/\alpha$ if $\alpha$ is odd. For $a > 0$, $a \neq 1$,

$$\Lambda_\tau(e^{-at}) = \frac{e^{-a\tau}}{2a}\left[\ln\left|\frac{a+1}{a-1}\right| - E_1(\tau - \tau a)\right] + \frac{E_1(\tau)}{2a} \tag{1-75}$$

In the above expression, the argument of the $E_1$ function may be negative. In this case, we have by analytic continuation (see Ref. 1, p. 228)

$$E_1(x) = -\gamma - \ln|x| - \sum_{k=1}^{\infty} \frac{|x|^k}{kk!}, \qquad (x < 0)$$

Similarly,

$$\Phi_\tau(1) = 2E_3(\tau) \tag{1-76}$$

$$\Phi_\tau(t) = \tfrac{4}{3} - 2E_4(\tau) \tag{1-77}$$

$$\Phi_\tau(t^p) = 2p!\left[\sum_{k=0}^{p} \frac{\tau^k}{k!}\delta_\alpha + (-1)^p E_{p+3}(\tau)\right] \tag{1-78}$$

where $\delta_\alpha = 0$ if $\alpha \equiv p + 2 - k$ is even, and $\delta_\alpha = 2/\alpha$ if $\alpha$ is odd.

Finally,

$$X_\tau(1) = \tfrac{4}{3} - 2E_4(\tau) \tag{1-79}$$

$$X_\tau(t) = \tfrac{4}{3}\tau + 2E_5(\tau) \tag{1-80}$$

Many other useful formulae of this type have been tabulated by Kourganoff (Ref. 23, p. 43 ff.).

Let us briefly explore the effects of the $\Lambda$- and $\Phi$-operators on a linear source function $S(t) = a + bt$. From equations (1-72) and (1-73) we may immediately write

$$\Lambda_\tau(a + bt) = a + b\tau + \tfrac{1}{2}[bE_3(\tau) - aE_2(\tau)] \tag{1-81}$$

Since the exponential integrals decay asymptotically as $e^{-x}/x$, it is evident from equation (1-81) that $\Lambda_\tau(S)$ strongly approaches the local value of $S(\tau)$ for $\tau \gg 1$; thus, the $\Lambda$-operator tends to reproduce the source function at depth. in contrast, at the surface, $E_2(0) = 1$, $E_3(0) = \frac{1}{2}$, and we obtain $\Lambda_0(S) = \frac{1}{2}a + \frac{1}{4}b$, which clearly is markedly different from the value $S(0) = a$. Note in particular that for $b = 0$, $S \equiv a$, $J(0) = \frac{1}{2}a = \frac{1}{2}S(0)$. Physically this reflects the fact that a point at the surface sees a hemisphere of empty space and a hemisphere in which, in this case, $S$ = constant = $a$; the average intensity is thus $\frac{1}{2}S$. When a gradient is present, $J(0)$ may be above or below this value depending upon how steep the gradient is and upon its sign. In a general way we may expect $\Lambda(S)$ to disagree with $S$ itself most strongly near the boundary $\tau = 0$. Let us now consider the effect of the $\Phi$-operator on a linear source function. From equations (1-76) and (1-77) we have

$$\Phi_\tau(a + bt) = 2aE_3(\tau) + \tfrac{4}{3}b - 2bE_4(\tau) \tag{1-82}$$

Here, for $\tau \gg 1$, $\Phi(S) \to \frac{4}{3}b$. Thus, the flux depends only upon the *gradient* of the source function at great depth. This shows quite clearly that the flux operator may be considered to be a *differencing* operator. At $\tau = 0$, $\Phi_0(S) = a + \frac{2}{3}b$. Clearly the surface flux will be the larger, the faster the source function grows inward. For $b = 0$, note that $F(0) = a = 2J(0)$, a result that will be used in later discussion.

## 1-5. The Condition of Radiative Equilibrium

Within the deep interior of a star, nuclear reactions release a flux of energy that diffuses outward, passes through those outermost layers that constitute the atmosphere of the star, and ultimately emerges as radiation. In all normal stars there are no sources or sinks of energy within the atmosphere itself; the atmosphere merely transports the total energy it receives outward. Thus, while the atmosphere may, for example, alter the frequency distribution of the radiation it receives or may alter the partitioning of energy between radiative and nonradiative modes, the energy flux as a whole is rigorously conserved. There are two basic modes of energy transport in a stellar atmosphere: radiative and convective (or some other hydrodynamical mode); conduction is ineffective and can be omitted. When all of the energy is transported by radiation, we have what is called *radiative equilibrium*; conversely, pure convective transport is called *convective equilibrium*. Whether or not radiative transport prevails over convection depends upon its efficiency and hence upon the stability of the atmosphere against convective motions.

The criterion determining the stability of radiative transport was first enunciated by K. Schwarzschild (see Ref. 26, p. 25) in another of the fundamental papers on the theory of radiative transfer. Schwarzschild was able to demonstrate convincingly that in the photospheric layers of the sun, the dominant mode of energy transport is radiative. Since his time a number of thorough analyses of the problem have been carried out; the results of some of these analyses are summarized by Unsöld (Ref. 36, p. 215 ff.) and Aller (Ref. 4, p. 449). The basic picture that emerges—for a star like the sun—is that radiative equilibrium obtains to optical depths of order unity, and below this hydrogen ionization causes the atmosphere to become unstable against convection. Convection zones below the outer radiative zone exist for all stars later than spectral type F5, approximately. Earlier than this, radiative equilibrium prevails throughout the entire outer envelope of a star. In this book we shall be concerned mainly with early-type stars, and accordingly we shall usually stress the radiative equilibrium regime. The theory of convective transport is unfortunately not well developed, but we will at least outline the so-called *mixing-length theory* (Chapter 6), which has been widely applied in astrophysics.

Let us now consider some of the mathematical implications of the requirement of radiative equilibrium. From the discussion in Section 1-2 we may write the total energy removed from the beam as

$$\int_0^\infty d\nu \oint k_\nu I_\nu \, d\omega = 4\pi \int_0^\infty k_\nu J_\nu \, d\nu \tag{1-83}$$

where $k_\nu$ is understood to be the total extinction coefficient. Similarly, the total energy replaced in the beam is

$$\oint d\omega \int_0^\infty j_\nu \, d\nu = 4\pi \int_0^\infty k_\nu S_\nu \, d\nu \tag{1-84}$$

where we have made use of equation (1-39). Thus, the condition of radiative equilibrium demands that at each point in the atmosphere

$$\int_0^\infty k_\nu J_\nu \, d\nu = \int_0^\infty k_\nu S_\nu \, d\nu \tag{1-85}$$

An alternative expression follows from equation (1-37) by integrating over all solid angles and frequency to obtain

$$\frac{1}{\rho}\frac{d}{dz}\left(\int_0^\infty d\nu \oint I_\nu \mu \, d\omega\right) = \oint d\omega \int_0^\infty j_\nu \, d\nu - \int_0^\infty d\nu \oint k_\nu I_\nu \, d\omega = 0 \tag{1-86}$$

where the last equality follows from equation (1-85). Now from the definition of the flux $F_\nu$, equation (1-6), we see that equation (1-86) reduces simply to

$$\frac{d}{dz}\left(\int_0^\infty F_\nu \, d\nu\right) = \frac{dF}{dz} = 0 \tag{1-87}$$

Hence, the condition of radiative equilibrium is equivalent to the requirement that the depth derivative of the total flux is zero, i.e., the total flux is constant with depth. Either equation (1-85) or equation (1-87) will be used to place constraints upon the mathematical models we construct.

Since the total flux is constant with depth, it is a uniquely specified parameter for each atmosphere. An equivalent quantity often employed to specify the flux is the *effective temperature*. If we were to make a small opening into a cavity in thermodynamic equilibrium at temperature $T$, the flux that would emerge would be

$$\mathscr{F}_{BB}(\nu) = 2\pi \int_{-1}^{1} I_\nu(\mu)\mu \, d\mu = 2\pi B_\nu(T) \int_0^1 \mu \, d\mu = \pi B_\nu(T) \tag{1-88}$$

so that the total flux is

$$\int_0^\infty \mathscr{F}_{BB}(\nu) \, d\nu = \pi \int_0^\infty B_\nu(T) \, d\nu = \sigma T^4 \tag{1-89}$$

where $\sigma$ is the well-known Stefan-Boltzmann constant

$$\sigma = \frac{2\pi^5 k^4}{15h^3c^2} = 5.669 \times 10^{-5} \text{ ergs cm}^{-2} \text{ sec}^{-1} \text{ deg}^{-4} \tag{1-90}$$

Although the radiation emergent from a star is by no means Planckian, it is, nevertheless, customary to define an effective temperature as that temperature which when substituted into equation (1-89) yields the actual flux in the stellar atmosphere. Thus

$$\frac{\sigma T_{\text{eff}}^4}{\pi} \equiv F = \int_0^\infty F_\nu \, d\nu \tag{1-91}$$

While $T_{\text{eff}}$ clearly has no direct physical significance, it is somewhat pictorial and provides a convenient parameter with which to characterize the atmosphere. Moreoever, the actual kinetic temperature $T$ will equal the effective temperature at some characteristic point in the atmosphere, usually of optical depth of order unity measured at frequencies where the opacity is lowest.

It is of interest to examine the implications of equation (1-85) in more detail. For simplicity let us employ the prototype source function given by equation (1-43). Substitution into equation (1-85) then yields

$$\int_0^\infty k_\nu J_\nu \, d\nu = \int_0^\infty \kappa_\nu J_\nu \, d\nu + \int_0^\infty \sigma_\nu J_\nu \, d\nu$$

$$= \int_0^\infty k_\nu S_\nu \, d\nu = \int_0^\infty \kappa_\nu B_\nu \, d\nu + \int_0^\infty \sigma_\nu J_\nu \, d\nu$$

which reduces to

$$\int_0^\infty \kappa_\nu J_\nu \, d\nu = \int_0^\infty \kappa_\nu B_\nu \, d\nu \tag{1-92}$$

Equation (1-92) demonstrates two very important aspects of the problem. First, we see that the scattering terms have canceled out. This is reasonable physically since scattering returns as much energy to the beam as it removes, though possibly redistributed in direction and frequency (aspects which we have ignored here for the sake of simplicity). Second, we see that the total thermal emission is set at a value determined by the mean intensity and hence by the solution of the transfer equation. It is this fact that leads to the coupling between the *local* value of the source function and the *overall* properties of the atmosphere, as mentioned earlier in our discussion of the formal nature of the solution of the transfer equation obtained in Section 1-4. Suppose, for a moment, that the opacity $\kappa_\nu$ were simply fixed once and for all. Then the integral on the right-hand side of equation (1-92) will be a monotone increasing function of temperature, for physically reasonable choices of $\kappa_\nu$. In this sense, the radiative equilibrium condition *determines* the local temperature at a value consistent with the radiation field in the atmosphere as a whole. [Note that in the non-LTE situation, equation (1-92) will need modification; we consider this point in Chapters 7 and 14.]

## 1-6. Asymptotic Form of the Transfer Equation

For the sake of generality, the development given above has been rather formal. To gain further insight into the problem, let us adopt here a somewhat different perspective and consider the nature of the radiation field and of the transfer equation at great depths in a semi-infinite atmosphere. If we are at depths far from the boundary of the atmosphere, we may then expect that the conditions of thermal equilibrium will be closely met and that $S_\nu$ will be the Planck function $B_\nu$. Suppose we consider some reference point $\tau$; we may then write a power-series expansion for the source function as:

$$S_\nu(t) = \sum_{n=0}^{\infty} \frac{(t-\tau)^n}{n!} \frac{d^n B_\nu}{d\tau^n} \tag{1-93}$$

Then substitution into equation (1-56) yields for $0 \le \mu \le 1$

$$\begin{aligned} I_\nu(\tau, \mu) &= \sum_{n=0}^{\infty} \frac{1}{n!} \frac{d^n B_\nu}{d\tau^n} \int_\tau^\infty (t-\tau)^n e^{-(t-\tau)/\mu} \frac{dt}{\mu} \\ &= \sum_{n=0}^{\infty} \frac{d^n B_\nu}{d\tau_n} \frac{1}{n!} \int_0^\infty x^n e^{-x/\mu} \frac{dx}{\mu} \\ &= \sum_{n=0}^{\infty} \mu^n \frac{d^n B_\nu}{d\tau^n} = B_\nu(\tau) + \mu \frac{dB_\nu}{d\tau} + \mu^2 \frac{d^2 B_\nu}{d\tau^2} + \cdots \end{aligned} \tag{1-94}$$

An expression for $I_\nu(\tau, \mu)$ when $-1 \le \mu \le 0$ may be obtained from substitution into equation (1-57); when this is done, one finds an expression that differs from equation (1-94) only by terms of order $e^{-\tau/\mu}$ so that if we consider the limit of great depth, we may apply the above result for both the incoming and outgoing radiation. By substitution into equation (1-9), we may calculate the mean intensity

$$J_\nu(\tau) = \frac{1}{2} \sum_{n=0}^{\infty} \frac{d^n B_\nu}{d\tau^n} \int_{-1}^{1} \mu^n \, d\mu$$

Clearly only even-order terms will survive the integration so that we obtain

$$J_\nu(\tau) = \sum_{n=0}^{\infty} \frac{1}{2n+1} \frac{d^{2n} B_\nu}{d\tau^{2n}} = B_\nu(\tau) + \frac{1}{3} \frac{d^2 B_\nu}{d\tau^2} + \cdots \tag{1-95}$$

Similarly, by substitution into equations (1-10) and (1-11), we find

$$H_\nu(\tau) = \sum_{n=0}^{\infty} \frac{1}{2n+3} \frac{d^{2n+1} B_\nu}{d\tau^{2n+1}} = \frac{1}{3} \frac{dB_\nu}{d\tau} + \frac{1}{5} \frac{d^3 B_\nu}{d\tau^3} + \cdots \tag{1-96}$$

and

$$K_\nu(\tau) = \sum_{n=0}^{\infty} \frac{1}{2n+3} \frac{d^{2n} B_\nu}{d\tau^{2n}} = \frac{1}{3} B_\nu(\tau) + \frac{1}{5} \frac{d^2 B_\nu}{d\tau^2} + \cdots \tag{1-97}$$

To estimate the convergence of the above series expressions, we approximate derivatives by appropriate differences. Thus, at least to order of magnitude we write

$$\left|\frac{d^n B_\nu}{d\tau^n}\right| \sim \frac{B_\nu}{\tau^n}$$

so that the ratio of successive terms in the series is of order

$$\left|\frac{d^{n+2} B_\nu}{d\tau^{n+2}}\right| \bigg/ \left|\frac{d^n B_\nu}{d\tau^n}\right| \sim \left(\frac{B_\nu}{\tau^{n+2}}\right) \bigg/ \left(\frac{B_\nu}{\tau^n}\right) \sim \frac{1}{\tau^2} \sim \frac{1}{(\langle\kappa\rho\rangle l)^2} \tag{1-98}$$

where $\langle\kappa\rho\rangle$ is an average opacity coefficient along the path length $l$. Thus the successive terms in the series will be very small, and we can expect rapid convergence at depth in the atmosphere. Clearly since $\kappa_\nu$ may differ from frequency to frequency, this convergence will occur at different physical depths for different frequencies—being the most rapid at those wavelengths where the material is opaque. We should also recall that we have explicitly neglected the effects of scattering; as we shall see in Chapters 6 and 12, the presence of scattering may alter the situation quite markedly. From the point of view of the star as a whole, we can take $l$ to be some significant fraction of a stellar radius, say $10^{10}$ cm, and since $\langle\kappa\rho\rangle$ will be $\sim 1$, we see from equation (1-98) that the convergence factor in each series will be of order $10^{-20}$! Thus it is clear that in the interior we may use just the first term of equations (1-95) through (1-97).

In the limit of large depth then we may write

$$I_\nu(\tau, \mu) \approx B_\nu(\tau) + \mu \frac{dB_\nu}{d\tau} \tag{1-99}$$

$$J_\nu(\tau) \approx B_\nu(\tau) \tag{1-100}$$

$$H_\nu(\tau) \approx \frac{1}{3}\frac{dB_\nu}{d\tau} \tag{1-101}$$

and

$$K_\nu(\tau) \approx \tfrac{1}{3} B_\nu(\tau) \tag{1-102}$$

We see here explicitly the important relation that

$$\lim_{\tau\to\infty} [K_\nu(\tau)/J_\nu(\tau)] = \tfrac{1}{3} \tag{1-103}$$

Equation (1-103) simply states that the radiation field is essentially isotropic; we had obtained this result earlier (Section 1-1) for strictly isotropic radiation. Of course, in the expansion for $I_\nu$ we carry along two terms, the second of which gives an anisotropy since we want in general a nonzero net flux.

In the limit of large optical depth, the transfer equation also reduces to a simple form. Writing equation (1-40),

$$\mu \frac{dI_\nu}{d\tau_\nu} = I_\nu - B_\nu$$

and taking the first moment, we obtain

$$\frac{dK_\nu}{d\tau_\nu} = H_\nu$$

and substituting from equation (1-102), we find

$$H_\nu = \frac{1}{3}\frac{dB_\nu}{d\tau_\nu} = -\frac{1}{3}\frac{1}{\kappa_\nu \rho}\frac{dB_\nu}{dr}$$

or

$$H_\nu = -\frac{1}{3}\left(\frac{1}{\kappa_\nu \rho}\frac{dB_\nu}{dT}\right)\frac{dT}{dr} \tag{1-104}$$

so that

$$F_\nu = -\frac{4}{3}\left(\frac{1}{\kappa_\nu \rho}\frac{dB_\nu}{dT}\right)\frac{dT}{dr} \tag{1-105}$$

These equations are referred to as the *diffusion approximation* because of their formal similarity to other kinds of diffusion equations where a flux is given by a diffusion coefficient times a gradient. In the present context the term $(4/3)[(1/\kappa_\nu \rho)(dB_\nu/dT)]$ may be regarded as a kind of *radiative conductivity*. (We shall refer again to this result in Section 2-2.) These results show the essential physical content of our previous demonstration (Section 1-4) that application of the $\Phi$-operator to a linear source function yields a flux at a depth that depends only upon the gradient term.

One extremely important result that we can immediately see from equation (1-105) is that the very fact that energy emerges from the star implies that there is a temperature gradient and that the temperature increases inward. This of course is consistent with our previous statement that the ultimate energy source in a star is thermonuclear energy-release at the center.

In an intuitive picture of diffusion one usually conceives of a slow leakage from a reservoir of large capacity. This notion is valid in the present case as well. Rewriting equation (1-99) by use of equation (1-101) and integrating over frequency, we have

$$I(\tau, \mu) \approx B(\tau) + \tfrac{3}{4} F\mu \tag{1-106}$$

Comparing the relative size of the two terms, we then see that

$$\frac{\text{Anisotropic term}}{\text{Isotropic term}} \sim \frac{3}{4}\frac{F}{B} = \frac{3}{4}\left(\frac{\sigma}{\pi} T_{\text{eff}}^4\right)\Big/\left(\frac{\sigma}{\pi} T^4\right) = \frac{3}{4}\left(\frac{T_{\text{eff}}}{T}\right)^4 \tag{1-107}$$

Clearly as we proceed to a great depth where $T \gg T_{\text{eff}}$, the anisotropy and the "leak" become ever smaller. From a slightly different physical point of view, we note that if $\pi F$ is the flux from differential volume of material and the photons emerge with velocity $c$, then the rate of energy flow is $\pi F/c$. We also recall that the energy density is given by $U = 4\pi J/c \approx 4\pi B/c$ so that an alternative restatement of equation (1-107) is

$$\frac{\text{Rate of energy leakage}}{\text{Energy content}} = \frac{(\pi F/c)}{(4\sigma T^4/c)} = \frac{1}{4}\left(\frac{T_{\text{eff}}}{T}\right)^4 \tag{1-108}$$

So again we verify our intuitive notion that at a great depth diffusion occurs in the sense that we described above. Simultaneously we see that at the surface where $T \approx T_{\text{eff}}$, essentially "free flow" of the radiant energy occurs.

## 1-7. The Classical Model Atmospheres Problem

Now that we have described the basic concepts that we shall employ, let us summarize the assumptions that define the central problem treated in this book: the *model atmospheres problem*. Let us attempt to bring the physical nature of the problem into sharp focus while we indicate clearly the kinds of data which are required for its solution. What we mean by the "model atmospheres problem" is the construction of numerical models that can provide an accurate estimate of both the run of the physical variables with depth in the atmosphere of a star and of the emergent spectrum—in the continuum and in the lines. Under the most general circumstances conceivable, this would be a problem of overwhelming difficulty; therefore, certain simplifications must necessarily be made. The particular choice of simplifications listed below define what may be called the *classical* model atmospheres problem.

(a) We assume that the atmospheres are stratified into *homogeneous plane-parallel layers*. The assumption of plane-parallel geometry merely implies that the "thickness" of the atmosphere is much less than the radius of the star. This is an adequate approximation for all but the most distended stellar envelopes. The assumption of homogeneity is a more severe simplification. For example, observations of the sun reveal considerable small-scale structures which show clearly that the horizontal layers are nonhomogeneous; only some of the most modern analyses have begun to account for these inhomogeneities. In the case of stars, we have no information whatever about inhomogeneities, and the assumption that none exist is at present a practical necessity—we can only hope that such models at least yield information about average conditions (in some ill-defined sense) even if the assumption is not in truth correct.

(b) We assume that the atmosphere is in a *steady state*, i.e., we neglect such phenomena as stellar pulsations, transient expanding envelopes, shocks, variable magnetic fields, heating by a binary component, and so on. We explicitly assume the transfer equation is independent of time and that atomic occupation numbers do not depend upon time. In a general way, the atomic occupation numbers follow from solutions of *statistical equilibrium equations* that account for the microprocesses that populate and depopulate each atomic level. The assumption of a steady state simply demands that total rate at which atoms leave a given level be exactly balanced by the total rate at which atoms enter that level—when all possible processes are taken into account. As we shall see in Chapter 5, this analysis provides a detailed description of the interaction between the gas field and the radiation field. A crude substitute for the steady-state statistical equilibrium assumption is the assumption of LTE.

(c) We assume that the atmosphere is in *hydrostatic equilibrium.* Thus we ignore all velocity fields and assume that the pressure stratification is such that it just balances the gravitational field.

(d) We assume that the atmosphere is in *radiative equilibrium.* Thus we neglect energy transport by convection or by other hydrodynamic phenomena in the very outer layers. As mentioned previously, this assumption is expected to be a good approximation for many stars.

Assumptions (b), (c), and (d) are clearly all related inasmuch as they all suppress consideration of hydrodynamical effects. The existence of velocity fields in the solar atmosphere is very well documented (see, e.g., Ref. 39, Chaps. 9 and 10); for stars the data naturally are less complete, but there is little doubt that mass motions play an important role in the atmospheres of many stars, for example, the supergiants. A *complete* theory of stellar atmospheres must be able to treat a wide variety of hydrodynamical problems and to specify the interchange of energy between radiative and nonradiative modes. At present, such a theory is not in hand, and its construction lies at the real frontier of research in this field of astrophysics.

Within the framework of the above assumptions, let us just sketch what is involved in calculating a model atmosphere. To begin as simply as possible, let us suppose that LTE prevails. Then the distribution of atoms among their bound states will depend upon the temperature $T$ while their distribution among various states of ionization will depend upon $T$ and the electron density $N_e$. Thus opacities which are determined by occupation numbers will depend upon $T$ and $N_e$, i.e., we may write $\kappa_\nu = \kappa_\nu(T, N_e)$ and $\sigma_\nu = \sigma_\nu(T, N_e)$. Similarly, the gas pressure will depend upon occupation numbers and hence upon $N_e$ and $T$: $p = p(T, N_e)$. Now if we know a temperature distribution $T(\tau)$ (where $\tau$ is the optical depth at some prechosen wavelength), the density structure of the atmosphere is determined by the condition of hydrostatic equilibrium. By integration of the hydrostatic equation, we may calculate step by step the pressure distribution $p(\tau)$ and the density distribution $N_e(\tau)$ and hence $\kappa_\nu$ and $\sigma_\nu$ at all frequencies. The transfer equation can then be solved since we can compute $\tau_\nu(\tau)$ from knowledge of the opacities and have estimates of the source function [e.g., that given by equation (1-43)]. From the solution of the transfer equation, we obtain $J_\nu(\tau_\nu)$ and $F_\nu(\tau_\nu)$. Finally, we must check the condition of radiative equilibrium and demand that equation (1-85) or equation (1-87) be satisfied. In general, we will find that our (essentially arbitrarily chosen) $T(\tau)$ will not yield a solution satisfying the radiative equilibrium condition. We must then find a means of altering $T(\tau)$, making use of the differences between our computed values of the integrated flux $F$ and its derivative $dF/d\tau$ and the desired values in such a way as to obtain radiative equilibrium. With this altered $T(\tau)$, we recompute the model and again check the radiative equilibrium requirement. This procedure is repeated until one is satisfied with the accuracy of the answer. In short, in LTE, the problem reduces to the determination of the appropriate temperature distribution $T(\tau)$. In the more general non-LTE case, the overall pocedure is similar, except that now the occupation numbers of atomic states depend not only upon $T$ and $N_e$ but also upon the radiation field. Thus it is no longer sufficient to regard the problem as one of determining the appropriate temperature distribution $T(\tau)$ but rather as one of specifying occupation numbers, temperatures, and radiation fields in such a way as to satisfy, self-consistently, the requirements of steady-state and energy equilibrium. This is a vastly more complex and difficult calculation. Happily, a number of powerful techniques have recently become available, and the problem appears to be in hand.

The program outlined above is obviously complicated and requires considerable mathematical and physical theory beyond that which has already been presented. We will discuss the necessary information in a systematic way in Chapters 3 to 5 and 9 and 10. Even before we do this, however, it is important to gain whatever additional insight we can from simplified problems. One such problem is the so-called *gray problem*, which is particularly instructive since it does not depend upon the detailed physics of the state of the gas while at the same time it demonstrates several basic points of transfer theory. Therefore, we will next turn our attention to the gray problem, then to a discussion of the atomic parameters, and then to an attack upon the model atmospheres problem for the continuum only. Subsequently, we will again consider questions of atomic physics with particular emphasis upon spectrum lines and discuss certain important aspects of line transfer problems. Finally, we will return to the complete model atmospheres problem for both lines and continua.

# 2 | The Gray Atmosphere

## 2-1. Statement of the Problem

* The gray atmosphere problem makes the simplifying approximation that the opacity is independent of frequency, i.e., $\kappa_\nu \equiv \kappa$. This assumption is of course highly unrealistic in many cases. Yet as we shall see in later chapters, the opacity in some stars (e.g., the sun) is not too far from being gray, and, in addition, it is possible to reduce the nongray problem to a certain extent to the gray problem by suitable choices of *mean opacities.* It is therefore of considerable importance to obtain the solution of the gray problem both because of the instructional value of solving a relatively simple transfer problem and because the gray solution can provide a valuable starting approximation in some nongray cases.

If we assume $\kappa_\nu \equiv \kappa$, then the standard equation of transfer for plane-parallel geometry is

$$\mu \frac{dI_\nu}{d\tau} = I_\nu - S_\nu \tag{2-1}$$

or by integrating over frequency and writing

$$I \equiv \int_0^\infty I_\nu \, d\nu \tag{2-2}$$

and

$$S \equiv \int_0^\infty S_\nu \, d\nu \tag{2-3}$$

we have

$$\mu \frac{dI}{d\tau} = I - S \tag{2-4}$$

Furthermore, the condition of radiative equilibrium requires

$$\int_0^\infty \kappa J_\nu \, d\nu = \int_0^\infty \kappa S_\nu \, d\nu \tag{2-5}$$

which because the opacity is gray simply states that $S = J$. Thus, equation (2-4) becomes

$$\mu \frac{dI}{d\tau} = I - J \tag{2-6}$$

which by use of equation (1-62) implies

$$J(\tau) = \Lambda_\tau[S(t)] = \Lambda_\tau[J(t)] = \frac{1}{2}\int_0^\infty J(t)E_1 |t - \tau| \, dt \tag{2-7}$$

Equation (2-7) is a linear integral equation for $J$ (or $S$), known as *Milne's equation*; the gray problem itself is sometimes referred to as the *Milne problem*. If we now introduce the additional hypothesis that the source function is due entirely to LTE thermal emission, we have $S_\nu = B_\nu$, which, from the radiative equilibrium condition implies that

$$J(\tau) = S(\tau) = B(\tau) = \frac{\sigma T^4}{\pi} \tag{2-8}$$

where we have made use of equation (1-89). Given the solution of the purely formal equation (2-7), the additional hypothesis of LTE allows us to associate a temperature with the radiative equilibrium radiation field via equation (2-8).

Several important results may be obtained immediately from equation (2-6). Now taking the zero and first order moments of this equation, we have

$$\frac{dH}{d\tau} = J - J = 0 \tag{2-9}$$

which implies $H$ is constant, and

$$\frac{dK}{d\tau} = H \tag{2-10}$$

which yields the *exact* integral

$$K(\tau) = H\tau + c = \tfrac{1}{4}F\tau + c \tag{2-11}$$

To make further progress, we must see if we can obtain a relation among the moments $J$ and $K$. This can be done if we recall that at a very great depth in the atmosphere, the radiation field must be very nearly isotropic; some anisotropy must exist of course in order to transport a nonzero flux. We showed previously (Section 1-6) that in this limit the radiation field can be represented with very good accuracy by an expression of the form

$$I(\mu) = I_0 + I_1\mu \tag{2-12}$$

where $I_1 \ll I_0$. Using this expression, we immediately obtain by direct calculation the following results:

$$J = \frac{I_0}{2}\int_{-1}^{1} d\mu + \frac{I_1}{2}\int_{-1}^{1} \mu\, d\mu = I_0 \tag{2-13}$$

$$H = \frac{I_0}{2}\int_{-1}^{1} \mu\, d\mu + \frac{I_1}{2}\int_{-1}^{1} \mu^2\, d\mu = \frac{I_1}{3} \tag{2-14}$$

and

$$K = \frac{I_0}{2}\int_{-1}^{1} \mu^2\, d\mu + \frac{I_1}{2}\int_{-1}^{1} \mu^3\, d\mu = \frac{I_0}{3} \tag{2-15}$$

so that for $\tau \gg 1$, $J = 3K$. Thus, as $\tau \to \infty$, the fact that $K(\tau) \to \frac{1}{4}F\tau$ implies

$$J(\tau) \to \tfrac{3}{4}F\tau \tag{2-16}$$

so that asymptotically the mean intensity varies linearly with depth. On general grounds we may expect the behavior of $J(\tau)$ to depart most from linearity near the surface. These results suggest that a reasonable general expression for $J(\tau)$ would be

$$J(\tau) = \tfrac{3}{4}F[\tau + q(\tau)] \tag{2-17}$$

where the function $q(\tau)$, known as the *Hopf function*, remains to be determined. From equation (2-7) it is clear that $q(\tau)$ must satisfy the relation

$$\tau + q(\tau) = \frac{1}{2}\int_{0}^{\infty} [t + q(t)]E_1\,|t - \tau|\, dt \tag{2-18}$$

We may immediately obtain one further result from equations (2-11) and (2-17). By subtraction and passing to the limit of great depth, we have

$$\lim_{\tau\to\infty}[\tfrac{1}{3}J(\tau) - K(\tau)] = \lim_{\tau\to\infty}[\tfrac{1}{4}F\tau + q(\tau) - \tfrac{1}{4}F\tau - C] = 0$$

so that

$$C = q(\infty) \tag{2-19}$$

and

$$K(\tau) = \tfrac{1}{4}F[\tau + q(\infty)] \tag{2-20}$$

A solution of the gray atmosphere problem consists of a specification of function $q(\tau)$; we shall derive both approximate expressions and the exact solution below. Before deriving these solutions, however, we must clarify the nature and extent of the connection between the gray and nongray problems.

## 2-2. Relation to the Nongray Problem: Mean Opacities

As we have mentioned above, the gray problem can be solved exactly. It is therefore important to inquire whether any connection exists between the gray and nongray problem, for if there is, it may provide valuable guidance in the latter more complicated case.

Let us first compare side by side the gray and nongray transfer equations. Starting with the transfer equation and calculating the zero and first order moments, we have in the nongray and gray cases, respectively,

$$\frac{-\mu}{\rho\kappa_\nu}\frac{dI_\nu}{dz} = I_\nu - S_\nu \quad \text{(2-21a)} \qquad \frac{-\mu}{\rho\kappa}\frac{dI}{dz} = I - J \quad \text{(2-21b)}$$

$$\frac{-1}{\rho\kappa_\nu}\frac{dH_\nu}{dz} = J_\nu - S_\nu \quad \text{(2-22a)} \qquad \frac{-1}{\rho\kappa}\frac{dH}{dz} = 0 \quad \text{(2-22b)}$$

$$\frac{-1}{\rho\kappa_\nu}\frac{dK_\nu}{dz} = H_\nu \quad \text{(2-23a)} \qquad \frac{-1}{\rho\kappa}\frac{dK}{dz} = H \quad \text{(2-23b)}$$

where variables without frequency subscripts are defined to be integrated quantities, i.e.,

$$I = \int_0^\infty I_\nu\, d\nu,\ J = \int_0^\infty J_\nu\, d\nu,\ H = \int_0^\infty H_\nu\, d\nu, \text{ and } K = \int_0^\infty K_\nu\, d\nu$$

We now ask whether it is possible to define a mean opacity $\bar{\kappa}$, which is some specified weighted mean of the monochromatic opacity, in such a way that the monochromatic transfer equation when integrated over frequency has exactly the same form as the gray equation. Several such definitions have been suggested.

### THE FLUX-WEIGHTED MEAN

Suppose we wish to define a mean opacity in such a way as to guarantee an exact correspondence between the integrated form of equation (2-23a) and the gray equation (2-23b). If we can construct such a mean, we could again write $K(\bar{\tau}) = H\bar{\tau} + C$, as we did in the gray case. Starting from equation (2-23a),

$$\frac{-1}{\rho}\frac{dK_\nu}{dz} = \kappa_\nu H_\nu$$

and integrating over frequencies, we have

$$\frac{-1}{\rho}\frac{dK}{dz} = \int_0^\infty \kappa_\nu H_\nu \, d\nu = \bar{\kappa}_F H \tag{2-24}$$

where the last equality yields the desired identification with equation (2-23b). Thus, clearly,

$$\bar{\kappa}_F \equiv \frac{\int_0^\infty \kappa_\nu H_\nu \, d\nu}{H} = \frac{\int_0^\infty \kappa_\nu F_\nu \, d\nu}{F} \tag{2-25}$$

The opacity $\bar{\kappa}_F$ defined in this manner is called the *flux-weighted mean.* Note however that this choice does not reduce the nongray problem completely to the gray problem since the monochromatic equation (2-22a) cannot be made to correspond to equation (2-22b) with this choice of $\bar{\kappa}$. We are also faced with the practical problem that $F_\nu$ is not known a priori, and therefore we cannot actually calculate $\bar{\kappa}_F$. This latter problem could however be overcome by iterative construction of models and calculation of $\bar{\kappa}_F$. Although we have not fully attained our desired goal, the fact that the flux-weighted mean preserves the $K$-integral is quite important. In addition, the use of this mean yields the correct value of the radiation pressure gradient. Recalling equation (1-35),

$$\frac{-1}{\rho}\frac{dp_R}{dz} = \frac{\pi}{c}\int_0^\infty (\kappa_\nu + \sigma_\nu) F_\nu \, d\nu$$

and substituting equation (2-25), generalized to include a scattering coefficient $\sigma_\nu$, we may write

$$\frac{dp_R}{d\bar{\tau}} = \frac{\dfrac{\pi}{c}\displaystyle\int_0^\infty (\kappa_\nu + \sigma_\nu) F_\nu \, d\nu}{\dfrac{1}{F}\displaystyle\int_0^\infty (\kappa_\nu + \sigma_\nu) F_\nu \, d\nu} = \frac{\pi F}{c} = \frac{\sigma T_{\text{eff}}^4}{c} \tag{2-26}$$

Thus we have a simple expression for the radiation pressure gradient; this is a fact of some practical importance in the computation of model atmospheres for early-type stars since in these objects the mechanical force exerted by radiation appreciably affects the condition of hydrostatic equilibrium.

### THE ROSSELAND MEAN

Alternatively, suppose we wish to guarantee the relation

$$\int_0^\infty H_\nu \, d\nu = H$$

We see that from equations (2-23a) and (2-23b) that to do this we must choose $\bar{\kappa}$ such that

$$\frac{-1}{\rho}\int_0^\infty \frac{1}{\kappa_\nu}\frac{dK_\nu}{dz}\,d\nu = \int_0^\infty H_\nu\,d\nu = H = \frac{-1}{\rho\bar{\kappa}}\frac{dK}{dz}$$

or equivalently

$$\frac{1}{\bar{\kappa}} = \frac{\displaystyle\int_0^\infty \frac{1}{\kappa_\nu}\frac{dK_\nu}{dz}\,d\nu}{\displaystyle\int_0^\infty \frac{dK_\nu}{dz}\,d\nu} \tag{2-27}$$

Again we must face the practical difficulty that $K_\nu$ is not known a priori, and hence the indicated calculation cannot be performed. The above expression can at least be approximated in the following way. At great depth in the atmosphere, $K_\nu \to \frac{1}{3}J_\nu$ while $J_\nu \to B_\nu$. Thus we may write

$$\frac{dK_\nu}{dz} \approx \frac{1}{3}\frac{\partial B_\nu}{\partial T}\frac{dT}{dz}$$

We may then define a mean opacity $\bar{\kappa}_R$ such that

$$\frac{1}{\bar{\kappa}_R} = \frac{\displaystyle\frac{1}{3}\frac{dT}{dz}\int_0^\infty \frac{1}{\kappa_\nu}\frac{\partial B_\nu}{\partial T}\,d\nu}{\displaystyle\frac{1}{3}\frac{dT}{dz}\int_0^\infty \frac{\partial B_\nu}{\partial T}\,d\nu} = \frac{\displaystyle\int_0^\infty \frac{1}{\kappa_\nu}\frac{\partial B_\nu}{\partial T}\,d\nu}{\displaystyle\frac{dB}{dT}}$$

or

$$\frac{1}{\bar{\kappa}_R} = \frac{\pi}{4\sigma T^3}\int_0^\infty \frac{1}{\kappa_\nu}\frac{\partial B_\nu}{\partial T}\,d\nu \tag{2-28}$$

The opacity $\bar{\kappa}_R$ is called the *Rosseland mean* in honor of its originator. Note that the harmonic nature of the averaging process gives highest weight to those regions in which the opacity is lowest, and, as a result, the greatest amount of radiation is transported—a very desirable feature. Again, the use of $\bar{\kappa}_R$ does not permit a strict correspondence between equations (2-22a) and (2-22b) and therefore does not strictly allow the nongray problem to be replaced by the gray problem. On the other hand, it is obvious that the approximations made here are precisely those introduced in the derivation of the diffusion approximation to the transfer equation (1-104)

$$H_\nu = \frac{-1}{3\rho\kappa_\nu}\frac{dB_\nu}{dT}\frac{dT}{dr}$$

and that the frequency-integrated form of this equation

$$H = -\frac{1}{3\rho\bar{\kappa}}\frac{dB}{dT}\frac{dT}{dr} \tag{2-29}$$

is exact if $\bar{\kappa}$ is chosen to be $\bar{\kappa}_R$. It is clear therefore why Rosseland mean opacities are employed throughout the interior and apply as well to at least the deeper layers in the atmosphere.

### THE PLANCK MEAN

Several other expressions for mean opacities may be chosen. For example, if we demand that the mean opacity be defined to yield the correct value of the thermal emission, then we should require

$$\bar{\kappa}_P = \frac{\int_0^\infty \kappa_\nu B_\nu(T)\,d\nu}{B(T)} = \frac{\pi}{\sigma T^4}\int_0^\infty \kappa_\nu B_\nu(T)\,d\nu \tag{2-30}$$

The opacity $\bar{\kappa}_P$ is known as the Planck mean; it has the advantage of being calculable immediately. On the other hand, $\bar{\kappa}_P$ does not allow a correspondence between equations (2-23a) and (2-23b) and therefore does not have some of the desirable features possessed by $\bar{\kappa}_F$ or $\bar{\kappa}_R$. Nevertheless, the Planck mean does possess certain additional significance.

In particular, near the surface of the star, the condition of radiative equilibrium has the most direct statement of its physical content via equations (1-85) and (2-5). Taken together, we see that these equations allow a correspondence between equations (2-22a) and (2-22b) provided that we can choose $\bar{\kappa}$ to satisfy the relation

$$\int_0^\infty (\kappa_\nu - \bar{\kappa})(J_\nu - B_\nu)\,d\nu = 0$$

This integral will be dominated by those frequencies at which $\kappa_\nu \gg \bar{\kappa}$. Now if we assume a linear expansion for $B_\nu$ on a $\bar{\tau}$-scale, then by application of the $\Lambda$-operator we find [see equation (1-81)]

$$J_\nu(\tau) - B_\nu(\tau) \approx -\frac{1}{2}B_\nu(\tau)E_2(\tau) + \left(\frac{\bar{\kappa}}{\kappa_\nu}\right)\left(\frac{dB_\nu}{d\bar{\tau}}\right)\left[\frac{1}{2}E_3(\tau) + \frac{1}{2}\tau E_2(\tau)\right]$$

The last term is least important when $\kappa_\nu \gg \bar{\kappa}$ so that $\bar{\kappa}$ should essentially fulfill the requirement that

$$\int_0^\infty \kappa_\nu B_\nu\,d\nu = \bar{\kappa}\int_0^\infty B_\nu\,d\nu$$

if we can neglect the factor $E_2(\tau)$, which is permissible for $\tau \ll 1$. This analysis suggests that near the surface, the Planck mean is, physically speaking, the most relevant mean value.

THE ABSORPTION MEAN

On the other hand, we might demand that the mean opacity yield a correct total for the amount of energy absorbed. Then we obtain the so-called *absorption mean*

$$\bar{\kappa}_J = \frac{\int_0^\infty \kappa_\nu J_\nu \, d\nu}{J} \tag{2-31}$$

Once again we cannot calculate $\bar{\kappa}_J$ a priori, and, in addition, it fails to allow a correspondence between the gray and nongray transfer equations.

SUMMARY COMMENTS ON MEAN OPACITIES

We have seen that no one of the mean opacities described above allows in itself a complete correspondence of the nongray problem to the gray problem. Yet insofar as they allow even a partial reduction, mean opacities can provide a useful first estimate of the temperature distribution in a stellar atmosphere if we assume, as a starting approximation, $T(\bar{\tau}) = T_{\text{gray}}(\bar{\tau})$ and then improve this estimate by an appropriate correction procedure. Indeed, we will find that certain mean opacities appear explicitly in some temperature-correction procedures. From a historical point of view we must recognize that before the advent of high-speed computers, the nongray atmospheres problem required far too much computation to permit a direct attack, and the use of the mean opacities $\bar{\kappa}_R$ and $\bar{\kappa}_P$ provided a practical method of approaching an otherwise intractable problem. In fact, the answers obtained in this way do not compare too unfavorably with more recent results despite the apparent crudeness of the approximation. We have mentioned only some of the more basic aspects of the properties of mean opacities; further information may be obtained from the discussions of Michard (*Ann. d'Ap.*, **12**, 291, 1949) and Kourganoff (Ref. 23, p. 234 ff.).

## 2-3. Approximate Solutions

THE EDDINGTON APPROXIMATION

We noted previously that in the deeper layers of the atmosphere, the expression $J = 3K$ is valid. Actually, this relation holds under a wider variety of circumstances than previously indicated. To demonstrate this fact, let us consider a few examples.

(a) Suppose $I(\mu)$ is expandable in *odd* powers of $\mu$ only, i.e.,

$$I(\mu) = I_0 + \sum_{k=1}^{\infty} I_{2k-1} \mu^{2k-1} \tag{2-32}$$

It is obvious that since all of the terms in $\mu$ are odd functions, they will integrate to zero on the interval $(-1, 1)$ when weighted by even powers of $\mu$. Therefore only the term $I_0$ contributes to either $J$ or $K$, and by direct calculation we again recover $J = 3K$.

(b) Suppose $I(\mu) = I_0$ for $0 \leq \mu \leq 1$, but $I(\mu) = 0$ for $\mu < 0$. This simple choice might be regarded as a schematic boundary condition at the surface. Then by direct calculation we find

$$J = \frac{1}{2} I_0 \int_0^1 d\mu = \frac{1}{2} I_0 \tag{2-33}$$

$$H = \frac{1}{2} I_0 \int_0^1 \mu \, d\mu = \frac{1}{4} I_0 = \frac{1}{2} J \tag{2-34}$$

and

$$K = \frac{1}{2} I_0 \int_0^1 \mu^2 \, d\mu = \frac{1}{6} I_0 = \frac{1}{3} J \tag{2-35}$$

so that again we have $J = 3K$.

(c) Suppose $I(\mu) = I_+$ for $0 \leq \mu \leq 1$ and $I(\mu) = I_-$ for $-1 \leq \mu \leq 0$. This is the so-called *two-stream model* first employed by Schuster (*Ap. J.*, **21**, 1, 1905, a paper of great historical importance in the development of the theory of radiative transfer). Then we readily find

$$J = \frac{1}{2} I_+ \int_0^1 d\mu + \frac{1}{2} I_- \int_{-1}^0 d\mu = \frac{1}{2}(I_+ + I_-) \tag{2-36}$$

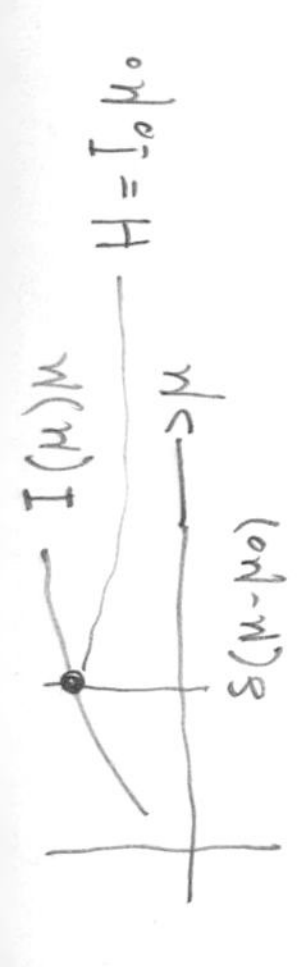


$$H = \frac{1}{2} I_+ \int_0^1 \mu \, d\mu + \frac{1}{2} I_- \int_{-1}^0 \mu \, d\mu = \frac{1}{4}(I_+ - I_-) \tag{2-37}$$

$$K = \frac{1}{2} I_+ \int_0^1 \mu^2 \, d\mu + \frac{1}{2} I_- \int_{-1}^0 \mu^2 \, d\mu = \frac{1}{6}(I_+ + I_-) = \frac{1}{3} J \tag{2-38}$$

so once again we have $J = 3K$.

(d) Finally, suppose $I(\mu) = \delta(\mu - \mu_0)$, which describes a strictly monodirectional beam. Then

$$J = I_0 \tag{2-39}$$

$$H = I_0 \mu_0 \tag{2-40}$$

and

$$K = I_0 \mu_0{}^2 \tag{2-41}$$

In this case we do *not* have $J = 3K$.

In view of considerations such as those listed above, Eddington made the simplifying approximation that $J = 3K$ *everywhere* in the atmosphere. Then the exact integral $K = \frac{1}{4}F\tau + C$ implies, in the Eddington approximation, $J_E(\tau) = \frac{3}{4}F\tau + C'$. To evaluate the constant $C'$, we may calculate the emergent flux and fit it to the desired value. Thus from equation (1-64) we have

$$F(0) = 2\int_0^\infty \left(\frac{3}{4}F\tau + C'\right)E_2(\tau)\,d\tau = 2C'E_3(0) + \frac{3}{4}F\left[\frac{4}{3} - 2E_4(0)\right] \quad (2\text{-}42)$$

so that using the fact that $E_n(0) = 1/(n-1)$ and requiring $F(0) \equiv F$ implies $C' = \frac{1}{2}F$. Thus we finally obtain

$$J_E(\tau) = \tfrac{3}{4}F(\tau + \tfrac{2}{3}) \quad (2\text{-}43)$$

In Eddington's approximation we clearly have $q(\tau) \equiv 2/3$.

From the radiative equilibrium requirement with the additional assumption of LTE, we may obtain the run of temperature with depth. Thus, combining equations (1-91), (2-8), and (2-43), we may write

$$B(\tau) = \frac{\sigma T^4}{\pi} = J_E(\tau) = \frac{3}{4}\frac{\sigma T_{\text{eff}}^4}{\pi}\left(\tau + \frac{2}{3}\right) \quad (2\text{-}44)$$

or

$$T^4 = \tfrac{3}{4}T_{\text{eff}}^4(\tau + \tfrac{2}{3}) \quad (2\text{-}45)$$

Equation (2-45) predicts that the boundary temperature is related to the effective temperature in the ratio $T_0/T_{\text{eff}} = (\frac{1}{2})^{1/4} = 0.841$. By way of comparison, the exact solution that $T_0/T_{\text{eff}} = (\sqrt{3}/4)^{1/4} = 0.8114$ so that in fact the Eddington approximation is fairly accurate. Note further that when $\tau = 2/3$, equation (2-45) predicts that $T = T_{\text{eff}}$. This result has given rise to the useful conceptual notion that the "effective depth" of continuum formation is $\tau = 2/3$. In fact, this is often a rather good estimate. In particular, we might note that a photon emitted at $\tau = 0.67$ has a chance of the order $e^{-0.67} \approx 0.5$ of emerging at the surface; this corresponds in a reasonable way with what we might intuitively identify as the place of continuum formation.

From equation (2-43) we may calculate directly the angular dependence of the emergent intensity. Substituting into equation (1-55), we have

$$I_E(0, \mu) = \frac{3}{4}F\int_0^\infty \left(\tau + \frac{2}{3}\right)e^{-\tau/\mu}\frac{d\tau}{\mu} = \frac{3}{4}F\left(\mu + \frac{2}{3}\right) \quad (2\text{-}46)$$

The center of the star's disc, as seen by the observer, corresponds to $\theta = 0$ or $\mu = 1$. If we take the ratio $I(0, \mu)/I(0, 1)$ we obtain the relative intensity from center to limb, which is referred to as the *limb-darkening law*. In the Eddington approximation

$$\frac{I_E(0, \mu)}{I_E(0, 1)} = \frac{3}{5}\left(\mu + \frac{2}{3}\right) \quad (2\text{-}47)$$

This result predicts the limb intensity to be 40% of the central intensity. Observations of the sun in the visual regions of the spectrum are actually in good agreement with this result, and in fact it was precisely this agreement that led K. Schwarzschild to propose the validity of radiative equilibrium in the outermost layers of the solar atmosphere.

Note in passing from equations (2-43) and (2-46) that $J_E(0) = \frac{1}{2}F$, and $I_E(0, 0) = \frac{1}{2}F$ so that in the Eddington approximation

$$J_E(0) = I_E(0, 0) \tag{2-48}$$

We shall show later that this result is *exact*. Anticipating the calculation of the exact solution, it is of interest to ask how accurate the Eddington solution is. One finds that the worst error occurs at the surface where $\Delta J/J = (J_E - J_{\text{exact}})/J_{\text{exact}} = 0.155$; a more complete tabulation is given by Kourganoff (Ref. 23, p. 89). This error is not surprising since the assumption $J = 3K$ cannot be expected to be completely accurate at the surface.

An improvement upon the Eddington solution can be obtained by using the relation that defines the gray problem, namely, equation 2-7:

$$J(\tau) = \Lambda_\tau[J(t)]$$

Writing $J_E{}^*$ as the new estimate of $J$, we have

$$J_E{}^*(\tau) = \Lambda_\tau[\tfrac{3}{4}F(\tau + \tfrac{2}{3})] = \tfrac{3}{4}F[\tau + \tfrac{1}{2}E_3(\tau) + \tfrac{2}{3} - \tfrac{1}{2}E_2(\tau)] \tag{2-49}$$

Observe that asymptotically $J_E{}^*(\tau) \to J_E(\tau)$ for $\tau \gg 1$. The largest difference. between $J_E{}^*$ and $J_E$ occurs at the surface, where we find $J_E{}^*(0)/J_E(0) = 7/8$. The new estimate of $T_0/T_{\text{eff}}$ is thus $(7/16)^{1/4} = 0.813$; $q(\infty)$ remains 2/3 while $q(0)$ drops from $2/3$ to $7/12 = 0.583$, in comparison with the exact value $q_{\text{exact}}(0) = 1/\sqrt{3} = 0.577$. By comparison with the exact solution we find that the error at the surface has been reduced to 1% and at worst is only about 1.6%; further comparisons are given by Kourganoff (Ref. 23, p. 117). The solution is thus much improved, particularly at the surface. Another application of the lambda operator is possible in principle but immediately introduces functions of the form $\Lambda_\tau[E_n(t)]$ which are cumbersome to compute [see Ref. 23, equations (14–50), (14–53), (37–41), and (37–42)].

By applying the X-operator to $J_E(\tau)$, we may calculate $K_E{}^*(\tau)$ and compare the ratio $\varphi(\tau) = K_E{}^*(\tau)/J_E{}^*(\tau)$ to the value of 1/3 assumed in the derivation of $J_E(\tau)$. We find

$$K_E{}^*(\tau) = \tfrac{1}{4}X_\tau[J_E(t)] = \tfrac{3}{16}F[\tfrac{4}{3}\tau + 2E_5(\tau) + \tfrac{8}{9} - \tfrac{4}{3}E_4(\tau)] \tag{2-50}$$

so that

$$\varphi(\tau) = \frac{\frac{1}{3}[\tau + \frac{2}{3} + \frac{3}{2}E_5(\tau) - E_4(\tau)]}{[\tau + \frac{2}{3} + \frac{1}{2}E_3(\tau) - \frac{1}{3}E_2(\tau)]} \tag{2-51}$$

We easily see that $\lim_{\tau\to\infty} \varphi(\tau) = 1/3$ as desired. Further, we find that $\varphi(0) = 17/42 = 0.405$ so that $K(0)/J(0) > 1/3$. It is easy to understand why this occurs. We note from equation (2-47) that $I(0, \mu)$ is always greater than $I(0, 0)$ which, by equation (2-48) equals $J(0)$. Therefore, in forming the weighted mean against $\mu$ used in computing $K$, values larger than $J(0)$ will dominate. Physically, this merely states that at small angles from the center of the disc, we see into deeper layers of the atmosphere and hence to larger values of the source function. These small angles (large $\mu$) also contribute most heavily to $K(0)$.

An improved estimate of the limb darkening may be obtained from $J_E^*(\tau)$ by direct calculation with equation (1-55); we obtain

$$I_E^*(0, \mu) = \frac{3}{4} F \int_0^\infty e^{-t/\mu} \left[ t + \frac{2}{3} + \frac{1}{2} E_3(t) - \frac{1}{3} E_2(t) \right] \frac{dt}{\mu}$$

$$= \frac{3}{4} F \left[ \frac{7}{12} + \frac{1}{2} \mu + \left( \frac{\mu}{3} + \frac{\mu^2}{2} \right) \ln \left( \frac{1 + \mu}{\mu} \right) \right] \tag{2-52}$$

Note that since $\lim_{\mu\to 0} (\mu \ln \mu) = 0$, we have $I_E^*(0, 0) = 7/16F = J_E^*(0)$ which shows that the improved solution again satisfies the (exact) relation given in equation (2-48). This limb-darkening law is an improvement over that obtained previously, particularly for small $\mu$, as may be seen in Table 2–1.

One final comment about the Eddington approximation should be made. Although the source function $J_E(\tau)$ was derived under the assumption that

TABLE 2-1. Comparison of Eddington, Improved Eddington, and Exact Limb-Darkening Laws

| $\mu$ | $\dfrac{I_E(0, \mu)}{F}$ | $\dfrac{I_E^*(0, \mu)}{F}$ | $\dfrac{I_{\text{exact}}(0, \mu)}{F}$ |
|---|---|---|---|
| 0.0 | 0.5000 | 0.4375 | 0.4330 |
| 0.1 | 0.5750 | 0.5439 | 0.5401 |
| 0.2 | 0.6500 | 0.6290 | 0.6280 |
| 0.3 | 0.7250 | 0.7095 | 0.7112 |
| 0.4 | 0.8000 | 0.7880 | 0.7921 |
| 0.5 | 0.8750 | 0.8653 | 0.8716 |
| 0.6 | 0.9500 | 0.9420 | 0.9501 |
| 0.7 | 1.0250 | 1.0185 | 1.0280 |
| 0.8 | 1.1000 | 1.0942 | 1.1053 |
| 0.9 | 1.1750 | 1.1702 | 1.1824 |
| 1.0 | 1.2500 | 1.2457 | 1.2591 |

the flux be constant, the flux computed from $J_E(\tau)$ is *not* constant. Thus, applying equation (1-64), we find

$$F_E(\tau) = \Phi_\tau[J_E(t)] = \tfrac{3}{4}F[\tfrac{4}{3} - 2E_4(\tau) + \tfrac{4}{3}E_3(\tau)] \tag{2-53}$$

or

$$\frac{F_E(\tau)}{F} = 1 + [E_3(\tau) - \tfrac{3}{2}E_4(\tau)] \tag{2-54}$$

Now at $\tau = 0$ we have $E_3(0) = \frac{1}{2}$ and $E_4(0) = \frac{1}{3}$, so equation (2-54) shows that $F_E(0)$ does indeed equal $F$. At greater depths the errors grow, reaching a maximum of about 2.7% near $\tau = 0.5$; more complete results are given by Kourganoff (Ref. 23, p. 89). If one were to compute the flux from $J_E{}^*(\tau)$ instead, much smaller errors would be obtained.

## THE LAMBDA-ITERATION METHOD

We found above that application of the lambda operator to the approximate Eddington solution greatly improved the accuracy near the upper boundary but left it unaltered at infinity. This fact is of importance both in the gray and nongray problems and merits attention. To be specific, suppose we have an estimate of $q(\tau)$ which differs from the true $q(\tau)$ only by a constant, i.e.,

$$q(\tau) = q_{\text{exact}}(\tau) + C \tag{2-55}$$

Then, application of the $\Lambda$-operator yields

$$\Lambda_\tau[t + q(t)] = \Lambda_\tau[t + q_{\text{exact}}(t) + C] = \tau + q_{\text{exact}}(\tau) + \Lambda_\tau(C) \tag{2-56}$$

so that the error in the solution after lambda iteration is

$$\Delta q(\tau) = C[1 - \tfrac{1}{2}E_2(\tau)] \tag{2-57}$$

We see that the error is halved at the boundary but remains virtually unaffected at depth. Indeed at $\tau = 1$, the error is diminished only by 20%. Physically, this is related to the fact that photons can propagate on the average only over an optical depth of order unity. Thus, if we start from a solution which in essence ignores the existence of the boundary and apply the $\Lambda$-operator, the presence of the boundary will be "felt" by the new solution only to optical depth unity; below this, the solution will remain "unaware" of the boundary. In principle, one can propagate the information to greater depth by successive applications of the $\Lambda$-operator; in fact, one can show (Ref. 38, p. 31) that $\lim_{n\to\infty} \Lambda^n(1) = 0$ so that in principle an initial error can be removed by iteration; in practice, the convergence is too slow to be of value. Another example showing the limited range of effectiveness of the

$\Lambda$-operator was devised by Unsöld (Ref. 36, p. 141). Suppose we have an approximate estimate of $J$ with an error $\Delta J$ of the form:

$$\begin{aligned} \Delta J &= 0 \quad \text{for} \quad \tau \leq \tau_1 \\ \Delta J &= 1 \quad \text{for} \quad \tau_1 \leq \tau \leq \tau_2 \\ \Delta J &= 0 \quad \text{for} \quad \tau_2 \leq \tau \end{aligned} \tag{2-58}$$

Then application of the $\Lambda$-operator yields

$$\begin{aligned} \Lambda(\Delta J) &= \tfrac{1}{2}[E_2(\tau_1 - \tau) - E_2(\tau_2 - \tau)] \quad &\text{for} \quad \tau \leq \tau_1 \\ \Lambda(\Delta J) &= 1 - \tfrac{1}{2}[E_2(\tau_2 - \tau) + E_2(\tau - \tau_1)] \quad &\text{for} \quad \tau_1 \leq \tau \leq \tau_2 \\ \Lambda(\Delta J) &= \tfrac{1}{2}[E_2(\tau - \tau_2) - E_2(\tau - \tau_1)] \quad &\text{for} \quad \tau_2 \leq \tau \end{aligned} \tag{2-59}$$

As is obvious by symmetry, the maximum value of $\Lambda(\Delta J)$ occurs at $(\tau_1 + \tau_2)/2$ and is

$$\|\Lambda(\Delta J)\| = 1 - E_2\left(\frac{\tau_2 - \tau_1}{2}\right)$$

which is clearly less than one. Thus successive iterations will tend to reduce the size of $\Delta J$ and will be the more effective the smaller the value of $(\tau_2 - \tau_1)$. Here again we see that the $\Lambda$-operator has an extremely short effective range, in fact, less than about $\Delta\tau = 1$.

### THE METHOD OF DISCRETE ORDINATES

The method to be described now furnishes both a means of obtaining approximate solutions to the gray problem and a method by which the exact solution can be determined. Substitution of equations (2-8) and (1-3) into equation (2-4) allows us to write the equation to be solved in the form

$$\mu \frac{dI}{d\tau} - I - \frac{1}{2}\int_{-1}^{1} I(\tau, \mu)\, d\mu \tag{2-60}$$

which is classified as an *integro-differential equation.* The essential difficulty in obtaining a solution is introduced by the presence of the integral over angle. However, definite integrals such as that in equation (2-60) may be performed numerically by replacing the integral with a *quadrature sum* evaluated at a finite set of points on the interval of integration. Thus we can write

$$\int_{-1}^{1} I(\tau, \mu)\, d\mu \approx \sum_{j=-n}^{n} a_j I(\tau, \mu_j) \tag{2-61}$$

The accuracy of the quadrature depends both upon the order $n$ and the distribution of the points on the interval. We might, for example, choose the three points $\mu = (-1, 0, 1)$, in which case we would have an ordinary *Simpson's rule* integration. A better choice than placing the points at equal intervals is to make use of a so-called *Gaussian quadrature* formula, in which the points are chosen to be zeros of Legendre polynomials. It would take us too far afield to discuss the construction and accuracy of quadrature formulae; a brief but excellent discussion has been given by Chandrasekhar (Ref. 9, Chap. 2). One important result that we will merely state is that while an $n$-point formula with equally spaced points is exact for polynomials of order $n - 1$ (for $n$ even) or $n$ (for $n$ odd), an $n$-point Gauss formula is exact for polynomials of order $2n - 1$. A variant form known as the *double-Gauss* formula, which is superior in some respects, was suggested by Sykes (*M. N.*, **111**, 377, 1951). One general property of all of these formulae is that $\mu_{-j} = -\mu_j$ while $a_{-j} = a_j$.

Having chosen a quadrature formula, we replace the integro-differential transfer equation by a *set* of 2n equations,

$$\mu_i \frac{dI_i}{d\tau} = I_i - \frac{1}{2} \sum_{j=-n}^{n} a_j I_j, \qquad (i = \pm 1, \ldots, \pm n) \tag{2-62}$$

In this way we have replaced the continuous radiation field by a finite set of *pencils*, each of which represents the value of $I(\mu)$ over a finite interval. On physical grounds it is quite reasonable to expect the solution to become exact as $n \to \infty$.

Observing that the system is linear and of the first order, we take a trial solution of the form $I_i = q_i \exp(-k\tau)$ where $g_i$ and $k$ are to be specified. Substituting into equation (2-62), we obtain

$$-\mu_i g_i k e^{-k\tau} = g_i e^{-k\tau} - \frac{1}{2} e^{-k\tau} \sum_{j=-n}^{n} a_j g_j \tag{2-63}$$

or

$$g_i(1 + \mu_i k) = \frac{1}{2} \sum a_j g_j = C \tag{2-64}$$

so that

$$g_i = \frac{C}{1 + \mu_i k} \tag{2-65}$$

If we use this value of $g_i$ and again substitute into the equation of transfer, we find

$$\frac{1}{2} \sum_{j=-n}^{n} \frac{a_j}{1 + \mu_j k} = 1 \tag{2-66}$$

This equation is an *eigenvalue equation* for $k$ and is known as the *characteristic equation*. Now recalling that $a_{-j} = a_j$ and $\mu_{-j} = -\mu_j$, we can add together the terms in $\pm j$ to obtain

$$\sum_{j=1}^{n} \frac{a_j}{1 - \mu_j^2 k^2} = 1 \tag{2-67}$$

We now wish to find the solution of equation (2-67).

If we observe that the integral $\frac{1}{2}\int_{-1}^{1} d\mu = 1$ is approximated by $\frac{1}{2}\sum a_j$ in the discrete ordinate scheme, then we see that

$$\sum_{j=-n}^{n} a_j = 1 \tag{2-68}$$

We may then see that $k^2 = 0$ is a solution of equation (2-67) since substitution leads directly to equation (2-68). There are an additional $(n - 1)$ nonzero solutions for $k^2$, as may be seen as follows.

Note that $k^2 = 1/\mu_j^2$ is a pole of the sum on the left-hand side of equation (2-67). For $k = (1/\mu_j) - \varepsilon$, the sum is greater than zero, and by making $\varepsilon$ arbitrarily small, it can be made arbitrarily large. For $k = (1/\mu_j) + \varepsilon$, the sum is less than zero and again can be made arbitrarily large (in absolute value). Let us define a characteristic function

$$T(k^2) \equiv 1 - \sum_{j=1}^{n} \frac{a_j}{1 - k^2 \mu_j^2} \tag{2-69}$$

which has the behavior sketched in Figure 2-1. From this figure we see that there must exist $(n - 1)$ nonzero roots for $k^2$, such that

$$\frac{1}{\mu_1^2} < k_1^2 < \frac{1}{\mu_2^2} < \cdots < k_{n-1}^2 < \frac{1}{\mu_n^2}$$

Note in passing that since the largest $\mu_i$ must be less than unity, the smallest nonzero $k$ must be greater than unity. Thus in all, there exist $2n - 2$ nonzero

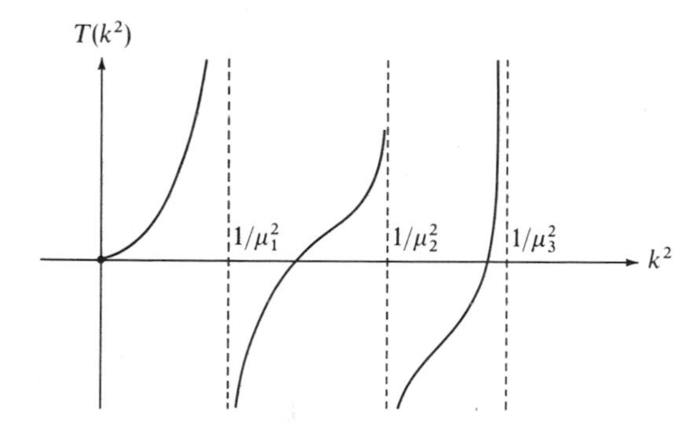


FIG. 2-1. Schematic behavior of characteristic function $T(k^2)$. One root occurs at $k^2 = 0$. Nonzero roots are bounded by poles at successive values of $1/\mu^2$.

values for the $k$'s in pairs of the form $\pm k_i$, where $i$ runs from 1 to $(n-1)$. The *general* solution for the system in equation (2-62) is therefore of the form

$$I_i(\tau) = \sum_{\alpha=1}^{n-1} \frac{L_\alpha e^{-k_\alpha \tau}}{1+\mu_i k_\alpha} + \sum_{\alpha=1}^{n-1} \frac{L_{-\alpha} e^{k_\alpha \tau}}{1-\mu_i k_\alpha} \tag{2-70}$$

We must now seek a particular solution corresponding to the root $k^2 = 0$. In view of the solution obtained in the Eddington approximation, as well as the general expectation that $J$ must become linear in $\tau$ at depth, we examine trial solutions of the form

$$I_i = b(\tau + q_i) \tag{2-71}$$

Inserting equation (2-71) into equation (2-62), we have

$$\mu_i b = b(\tau + q_i) - \frac{1}{2} b\left(\tau \sum_{j=-n}^{n} a_j + \sum_{j=-n}^{n} a_j q_j\right)$$

which reduces to

$$q_i = \mu_i + \frac{1}{2} \sum_{j=-n}^{n} a_j q_j$$

Now observing that

$$\sum_{j=-n}^{n} a_j \mu_j = \int_{-1}^{1} \mu \, d\mu = 0$$

we see that a simple solution for $q_i$ is

$$q_i = Q + \mu_i$$

so that the particular solution for $k = 0$ is

$$I_i(\tau) = b(\tau + Q + \mu_i)$$

and the complete solution is

$$I_i(\tau) = b\left[\tau + Q + \mu_i + \sum_{\alpha=1}^{n-1} \frac{L_\alpha e^{-k_\alpha \tau}}{1 + k_\alpha \mu_i} + \sum_{\alpha=1}^{n-1} \frac{L_{-\alpha} e^{k_\alpha \tau}}{1 - k_\alpha \mu_i}\right] \tag{2-72}$$

We must now specify the $2n$ unknown coefficients $Q$, $b$, and $L_{\pm\alpha}$. This is done by application of the boundary conditions. In the case of a *finite* atmosphere of total optical depth $T$, we have $2n$ boundary conditions of the form

$$I_{-i}(0) \equiv f(\mu_i) = b\left(Q - \mu_i + \sum_{\alpha=1}^{n-1} \frac{L_\alpha}{1 - k_\alpha \mu_i} + \sum_{\alpha=1}^{n-1} \frac{M_\alpha e^{-k\alpha T}}{1 + k_\alpha \mu_i}\right) \tag{2-73}$$

and

$$I_{+i}(T) \equiv g(\mu_i) = b\left(Q + T + \mu_i + \sum_{\alpha=1}^{n-1} \frac{L_\alpha e^{-k\alpha T}}{1 + k_\alpha \mu_i} + \sum_{\alpha=1}^{n-1} \frac{M_\alpha}{1 - k_\alpha \mu_i}\right) \tag{2-74}$$

where we have defined $M_\alpha \equiv L_\alpha \exp(k_\alpha T)$ to improve the numerical condition of the equations. The solution of these sets of linear equations is straightforward and can be carried out by standard numerical techniques.

In the case of a semi-infinite atmosphere in radiative equilibrium, we have the boundary condition that $I_{-i}(0) = 0$ and that $I(\tau)$ must not diverge exponentially as $\tau \to \infty$. The latter constraint is satisfied by demanding

$$L_{-\alpha} \equiv 0 \tag{2-75}$$

while the former requires

$$Q + \sum_{\alpha=1}^{n-1} \frac{L_\alpha}{1 - k_\alpha \mu_i} - \mu_i = 0 \tag{2-76}$$

Solution of equations (2-76) yields the $n$ unknowns $Q$ and $L_\alpha$. In addition, we require that the flux must equal the nominal flux $F$. Thus we demand that

$$F \equiv 2\int_{-1}^{1} I(\mu)\mu \, d\mu = 2\sum_{j=-n}^{n} a_j \mu_j I_j \tag{2-77}$$

Substituting equations (2-72), we have

$$F = 2b\left[(\tau + Q)\sum_{j=-n}^{n} a_j \mu_j + \sum_{j=-n}^{n} a_j \mu_j^2 + \sum_{\alpha=1}^{n-1} L_\alpha e^{-k_\alpha \tau} \sum_{j=-n}^{n} \frac{a_j \mu_j}{1 + \mu_j k_\alpha}\right] \tag{2-78}$$

Now the first sum is zero and the second is 2/3 while the third sum may be written

$$\frac{1}{k_\alpha}\sum_{j=-n}^{n} a_j\left(1 - \frac{1}{1 + \mu_j k_\alpha}\right) = \frac{2}{k_\alpha}\left(1 - \frac{1}{2}\sum_{j=-n}^{n} \frac{a_j}{1 + \mu_j k_\alpha}\right) = 0 \tag{2-79}$$

since the $k$'s are solutions of the characteristic equation. Thus we find that $b = \frac{3}{4}F$, as might have been expected on the basis of the Eddington solution. Note also that the quadrature calculation of $F$ yields a *constant flux* automatically. Finally, the complete solution in the semi-infinite case is

$$I_i(\tau) = \frac{3}{4} F\left(\tau + Q + \mu_i + \sum_{\alpha=1}^{n-1} \frac{L_\alpha e^{-k_\alpha \tau}}{1 + k_\alpha \mu_i}\right) \tag{2-80}$$

We may now compute the mean intensity $J(\tau) \equiv B(\tau)$ by the quadrature sum

$$J(\tau) = \frac{1}{2}\sum_{j=-n}^{n} a_j I_j = \frac{3}{4} F\left[(\tau + Q)\frac{1}{2}\sum_{j=-n}^{n} a_j + \frac{1}{2}\sum_{j=-n}^{n} a_j \mu_j \right.$$
$$\left. + \sum_{\alpha=1}^{n-1} L_\alpha e^{-k_\alpha \tau} \frac{1}{2}\sum_{j=-n}^{n} \frac{a_j}{1 + \mu_j k_\alpha}\right] \tag{2-81}$$

Making use of the characteristic equation, this reduces to

$$J(\tau) = \frac{3}{4} F\left[\tau + Q + \sum_{\alpha=1}^{n-1} L_\alpha e^{-k_\alpha \tau}\right] \tag{2-82}$$

Thus the discrete-ordinate solution for $q(\tau)$ is

$$q(\tau) = Q + \sum_{\alpha=1}^{n-1} L_\alpha e^{-k_\alpha \tau} \tag{2-83}$$

Clearly, $q(\infty) \equiv Q$. Before discussing several general properties of the solution, let us merely mention a few numerical results of the method.

Chandrasekhar (*Ap. J.*, **100**, 76, 1944) has carried out the solution for $n \leq 4$; he obtains the following results:

$n = 1$: $q(\tau) = 1/\sqrt{3}$ (exact value for $\tau = 0$)
$n = 2$: $q(\tau) = 0.694025 - 0.116675 \exp(-1.97203\ \tau)$
$n = 3$: $q(\tau) = 0.703899 - 0.101245 \exp(-3.20295\ \tau) - 0.02530 \exp(-1.22521\ \tau)$
$n = 4$: $q(\tau) = 0.70692 - 0.08392 \exp(-4.45808\ \tau) - 0.03619 \exp(-1.59178\ \tau) - 0.00946 \exp(-1.10319\ \tau)$

The error in $J(\tau)$, compared with the exact solution, is about 2% for $n = 4$; a more detailed comparison is given by Kourganoff (Ref. 23, p. 109). This error is a bit larger than one might have expected and is in large part due to the choice of a standard Gaussian quadrature formula. Sykes (*M. N.*, **111**, 377, 1951) used the so-called double-Gauss formula and obtained a solution accurate to 0.6% even for $n = 2$. The main importance of the discrete-ordinate method is not that it is amazingly accurate in low orders of approximation (it is not) but that it can be generalized to yield the exact solution, as will be shown in Section 2-4 below.

Let us now establish several important relations that will be of use later. Consider first the characteristic equation (2–69):

$$T(k^2) \equiv 1 - \sum_{j=1}^{n} \frac{a_j}{1 - k^2 \mu_j^2}$$

To simplify the notation we define $x = 1/k$ and $X = 1/k^2$. We previously noted that the smallest nonzero $k$ is greater than unity; hence the largest $X$ must be less than unity. We may now write two equivalent forms for the characteristic function, namely,

$$T(X) \equiv 1 - X \sum_{j=1}^{n} \frac{a_j}{X - \mu_j^2} = \sum_{j=1}^{n} a_j - X \sum_{j=1}^{n} \frac{a_j}{X - \mu_j^2} = \sum_{j=1}^{n} \frac{a_j \mu_j^2}{\mu_j^2 - X} \tag{2-84}$$

We may clear $T(X)$ of fractions by multiplying through by $\prod_{j=1}^{n}(\mu_j^2 - X)$. Then we obtain the function $P(X)$,

$$P(X) = \prod_{j=1}^{n}(\mu_j^2 - X)T(X) = \sum_{i=1}^{n} a_i \mu_i^2 \prod_{j \neq i}^{n}(\mu_j^2 - X) \tag{2-85}$$

which clearly is a polynomial of order $(n - 1)$ in $X$. Now we know that $T(X)$ has the $(n - 1)$ roots $X_1 = 1/k_1^2, \ldots, X_{n-1} = 1/k_{n-1}$, so that the polynomial $P(X)$ must be of the form

$$P(X) = C(X - X_1) \cdots (X - X_{n-1}) \tag{2-86}$$

To evaluate the constant, we note that in equation (2-85) the coefficient of the term in $X^{n-1}$ can be identified as

$$(-1)^{n-1} \sum_{i=1}^{n} a_i \mu_i^2 = \frac{(-1)^{n-1}}{3} \tag{2-87}$$

In equation (2-86), the coefficient of $X^{n-1}$ is simply $C$ itself. Thus we have

$$P(X) = \tfrac{1}{3}(X_1 - X) \cdots (X_{n-1} - X) \tag{2-88}$$

or

$$T(X) = \frac{\frac{1}{3}\prod_{i=1}^{n-1}(X_i - X)}{\prod_{j=1}^{n}(\mu_j^2 - X)} \tag{2-89}$$

An important relation may also be established between the quadrature points $\mu_i$ and roots $k_i$. Thus, using equations (2-84) and (2-89), we may write

$$\frac{1}{3}\prod_{j=1}^{n-1}(X_j - X) = T(X)\prod_{j=1}^{n}(\mu_j^2 - X)$$

$$= \prod_{j=1}^{n}(\mu_j^2 - X) + X\sum_{i=1}^{n} a_i \prod_{j \neq i}^{n}(\mu_j^2 - X) \tag{2-90}$$

Now observing that the term on the right which is independent of $X$ is $\prod_{j=1}^{n}\mu_j^2$ while that on the left is $\frac{1}{3}\prod_{j=1}^{n-1} X_j$, we must have

$$\mu_1\mu_2 \cdots \mu_n k_1 k_2 \cdots k_{n-1} = \frac{1}{\sqrt{3}} \tag{2-91}$$

We will make use of this result at a later time.

Now consider the emergent intensity $I(0, \mu)$. Let us define a function $S(\mu)$ such that

$$S(\mu_i) = Q - \mu_i + \sum_{\alpha=1}^{n-1} \frac{L_\alpha}{1 - k_\alpha \mu_i} \tag{2-92}$$

The surface boundary conditions in equation (2-76) then may be written

$$I(0, -\mu_i) = 0 = \tfrac{3}{4}F\, S(\mu_i) \tag{2-93}$$

We now assume that we may generalize $S(\mu)$ to apply at all values of $\mu$ so that

$$I(0, \mu) = \tfrac{3}{4}F\,S(-\mu) \tag{2-94}$$

for $\mu \geq 0$. Note that with this generalization, $I(0, -\mu)$ is *not* $\equiv 0$ but will in general have nonzero values for $-\mu \neq -\mu_i$. By using $S(\mu)$, we can obtain an expression for $I(0, \mu)$ that does not involve explicitly the constants $L_\alpha$ and $Q$. Let

$$R(\mu) = \prod_{\alpha=1}^{n-1} (1 - k_\alpha \mu) \tag{2-95}$$

Then the product

$$R(\mu)S(\mu) = (Q - \mu)\prod_{\alpha=1}^{n-1} (1 - k_\alpha \mu) + \sum_{\alpha=1}^{n-1} L_\alpha \prod_{i \neq \alpha}^{n-1} (1 - k_i \mu) \tag{2-96}$$

is clearly a polynomial of order $n$ in $\mu$. But since $S(\mu)$ has the $n$ roots $\mu_1, \ldots, \mu_n$, this polynomial can differ at most by a constant factor from the polynomial $(\mu - \mu_1) \ldots (\mu - \mu_n)$. To find the constant factor, we merely note that the coefficient of the term $\mu_\nu$ on the right-hand side of equation (2-96) is $(-1)^n k_1 \ldots k_{n-1}$, which is therefore the desired factor. Thus we have

$$S(\mu) = \frac{\prod_{i=1}^{n-1} k_i \prod_{i=1}^{n} (\mu_i - \mu)}{\prod_{i=1}^{n-1} (1 - k_i \mu)} = \frac{\prod_{i=1}^{n} (\mu_i - \mu)}{\prod_{i=1}^{n-1} (x_i - \mu)} \tag{2-97}$$

Therefore we may rewrite equation (2-94) as

$$I(0, \mu) = \frac{3}{4} F \frac{\prod_{i=1}^{n} (\mu + \mu_i)}{\prod_{i=1}^{n-1} (x_i + \mu_i)} \tag{2-98}$$

It is customary to define a limb-darkening function $H(\mu) = I(0, \mu)/I(0, 0)$. Using equation (2-98), we have

$$H(\mu) = \frac{\prod_{i=1}^{n} \left(1 + \dfrac{\mu}{\mu_i}\right)}{\prod_{i=1}^{n-1} (1 + \mu k_i)} \tag{2-99}$$

A very important general result can be obtained at this point. If we identify $f = \lim_{n \to \infty} f_n$, where $f_n$ is any function computed in the $n$th discrete-ordinate approximation with the exact solution for $f$ in the gray problem, we can now show that $q(0) = 1/\sqrt{3}$ is the exact value. First, note that in the $n$th approximation

$$J_n(0) = \frac{3}{4} F \left[ Q + \sum_{\alpha=1}^{n-1} L_\alpha \right] \tag{2-100}$$

while equations (2-92) and (2-93) imply

$$I_n(0, 0) = \frac{3}{4} F\left[Q + \sum_{\alpha=1}^{n-1} L_\alpha\right] = J_n(0) \tag{2-101}$$

*independent of the order n.* Thus we conclude that in the *exact* solution

$$J(0) = I(0, 0) \tag{2-102}$$

a result we have previously quoted [equation (2-48)] in our discussion of the Eddington approximation. Note further from equation (2-98) that

$$I_n(0, 0) = \frac{\frac{3}{4}F\prod_{i=1}^{n} \mu_i}{\prod_{i=1}^{n-1} x_i} = \frac{3}{4} F\mu_1 \cdots \mu_n k_1 \cdots k_{n-1} \tag{2-103}$$

Now using equation (2-91) we immediately see that *independent of the order n*

$$I_n(0, 0) = \frac{\sqrt{3}}{4} F \tag{2-104}$$

which implies that in the exact solution

$$I(0, 0) = \frac{\sqrt{3}}{4} F \tag{2-105}$$

and in light of equation (2-102),

$$J(0) = \frac{\sqrt{3}}{4} F \tag{2-106}$$

But by the definition of $q(\tau)$ given in equation (2-17), we have $J(0) = \frac{3}{4}Fq(0)$ so that the *exact* value of $q(0)$ at the surface is

$$q(0) = \frac{1}{\sqrt{3}} \tag{2-107}$$

Finally, using equation (2-105), we may rewrite equation (2-99) as

$$I(0, \mu) = \frac{\sqrt{3}}{4} FH(\mu) \tag{2-108}$$

As a final step in the analysis we may now find explicit expressions for the $L_\alpha$'s and $Q$. From equation (2-92) we have

$$(1 - k_\alpha \mu)S(\mu) = (1 - k_\alpha \mu)(Q - \mu) + L_\alpha + \sum_{i\neq\alpha}^{n-1} \frac{L_i(1 - \mu k_\alpha)}{(1 - \mu k_i)} \tag{2-109}$$

On the other hand, from equation (2-97) we see that

$$(1 - k_\alpha \mu)S(\mu) = \frac{k_\alpha \prod_{i=1}^{n} (\mu_i - \mu)}{\prod_{i\neq\alpha}^{n-1} (x_i - \mu)} \tag{2-110}$$

Thus, taking the limit as $\mu \to 1/k_\alpha$ we find from equations (2-109) and (2-110)

$$L_\alpha = k_\alpha \frac{\prod_{i=1}^{n}\left(\mu_i - \frac{1}{k_\alpha}\right)}{\prod_{i \neq \alpha}^{n-1}\left(\frac{1}{k_i} - \frac{1}{k_\alpha}\right)} \tag{2-111}$$

To calculate $Q$, we see from equation (2-92) that

$$S(\mu)\prod_{i=1}^{n-1}(x_i - \mu) = (Q - \mu)\prod_{i=1}^{n-1}(x_i - \mu) + \sum_{\alpha=1}^{n-1} x_\alpha L_\alpha \prod_{i \neq \alpha}^{n-1}(x_i - \mu) \tag{2-112}$$

Substituting equation (2-97) for $S(\mu)$, we have

$$\prod_{i=1}^{n}(\mu_i - \mu) = (Q - \mu)\prod_{i=1}^{n-1}(x_i - \mu) + O(\mu^{n-2}) \tag{2-113}$$

Expanding the right-hand side, we have

$$(-1)^n\mu^n + (-1)^{n-1}\mu^{n-1}\left(Q + \sum_{i=1}^{n-1} x_i\right) + \cdots$$

while on the left-hand side we have

$$(-1)^n\mu^n + (-1)^{n-1}\mu^{n-1}\sum_{i=1}^{n}\mu_i + \cdots$$

Equating the coefficients of the terms in $\mu^{n-1}$, we thus find

$$Q = \sum_{i=1}^{n}\mu_i - \sum_{i=1}^{n-1}\frac{1}{k_i} \tag{2-114}$$

We have now completed our discussion of the use of the discrete-ordinate method to obtain approximate solutions of the gray problem; let us now consider the exact solution.

## 2-4. The Exact Solution

We shall obtain the exact solution by passing to the limit as $n \to \infty$ in the discrete ordinate approximation; our discussion will parallel that of Kourganoff (Ref. 23, Chap. 6). In principle, alternative approaches are possible, for example, using Laplace transformations, but the one we have chosen is the least complicated (and even then is complicated enough!).

In our previous work we obtained the exact value for $q(0)$; let us now consider the determination of $q(\infty)$. In principle, we can obtain $q(\infty)$ by noting that in the $n$th approximation,

$$\lim_{\tau \to \infty} q_n(\tau) = \lim_{\tau \to \infty}\left(Q + \sum_{\alpha=1}^{n-1} L_\alpha e^{-k_\alpha \tau}\right) = Q \tag{2-115}$$

But $Q$ is known only in terms of the integration points and characteristic roots, as shown in equation (2-114) which clearly is not a suitable form for generalization, since we cannot directly calculate the limiting values of the sums as $n \to \infty$. Alternatively, if we recall that $K(\tau) = H \cdot [\tau + q(\infty)]$, we can also write

$$q(\infty) = \frac{K(0)}{H} = \frac{\int_0^1 I(0, \mu)\mu^2\, d\mu}{\int_0^1 I(0, \mu)\mu\, d\mu} = \frac{\int_0^1 H(\mu)\mu^2\, d\mu}{\int_0^1 H(\mu)\mu\, d\mu} \tag{2-116}$$

Thus, if we can obtain the exact form of the limb-darkening law $H(\mu)$, we can obtain $q(\infty)$ by a direct quadrature. Let us pursue this approach further. Our previous results for $I(0, \mu)$ were,

$$I(0, \mu) = \frac{3}{4} F \left[ Q + \mu + \sum_{\alpha=1}^{n-1} \frac{L_\alpha}{1 + k_\alpha \mu} \right] \tag{2-117}$$

from equations (2-92) and (2-94), or, from equation (2-98),

$$I(0, \mu) = \frac{3}{4} F \frac{\prod_{i=1}^{n} (\mu + \mu_i)}{\prod_{i=1}^{n-1} (x_i + \mu)}$$

Clearly neither one is in a form to yield in an obvious way the exact expression when we pass to the limit as $n \to \infty$. Observe, however, that $I(0, \mu)$ is related to the $n$th approximation of the characteristic function $T(\mu)$ from equations (2-68) and (2-89)

$$T(\mu) = 1 + \mu^2 \sum_{j=1}^{n} \frac{a_j}{\mu_j^2 - \mu^2} = \frac{1}{3} \frac{\prod_{i=1}^{n-1} (x_i^2 - \mu^2)}{\prod_{i=1}^{n} (\mu_i^2 - \mu^2)} \tag{2-118}$$

since by comparing equations (2-98) and (2-118), we can write

$$I(0, \mu) I(0, -\mu) = \frac{3}{16} F^2 \frac{1}{T(\mu)} \tag{2-119}$$

or

$$H(\mu) H(-\mu) = \frac{1}{T(\mu)} \tag{2-120}$$

Now the generalization of $T(\mu)$ can be written simply by replacing the quadrature sum in equation (2-118) by its corresponding integral, namely

$$T(\mu) = 1 + \mu^2 \int_0^1 \frac{dz}{z^2 - \mu^2} \tag{2-121}$$

Note, however, that this integral has a pole on the range of integration so that special care must be taken in its evaluation. Indeed, the $n$th approximation to $T(\mu)$ has $2n$ poles, at $\pm\mu_i$, on the range $(-1, 1)$. To handle this problem, it is necessary to consider the *analytic continuation* of $T$ obtained by allowing $\mu$ to become a complex variable. To avoid the singularities in the range $(-1, 1)$ on the real axis, we insert a *branch cut* there. The analytic continuation of $T(\mu)$ can be made completely definite by choosing the correct sheet of the generalized function; we make this choice such that $T$ is real when $\mu$ is real and greater than one.

Thus, integrating equation (2-121), we have

$$T(\mu) = 1 + \frac{1}{2}\mu\left(\int_0^1 \frac{dz}{z-\mu} - \int_0^1 \frac{dz}{z+\mu}\right)$$

$$= 1 + \frac{1}{2}\mu[\text{Log}(1-\mu) - \text{Log}(-\mu)] - \frac{1}{2}\mu[\text{Log}(1+\mu) - \text{Log}(\mu)] \tag{2-122}$$

where "Log" is the complex logarithm function

$$\text{Log}(z) = \text{Log}(\rho e^{i\theta}) = \ln \rho + i(\theta + 2\pi k) \tag{2-123}$$

while "ln" denotes the ordinary logarithm function. Now let $\mu = \rho \exp(i\theta)$, $-\mu = \rho \exp[i(\theta+\pi)]$, $\mu + 1 = \rho' \exp(i\theta')$, $1 - \mu = \rho'' \exp(i\theta'')$ (see Figure 2-2). Note that $\theta'' < 0$ as shown in the figure. Then we may rewrite equation (2-122) as

$$T(\mu) = 1 + \mu \ln \frac{|1-\mu|}{|1+\mu|} + \frac{1}{2}\mu(\theta'' - \theta' + \pi) \tag{2-124}$$

But we see that $\theta'' - \theta' = \pi$ when $\mu$ is real and greater than unity, so this is the desired branch of the function; alternatively we can write

$$T(\mu) = 1 - \frac{1}{2}\mu \ln\left(\frac{\mu+1}{\mu-1}\right) \tag{2-125}$$

with the understanding that the $\mu$-plane is cut on the range $(-1, 1)$ along the real axis. We must now express $H(\mu)$ in terms of $T(\mu)$. In the $n$th approximation we may write

$$\ln H_n(\mu) = \frac{1}{2}\ln 3 + \sum_{j=1}^{n} \ln(\mu_j + \mu) - \sum_{j=1}^{n-1} \ln(x_j + \mu) \tag{2-126}$$

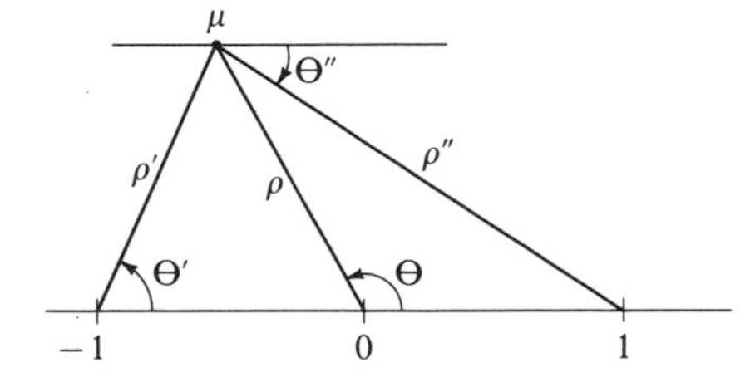


FIG. 2-2. Definition of auxiliary quantities in derivation of $T(\mu)$.

and

$$\ln T_n(\mu) = -\ln 3 - \sum_{j=1}^{n} \ln(\mu_j^2 - \mu^2) + \sum_{j=1}^{n-1} \ln(x_j^2 - \mu^2) \tag{2-127}$$

We must now find a relation between functions of the form $\ln(\alpha + \mu)$ and $\ln(\alpha^2 - \mu^2)$. An elegant way to do this was found by Chandrasekhar. Let $\alpha$ be such that $\text{Re}(\alpha) > 0$, and assume for the moment $\text{Re}(\mu)$ is also $> 0$. Then consider the function

$$f(\omega) = \frac{\mu \ln(\alpha + \omega)}{(\omega^2 - \mu^2)} \tag{2-128}$$

integrated around the contour $C$ shown in Figure 2-3. In passing to the limit as $R \to \infty$, the contribution on the semicircular part of the contour $\to 0$, and only the pole at $\omega = +\mu$ contributes to the contour integral. Thus by Cauchy's theorem

$$\oint_C f(z)\,dz = 2\pi i \sum \text{(residues of } f \text{ at poles within } C) \tag{2-129}$$

we have

$$\frac{1}{2\pi i} \oint f(\omega)\,d\omega = \frac{1}{2\pi i} \int_{+i\infty}^{-i\infty} \frac{\mu \ln(\alpha + \omega)}{\omega^2 - \mu^2}\,d\omega = \frac{1}{2} \ln(\alpha + \mu) \tag{2-130}$$

Similarly, computing the integral around the contour $C'$ shown in Figure 2-4 and again passing to the limit as $R \to \infty$, we have

$$\frac{1}{2\pi i} \int_{-i\infty}^{i\infty} \frac{\mu \ln(\alpha - \omega)}{\omega^2 - \mu^2}\,d\mu = -\frac{1}{2} \ln(\alpha + \mu) \tag{2-131}$$

since the pole is at $\omega = -\mu$. Subtracting equation (2-130) from equation (2-131), we find

$$-\ln(\alpha + \mu) = \frac{1}{2\pi i} \int_{-i\infty}^{i\infty} \frac{\mu \ln(\alpha^2 - \omega^2)}{\omega^2 - \mu^2}\,d\omega, \ [Re(\alpha) > 0,\ Re(\mu) > 0] \tag{2-132}$$

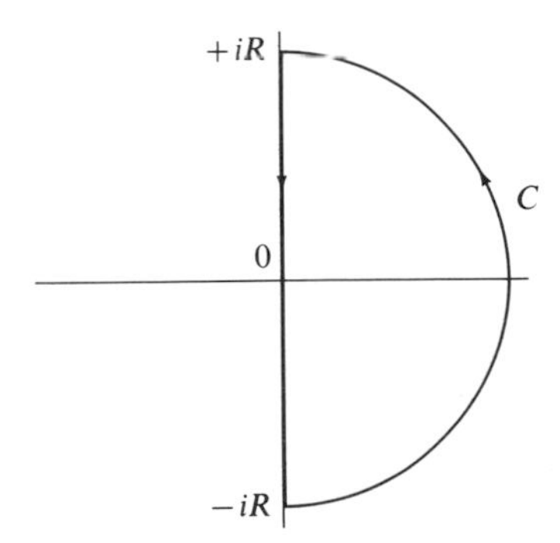


FIG. 2-3. Contour used in derivation of equation (2-130).

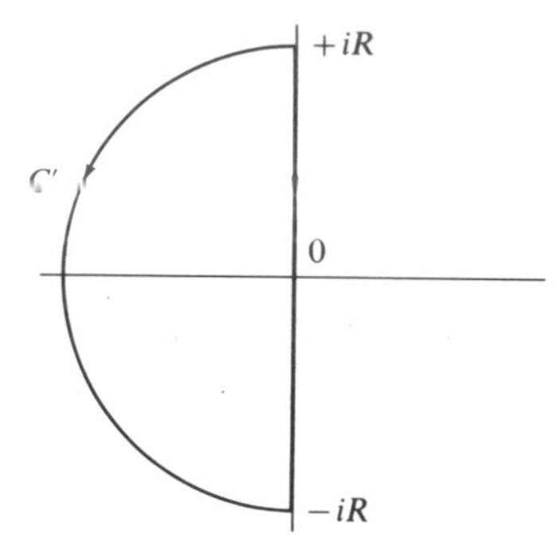


FIG. 2-4. Contour used in derivation of equation (2-131).

By an analogous argument using the function $\mu/(\omega^2 - \mu^2)$, one can show

$$\frac{1}{2\pi i}\int_{-i\infty}^{i\infty} \frac{\mu\, d\omega}{\omega^2 - \mu^2} = -\frac{1}{2}, \; [Re(\mu) > 0] \tag{2-133}$$

Making use of these identities, we have from equations (2-126) and (2-127)

$$\begin{aligned}\frac{1}{2\pi i}\int_{-i\infty}^{i\infty} &\frac{\mu \ln T_n(\omega)\, d\omega}{\omega^2 - \mu^2}\, d\omega \\ &= (-\ln 3)\frac{1}{2\pi i}\int_{-i\infty}^{i\infty}\frac{\mu\, d\omega}{\omega^2 - \mu^2} - \frac{1}{2\pi i}\sum_{j=1}^{n}\int_{-i\infty}^{i\infty}\frac{\mu \ln(\mu_j^2 - \omega^2)\, d\omega}{\omega^2 - \mu^2} \\ &\quad + \frac{1}{2\pi i}\sum_{j=1}^{n-1}\int_{-i\infty}^{i\infty}\frac{\mu \ln(x_j^2 - \omega^2)\, d\omega}{\omega^2 - \mu^2} \\ &= \frac{1}{2}\ln 3 + \sum_{j=1}^{n}\ln(\mu_j + \mu) - \sum_{j=1}^{n-1}\ln(x_j + \mu) \\ &\equiv \ln H_n(\mu)\end{aligned} \tag{2-134}$$

Thus

$$\ln H(\mu) = \frac{1}{2\pi i}\int_{-i\infty}^{i\infty}\frac{\mu \ln T(\omega)}{\omega^2 - \mu^2}\, d\omega, \; [Re(\mu) > 0] \tag{2-135}$$

This integral in the complex plane can be reduced to a simple real integral by the substitution $\omega = i \cot \theta$. First, with this substitution, equation (2-125) allows us to write

$$T(\omega) = 1 - \theta \cot \theta \tag{2-136}$$

Then inserting this expression into equation (2-135), we obtain

$$\ln H(\mu) = -\frac{\mu}{\pi}\int_0^{\pi/2}\frac{\ln(1 - \theta \cot \theta)}{\cos^2\theta + \mu^2 \sin^2\theta}\, d\theta \tag{2-137}$$

or

$$H(\mu) = \exp\left[-\frac{\mu}{\pi}\int_0^{\pi/2} \frac{\ln(1 - \theta \cot \theta)}{\cos^2 \theta + \mu^2 \sin^2 \theta}\, d\theta\right] \tag{2-138}$$

From this equation, $H(\mu)$ can be calculated directly. Numerical values of the exact solution for $I(0, \mu)/F = \sqrt{3}\, H(\mu)/4$ have been given already in Table 2-1. Alternative forms of $H(\mu)$ are given by Kourganoff (Ref. 23, p. 186). From the expression for $H(\mu)$, one can now obtain an expression for $q(\infty)$, namely,

$$q(\infty) = \frac{\int_0^1 H(\mu)\mu^2\, d\mu}{\int_0^1 H(\mu)\mu\, d\mu} = \frac{6}{\pi^2} + \frac{1}{\pi}\int_0^{\pi/2}\left(\frac{3}{\theta^2} - \frac{1}{1 - \theta \cot \theta}\right) d\theta \tag{2-139}$$

which yields, by direct numerical integration, $q(\infty) = 0.71044609\ldots$.

Finally, we may derive an expression for $q(\tau)$ over the entire range of optical depth in terms of the function $H(\mu)$. In the $n$th approximation we have

$$q_n(\tau) = Q + \sum_{\alpha=1}^{n-1} L_\alpha e^{-k_\alpha \tau} = q_n(\infty) + \sum_{\alpha=1}^{n-1} L_\alpha e^{-k_\alpha \tau} \tag{2-140}$$

We now need to find the limiting form for the sum term as $n \to \infty$. We proceed by noting that from equations (2-108) and (2-117), the function

$$\varphi(\mu) \equiv \frac{H_n(\mu)e^{\tau/\mu}}{\sqrt{3}(-\mu)} = \frac{(\mu + Q)e^{\tau/\mu}}{-\mu} + \sum_{\alpha=1}^{n-1} \frac{x_\alpha}{-\mu}\frac{L_\alpha e^{\tau/\mu}}{\mu + x_\alpha} \tag{2-141}$$

has $(n - 1)$ poles at $\mu = -x_\alpha = -1/k_\alpha$. If we perform an integration on a contour containing these poles, we obtain the sum of the residues at these points, namely $\sum L_\alpha e^{-k_\alpha \tau}$, which is precisely the sum appearing in equation (2-140). Thus, making use of equations (2-120) and (2-141), we have

$$\sum_{\alpha=1}^{n-1} L_\alpha e^{-k_\alpha \tau} = \frac{1}{2\pi i}\oint_C \varphi(\mu)\, d\mu = \frac{1}{2\sqrt{3}\,\pi i}\oint \frac{e^{\tau/\mu}\, d\mu}{(-\mu)T(\mu)H(-\mu)} \tag{2-142}$$

In choosing the contour $C$, we exclude the singular point at zero. We noted previously that the points $x_\alpha$ all lie on the range (0, 1), so we must calculate the integral around the contour encircling the range $(-1, -\varepsilon)$ (see Figure 2-5). Along the upper edge of the cut $(-1, 0)$, we see from Figure 2-2 that $\theta' = 0$, $\theta'' = 0$, $\mu = -u + i0$; thus equation (2-142) becomes

$$T_A(\mu) = 1 - \frac{u}{2}\ln\left(\frac{1 + u}{1 - u}\right) - \frac{1}{2}i\pi u \tag{2-143}$$

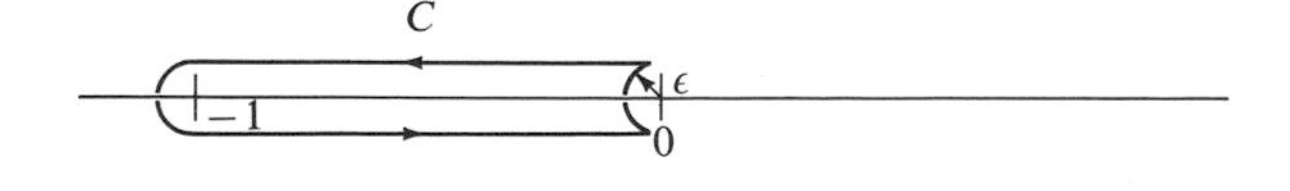


FIG. 2-5. Contour used in derivation of equation (2-146).

Passing counterclockwise around the point $-1$, we see that $\theta' = 2\pi$ and $\theta'' = 0$ so that equation (2-124) now becomes

$$T_B(\mu) = 1 - \frac{u}{2}\ln\left(\frac{1+u}{1-u}\right) + \frac{1}{2}i\pi u \tag{2-144}$$

Thus, substituting into equation (2-142) and passing to the limit as $\varepsilon \to 0$, we have

$$\begin{aligned}\sum_{\alpha=1}^{n-1} L_\alpha e^{-k_\alpha \tau} &= \lim_{\varepsilon\to 0} \frac{-1}{2\sqrt{3}\pi i}\int_\varepsilon^1 \frac{e^{-\tau/u}}{uH(u)}\left[\frac{1}{T_A(u)} - \frac{1}{T_B(u)}\right] du \\ &= \lim_{\varepsilon\to 0}\frac{-1}{2\sqrt{3}}\int_\varepsilon^1 \frac{e^{-\tau/u}}{H(u)}\left\{\left[1 - \frac{u}{2}\ln\left(\frac{1+u}{1-u}\right)\right]^2 + \frac{1}{4}\pi^2 u^2\right\}^{-1} du\end{aligned} \tag{2-145}$$

or

$$\sum_{\alpha=1}^{n-1} L_\alpha e^{-k_\alpha \tau} = \frac{-1}{2\sqrt{3}}\int_0^1 \frac{e^{-\tau/u}\,du}{H(u)Z(u)} \tag{2-146}$$

where

$$Z(u) \equiv \left[1 - \frac{u}{2}\ln\left(\frac{1+u}{1-u}\right)\right]^2 + \frac{1}{4}\pi^2 u^2 \tag{2-147}$$

Finally, we may write

$$J(\tau) = \frac{3}{4}F\left[\tau + q(\infty) - \frac{1}{2\sqrt{3}}\int_0^1 \frac{e^{-\tau/u}\,du}{H(u)Z(u)}\right] \tag{2-148}$$

This solution was first obtained by Mark (*Phys. Rev.*, **72**, 558, 1947). The exact $q(\tau)$ may thus be obtained directly by numerical integration; a few values of the exact solution are listed in Table 2-2.

TABLE 2-2. The Exact Solution for $q(\tau)$

| $\tau$ | $q(\tau)$ |
|---|---|
| 0.0 | 0.5773 |
| 0.1 | 0.6279 |
| 0.2 | 0.6495 |
| 0.4 | 0.6731 |
| 0.6 | 0.6858 |
| 0.8 | 0.6935 |
| 1.0 | 0.6985 |
| 1.5 | 0.7051 |
| 2.0 | 0.7079 |
| $\infty$ | 0.710446 |

## 2-5. Emergent Flux from a Gray Atmosphere

The physical assumption at the basis of the gray atmospheres problem is that the opacity is independent of frequency. This assumption allows one to make the identification $J = S$ through the requirement of radiative equilibrium and to reduce the problem to that of obtaining the solution of equation (2-7). If in addition we assume that LTE prevails, then as shown in equation (2-8) we may write

$$B(\tau) = \frac{\sigma T^4}{\pi} = J(\tau)$$

so that using equations (1-91) and (2-17), we clearly may write

$$T^4 = \tfrac{3}{4} T_{\text{eff}}^4 [\tau + q(\tau)] \tag{2-149}$$

Equation (2-149) gives the run of temperature with depth. Since we assume LTE, the source function will be the Planck function, $B_\nu(T)$. Since we know the frequency dependence of the source function, we may determine the frequency dependence of the flux from the formula

$$F_\nu(\tau) = 2 \int_\tau^\infty B_\nu[T(\tau)] E_2(t - \tau)\, dt - 2 \int_0^\tau B_\nu[T(t)] E_2(\tau - t)\, dt \tag{2-150}$$

Now the temperature enters the Planck function

$$B_\nu(T) = \frac{2h\nu^3}{c^2} (e^{h\nu/kT} - 1)^{-1}$$

only in the combination $(h\nu/kT)$, and from equation (2-149) we note that $T/T_{\text{eff}}$ is a unique function of depth, say $1/p(\tau)$. We may therefore simplify equation (2-150) if we introduce the parameter

$$\alpha \equiv \frac{h\nu}{kT_{\text{eff}}} \tag{2-151}$$

and use the corresponding flux

$$F_\alpha(\tau) \equiv F_\nu(\tau) \frac{d\nu}{d\alpha} \tag{2-152}$$

Combining equations (2-150) through (2-152), we find

$$\frac{F_\alpha(\tau)}{F} = \left(\frac{4\pi k^4}{h^3 c^2 \sigma}\right) \alpha^3 \left\{ \int_\tau^\infty \frac{E_2(t - \tau)\, dt}{\exp[\alpha p(t)] - 1} - \int_0^\tau \frac{E_2(\tau - t)\, dt}{\exp[\alpha p(t)] - 1} \right\} \tag{2-153}$$

The term in the brackets is a simple function of $\alpha$ and $\tau$ and only may be calculated once and for all. A detailed tabulation of $F_\alpha(\tau)/F$ has been given by Chandrasekhar (Ref. 9, p. 295). A plot of this function is shown in Figure 2-6. This figure shows clearly the marked degradation of photon energies as

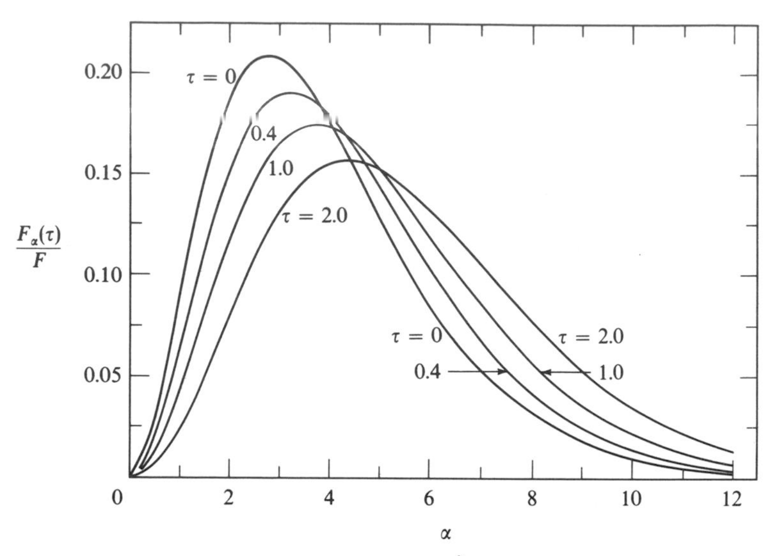


FIG. 2-6. Frequency distribution of flux in a gray atmosphere. (From S. Chandrasekhar, *Ap. J.*, **101**, 328, 1945; by permission.)

they transfer from depth to the surface; in particular, the most common photon energy at $\tau = 0$ is only about 75% of that at $\tau = 1$. This is a characteristic of the requirement of radiative equilibrium, which leads to a monotonic increase of the temperature with depth.

## 2-6. Small Departures from Grayness

It is possible to account for small departures from grayness—at least approximately—by use of an appropriate mean opacity and thus greatly extend the usefulness of the results obtained for a gray atmosphere. Suppose that the frequency spectrum of the opacity does not change with depth and that we can write

$$\kappa_\nu = \bar{\kappa}_c(1 + \beta_\nu) \tag{2-154}$$

where

$$\bar{\kappa}_c \equiv \int_0^\infty \kappa_\nu F_\nu^{(1)}\, d\nu \tag{2-155}$$

In equation (2-155), $F_\nu^{(1)}$ denotes the flux given by the gray atmosphere; this mean opacity is known as the *Chandrasekhar mean.* Unlike the flux mean, equation (2-25), the Chandrasekhar mean can be computed straightaway since $F_\nu^{(1)}$ is a known function. The usefulness of this choice may be seen as follows. The appropriate transfer equation in the nongray case (again assuming LTE) is

$$\mu \frac{dI_\nu}{d\bar{\tau}} = \frac{\kappa_\nu}{\bar{\kappa}}(I_\nu - B_\nu) = (1 + \beta_\nu)(I_\nu - B_\nu) \tag{2-156}$$

If the departures $\beta_\nu$ are regarded as small, we may solve equation (2-156) by successive approximations. Thus if we assume $\beta_\nu \equiv 0$, we may write

$$\mu \frac{dI_\nu^{(1)}}{d\bar{\tau}} = I_\nu^{(1)} - B_\nu \tag{2-157}$$

which is simply the equation for the gray problem itself, whose solution is already known. To obtain a second approximation, we may write

$$\begin{aligned} \mu \frac{dI_\nu^{(2)}}{d\bar{\tau}} &= I_\nu^{(2)} - B_\nu + \beta_\nu(I_\nu^{(1)} - B_\nu) \\ &= I_\nu^{(2)} - B_\nu + \mu\beta_\nu \frac{dI_\nu^{(1)}}{d\bar{\tau}} \end{aligned} \tag{2-158}$$

where we have substituted from equation (2-157). The requirement of radiative equilibrium for equation (2-158) is that

$$J^{(2)} - B + \frac{d}{d\bar{\tau}} \int_0^\infty \beta_\nu F_\nu^{(1)}\, d\nu = 0 \tag{2-159}$$

But note that equation (2-155) implies that

$$\bar{\kappa}_c F = \bar{\kappa}_c \int_0^\infty (1 + \beta_\nu)F_\nu^{(1)}\, d\nu = \bar{\kappa}_c F + \bar{\kappa}_c \int_0^\infty \beta_\nu F_\nu^{(1)}\, d\nu$$

or that

$$\int_0^\infty \beta_\nu F_\nu^{(1)}\, d\nu \equiv 0 \tag{2-160}$$

Thus, on the Chandrasekhar mean depth scale, the radiative equilibrium requirement in the nongray case (approximated by the above procedure) is simply

$$J^{(2)} = B(\bar{\tau})$$

which is equivalently to simply using the gray solution on the mean optical depth scale, i.e., in the second approximation we again have

$$T^4 = \tfrac{3}{4}T_{\text{eff}}^4[\bar{\tau} + q(\bar{\tau})]$$

The method of obtaining higher approximations has been described by Chandrasekhar (Ref. 9, p. 296 ff.).

Within this stage of approximation we may now compute the emergent intensity as

$$I_\nu(0, \mu) = \int_0^\infty B_\nu[T(\bar{\tau})]\exp(-\beta_\nu\bar{\tau}/\mu)\frac{\beta_\nu d\bar{\tau}}{\mu} \tag{2-161}$$

and

$$F_\nu(0) = 2\int_0^\infty B_\nu[T(\bar{\tau})]\, E_2(\beta_\nu\bar{\tau})\beta_\nu d\bar{\tau} \tag{2-162}$$

As we did in the gray case, we introduce the parameter $\alpha = h\nu/kT_{\text{eff}}$ and write

$$B_\nu(T) = B_\nu(T_0)\left[\frac{\exp(\alpha T_{\text{eff}}/T_0) - 1}{\exp[\alpha p(\tau)] - 1}\right] = B_\nu(T_0)b_\alpha(\tau) \tag{2-163}$$

so that we can reduce equations (2-161) and (2-162) to the parameterized forms

$$I_\nu(0, \mu) = B_\nu(T_0)\mathcal{J}(\alpha, \beta_\nu/\mu) \tag{2-164}$$

and

$$F_\nu(0) = B_\nu(T_0)\mathcal{F}(\alpha, \beta_\nu) \tag{2-165}$$

where

$$\mathcal{J}(\alpha, \beta) \equiv \int_0^\infty b_\alpha(\tau)\exp(-\beta\tau)\beta\, d\tau \tag{2-166}$$

and

$$\mathcal{F}(\alpha, \beta) \equiv 2\int_0^\infty b_\alpha(\tau)E_2(\beta\tau)\beta\, d\tau \tag{2-167}$$

Tables of the functions $\mathcal{J}(\alpha, \beta)$ and $\mathcal{F}(\alpha, \beta)$ have been given by Chandrasekhar (Ref. 9, pp. 306–7).

The functions $\mathcal{J}(\alpha, \beta)$ and $\mathcal{F}(\alpha, \beta)$ described here have played an important role in the development of the theory of stellar atmospheres. By analyzing the observational material available for the sun, Münch (*Ap. J.*, **102**, 385, 1945) was able to find values of $\beta_\nu$ which best reproduced the observed fluxes and limb darkening. These were shown to be compatible with the variation due to absorption by the negative hydrogen ion, $H^-$, as computed by Chandrasekhar. Analyses such as these led to the conclusive identification of $H^-$ as a major opacity source in the solar atmosphere. We shall discuss opacities at greater length in Chapter 4.

# 3 | The Equation of State

Stellar atmospheres are regions of relatively high temperature and low density. Therefore, the gas consists mainly of single atoms, ions, and free electrons; only in the cooler stars does molecule formation occur. Because of the low densities, the gas may always be treated as a perfect gas. To obtain an expression for the equation of state, we must specify the *occupation numbers* in each atomic level (whether a bound level or an ionization state). A knowledge of these occupation numbers allows computation of gas pressures, densities, and opacities.

In this chapter we shall consider primarily the LTE equation of state and merely introduce some aspects of the non-LTE case, deferring a complete discussion until Chapter 5.

## 3-1. LTE Occupation Numbers

The assumption of LTE implies that the state of the gas, i.e., the distribution of atoms over bound and free levels, may be specified completely in terms of two thermodynamic variables via the well-known equilibrium equations of thermodynamics. A typical choice of variables might be the absolute temperature $T$ and the total particle density $N$; we shall find it more convenient to use the equivalent set of $T$ and the electron density $N_e$. We wish to stress here that LTE provides only a *local* description of the state of the gas. That is to say, given the local variables $T$ and $N_e$, one assumes that he may uniquely calculate the above-mentioned distributions without further reference to the physical ensemble in which the element of material under consideration is found. Thus, it is assumed that it is irrelevant whether the element of material we consider is contained within an equilibrium cavity (the classical *hohlraum*), in an atmosphere with a strong radiation field (or no radiation field), or in the exhaust of a space vehicle. There is, in short, no coupling allowed from one element of the gas to another (except as may be imposed by additional constraints, e.g., energy balance and hydrostatic equilibrium). Moreover, in LTE, the absolute temperature $T$ has a quite general significance. The *same* $T$ applies in the calculation of the velocity distribution functions of atoms, ions, and electrons; the distribution of atoms and ions via the Boltzmann and Saha equations; and the thermally emitted photon distribution via the Planck function.

Thus we see that the full implications of the LTE assumption are quite sweeping. It is this broad nature that makes the assumption so effective in reducing the complexity of the equations and at the same time so difficult to justify physically.

## MAXWELL-BOLTZMANN STATISTICS

In thermodynamic equilibrium at temperature $T$, atoms are distributed among their bound levels according to the *Boltzmann distribution function.* Consider an atom in an excited level $i$ that has an excitation energy $\chi_i$ relative to the ground level. Since this level may have degenerate sublevels (e.g., $2J \times 1$ $m$-states in the absence of a magnetic field), let us assign to it a statistical weight $g_i$. Let the index $j$ denote the ionization state of the atoms ($j = 0$ for neutrals, 1 for singly ionized, etc.). Then, according to the Boltzmann law, the population of the excited level may be written

$$\frac{N_{i,j}}{N_{0,j}} = \frac{g_i}{g_0}\, e^{-\chi_i/kT} \tag{3-1}$$

where the subscript 0 denotes the ground level. In our work, we usually will want the number of atoms in a specific excitation and ionization state relative to the *total* number in that ionization state. We may calculate this total by simply summing equation (3-1) over all levels to obtain

$$N_j = \sum_i N_{i,j} = \left(\frac{N_{0,j}}{g_0}\right) \sum_i g_i e^{-\chi_i/kT} = \left(\frac{N_{0,j}}{g_0}\right) U_j(T) \tag{3-2}$$

where

$$U_j(T) \equiv \sum_i g_i e^{\chi_i/kT} \tag{3-3}$$

is called the *partition function.* Then we may write

$$\frac{N_{i,j}}{N_j} = \frac{g_i e^{-\chi_i/kT}}{U_j(T)} \tag{3-4}$$

The partition function is tedious to compute, and for some atoms our knowledge of the term structure is so incomplete that serious problems arise in its calculation. Another problem with partition functions is that the sum, if taken over *all* bound levels as indicated formally, must diverge. This is clear from the fact that there is an infinite number of bound levels, and the smallest that any one of the terms in the sum can be is $\exp(-\chi_{\text{ion}}/kT)$, which is nonzero. Actually, this is not a genuine physical problem since in reality the highest bound levels cannot exist because of perturbations of neighboring atoms and ions. As a simple estimate of this effect, we might suppose that an electron will become unbound whenever it is closer to a neighboring atom than to the atom under consideration. Then, for a hydrogen-like atom, the highest quantum number to be bound is given by

$$n^2 = \frac{Zr_0}{2a_0} \tag{3-5}$$

where $r_0$ is the mean interatomic spacing in the gas, $a_0$ is the Bohr radius, and $Z$ is the nuclear charge. Now typically in a stellar atmosphere $N \sim 10^{15}$ cm$^{-3}$ so that $r_0/2 \approx 3 \times 10^{-6}$ cm while $a_0 = 5 \times 10^{-9}$ cm; so from equation (3-5) we find $n = 25\, Z^{1/2}$. Clearly, the sum will be finite. A more accurate treatment (to be discussed further in Chapter 9) considers the atom to be immersed in a plasma of electrons and ions and calculates the potential due to their collective interaction. If this potential is large enough, than an electron may be lifted from its bound level and imbedded in the plasma with a net release of energy. This process will remove bound electrons from all levels above some critical value below the normal level of the continuum and is sometimes referred to as *pressure ionization*. One finds that

$$\Delta\chi = \frac{2.5 p_e^{1/2}}{T}\ \text{eV} \tag{3-6}$$

where $p_e$ is the electron pressure

$$p_e = N_e kT \tag{3-7}$$

For hydrogen, the corresponding quantum number of the last bound level is given by

$$n^2 = \frac{5.5T}{p_e^{1/2}} \tag{3-8}$$

Again equation (3-8) provides a means of truncating the partition function sum.

Actually, the question of partition functions is somewhat academic because the partition function in the denominator of the excitation equilibrium equation is usually canceled by the same partition function appearing in the numerator of the ionization equilibrium (see below). A fair approximation is to estimate $U$ by the sum over only the lowest bound states. Numerical results for partition functions of several atoms and ions have been tabulated by Aller (Ref. 4, pp. 115–17).

### THE SAHA EQUATION

Above the discrete, bound eigenstates of an atom, there exists a *continuum* of levels in which the electron is unbound from the atom and has a nonzero kinetic energy. The energy above the ground level at which this continuum begins is called the *ionization potential* $\chi_I$. The relative numbers of atoms with electrons in bound and free states can be computed by a simple extension of the Boltzmann formula, which leads to the *Saha ionization formula.*

Consider a unit process in which an atom in the ground level is ionized, resulting in an ion in the ground level plus a free electron in the continuum moving at velocity $v$. The energy required to carry out this process is $\chi_I + \frac{1}{2}mv^2$. Denote the statistical weight of the initial state as $g_{0,0}$, the ground-term statistical weight of the atom, and write the statistical weight of the final state as $g = g_{\text{ion}} \times g_{\text{electron}}$. If we use $N_{0,1}(v)$ to denote the number of ions in the ground level with a free electron in the velocity range $(v, v + dv)$, we may write from the Boltzmann equation (3-1)

$$\frac{N_{0,1}(v)}{N_{0,0}} = \frac{g}{g_{0,0}} \exp[-(\chi_I + \tfrac{1}{2}mv^2)/kT] \tag{3-9}$$

Now $g_{\text{ion}}$ is simply $g_{0,1}$, the statistical weight of the ground level of the ion. The factor $g_{\text{electron}}$ is to be identified with the number of phase space elements available to the electron, which, according to quantum statistics, is

$$g_{\text{electron}} = \frac{2dq_1\, dq_2\, dq_3\, dp_1\, dp_2\, dp_3}{h^3} \tag{3-10}$$

The factor of 2 enters because of the two possible orientations of the electron spin. If the space volume element $dq_1 dq_2\, dq_3$ is chosen to contain exactly one electron (since we are considering a unit process), we may substitute $dq_1\, dq_2\, dq_3 = 1/N_e$. The momentum volume element is best written in polar coordinates as $dp_1\, dp_2\, dp_3 = 4\pi p^2\, dp = 4\pi m^3 v^2\, dv$. Thus, combining these expressions we may rewrite equation (3-9) as

$$\frac{N_{0,1}(v)}{N_{0,0}} = \frac{8\pi m^3}{h^3} \frac{g_{0,1}}{N_e g_{0,0}} e^{-(\chi_I + \frac{1}{2}mv^2)/kT} v^2\, dv \tag{3-11}$$

Now, summing over all final states by integrating over the electron velocity distribution, we have

$$\frac{N_{0,1}N_e}{N_{0,0}} = \frac{8\pi m^3}{h^3} \frac{g_{0,1}}{g_{0,0}} e^{-\chi_I/kT} \left(\frac{2kT}{m}\right)^{3/2} \int_0^\infty e^{-x^2} x^2\, dx \tag{3-12}$$

or

$$\frac{N_{0,1}N_e}{N_{0,0}} = \left(\frac{2\pi mkT}{h^2}\right)^{3/2} \frac{2g_{0,1}}{g_{0,0}} e^{-\chi_I/kT} \tag{3-13}$$

Thus far we have considered both the atom and the ion to exist in their ground levels; we may extend equation (3-13) to include the distribution of the atoms and ions over all levels using equation (3-2) to write

$$\frac{N_{0,0}}{N_0} = \frac{g_{0,0}}{U_0(T)} \tag{3-14}$$

and

$$\frac{N_{0,1}}{N_1} = \frac{g_{0,1}}{U_1(T)} \tag{3-15}$$

so that substitution into equation (3-13) yields

$$\frac{N_1 N_e}{N_0} = \left(\frac{2\pi mkT}{h^2}\right)^{3/2} \frac{2U_1(T)}{U_0(T)} e^{-\chi_I/kT} \tag{3-16}$$

This equation is known as *Saha's equation* and gives the relation between two successive states of ionization in thermodynamic equilibrium. Since we have made no explicit reference to the degree of ionization of the initial atom, we may regard the Saha relation as valid between any two ionization states; thus we may write, in general,

$$\frac{N_{j+1}N_e}{N_j} = \left(\frac{2\pi mkT}{h^2}\right)^{3/2} \frac{2U_{j+1}(T)}{U_j(T)} e^{-\chi_I/kT} \tag{3-17}$$

More generally, one wishes to compute the number of atoms of a particular type in a specific ionization state relative to the total number of atoms of that type in all ionization states, i.e., the fraction

$$\begin{aligned} f_j &= \frac{N_j}{N} \\ &= \frac{(N_j/N_{j-1})\cdots(N_1/N_0)}{1 + (N_1/N_0) + (N_2/N_1)(N_1/N_0) + \cdots + (N_n/N_{n-1})\cdots(N_1/N_0)} \end{aligned} \tag{3-18}$$

We may calculate this fraction by repeated application of Saha's equation; for brevity define $r_{j+1} = N_{j+1}/N_j$ as computed from equation (3-17). Then clearly

$$f_j = \frac{\prod_{k=1}^{j} r_k}{\sum_{k=0}^{n} \prod_{l=0}^{k} r_l} \tag{3-19}$$

where we adopt the convention that $r_0 \equiv 1$.

## 3-2. The LTE Equation of State

Let us summarize here some of the practical aspects of calculating the LTE equation of state in stellar atmospheres.

### IONIZING GASES

The main relevance of the Saha equation in astrophysical calculation is in determining fractions of atomic species in various states of ionization and the numbers of free electrons they contribute to the plasma. We normally write the total pressure in the gas as the sum of partial pressures due to each constituent; for example, for a mixture of atoms, ions and electrons we may write

$$p_g = p_{\text{atoms}} + p_{\text{ions}} + p_{\text{electrons}} = (N_A + N_I + N_e)kT$$
$$= (N_{\text{nuclei}} + N_e)kT = p_N + p_e \tag{3-20}$$

Normal stellar atmospheres consist of a mixture of elements. Hydrogen is by far the most abundant constituent, and helium is next most abundant with $N(\text{He})/N(\text{H}) \approx 0.1$. In addition, there are heavier elements with much smaller abundances [see, e.g., the tabulation by Goldberg, Müller, and Aller (*Ap. J. Supp. No. 45*, **5**, 1, 1960) for the abundances of elements in the solar atmosphere]. At solar temperatures the hydrogen in the atmosphere is largely neutral, and electrons are contributed mainly by the heavier elements with lower ionization potentials (commonly called "metals" in astrophysical applications), such as Na, Mg, Al, Si, Ca, and Fe. At higher temperatures, characteristic of the A and B stars, hydrogen begins to ionize strongly and is the dominant source of electrons. At very high temperature, characteristic of O and B stars, helium ionizes strongly and makes an appreciable contribution to the number of electrons.

For the present, suppose that no molecule formation occurs. Consider a mixture of elements and let us define $\alpha_i$ as equal to the abundance of element $i$ relative to hydrogen $= N_i/N_{\text{H}}$, $A_i$ as equal to the atomic weight of element $i$, and $f_{i,j}$ as equal to the fraction of element $i$ in ionization state $j$. Then, summing over all ions of all elements, we may write

$$\frac{p_e}{p_N} = \frac{N_e}{N_N} = \frac{\sum_i N_i \sum_j j f_{i,j}}{\sum_i N_i} = \frac{\sum_i \alpha_i \sum_j j f_{i,j}}{\sum_i \alpha_i} \tag{3-21}$$

Thus

$$p_g = p_N + p_e = p_e\left[1 + \frac{\sum_i \alpha_i}{\sum_i \alpha_i \sum_j j f_{i,j}}\right] \tag{3-22}$$

Under the assumption of LTE, knowledge of $p_e$ and $T$ yields $f_{i,j}$ from equation (3-19), and we therefore can compute $p_g(T, P_e)$ directly from equation

(3-22). Tabulations of $p_g(T, p_e)$ have been given, for example, by Aller (Ref. 13, pp. 235–36).

As we shall see in Chapter 6, the usual situation in the construction of LTE model atmospheres is that from the integration of the equation of hydrostatic equilibrium, we obtain $p_g$ at a given depth point. The quantity then needed for other computations (e.g., of the opacity) is $p_e(T, p_g)$. For a pure hydrogen atmosphere this inverse computation is simple. For brevity rewrite equation (3-16) as

$$N_0 = N_1 N_e \, \Phi(T) \tag{3-23}$$

Then making use of the fact that for pure hydrogen $N_{\text{ion}}$ must equal $N_e$ (charge conservation), we may write

$$\frac{p_g}{kT} = N_0 + N_1 + N_e = 2N_e + N_e^2 \Phi(T) \tag{3-24}$$

Solving this quadratic equation for $N_e$, we have

$$p_e(p_g, T) = N_e kT = \left\{ \left[ 1 + \left( \frac{p_g}{4kT} \right) \Phi(T) \right]^{1/2} - 1 \right\} kT/\Phi(T) \tag{3-25}$$

When the gas is comprised of several constituents, we can no longer write a simple formula such as equation (3-25). In this more general case, one may derive the appropriate value of $p_e$ either by assuming several trial values of $p_e$, computing the corresponding $p_g(T, P_e)$ and searching numerically to find the correct value, or by precomputing a table of $p_g(T, p_e)$ and interpolating backward to obtain $p_e$. It is instructive to consider certain limiting forms of the ratio $p_g/p_e$ that can be derived in simple cases. Suppose the gas consists only of hydrogen and one metal. Let the hydrogen/metal abundance be denoted as $A$, assumed to be much greater than unity. Let $X_{\text{H}}$ be the fraction of hydrogen that is ionized and $X_m$ be the fraction of the metal that is ionized. Then the number of particles of all types is

$$N = N_{\text{H}} + \frac{N_{\text{H}}}{A} + X_{\text{H}} N_{\text{H}} + \frac{X_m N_{\text{H}}}{A} \tag{3-26}$$

while the number of electrons only is

$$N_e = X_{\text{H}} N_{\text{H}} + \frac{X_m N_{\text{H}}}{A} \tag{3-27}$$

so that

$$\frac{p_g}{p_e} = \frac{1 + X_{\text{H}} + (1 + X_m)/A}{X_{\text{H}} + X_m/A} \tag{3-28}$$

For high enough temperatures, $X_{\rm H} \to 1$, and since $A \gg 1$,

$$\frac{p_g}{p_e} \to 2 \tag{3-29}$$

For intermediate temperatures $1/A \ll X_{\rm H} \ll 1$ while $X_m \approx 1$ so that

$$\frac{p_g}{p_e} \to \frac{1}{X_{\rm H}} \tag{3-30}$$

For low enough temperatures, $X_{\rm H} \approx 0$ while $X_m$ remains finite so that

$$\frac{p_g}{p_e} \to \frac{A}{X_m} \tag{3-31}$$

We see then that at high temperatures the metal abundance in the atmosphere is essentially irrelevant in determining the relation $p_g(T, p_e)$ while at low temperatures it plays an important role. We also note that in the cooler stars, the opacity will be due to the ion $H^-$ over a wide range of physical conditions. This opacity is in turn proportional to $p_e$ (see Chapter 4) so that in these stars the metal abundance plays a role in fixing the opacity as well.

### MOLECULE FORMATION

At sufficiently low temperatures, atoms in a gas will form molecules. Since hydrogen is the most abundant element in stellar atmospheres, the formation of $H_2$ is particularly important. In addition, at these low temperatures, the negative ion $H^-$, which is an important source of opacity, acts as a sink for free electrons, and the electron pressure is determined by a delicate balance between those electrons donated by the metals and those attached by $H^-$. Let us therefore work out the LTE equation of state in the case where we must consider hydrogen as an atom, an ion (whether positive or negative,) and a molecule.

The usual notation for treating molecule formation in the reaction $A + B \rightleftarrows AB$ is to introduce a dissociation coefficient $K(AB)$, defined so that

$$\frac{p(A)p(B)}{p(AB)} = K(AB) \tag{3-32}$$

Equation (3-32) is analogous to Saha's equation, which clearly can be cast into the corresponding form

$$\frac{p(X^+)p_e}{p(X)} = K(X) \tag{3-33}$$

The dissociation coefficient contains a Boltzmann factor involving the molecular dissociation energy and a partition function accounting for the various modes of rotational and vibrational excitation (see, e.g., Dolan, *Ap. J.*, **142**, 1621, 1965). Numerical values for $K(AB)$ are given by Aller (Ref. 4, p. 132) for several diatomic molecules. A convenient formula for $K(H_2)$ has been given by Vardya (*Ap. J.*, **133**, 107, 1961).

Consider now the case where we allow for hydrogen molecule formation, but assume all other elements merely ionize. We define

$$f_1 \equiv n_{\text{H}}/(n_{\text{H}} + n_{\text{H}^+} + n_{\text{H}^-} + 2n_{\text{H}_2}) = n_{\text{H}}/N_{\text{H}} \tag{3-34}$$

$$f_2 \equiv n_{\text{H}^+}/N_{\text{H}} \tag{3-35}$$

$$f_3 \equiv n_{\text{H}^-}/N_{\text{H}} \tag{3-36}$$

$$f_4 \equiv n_{\text{H}_2}/N_{\text{H}} \tag{3-37}$$

and

$$f_e \equiv N_e/N_{\text{H}} \tag{3-38}$$

In equations (3-34) through (3-38), $n_{\text{H}}$, $n_{\text{H}^+}$, etc., denote the densities of the particular form of hydrogen indicated while $N_{\text{H}}$ denotes the density of hydrogen in *all* forms. Then from Saha's equation as expressed by equation (3-33), we may write

$$\frac{p(\text{H}^+)p_e}{p(\text{H})} = K(\text{H})$$

so that using equations (3-34) and (3-35) we have

$$\frac{f_2}{f_1} = \frac{K(\text{H})}{p_e} \equiv G_2 \tag{3-39}$$

Similarly, from the expression

$$\frac{p(\text{H})p_e}{p(\text{H}^-)} = K(\text{H}^-)$$

we may obtain

$$\frac{f_3}{f_1} = \frac{p(\text{H}^-)}{p(\text{H})} = \frac{p_e}{K(\text{H}^-)} \equiv G_3 \tag{3-40}$$

while the expression

$$\frac{p(\text{H})p(\text{H})}{p(\text{H}_2)} = K(\text{H}_2)$$

implies

$$\frac{f_4}{f_1} = \frac{p(H_2)}{p(H)} = \frac{p(H)}{K(H_2)} = \left(\frac{f_1}{f}\right)\frac{p_e}{K(H_2)} = \frac{f_1 G_4}{f_1} \tag{3-41}$$

In addition, conservation of mass demands that

$$f_1 + f_2 + f_3 + 2f_4 = 1 \tag{3-42}$$

while conservation of charge requires

$$f_e = f_2 - f_3 + \sum_{i \neq H} \alpha_i \sum_j j f_{i,j} \equiv f_2 - f_3 + Q \tag{3-43}$$

Eliminating $f_2$, $f_3$, $f_4$, and $f_e$ in terms of $f_1$, we may reduce the system of equations (3-39) through (3-43) to a single equation in $f_1$, namely,

$$[2G_4 + (1 + G_2 + G_3)(G_2 - G_3)]f_1^2 - [(G_2 - G_3) - Q(1 + G_2 + G_3)]f_1 - Q = 0 \tag{3-44}$$

which, when written in the standard form, $af_1^2 + bf_1 + c = 0$ yields $f_1$ directly as

$$f_1 = [-b + (b^2 + 4ac)^{1/2}]/2a \tag{3-45}$$

Substitution into equations (3-39) and (3-40) immediately yields $f_2$ and $f_3$, and $f_4$ follows from equation (3-42). Finally, we may write

$$p_g = p_e\left(1 + \frac{f_1 + f_2 + f_3 + f_4 + \sum_{i \neq H} \alpha_i}{f_2 - f_3 + Q}\right) \tag{3-46}$$

in analogy to equation (3-22).

Extensive tables of gas pressure as a function of temperature and electron pressure, allowing for hydrogen ionization and molecule formation, have been computed by Vardya (*Ap. J.*, **133**, 107, 1961). One interesting result is shown in Figure 3-1. The ordinate gives the ratio of the number of free electrons contributed by hydrogen to those contributed by the metals. We see that at high enough temperature and low enough density, hydrogen acts as an electron donor. In contrast, at low temperature and high density, $H^-$ forms in substantial quantities, and hydrogen acts as a sink of free electrons. Vardya also gives results that show clearly which of the metals acts as the main electron donor at each temperature and pressure.

In the atmospheres of the cooler stars, numerous diatomic and polyatomic molecules form in addition to $H_2$. These do not play a major role in determining the total pressure, which is dominated by the behavior of hydrogen, but do influence both the electron pressure and opacities. For example, the molecules $H_2O$ and CO are major opacity sources in the atmospheres of the M stars (see Yamashita, *P. A. S. Japan*, **14**, 390, 1962). The solution of the combined dissociation equations for several molecular species

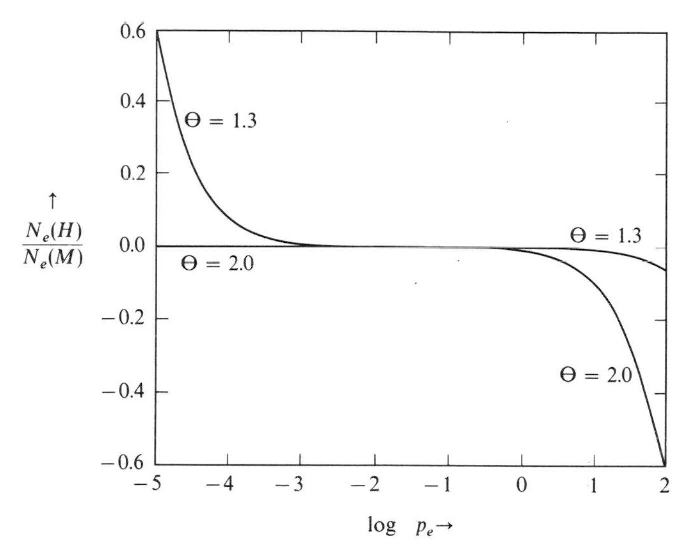


FIG. 3-1. Ratio of free electrons donated by hydrogen to those donated by metals at low temperatures. Note *negative* values for $\theta = 2.0$ ($T = 2500°$K); these imply that hydrogen acts as a sink of free electrons in the formation of $H^-$. (From M. S. Vardya, *Ap. J.*, **133**, 107, 1961; by permission.)

is naturally more complicated and, in general, must be carried out numerically; moreover, since the equations are nonlinear, some form of iteration scheme is usually required. Graphs of the relative abundances of many diatomic and polyatomic molecules for specified physical conditions and abundances of elements are given, for example, by Dolan (*Ap. J.*, **142**, 1621, 1965) and by Aller (Ref. 4, pp. 135–38).

THE DENSITY

The density of the stellar material in (gm/cm$^3$) is sometimes desired. In general, we may write

$$\rho = \sum_i m_i N_i = \frac{m_{\mathrm{H}} N_{\mathrm{H}}}{A_{\mathrm{H}}} \sum \alpha_i A_i \tag{3-47}$$

But from the definition $\alpha = N/N_{\mathrm{H}}$ it is clear that

$$N_{\mathrm{H}} = N_N \Big/ \sum_i \alpha_i \tag{3-48}$$

while from equation (3-20) we have

$$N_N = (p_g - p_e)/kT \tag{3-49}$$

Thus, combining equations (3-47), (3-48), and (3-49), we can write

$$\rho = \frac{(p_g - p_e)\mu m_{\mathrm{H}}}{kT} \tag{3-50}$$

where

$$\mu \equiv \frac{\sum_i \alpha_i A_i}{A_{\mathrm{H}} \sum_i \alpha_i} \tag{3-51}$$

As defined in equation (3-51), $\mu$ gives the number of atomic mass units per nucleus and is called the *mean molecular weight*. When molecule formation takes place, we can no longer calculate $N_N$ directly from $p_g$ and $p_e$ because in some cases, 2 or 3 atoms are bound together into a single and contribute only once to $p_g$ while contributing more to the mass density. We might, instead, employ equation (3-38) to write $N_{\mathrm{H}} = N_e/f_e = p_e/(f_e kT)$. Then substituting into equation (3-47) we have

$$\rho = \frac{p_e \mu' m_{\mathrm{H}}}{f_e kT} \tag{3-52}$$

where now

$$\mu' \equiv \frac{\sum_i \alpha_i A_i}{A_{\mathrm{H}}} \tag{3-53}$$

As defined in equation (3-53), $\mu'$ gives the number of grams of stellar material per gram of hydrogen.

## 3-3. The Non-LTE Equation of State

As we have seen in the preceding sections, under the assumption of LTE, complete knowledge of the distribution of atoms among their possible states of excitation and ionization follows from the local values of the two thermodynamic variables $T$ and $N_e$. In this sense, LTE is a local theory and does not contain any reference to the coupling between the radiation field and the state of the gas.

As we explained in Chapter 1, a more general assumption than LTE is one which considers the atomic populations to be in a steady state, with occupation numbers specified by the solution of equations of statistical equilibrium. We shall develop these equations in Chapter 5; it is not appropriate to do so here because we shall require results yet to be obtained in Chapter 4. Let us merely emphasize again, at this point, that the approach of using statistical equilibrium equations provides a complete microscopic picture of the interaction between the gas and the radiation field. As we previously stated, a complete solution requires that the radiation field (resulting, in the last analysis, from transfer with source and sink terms determined by the occupation numbers in the gas) be consistent with the occupation numbers (determined, in turn, by interaction with the radiation field). Since the radiation field is the result of a transfer process in the atmosphere, the state of the gas at a given point now depends upon the state of the gas in the ensemble as a whole; in this sense, the non-LTE equation of state is *nonlocal.*

If values of $T$, $N_e$, and the radiation field, characterized for this purpose by the mean intensity $J_\nu$, are given (with the understanding that the self-consistency requirements are also met somehow), then the non-LTE equation of state may formally be written as

$$N_i = f(T, N_e, J_\nu) \tag{3-54}$$

where $N_i$ denotes the population of some particular excitation and ionization state. It should be noted here that one must now be careful to specify the meaning of the temperature $T$. Since the Saha-Boltzmann relations are no longer valid, and the photon distribution of the source function is no longer given by Planck's law, $T$ can no longer be considered an absolute thermodynamic variable. We shall show in Chapter 5, however, that the velocity distributions of electrons, atoms, and ions remain Maxwellian and can be characterized by a single *kinetic temperature*; it is in this sense (only!) that the temperature retains its meaning. Since continuum states (i.e., electrons in the continuum) retain a description by a thermodynamic equilibrium distribution function, they may be said to be in LTE at the local value of $T$, but one must remember that this statement is made only in the sense we have prescribed here.

A convenient notational device sometimes employed to distinguish the non-LTE equations from the LTE equations is to write, following Menzel and his coworkers (see Ref. 25),

$$b_i \equiv N_i(T, N_e, J_\nu)/N_i^*(T, N_e) \tag{3-55}$$

where $N$ denotes the actual population of the level while $N_i^*$ denotes the population of the level that would be found for the given values of $T$, $N_{\text{ion}}$, and $N_e$ if the Saha-Boltzmann relations remained valid

As an example of this notation, consider atomic hydrogen. Combining equations (3-4), (3-16), and (3-55), we may write

$$\begin{aligned} N_{i,0} &= b_i N_{\text{ion}} N_e \left(\frac{2\pi mkT}{h^2}\right)^{-3/2} \frac{g_i}{2} \exp\left[\frac{(\chi_I - \chi_i)}{kT}\right] \\ &\equiv b_i N_{\text{ion}} N_e \varphi_i(T) \end{aligned} \tag{3-56}$$

where we have made use of the fact that $U_1 = 1$ for hydrogen. Summing over all bound states, we may write

$$\frac{N_{i,0}}{N_0} = \frac{b_i g_i \exp(-\chi_i/kT)}{U_0} \tag{3-57}$$

where now the partition function is generalized to mean

$$U_0 = \sum_i b_i g_i \exp(-\chi_i/kT) \tag{3-58}$$

Let us now turn to the question of specifying the opacity, both for bound-bound and bound-free transitions. We will then be in a position to write down the equations of statistical equilibrium that actually determine the non-LTE equation of state.

# 4 | The Opacity

In this chapter we shall consider the quantum mechanical basis for the calculation of atomic absorption cross-sections and discuss how these are combined to give the opacity of stellar material. The presentation is meant to be self-contained, but limitations of space require that knowledge of some of the basic principles of quantum mechanics be presupposed.

## 4-1. The Einstein Relations for Bound-Bound Transitions

Let us first consider the absorption and emission of radiation by an atom in a transition between two bound states. (Note: In this section we shall refer to all processes in which a photon is removed from the beam as "absorption" since the distinction made in Chapter 1 between absorption and scattering is not important in the present context.) We assume that the lower state, $i$, has statistical weight $g_i$, and the upper state, $j$, statistical weight $g_j$. There are three basic processes involved, which are usually described in terms of rate coefficients first introduced by Einstein (*Phys. Z.*, **18**, 121, 1917). The first process is the *direct absorption* of radiation, leading to an upward transition from level $i$ to level $j$. Einstein wrote the rate at which this process occurs as

$$N_i(\nu)R_{ij}\frac{d\omega}{4\pi} = N_i(\nu)B_{ij}I_\nu\frac{d\omega}{4\pi} \tag{4-1}$$

where $N_i(\nu)$ is the number of atoms per $cm^3$ in state $i$ that can absorb radiation at frequencies on the range $(\nu, \nu + d\nu)$. In general, the spectrum line corresponding to the transition will not be sharp, but because of the perturbations exerted by nearby atoms and ions, as well as the finite lifetime of the upper state, it will have a spread in frequency which we can characterize by an *absorption profile*, $\varphi_\nu$, normalized such that

$$\int_0^\infty \varphi_\nu \, d\nu = 1 \tag{4-2}$$

Thus if the total number of atoms in state $i$ is $N_i$, the number of atoms capable of absorbing radiation at frequency $\nu$ is

$$N_i(\nu) = N_i \varphi_\nu \tag{4-3}$$

In making the transition from level $i$ to level $j$, the atom absorbs photons of energy

$$h\nu_{ij} = E_j - E_i \tag{4-4}$$

The rate at which energy is removed from the beam may now be obtained by combining equations (4-1) through (4-4) to yield

$$\rho a_\nu I_\nu = N_i B_{ij} \frac{h\nu_{ij}}{4\pi} \varphi_\nu I_\nu \tag{4-5}$$

where $a_\nu$ denotes a macroscopic absorption coefficient as defined in Chapter 1.

For atoms returning from level $j$ to level $i$, two processes are possible. The first of these is a *spontaneous transition* with the emission of a photon. Writing the probability of spontaneous emission per unit time as $A_{ji}$, the rate of emission of energy is

$$\rho j_\nu(\text{spontaneous}) = N_j A_{ji} \frac{h\nu_{ij}}{4\pi} \psi_\nu \tag{4-6}$$

where the *emission profile* $\psi_\nu$ is normalized such that

$$\int_0^\infty \psi_\nu \, d\nu = 1 \tag{4-7}$$

The other return process possible is a transition *induced by the radiation field.* The energy emitted may be written in terms of a coefficient $B_{ji}$ so that

$$\rho j_\nu(\text{induced}) = N_j B_{ji} \frac{h\nu_{ij}}{4\pi} I_\nu \psi_\nu \tag{4-8}$$

where we have made use of the fact that one may show, on quite general grounds (Ref. 11, Chap. 10), that the induced emission profile is the same as that for spontaneous emission. An important distinction to note between spontaneous and induced emission, however, is that spontaneous emission takes place *isotropically* while *induced emission has the same angular distribution as* $I_\nu$. It is in this sense that induced emissions may be thought of as *negative absorptions.*

The coefficients $A_{ji}$, $B_{ji}$, and $B_{ij}$ are simply related, as can be shown by calculating the rate of absorption and emission in thermodynamic equilibrium. In this case we may write

$$I_\nu \equiv B_\nu$$

while

$$\frac{N_j^*}{N_i^*} = \frac{g_j}{g_i} \exp(-h\nu_{ij}/kT)$$

and

$$N_i^* B_{ij} B_\nu = N_j^* A_{ji} + N_j^* B_{ji} B_\nu \tag{4-9}$$

where we have integrated over $\nu$ and used the fact that $B_\nu$ changes negligibly over the line width. Solving for $B_\nu$, we have

$$B_\nu = \frac{N_j^* A_{ji}}{N_i^* B_{ij} - N_j^* B_{ji}} = \frac{(A_{ji}/B_{ji})}{\left(\dfrac{g_i B_{ij}}{g_j B_{ji}}\right) \exp(h\nu_{ij}/kT) - 1} \tag{4-10}$$

But we know that the correct expression for the Planck function is

$$B_\nu \quad \frac{2h\nu^3}{c^2} \frac{1}{\exp(h\nu/kT) - 1}$$

so that to make the expressions correspond, we clearly must have

$$A_{ji} = \frac{2h\nu^3}{c^2} B_{ji} \tag{4-11}$$

and

$$g_i B_{ij} = g_j B_{ji} \tag{4-12}$$

Note that while these relations were derived for ease from a thermodynamic equilibrium argument, the Einstein coefficients must for physical reasons be properties of the atom only and independent of the nature of the radiation field. Therefore, we must conclude that equations (4-11) and (4-12) are true in general. It is of interest to note that historically the kind of argument given above led to the realization of the necessity that a stimulated emission process must occur in nature—a fact not intuitively obvious at first sight.

The microscopic rate formulation described above may immediately be incorporated into the equation of transfer. If *only* the bound-bound process occurs, the appropriate transfer equation is

$$\mu \frac{dI_\nu}{dz} = \frac{h\nu_{ij}}{4\pi} [N_j A_{ji} \psi_\nu - (N_i B_{ij} \varphi_\nu - N_j B_{ji} \psi_\nu) I_\nu] \tag{4-13}$$

Here we have followed the usual practice of grouping together all terms involving $I_\nu$. In this way one can define a line absorption coefficient corrected for stimulated emission, namely,

$$\begin{aligned} l_\nu &= N_i B_{ij} \frac{h\nu_{ij}}{4\pi} \varphi_\nu \left(1 - \frac{N_j B_{ji} \psi_\nu}{N_i B_{ij} \varphi_\nu}\right) \\ &= N_i B_{ij} \frac{h\nu_{ij}}{4\pi} \varphi_\nu \left(1 - \frac{\psi_\nu}{\varphi_\nu} \frac{b_j}{b_i} e^{-h\nu_{ij}/kT}\right) \end{aligned} \tag{4-14}$$

and a line source function

$$\begin{aligned} S_\nu &= \frac{N_j A_{ji} \psi_\nu}{N_i B_{ij} \varphi_\nu - N_j B_{ji} \psi_\nu} = \frac{(A_{ji}/B_{ji})}{\dfrac{N_i}{N_j} \dfrac{B_{ij}}{B_{ji}} \dfrac{\varphi_\nu}{\psi_\nu} - 1} \\ &= \frac{2h\nu^3}{c^2} \frac{1}{\left(\dfrac{b_i}{b_j} \dfrac{\varphi_\nu}{\psi_\nu}\right) \exp(h\nu_{ij}/kT) - 1} \end{aligned} \tag{4-15}$$

Then the transfer equation reduces to its standard form

$$-\frac{\mu}{l_\nu} \frac{dI_\nu}{dz} = I_\nu - S_\nu \tag{4-16}$$

A simplifying assumption that may be made and which is valid in many cases of astrophysical interest is that $\varphi_\nu \equiv \psi_\nu$ (see Chapter 10). Then equations (4-14) and (4-15) simplify to

$$l_\nu = N_i B_{ij} \frac{h\nu_{ij}}{4\pi} \varphi_\nu \left(1 - \frac{b_j}{b_i} e^{-h\nu_{ij}/kT}\right) \tag{4-17}$$

and

$$S_\nu = \frac{2h\nu^3}{c^2} \frac{1}{\dfrac{b_i}{b_j} \exp(h\nu_{ij}/kT) - 1} \tag{4-18}$$

Aside from the approximation $\psi_\nu = \varphi_\nu$ mentioned above, these equations are completely general expressions for a single line without an overlapping transition of any kind.

In the case of LTE, we know that by definition $b_i = b_j = 1$, and we have for the absorption coefficient

$$l_\nu = N_i B_{ij} \frac{h\nu_{ij}}{4\pi} [1 - \exp(-h\nu_{ij}/kT)] \tag{4-19}$$

The factor $[1 - \exp(-h\nu/kT)]$ is usually referred to as the *correction factor for stimulated emission.* As is evident from equation (4-17), this form of the correction factor is valid in LTE only. Similarly, in LTE the source function becomes

$$S_\nu = \left(\frac{2h\nu^3}{c^2}\right) \frac{1}{\exp(h\nu/kT) - 1} = B_\nu$$

as expected, if we understand that the opacity appearing in the Kirchhoff-Planck relation [equation (1-21)] included the correction for stimulated emission.

Equation (4-18) contains implicitly the solution of the statistical equilibrium equations, inasmuch as it makes reference to $b$-factors for the levels involved. We shall therefore on occasion refer to this as the *implicit form* of the source function. An alternative approach is to introduce the solution of the statistical equilibrium equations into the source function explicitly by direct analytical substitution, yielding what we shall refer to as an *explicit form.* As we shall see in Chapters 7, 12, 13, and 14, the latter approach is by far the more powerful and useful.

## 4-2. The Calculation for Transition Probabilities

Let us now turn our attention to the problem of a calculation of the Einstein coefficients. Specifically, we will derive the direct absorption probability $B_{ij}$ since $B_{ji}$ and $A_{ji}$ can be obtained from $B_{ij}$ by use of equations (4-11) and (4-12). This computation may be made on three successively more accurate levels of approximation, as follows.

(a) Classical Atom and Electromagnetic Field. The electron is considered to be a damped harmonic oscillator driven by the electromagnetic field. The absorption coefficient that is derived is dimensionally correct but quantitatively may be wrong by orders of magnitude.

(b) Quantum-Mechanical Atom and Classical Electromagnetic Field. Here we can derive correct values of $B_{ij}$ and $B_{ji}$, but $A_{ji}$ does not appear in this formulation (although it is still correctly given by the Einstein relations).

(c) Quantum Mechanical Atom and Quantized Electromagnetic Field. Here the correct results are yielded automatically for all three rates, and this approach represents the complete theory.

In this section, we will carry through the calculation by method (a) for historical interest and general background and by method (b) to obtain the correct expression for $B_{ij}$. Application of method (c) is somewhat more complicated and really does not need to be carried through here if we are satisfied to use the Einstein relations. A complete discussion of the third method may be found in texts on quantum mechanics (e.g., Ref. 11, Chap. 10).

### THE CLASSICAL OSCILLATOR

The basic equations of classical electromagnetic theory are Maxwell's equations (see, e.g., Ref. 30, Chap. 9)

$$\nabla \cdot \mathbf{D} = \rho \tag{4-20}$$

$$\nabla \times \mathbf{E} = -\frac{\partial \mathbf{B}}{\partial t} \tag{4-21}$$

$$\nabla \cdot \mathbf{B} = 0 \tag{4-22}$$

$$\nabla \times \mathbf{H} - \frac{\partial \mathbf{D}}{\partial t} = \mathbf{j} \tag{4-23}$$

where $\mathbf{E}$ and $\mathbf{H}$ are the electric and magnetic fields, $\mathbf{B} \equiv \mu_0 \mathbf{H}$, $\mathbf{D} \equiv \varepsilon_0 \mathbf{E}$, $\rho$ is the charge density, and $\mathbf{j}$ is the current density, given by $\mathbf{j} = \rho \mathbf{v}$ where $\mathbf{v}$ is the velocity of the charges. These equations are in rationalized MKS units; in these units $c = \mu_0^{-1}\varepsilon_0^{-1}$, where $c$ is the velocity of light. It is convenient to introduce a scalar and vector potential, $\varphi$ and $\mathbf{A}$, such that

$$\mathbf{B} = \nabla \times \mathbf{A} \tag{4-24}$$

which satisfies equation (4-22), and

$$\mathbf{E} = -\nabla\varphi - \frac{\partial \mathbf{A}}{\partial t} \tag{4-25}$$

which satisfies equation (4-21). Since $\mathbf{B}$ is given by the *curl* of $\mathbf{A}$, one is free to define the *divergence* of $\mathbf{A}$ arbitrarily. One of the most convenient choices, first used by Lorentz, is

$$\nabla \cdot \mathbf{A} = -\frac{1}{c^2}\frac{\partial \varphi}{\partial t} \tag{4-26}$$

With this choice, Maxwell's equations reduce to linear wave equations

$$\frac{1}{c^2}\frac{\partial^2 \mathbf{A}}{\partial t^2} - \nabla^2 \mathbf{A} = \mu_0 \rho \mathbf{v} \tag{4-27}$$

$$\frac{1}{c^2}\frac{\partial^2 \varphi}{\partial t^2} - \nabla^2 \varphi = \frac{\rho}{\varepsilon_0} \tag{4-28}$$

The solutions of these equations can be written in the form of retarded potentials (see, e.g., Ref. 30, Chap. 14),

$$\mathbf{A}(\mathbf{r}, t) = \frac{\mu_0}{4\pi} \int \frac{\rho' \mathbf{v}'}{|\mathbf{r} - \mathbf{r}'|} \, d\tau' \tag{4-29}$$

and

$$\varphi(\mathbf{r}, t) = \frac{1}{4\pi\varepsilon_0} \int \frac{\rho'}{|\mathbf{r} - \mathbf{r}'|} \, d\tau' \tag{4-30}$$

where $\rho'$ and $\mathbf{v}'$ are evaluated at position $\mathbf{r}'$, and at time

$$t' = t - \frac{1}{c} |\mathbf{r} - \mathbf{r}'| \tag{4-31}$$

These retarded potentials are generalizations of the usual coulomb and ampere laws.

One of the most important solutions to Maxwell's equations is that of a plane electromagnetic wave propagating in vacuum (no charges present so that $\rho - 0$). The solution to the wave equations yields transverse wave of the form

$$\mathbf{E} = E_0 \cos\left[\omega\left(t - \frac{z}{c}\right)\right]\hat{\mathbf{i}} \tag{4-32}$$

and

$$\mathbf{H} = H_0 \cos\left[\omega\left(t - \frac{z}{c}\right)\right]\hat{\mathbf{j}} \tag{4-33}$$

where

$$H_0 = c\,\varepsilon_0\, E_0 \tag{4-34}$$

These waves propagate in the direction $\hat{\mathbf{k}}$, orthogonal to $\hat{\mathbf{i}}$ and $\hat{\mathbf{j}}$. The rate of flow of energy per $cm^2$ is given by Poynting's vector

$$\mathbf{S} = \mathbf{E} \times \mathbf{H} = c\varepsilon_0\, E_0{}^2 \cos^2\left[\omega\left(t - \frac{z}{c}\right)\right]\hat{\mathbf{k}} \tag{4-35}$$

If we average this energy flow over a cycle, the mean rate of flow is given by

$$\overline{S(\omega)} = \tfrac{1}{2}\, c\varepsilon_0\, E_0{}^2 \tag{4-36}$$

where $\overline{S(\omega)}$ denotes the energy per unit *circular* frequency interval $d\omega$. Equation (4-36) gives the flux of energy transported by the mono-directional beam through a unit area oriented normal to its direction of propagation. In terms of a specific intensity we could equivalently write

$$I(\omega) = \frac{1}{2} c\varepsilon_0 E_0^2 \frac{\delta(\mu - 1)}{2\pi} \tag{4-37}$$

or, in *cgs* units, which will be more convenient for our later work,

$$I(\omega) = \frac{c}{8\pi} E_0^2 \frac{\delta(\mu - 1)}{2\pi} \tag{4-38}$$

Then, from equation (1-6) we have

$$F(\omega) = 2\pi \int_{-1}^{1} I(\mu)\mu \, d\mu = \frac{c}{8\pi} E_0^2 \tag{4-39}$$

in agreement with equation (4-36).

The potential due to a moving point (or near-point) charge can be calculated from equations (4-29) and (4-30), but a bit of care is required. In particular, since the charge densities are to be evaluated at *retarded* times, we should consider the integration process as the result of an information-gathering sphere, moving with velocity $c$, as it approaches the observer. If the charges are stationary, they give exactly the right contribution to the integration process. If the charge follows the sphere, it contributes too much to the integral and would result in too large a measured charge for the particle. The excess charge that would be measured by the observer at position $\mathbf{r}$ relative to the charge is

$$\rho' \frac{\mathbf{v}' \cdot \mathbf{r}}{r} \, dA \, dt = \rho' \frac{\mathbf{v}' \cdot \mathbf{r}}{r} \, dA \frac{dr}{c} = \rho' \frac{\mathbf{v}' \cdot \mathbf{r}}{r} \frac{d\tau'}{c}$$

Thus the part of the charge actually contributed by the volume element is

$$de = \rho' \left(1 - \frac{\mathbf{v}' \cdot \mathbf{r}}{cr}\right) d\tau' \tag{4-40}$$

and the appropriate retarded density to substitute into equations (4-29) and (4-30) for the potential is

$$\frac{\rho' \, d\tau'}{r} = \frac{de}{r'\left(1 - \dfrac{\mathbf{v}' \cdot \mathbf{r}}{cr}\right)} \tag{4-41}$$

Then, integrating over all volume containing the charge (which is essentially a point), we find

$$\varphi = \frac{1}{4\pi\varepsilon_0} \left( \frac{e}{r - \dfrac{\mathbf{r} \cdot \mathbf{v}'}{c}} \right) \tag{4-42}$$

and

$$\mathbf{A} = \frac{\mu_0}{4\pi}\left(\frac{e\mathbf{v}'}{r - \dfrac{\mathbf{r}\cdot\mathbf{v}'}{c}}\right) \tag{4-43}$$

The calculation of the electric and magnetic fields from these equations is complicated by the necessity of connecting derivatives in the observer's frame with those in the moving frame of the charge. When this is done (see, e.g., Ref. 30, Chap. 20), we obtain

$$\frac{4\pi\varepsilon_0 \mathbf{E}}{e} = \frac{1}{s^3}\left(\mathbf{r} - \frac{r}{c}\mathbf{v}\right)\left(1 - \frac{v^2}{c^2}\right) + \frac{1}{c^2 s^3}\left\{\mathbf{r} \times \left[\left(\mathbf{r} - \frac{r}{c}\mathbf{v}\right) \times \dot{\mathbf{v}}\right]\right\} \tag{4-44}$$

and

$$\frac{4\pi\mathbf{B}}{\mu_0 e} = \frac{1}{s^3}\left(1 - \frac{v^2}{c^2}\right)(\mathbf{v} \times \mathbf{r}) + \left(\frac{1}{crs^3}\right)\mathbf{r} \times \left\{\mathbf{r} \times \left(\left[\mathbf{r} - \frac{r}{c}\mathbf{v}\right) \times \dot{\mathbf{v}}\right]\right\} \tag{4-45}$$

where

$$s = r - \frac{\mathbf{r}\cdot\mathbf{v}}{c} \tag{4-46}$$

Note that in equations (4-42) and (4-45) the first term is proportional to $r^{-2}$ while the second term varies as $r^{-1}$. The latter will obviously dominate at large distances where our observations will be made, so we retain it while neglecting the other. In addition, in our present work we will be considering a harmonic oscillator in which $\dot{\mathbf{v}}$ is parallel to $\mathbf{v}$ so that terms of the form $\mathbf{v} \times \dot{\mathbf{v}}$ are zero. Simplifying equations (4-44) and (4-45), we then obtain

$$\frac{4\pi\varepsilon_0 \mathbf{E}}{e} = \frac{1}{c^2 s^3}\mathbf{r} \times (\mathbf{r} \times \dot{\mathbf{v}}) = \frac{r^2 \dot{v} \sin\theta}{c^2 s^3}\hat{\mathbf{i}} \tag{4-47}$$

and

$$\frac{4\pi\mathbf{B}}{\mu_0 e} = \left(\frac{1}{cs^3 r}\right)\mathbf{r} \times [\mathbf{r} \times (\mathbf{r} \times \dot{\mathbf{v}})] = \frac{r}{cs^3}(\dot{\mathbf{v}} \times \mathbf{r}) = \frac{r^2 \dot{v} \sin\theta}{cs^3}\hat{\mathbf{j}} \tag{4-48}$$

where $\theta$ is the angle between $\mathbf{r}$ and $\dot{\mathbf{v}}$. If, in addition, we assume that $v \ll c$ so that $r \approx s$, then we may write

$$\mathbf{E} = \frac{e}{4\pi\varepsilon_0}\frac{\dot{v}\sin\theta}{rc^2}\cdot\hat{\mathbf{i}} \tag{4-49}$$

and

$$\mathbf{B} = \frac{\mu_0 e}{4\pi}\frac{\dot{v}\sin\theta}{rc}\hat{\mathbf{j}} \tag{4-50}$$

Then from equation (4-35), the power radiated per cm² is

$$\mathbf{S} = \mathbf{E} \times \mathbf{H} = \frac{e^2 \dot{v}^2 \sin^2 \theta}{16\pi^2 \varepsilon_0 c^3 r^3} \mathbf{r} \tag{4-51}$$

Finally, integrating over a sphere of radius $r$ and writing $\sin^2 \theta = (1 - \mu^2)$, the total power radiated in all directions is

$$P = \frac{e\dot{v}^2}{16\pi^2 \varepsilon_0 c^3} 2\pi \int_{-1}^{1} (1 - \mu^2)\, d\mu = \frac{e^2 \dot{v}^2}{6\pi\varepsilon_0 c^3} \tag{4-52}$$

or, in cgs units

$$P = \frac{2e\dot{v}^2}{3c^3} \tag{4-53}$$

Now for a harmonic oscillator, we have

$$x = x_0 \cos \omega t$$

so that

$$\dot{v} = -\omega^2 x_0 \cos \omega t$$

Substituting into equation (4-53) and averaging over a period, we find

$$\overline{P(\omega)} = \frac{e^2 x_0{}^2 \omega^4}{3c^3} \tag{4-54}$$

where we have made use of the fact that $\overline{\cos^2 \omega t} = 1/2$. Because the harmonic oscillator is radiating away energy, the oscillation will eventually decay. This is the same result that would be obtained for a damping force acting on the oscillator. We can calculate this effective damping force by assuming that the rate of work done against the force accounts for the rate of energy loss of the oscillator. Thus, from equation (4-53) we write

$$\mathbf{F}_{\text{rad}} \cdot \mathbf{v} + \frac{2e^2 \dot{v}^2}{3c^3} = 0 \tag{4-55}$$

Then

$$\int_{t_1}^{t_2} \mathbf{F}_{\text{rad}} \cdot \mathbf{v}\, dt + \frac{2e^2}{3c^3} \left( \Big|_{t_1}^{t_2} \mathbf{v} \cdot \dot{\mathbf{v}} - \int_{t_1}^{t_2} \ddot{\mathbf{v}} \cdot \mathbf{v}\, dt \right) = 0 \tag{4-56}$$

Over a single cycle, the integrated term vanishes; so on the average we must have

$$\overline{\mathbf{F}}_{\text{rad}} = \frac{2e^2 \ddot{\mathbf{v}}}{3c^3} \tag{4-57}$$

To a good order of approximation we may calculate $\ddot{\mathbf{v}}$ from its value for the undamped harmonic oscillator, namely,

$$\ddot{\mathbf{v}} = -\omega_0^2 \mathbf{v}$$

and thus we can write

$$\overline{\mathbf{F}}_{\text{rad}} = -m\gamma\mathbf{v} \tag{4-58}$$

where

$$\gamma = \frac{2}{3}\frac{e^2\omega_0^2}{mc^3} \tag{4-59}$$

This constant is referred to as the *classical damping constant* because of the formal resemblance of the radiation reaction force to a viscous damping term.

Let us now calculate the scattering coefficient for a classical electron oscillator driven by an electromagnetic field. Since in the classical picture the interaction is a pure scattering process, we can calculate the energy scattered out of the beam by calculating the energy radiated by a dipole driven by the field. The equation of motion for an oscillator driven by a field of frequency $\omega$ is

$$m(\ddot{\mathbf{x}} + \omega_0^2 \mathbf{x}) = e\mathbf{E}_0 e^{i\omega t} - m\gamma\dot{\mathbf{x}} \tag{4-60}$$

Writing a trial solution for $\mathbf{x}$ which is proportional to $\exp(i\omega t)$, we find for the steady-state solution

$$\mathbf{x} = Re\left[\frac{e}{m}\frac{\mathbf{E}_0\, e^{i\omega t}}{(\omega^2 - \omega_0^2) + i\gamma\omega}\right] \tag{4-61}$$

and

$$\ddot{\mathbf{x}} = Re\left[-\frac{e\omega^2}{m}\frac{\mathbf{E}_0\, e^{i\omega t}}{(\omega^2 - \omega_0^2) + i\gamma\omega}\right] \tag{4-62}$$

Thus, substituting into equation (4-53) and averaging, we have

$$\overline{P(\omega)} = \frac{2e^2}{3c^3}\overline{\ddot{\mathbf{x}}^2} = \frac{e^4\omega^4}{3m^2c^3}\frac{E_0^2}{(\omega^2 - \omega_0^2)^2 + \gamma^2\omega^2} \tag{4-63}$$

This is to be identified with the total energy scattered from the beam. Assuming, for simplicity, that the scattering cross-section $\sigma$ is isotropic and making use of equation (4-38), we find

$$\overline{P(\omega)} = \sigma(\omega)\oint I\, d\Omega = \sigma\left(\frac{cE_0^2}{8\pi}\right)\int_0^{2\pi} d\varphi \int_{-1}^{1}\frac{\delta(\mu - 1)}{2\pi}\, d\mu = \sigma\frac{cE_0^2}{8\pi} \tag{4-64}$$

By comparing equations (4-63) and (4-64), we see that the scattering coefficient is

$$\sigma(\omega) = \frac{8\pi e^4}{3m^2c^4}\left[\frac{\omega^4}{(\omega^2 - \omega_0{}^2)^2 + \gamma^2\omega^2}\right] \tag{4-65}$$

We may simplify this expression by noting that $\sigma(\omega)$ is a sharply peaked function in the neighborhood $\omega \approx \omega_0$. To a good approximation we can replace $\omega^2 \approx 2\omega(\omega - \omega_0)$ and from equation (4-59) substitute for $\gamma$ to find

$$\sigma(\omega) = \frac{\pi e^2}{mc}\left[\frac{\gamma}{(\omega - \omega_0)^2 + (\gamma/2)^2}\right] \tag{4-66}$$

If we integrate equation (4-66) over all frequencies, we obtain the total scattering coefficient

$$\begin{aligned}\sigma_{\text{tot}} &= \int_0^\infty \sigma\, d\nu = \frac{\pi e^2}{mc}\int_0^\infty \frac{(\gamma/4\pi^2)\, d\nu}{(\nu - \nu_0)^2 + (\gamma/4\pi)^2} \\ &= \left(\frac{\pi e^2}{mc}\right)\frac{1}{\pi}\int_{-\infty}^\infty \frac{dx}{1 + x^2} = \frac{\pi e^2}{mc}\end{aligned} \tag{4-67}$$

where we have written $x = 4\pi(\nu - \nu_0)/\gamma$ and observed that $-4\pi\nu_0/\gamma = -\infty$, for all practical purposes. This total cross-section gives a measure of the efficiency with which energy is removed from the beam. The factor in brackets in equation (4-66) is thus a normalized profile function known as a *Lorentz profile*. For our present purposes, we shall confine attention to the integrated cross-section alone; the profile function will be discussed in detail in Chapter 9. The classical result predicts a unique scattering efficiency for all transitions; this is not surprising insofar as no reference is made in the theory to the structure of the levels between which the transition occurs. The quantum mechanical treatment shows that scattering cross-sections may in fact differ greatly for various transitions. A customary way of writing the quantum mechanical result for the total cross-section is

$$\sigma_{\text{tot}} = \frac{\pi e^2}{mc} f_{ij} \tag{4-68}$$

where $f_{ij}$ is called the *oscillator strength* of the transition. In rough pictorial terms, $f_{ij}$ may be thought of as giving the "effective number" of classical oscillators involved in the transition under consideration; only for the strongest transitions does $f_{ij}$ approach unity. In terms of $B_{ij}$ we can write

$$\frac{\pi e^2}{mc} f_{ij} = B_{ij}\frac{h\nu_{ij}}{4\pi} \tag{4-69}$$

QUANTUM MECHANICAL CALCULATION

Let us now consider the calculation of $B_{ij}$ when the atom is treated according to quantum mechanics and the radiation field according to classical electrodynamics. The atomic structure is described by a *wave function* $\psi(\mathbf{r}_1, \mathbf{r}_2, \ldots, \mathbf{r}_N, t)$ where $\mathbf{r}_1$, etc., are the positions of the bound atomic electrons. The quantity $\psi\psi^* \, d\mathbf{r}_1 \ldots d\mathbf{r}_N$ may be interpreted as the probability of finding the atom with the electrons in the volume element $\mathbf{r}_1$ to $\mathbf{r}_1 + d\mathbf{r}_1, \ldots,$ $\mathbf{r}_N$ to $\mathbf{r}_N + d\mathbf{r}_N$. These wave functions are solutions of *Schrödinger's equation*

$$H\psi = i\hbar \frac{\partial \psi}{\partial t} \tag{4-70}$$

where $H$ is the total *Hamiltonian* of the system in operator form (see, e.g., Ref. 27, Chaps. 8 and 9, for a detailed discussion of mathematical expressions for the Hamiltonian). The Hamiltonian operator is to be constructed according to the rule

$$H(q_i, p_i) \rightarrow H\left(q_i, \frac{\hbar}{i} \frac{\partial}{\partial q_i}\right)$$

where $q_i$ and $p_i$ are space coordinates and momenta, respectively. The atomic system has certain *stationary states* (or *eigenstates*) in which the total energy is constant. Thus, if $H_A$ is the Hamiltonian of an atom which is in some stationary state $j$, then

$$H_A \psi_j = i\hbar \frac{\partial \psi_j}{\partial t} = E_j \psi_j \tag{4-71}$$

which implies

$$\psi_j(t) = \psi_j(0) \exp(-iE_j t/\hbar) \tag{4-72}$$

We may thus write the general solution in the form

$$\psi_j(\mathbf{r}, t) = \varphi_j(\mathbf{r}) \exp(-iE_j t/\hbar) \tag{4-73}$$

The *time-independent* solutions $\varphi_j$ satisfy the equation

$$H_A \varphi_j = E_j \varphi_j \tag{4-74}$$

and are orthogonal in the sense that

$$\int \varphi_i^* \varphi_j \, d\tau \equiv \langle \varphi_i^* | \varphi_j \rangle = \delta_{ij} \tag{4-75}$$

A general state of the system at time $t = 0$ can be expanded in terms of the eigenstates (which form a complete set) by writing

$$\varphi = \sum_j a_j \varphi_j \tag{4-76}$$

Then the probability of measuring the system in this general state in a state $j$ is $a_j{}^*a_j = |a_j|^2$. At an arbitrary time $t$, we can write the general state as

$$\psi(t) = \sum_j a_j(t)\psi_j(t) \tag{4-77}$$

and again the probability of finding the system in state $j$ is $|a_j(t)|^2$. If the atom is unperturbed (i.e., $H \equiv H_{\text{atom}}$), then of course the $a$'s are independent of time. If, however, the atom is perturbed with some potential $V$, then the $a$'s will in general change with time, which can be interpreted as the atom undergoing *transitions* from one state to another. An example of such a perturbation is that exerted by an electromagnetic field upon the atomic electrons. In the lowest order of approximation we can assume that the atom is in a uniform, time-varying electromagnetic field, $\mathbf{E} = E_0 \cos \omega t\, \hat{\mathbf{i}}$, and that the potential of the electrons in this field is given by the expression

$$V = \sum_{i=1}^{N} e\, \mathbf{E} \cdot \mathbf{r}_i = E_0 \cos \omega\, \hat{\mathbf{i}} \cdot \mathbf{d} \tag{4-78}$$

where $\mathbf{d}$ is the dipole moment defined as

$$\mathbf{d} = e \sum_{i=1}^{N} \mathbf{r}_i \tag{4-79}$$

With a perturbing potential Schrödinger's equation becomes

$$(H_A + V)\psi = i\hbar \frac{\partial \psi}{\partial t} \tag{4-80}$$

Substituting equation (4-77) for $\psi$, we have

$$(H_A + V) \sum_n a_n(t)\psi_n = i\hbar \sum_n \dot{a}_n \psi_n + i\hbar \sum_n a_n \frac{\partial \psi_n}{\partial t} \tag{4-81}$$

In view of equation (4-71), equation (4-81) reduces to

$$i\hbar \sum_n \dot{a}_n \psi_n = \sum_n a_n V \psi_n \tag{4-82}$$

We may isolate a particular term $\dot{a}_m$ by using the orthogonality of the $\varphi$'s. Thus we multiply through by $\psi_m{}^*$ and integrate over all space. We then have

$$\begin{aligned} i\hbar \sum_n \dot{a}_n \exp[i(E_m - E_n)t/\hbar]\langle \varphi_m{}^* \,|\, \varphi_n \rangle \\ = \sum_n a_n(t) \exp[i(E_m - E_n)t/\hbar]\langle \varphi_m{}^* |\, V \,|\varphi_n \rangle \end{aligned} \tag{4-83}$$

But from equation (4-75),

$$\langle \varphi_m{}^* \,|\, \varphi_n \rangle = \delta_{mn}$$

and writing

$$\omega_{mn} \equiv (E_m - E_n)/\hbar \tag{4-84}$$

and

$$V_{mn} \equiv \langle \varphi_m^* | V | \varphi_n \rangle \tag{4-85}$$

equation (4-83) reduces to

$$\dot{a}_m(t) = \frac{1}{i\hbar} \sum_n a_n e^{i\omega_{mn}t} V_{mn} \tag{4-86}$$

For the perturbing potential given by equation (4-78), we readily find

$$\begin{aligned} V_{mn} &= E_0 \cos \omega t \, \hat{\mathbf{i}} \langle \varphi_m^* | \mathbf{d} | \varphi_n \rangle \\ &= E_0 \cos \omega t \, \hat{\mathbf{i}} \cdot \mathbf{d}_{mn} \equiv 2h_{mn} \cos \omega t \\ &= h_{mn}(e^{i\omega t} + e^{-i\omega t}) \end{aligned} \tag{4-87}$$

The quantities $\mathbf{d}_{mn}$ are referred to as *matrix elements* of the dipole moment. Thus, substituting into equation (4-86),

$$\dot{a}_m(t) = \frac{1}{i\hbar} \sum_n a_n(t) h_{mn} e^{i\omega_{mn}t}(e^{i\omega t} + e^{-i\omega t}) \tag{4-88}$$

We now make the simplifying assumption that at time $t = 0$, the atom is in some definite eigenstate $k$, and we consider a time interval $T$ so short that this state is not appreciably depopulated. That is, at $t = 0$, we assume $a_k(0) = 1$ and $a_n(0) = 0$ for all $n \neq k$. Moreover, we choose $T$ such that $a_k(t) \approx 1$ for all $t \leq T$. Then the sum in equation (4-88) may be replaced by a single term

$$\dot{a}_m(t) = \frac{1}{i\hbar} h_{mk} e^{i\omega_{mk}}(e^{i\omega t} + e^{-i\omega t}) \tag{4-89}$$

Integrating this expression with respect to time, we have

$$a_m(t) = \frac{h_{mk}}{i\hbar} \left\{ \frac{\exp[i(\omega_{mk} - \omega)t] - 1}{(\omega_{mk} - \omega)} + \frac{\exp[i(\omega_{mk} + \omega)t] - 1}{(\omega_{mk} + \omega)} \right\} \tag{4-90}$$

Since we are interested in absorption processes, we choose $E_m > E_k$ so that $\omega_{mk} > 0$. From the denominator of the first term of equation (4-90), we see that the main contribution will come when $\omega \approx \omega_{mk}$. We can thus clearly neglect the second term in comparison with the first. Then, writing $x = \omega - \omega_{mk}$, we have

$$|a_m(t)|^2 = \frac{h_{mk}^2}{\hbar^2 x^2} 4 \sin^2\left(\frac{xt}{2}\right) = \frac{E_0{}^2 |\hat{\mathbf{i}} \cdot \mathbf{d}_{mk}|^2}{4\hbar^2} \frac{4 \sin^2(xt/2)}{x^2} \tag{4-91}$$

Now this rate must be summed over all frequencies. Suppose the absorption line has a profile $\varphi_\nu$, and suppose that over a frequency range $X$, much wider than the range of $\varphi_\nu$, the intensity of radiation is constant and has the value $\bar{J}_\omega$. From equation (4-38), we see that the energy density in the beam [as defined by equation (1-15)] is simply

$$\frac{4\pi}{c}\bar{J}_\omega = \frac{E_0{}^2}{8\pi} \tag{4-92}$$

Then the total transition rate is

$$|a_m(t)|^2 = \frac{8\pi^2 \bar{J}_\omega}{\hbar^2 c} |\hat{\mathbf{i}} \cdot \mathbf{d}_{mk}|^2 2t \int_{-X}^{X} \frac{\sin^2(xt/2)\, d(xt/2)}{(xt/2)^2} \tag{4-93}$$

Now if $Xt \gg 1$, the limits on the integral in equation (4-93) may formally be extended to infinity; this is usually a good approximation since a characteristic transition time is of order $10^{-8}$ sec while $\omega$ is of order $10^{15}$ sec$^{-1}$. But from standard tables we find

$$\int_{-\infty}^{\infty} \frac{\sin^2 x}{x^2}\, dx = \pi \tag{4-94}$$

so that

$$|a_m(t)|^2 = \frac{8\pi^2}{c\hbar^2} 2\pi \bar{J}_\omega |\hat{\mathbf{i}} \cdot \mathbf{d}_{mk}|^2 t \tag{4-95}$$

Now $J_\nu\, d\nu = J_\omega\, d\omega$, so we may write $\bar{J}_\nu = 2\pi \bar{J}_\omega$. Also the *transition rate* is simply the probability the transition has occurred per unit time so that from equation (4-95)

$$R_{mk} = \frac{8\pi^2}{c\hbar^2} |\hat{\mathbf{i}} \cdot \mathbf{d}_{mk}|^2 \bar{J}_\nu \tag{4-96}$$

On the other hand, from equations (4-1) and (4-3) we have

$$R_{mk} = B_{mk} \int_0^\infty J_\nu \varphi_\nu\, d\nu = B_{mk} \bar{J}_\nu \int_0^\infty \varphi_\nu\, d\nu = B_{mk} \bar{J}_\nu \tag{4-97}$$

Thus it is clear that

$$B_{mk} = \frac{8\pi^2}{\hbar^2 c} |\hat{\mathbf{i}} \cdot \mathbf{d}_{mk}|^2 \tag{4-98}$$

In general we will be interested in the absorptivity of bulk material. If we assume the atoms are oriented at random relative to the beam of radiation, then we must have

$$\overline{|\hat{\mathbf{i}} \cdot \mathbf{d}_{mk}|^2} = d_{mk}^2 \overline{\cos^2 \theta} = \tfrac{1}{3} d_{mk}^2 \tag{4-99}$$

so that we may write, finally,

$$B_{mk} = \frac{8\pi^2 d_{mk}^2}{3c\hbar^2} \tag{4-100}$$

or

$$B_{mk} = \frac{8\pi^2 e^2 r_{mk}^2}{3c\hbar^2} \tag{4-101}$$

where

$$r_{mk}^2 \equiv \left| \langle m | \sum_{i=1}^{N} \mathbf{r}_i | k \rangle \right|^2 \tag{4-102}$$

It is interesting to compare the quantum mechanical result with the classical result. For simplicity, suppose the levels are nondegenerate so that all statistical weight factors are unity. Then, substituting equation (4-101) into equation (4-11), we obtain the spontaneous emission rate

$$A_{ji} = \frac{64\pi^4 e^2 \nu^3}{3hc^3} r_{ij}^2 \tag{4-103}$$

and the rate of energy emission

$$A_{ji}\, h\nu = \frac{64\pi^4 e^2 \nu^4}{3c^3} r_{ij}^2 = \frac{e^2 \omega^4 (2r_{ij})^2}{3c^3} \tag{4-104}$$

which, as we see from equation (4-54), is merely the classical formula with $x_0$ replaced with $2r_{ij}$. We see clearly from equation (4-102) that both the absorption and emission rate must depend on the structure of the initial and final states of the transition.

Let us now suppose that the upper and lower levels of the line are degenerate. Then the total amount of energy emitted in the line is simply

$$E_{ji} = \frac{64\pi^4 \nu^4}{3c^3} \sum_{i,j} d_{ij}^2 \tag{4-105}$$

the sum being carried over all substates of the upper and lower levels. Because it is always necessary to sum over all initial and final states, it is customary to define a *line strength* $S$ such that

$$S(i, j) \equiv \sum_{i,j} d_{ij}^2 \tag{4-106}$$

which is symmetrical in $i$ and $j$. Then we may write

$$g_j A_{ji} = \frac{64\pi^4 \nu^3}{3hc^3} S(i, j) = \frac{2h\nu^3}{c^2} g_j B_{ji} \tag{4-107}$$

or

$$g_i B_{ij} = g_j B_{ji} = \frac{32\pi^4}{3h^2c} S(i, j) \tag{4-108}$$

Recall that in classical theory the integrated absorption coefficient is given by equation (4-69) so that on substitution into equation (4-108) we have

$$f_{ij} = \frac{h\nu mc}{4\pi^2 e^2} B_{ij} = \frac{8\pi^2 m\nu}{3he^2} \frac{S(i, j)}{g_j} \tag{4-109}$$

We can also define an *emission oscillator strength* $f_{ij}$ as

$$g_j f_{ji} = -g_i f_{ij} \tag{4-110}$$

where the minus sign is written to denote emission. Finally, we note that since $S(i, j)$ is defined in terms of sums over all substates, a simple rule can be written for the total oscillator strength of a level in terms of the oscillator strengths for transitions among sublevels. Let $n'$ be the principal quantum number of the lower level, and label each sublevel with $l'$; let $n$ and $l$ correspond to the upper sublevels. Then we may write simply

$$f(n', n) = \frac{1}{g_{n'}} \sum_{l'} g_{n'l'} f(n', l'; n, l) = \frac{\sum_{l'} g_{n'l'} f(n', l'; n, l)}{\sum_{l'} g_{n'l'}} \tag{4-111}$$

## APPLICATION TO HYDROGEN

Because hydrogen is both the simplest atomic structure and the most abundant element astrophysically, it is important to be familiar with its quantum mechanical properties in some detail. Let us therefore work out explicit expressions for hydrogen $f$-values since such calculations illustrate the general approach and are very instructive.

Schrödinger's equation for hydrogen may be written (Ref. 27, p. 186)

$$\frac{\hbar^2}{2\mu_H} \nabla^2\psi + \left(\frac{e^2}{r} + E\right)\psi = 0 \tag{4-112}$$

where

$$\frac{1}{\mu_H} = \frac{1}{m_H} + \frac{1}{m_e} \tag{4-113}$$

In spherical coordinates, one assumes a solution of the form

$$\psi(r, \theta, \varphi) = R(r)\Theta(\theta)\Phi(\varphi) \tag{4-114}$$

which allows equation (4-112) to be decomposed into the three separated equations

$$\frac{1}{r^2}\frac{d}{dr}\left(r^2\frac{dR}{dr}\right)+\left[E+\frac{2}{r}-\frac{l(l+1)}{r^2}\right]R=0 \tag{4-115}$$

$$\frac{1}{\sin\theta}\frac{d}{d\theta}\left(\sin\theta\frac{d\Theta}{d\theta}\right)+\left[l(l+1)-\frac{m^2}{\sin^2\theta}\right]\Theta=0 \tag{4-116}$$

$$\frac{d^2\Phi}{d\varphi^2}+m^2\Phi=0 \tag{4-117}$$

where radial distances are now expressed in units of the Bohr radius

$$a_0=\frac{h^2}{4\pi^2e^2\mu_{\rm H}} \tag{4-118}$$

and energies in units of the Rydberg energy

$$\mathscr{R}=\frac{2\pi^2e^4\mu_{\rm H}}{h^2} \tag{4-119}$$

The solution of equation (4-117) yields immediately

$$\Phi(\varphi)-\exp(\pm im\varphi) \tag{4-120}$$

The solution of equation (4-116) can be expressed in terms of associated Legendre functions, $P_l^{|m|}(\mu)$. These functions are orthogonal and have the property

$$\int_{-1}^{1}P_l^m(\mu)P_{l'}^m(\mu)\,d\mu=\frac{2}{2l+1}\frac{(l+m)!}{(l-m)!}\delta_{ll'} \tag{4-121}$$

(see, e.g., Ref. 1, p. 338). We may thus take the solution of equation (4-16) to be

$$\Theta_l^m(\theta)=\left[\frac{2l+1}{2}\frac{(l-m)!}{(l+m)!}\right]^{1/2}(-1)^{(m+|m|)/2}P_l^{|m|}(\cos\theta) \tag{4-122}$$

where $|m|\le l$. The factor involving $(-1)$ is inserted to preserve certain phase relations. All the angular factors may be grouped together into a *spherical harmonic*

$$Y_l^m(\theta,\varphi)=\left[\frac{2l+1}{4\pi}\frac{(l-m)!}{(l+m)!}\right]^{1/2}(-1)^{(m+|m|)/2}P_l^{|m|}(\cos\theta)e^{im\varphi} \tag{4-123}$$

These functions are orthogonal and are normalized such that

$$\int_0^{2\pi}d\varphi\int_0^{\pi}Y_l^m(\theta,\varphi)^*\,Y_{l'}^{m'}(\theta,\varphi)\sin\theta\,d\theta=\delta_{ll'}\,\delta_{mm'} \tag{4-124}$$

Important expansion relations which we will need later are (see, e.g., Ref. 1, pp. 333–34)

$$\mu\Theta_l^m(\mu) = \left[\frac{(l+m)(l-m)}{(2l-1)(2l+1)}\right]^{1/2} \Theta_{l-1}^m(\mu) + \left[\frac{(l+m+1)(l-m+1)}{(2l+3)(2l+1)}\right]^{1/2} \Theta_{l+1}^m(\mu) \tag{4-125}$$

and

$$(1-\mu^2)^{1/2}\Theta_l^m(\mu) = \left[\frac{(l+m+1)(l+m+2)}{(2l+1)(2l+3)}\right]^{1/2} \Theta_{l+1}^{m+1}(\mu) - \left[\frac{(l-m-1)(l-m)}{(2l+1)(2l-1)}\right]^{1/2} \Theta_{l-1}^{m+1}(\mu) \tag{4-126}$$

The solution of the radial equation can be expressed in terms of associated Laguerre polynomials as (see, e.g., Ref. 27, Chap. 10)

$$R_{nl}(r) = \left\{\frac{k^2(n-l-1)!}{n^2[(n+l)!]^3}\right\}^{1/2} (kr)^l e^{-kr/2} L_{n-l-1}^{2l+1}(kr) \tag{4-127}$$

where $l \leq n$ and $k \equiv 2/na_0$. These functions are normalized such that

$$\int_0^\infty R_{nl}^2(r) r^2\, dr = 1 \tag{4-128}$$

when $r$ is measured in units of $a_0$. Frequently, it is convenient to define another radial function

$$P_{nl}(r) \equiv rR_{nl}(r) \tag{4-129}$$

The energy eigenvalues depend only upon the quantum number $n$ via the well-known Rydberg formula

$$E_n = -\frac{\mathcal{R}}{n^2} \tag{4-130}$$

There are a total of four quantum numbers specifying a given state: $n$, the principal quantum number, which characterizes energy; $l$, the azimuthal quantum number, which characterizes angular momentum; $m$, the magnetic quantum number, which characterizes the projection of the angular momentum along a preferred axis (taken to be the $z$-axis); and $s$, the spin quantum number of an electron, which is equal to $\pm 1/2$. In most atomic systems, the energies of different $(n, l)$ states are different; hydrogen is a special case in this respect. The total number of hydrogen sublevels degenerate at a given energy is easily calculated. At each $l$, there are $(2l + 1)$ degenerate $m$-states;

also, since the spin can be either $\pm 1/2$, there are $2(2l+1)$ degenerate $(m, s)$ states. Summing over $l$, we find the number of degenerate states is

$$2\sum_{l=0}^{n-1}(2l+1) = 2n^2 \tag{4-131}$$

Thus

$$g_{n,l} = 2(2l+1) \tag{4-132}$$

and

$$g_n = 2n^2 \tag{4-133}$$

Let us now calculate the matrix elements of the electric dipole moment for hydrogen. Consider transitions from the state $n'$, $l'$, $m'$ to the state $n$, $l$, $m$. We wish to compute $|\langle nlm|\, e\mathbf{r}\, |n'l'm'\rangle|^2$ where $\mathbf{r} = x\hat{\mathbf{i}} + y\hat{\mathbf{j}} + z\hat{\mathbf{k}}$. We may write quite generally

$$|\langle nlm|\, e\mathbf{r}\, |n'l'm'\rangle|^2 = \tfrac{1}{2}|\langle nlm|\, e(x+iy)\, |n'l'm'\rangle|^2 + \tfrac{1}{2}|\langle nlm|\, e(x-iy)\, |n'l'm'\rangle|^2 + |\langle nlm|\, ez\, |n'l'm'\rangle|^2 \tag{4-134}$$

and note that

$$x = r\sin\theta\cos\varphi \tag{4-135}$$

$$y = r\sin\theta\sin\varphi \tag{4-136}$$

and

$$z = r\cos\theta \tag{4-137}$$

so that

$$x + iy = r\sin\theta e^{i\phi} \tag{4-138}$$

and

$$x - iy = r\sin\theta e^{-i\phi} \tag{4-139}$$

For brevity we will write $\mu = \cos\theta$. Now consider the matrix element

$$\langle nlm|\, ez\, |n'l'm'\rangle = ea_0 \int_0^\infty rP_{nl}(r)P_{n'l'}(r)\, dr \int_{-1}^{1} \Theta_l^m(\mu)\Theta_{l'}^{m'}(\mu)\, d\mu \times \int_0^{2\pi} e^{i(m-m')\varphi}\, d\varphi \tag{4-140}$$

We see immediately from the last integral that this matrix element is zero unless $m' = m$. Also we may calculate

$$\langle nlm|\, e(x \pm iy)\, |n'l'm'\rangle = ea_0 \cdots \int_0^{2\pi} e^{i(m'-m+1)}\, d\varphi \tag{4-141}$$

and here we see that the matrix element is zero unless $m = m' \pm 1$, the sign being chosen to agree with the sign of $(x \pm iy)$. These results are true fairly generally and are not restricted merely to hydrogen. The rule $\Delta m = \pm 1$ or 0 for a transition with nonzero probability is called a *spectroscopic selection rule*. Thus in evaluating the matrix element in equation (4-134), we need only consider terms of the form

$$|\langle nlm'|\,e\mathbf{r}\,|n'l'm'\rangle|^2 = \langle nlm'|\,ez\,|n'l'm\,\rangle^2$$
$$= \sigma^2 e^2\left(\int_{-1}^{1} \Theta_l^{m'}(\mu)\Theta_{l'}^{m'}(\mu)\mu\,d\mu\right)^2 \tag{4-142}$$

$$|\langle nlm'+1|\,e\mathbf{r}\,|n'l'm'\rangle|^2 = \tfrac{1}{2}|\langle nlm'+1|\,e(x+iy)\,|n'l'm'\rangle|^2$$
$$= \frac{\sigma^2 e^2}{2}\left(\int_{-1}^{1} \Theta_l^{m'+1}(\mu)\Theta_{l'}^{m'}(\mu)(1-\mu^2)^{1/2}\,d\mu\right)^2 \tag{4-143}$$

and

$$|\langle nlm'-1|\,e\mathbf{r}\,|n'l'm'\rangle|^2 = \tfrac{1}{2}|\langle nlm'-1|\,e(x-iy)\,|n'l'm'\rangle|^2$$
$$= \frac{\sigma^2 e^2}{2}\left(\int_{-1}^{1} \Theta_l^{m'-1}(\mu)\Theta_{l'}^{m'}(\mu)(1-\mu^2)^{1/2}\,d\mu\right) \tag{4-144}$$

where we have defined

$$\sigma^2 = a_0{}^2\left(\int_0^{\infty} P_{n'l'}(r)P_{nl}(r)r\,dr\right)^2 \tag{4-145}$$

Consider first equation (4-142). Recall that from equation (4-125) that we may write

$$\mu\Theta_l{}^m = C_{l-1}^m\Theta_{l-1}^m + C_{l+1}^m\Theta_{l+1}^m$$

so that we can guarantee that

$$\int_{-1}^{1} \Theta_l^{m'}\Theta_{l'}^{m'}\mu\,d\mu \neq 0$$

only if $l = l' \pm 1$. This rule is again quite general (for dipole transitions) and leads to the spectroscopic selection rule that $\Delta l = \pm 1$. To be specific, suppose that $l = l' + 1$. Then

$$\int_{-1}^{1} \Theta_{l'+1}^{m'}\Theta_{l'}^{m'}\mu\,d\mu = C_{l'+1}^m \int_{-1}^{1} (\Theta_{l'+1}^{m'})^2\,d\mu$$
$$= \left[\frac{(l'+m'+1)(l'-m'+1)}{(2l'+3)(2l'+1)}\right]^{1/2} \tag{4-146}$$

Thus

$$|\langle nl'+1m'|\,e\mathbf{r}\,|n'l'm'\rangle|^2 = \frac{e^2\sigma^2(l'+m'+1)(l'-m'+1)}{(2l'+1)(2l'+3)} \tag{4-147}$$

Suppose now $l = l' - 1$. Then

$$\int_{-1}^{1} \Theta_{l'-1}^{m'} \Theta_{l'}^{m'} \mu \, d\mu = C_{l'-1}^{m'} \int_{-1}^{1} (\Theta_{l'-1}^{m'})^2 d\mu = \left[\frac{(l' + m')(l' - m')}{(2l' + 1)(2l' - 1)}\right]^{1/2} \tag{4-148}$$

Thus

$$|\langle nl' - 1m'|\, e\mathbf{r}\, |n'l'm'\rangle|^2 = \frac{e^2\sigma^2(l' + m')(l' - m')}{(2l' + 1)(2l' - 1)} \tag{4-149}$$

Now consider equation (4-143). Recalling from equation (4-126) that

$$(1 - \mu^2)^{1/2}\Theta_l{}^m = C_{l+1}^{m+1}\Theta_{l+1}^{m+1} + C_{l-1}^{m+1}\Theta_{l-1}^{m+1}$$

we again find that nonzero matrix elements occur only for $l = l' \pm 1$ and that

$$|\langle nl' + 1m' + 1|\, e\mathbf{r}\, |n'l'm'\rangle|^2 = \frac{e^2\sigma^2}{2} \frac{(l' + m' + 1)(l' + m' + 2)}{(2l' + 1)(2l' + 3)} \tag{4-150}$$

and

$$|\langle nl' - 1m' + 1|\, e\mathbf{r}\, |n'l'm'\rangle|^2 = \frac{e^2\sigma^2}{2} \frac{(l' - m' - 1)(l' - m')}{(2l' + 1)(2l' - 1)} \tag{4-151}$$

Finally, by an analogous argument from equation (4-144), we have

$$|\langle nl' + 1m' - 1|\, e\mathbf{r}\, |n'l'm'\rangle|^2 = \frac{e^2\sigma^2}{2} \frac{(l' - m' + 2)(l' - m' + 1)}{(2l' + 1)(2l' + 3)} \tag{4-152}$$

and

$$|\langle nl' - 1m' - 1|\, e\mathbf{r}\, |n'l'm'\rangle|^2 = \frac{e^2\sigma^2}{2} \frac{(l' + m')(l' + m' - 1)}{(2l' - 1)(2l' + 1)} \tag{4-153}$$

To find the oscillator strength $f(n', l'; n, l)$, we must sum the above matrix elements over $m$ and $m'$. Thus, if we let $l = l' + 1$, we compute first the sum over $m$ and find

$$\begin{aligned}
&|\langle nl' + 1m|\, e\mathbf{r}\, |n'l'm'\rangle|^2 \\
&\quad = |\langle nl' + 1m' + 1|\, e\mathbf{r}\, |n'l'm'\rangle|^2 \\
&\qquad + |\langle nl' + 1m'|\, er\, |n'l'm'\rangle|^2 + |\langle nl' + 1m' - 1|\, e\mathbf{r}\, |n'l'm'\rangle|^2 \\
&\quad = \frac{e^2\sigma^2}{(2l + 1)(2l + 3)} \left[\frac{1}{2}(l' + m' + 1)(l' + m' + 2)\right. \\
&\qquad \left. + (l' + m' + 1)(l' - m' + 1) + \frac{1}{2}(l' - m' + 2)(l' - m' + 1)\right]
\end{aligned}$$

or

$$\sum_m |\langle nl'+1m|\,e\mathbf{r}\,|n'l'm'\rangle|^2 = \frac{e^2\sigma^2(l'+1)}{(2l'+1)} \tag{4-154}$$

a result independent of $m'$. Similarly, we find

$$\sum_m |\langle nl'-1m|\,e\mathbf{r}\,|n'l'm'\rangle|^2 = \frac{e^2\sigma^2 l'}{(2l'+1)} \tag{4-155}$$

also independent of $m'$. In short, we deduce from equations (4-154) and (4-155) that

$$\sum_m |\langle nlm|\,e\mathbf{r}\,|n'l'm'\rangle|^2 = \frac{e^2\sigma^2 \max(l, l')}{(2l'+1)} \tag{4-156}$$

where $l = l' \pm 1$. Now the sum over $m'$ simply extends over the $(2l'+1)$ values $(-l', \ldots, 0, \ldots, l')$, so from equation (4-156) we have

$$\sum_{m'}\sum_m |\langle nlm|\,e\mathbf{r}\,|n'l'm'\rangle|^2 = e^2\sigma^2 \max(l, l') \tag{4-157}$$

Thus the line strength $S(n', l'; n, l)$ is given by

$$S(n', l'; n, l) = \sum_s\sum_{m'}\sum_m |\langle nlm|\,e\mathbf{r}\,|n'l'm'\rangle|^2 = 2e^2\sigma^2 \max(l, l') \tag{4-158}$$

where the factor of 2 comes from the two possible spin states. Alternatively, from equation (4-145)

$$S(n', l'; n, l) = 2e^2 {a_0}^2 \max(l, l')\left(\int_0^\infty P_{n'l'}(r)P_{nl}(r)r\,dr\right)^2 \tag{4-159}$$

In terms of oscillator strengths, it follows from equation (4-109) that

$$\begin{aligned} f(n', l'; n, l) &= \frac{8\pi^2 m\nu}{3he^2}\,\frac{2e^2\sigma^2 \max(l, l')}{2(2l'+1)} \\ &= \frac{1}{3}\left(\frac{1}{n'^2} - \frac{1}{n^2}\right)\frac{\max(l, l')}{(2l'+1)}\left(\int_0^\infty P_{n'l'}(r)P_{nl}(r)r\,dr\right)^2 \end{aligned} \tag{4-160}$$

where we have made use of the definition of $a_0$ and the Rydberg energy. Finally, we can write

$$A(n, l; n', l') = \frac{64\pi^4\nu^3}{3hc^3}\,\frac{\max(l, l')}{(2l+1)}\,e^2\sigma^2 \tag{4-161}$$

Extensive tables of $\sigma^2$ for hydrogen have been published by Green, Rush, and Chandler (*Ap. J. Supp. No. 26*, **3**, 37, 1957). The values of $\sigma^2$ may easily be computed for simple cases but in general are difficult to evaluate; a general expression for the integral was first derived by Gordon (*Ann. Phys.*, **2**, 1031,

1929), who showed that

$$\left(\int_0^\infty P_{n'l'-1}P_{nl}\, r\, dr\right)^2 = \left\{\frac{(-1)^{n'-l}}{4(2l-1)!}\left[\frac{(n+l)!(n'+l-1)!}{(n-l-1)!(n'-l')!}\right]^{1/2} \times \frac{(4nn')^{l+1}(n-n')^{n+n'-2l-2}}{(n+n')^{n+n'}} \times \left[F\left(-n+l+1,\, -n'+l,\, 2l,\, \frac{4nn'}{(n'-n)^2}\right) - \left(\frac{n-n'}{n+n'}\right)^2 F\left(-n+l-1,\, -n'+l,\, 2l,\, \frac{-4nn'}{(n-n')^2}\right)\right]\right\}^2 \tag{4-162}$$

where $F(a, b, c, x)$ is the hypergeometric function

$$F(a, b, c, x) = 1 + \frac{ab}{c}x + \frac{a(a+1)b(b+1)}{2!c(c+1)}x^2 + \cdots \tag{4-163}$$

In equation (4-162) *only*, $n$ denotes the quantum number of the state with the larger $l$-value, and $n'$ the number of the state with the smaller $l$-value. Since the $l$-states are normally degenerate, we often desire $f$-values for the entire transition $n' \to n$. We easily find from equations (4-149) and (4-160)

$$S(n', n) = 2e^2a_0^2\left[\sum_{l'=1}^{n'-1} l'\left(\int_0^\infty P_{n'l'}P_{nl'-1}\, r\, dr\right)^2 + \sum_{l'=0}^{n'-1}(l'+1)\left(\int_0^\infty P_{n'l'}P_{nl'+1}\, r\, dr\right)^2\right] \tag{4-164}$$

$$f(n', n) = \frac{1}{3n'^2}\left(\frac{1}{n'^2} - \frac{1}{n^2}\right)\left[\sum_{l'=1}^{n'-1} l'\left(\int_0^\infty P_{n'l'}P_{nl'-1} r\, dr\right)^2 + \sum_{l'=0}^{n'-1}(l'+1)\left(\int_0^\infty P_{n'l'}P_{nl'+1} r\, dr\right)^2\right] \tag{4-165}$$

and

$$A(n, n') = \frac{64\pi^4\nu^3}{3hc^3}\frac{S(n', n)}{2n^2} \tag{4-166}$$

An explicit form for $f(n', n)$ has been given by Menzel and Pekeris (*M. N.*, **96**, 77, 1935), namely,

$$f(n', n) = \frac{32}{3}n^4n'^2\frac{(n-n')^{2n+2n'-4}}{(n+n')^{2n+2n'+3}}\left\{\left[F\left(-n',\, -n+1,\, 1,\, \frac{-4n'n}{(n-n')^2}\right)\right]^2 - \left[F\left(-n'+1,\, -n,\, 1,\, \frac{-4n'n}{(n-n')^2}\right)\right]^2\right\} \tag{4-167}$$

The factor in curly brackets is customarily abbreviated as $\Delta(n', n)$. Extensive tables of $f(n' n)$ are given both by Menzel and Pekeris (*M. N.*, **96**, 77, 1935) and by Green, Rush, and Chandler (*Ap. J. Supp. No. 26*, **3**, 37, 1957).

A very convenient form for hydrogen oscillator strength is to express them in terms of the value derived by Kramers (*Phil. Mag.*, **44**, 836, 1923), using a semiclassical analysis, i.e.,

$$f_K(n', n) = \frac{32}{3\pi\sqrt{3}} \left( \frac{1}{n'^2} - \frac{1}{n^2} \right)^{-3} \frac{1}{n^3 n'^5} \tag{4-168}$$

Then we may write

$$f(n', n) = g_I(n', n) f_K(n', n) \tag{4-169}$$

where the function $g_I$ is called the *Gaunt factor*. From equations (4-167) through (4-169) we must have

$$g_I(n', n) = \pi\sqrt{3} \frac{nn'}{(n - n')} \left[ \frac{n - n'}{n + n'} \right]^{2n' + 2n} \Delta(n', n) \tag{4-170}$$

A table of $g_I(n' n)$ has been given by Baker and Menzel (*Ap. J.*, **88**, 52, 1938). The introduction of the Gaunt factor is a convenient formalism for treating continuum transitions, as we shall see below. We note from equation (4-168) that oscillator strengths decrease rapidly as one progresses up a series with a given lower state.

## TRANSITION PROBABILITIES FOR OTHER LIGHT ELEMENTS

(a) Hartree-Fock Method. When more than one electron is present in the atom, we can no longer solve the wave equation in closed form, and approximations must be made. The actual Hamiltonian for an $N$ electron system is

$$H = \frac{-\hbar^2}{2m} \sum_{i=1}^{N} \nabla_i^2 - \sum_{i=1}^{N} \frac{Ze^2}{r_i} + \sum_{\substack{\text{all pairs} \\ i, j}} \frac{e^2}{|\mathbf{r}_i - \mathbf{r}_j|} \tag{4-171}$$

The first term represents the kinetic energy of the electrons, the second their electrostatic potential with the nucleus of charge $Z$, and the third their mutual coulomb repulsion. It is this last term that introduces the principal difficulties.

One of the most important methods of deriving approximate wave functions is *Hartree's self-consistent field method.* In this method, the sum over electron pairs is replaced for each electron by its spherical average. An excellent description of how this replacement is made has been given by Slater (Ref. 31, Chaps. 3 and 9). In this way, each electron moves in a potential that depends only upon its distance from the nucleus, and we may make the correspondence

$$\sum_{\substack{\text{all pairs} \\ (i,j)}} \frac{e^2}{|\mathbf{r}_i - \mathbf{r}_j|} \rightarrow \sum_i V_i(r_i) \tag{4-172}$$

This results in the approximation of the actual potential by a *central field.* When the potential in which each electron moves depends only upon the distance, the angular factors appearing in the Schrödinger equation can be separated out in exactly the same way as for hydrogen, and we can write for each electron

$$U_i(r, \theta, \varphi; n, l, m, s) = \frac{1}{r} P_{nl}(r) Y_l^m(\theta, \varphi) X(s) \tag{4-173}$$

where the normalizations given in equations (4-124) and (4-173) still apply. These functions are called *electron orbitals.* The radial equation for each orbital is of the form ($r$ in units of $a_0$, $E$ in Rydbergs)

$$\frac{d^2 P_{nl}}{dr^2} + \left[\frac{2Z_{\text{eff}}(r)}{r} + E_{nl} - \frac{l(l+1)}{r^2}\right] P_{nl} = 0 \tag{4-174}$$

Here $Z_{\text{eff}}(r)$ is the "effective nuclear charge" as shielded by the other electrons, making use of the central field approximation in equation (4-172). We now consider the atom to be made up of $N$ such orbitals and use these to construct the wave function for the entire configuration. Because of the Pauli exclusion principle, the set of four numbers ($n$, $l$, $m$, $s$) characterizing each orbital cannot be identical for any two orbitals. Also, in constructing the wave function of the atom, an important property that must be conserved is that it should be antisymmetric under the interchange of the coordinates of any two electrons. In practice these conditions may be met by writing the complete wave function as the determinant

$$\varphi(\mathbf{r}_1, \ldots, \mathbf{r}_N) = \frac{1}{\sqrt{N!}} \begin{vmatrix} u_1(\alpha) & u_1(\beta) & \ldots & u_1(\nu) \\ u_2(\alpha) & u_2(\beta) & \ldots & u_2(\nu) \\ \vdots & & & \\ u_N(\alpha) & u_N(\beta) & \ldots & u_N(\nu) \end{vmatrix} \tag{4-175}$$

where numbers 1, 2, ..., $N$ denote the orbitals of electrons 1, 2, etc., while $\alpha, \beta, \ldots, \nu$ stand for the space and spin coordinates of electrons $\alpha, \ldots, \nu$, respectively. This determinant is called a *Slater determinant* (Ref. 31, Chap. 12).

The solution for the wave functions is carried out iteratively. Thus $Z_{\text{eff}}(r)$ depends in an involved way upon integrals over the electron orbitals but in turn determines those orbitals. Thus we must start with an approximate set of orbitals, compute $Z_{\text{eff}}$, solve for the $P_{nl}$'s, recompute $Z_{\text{eff}}$, and iterate until the procedure converges. The calculations are time-consuming and laborious but are within the capabilities of modern computers, and a considerable number of wave functions are now available for a wide variety of atomic configurations.

A specific term in an atomic spectrum can be characterized by certain quantum numbers describing the atom as a whole. In light atoms, these quantum numbers describe the total orbital angular momentum **L**, the total spin **S**, and the total angular momentum **J**, which is the vector sum of **L** and **S**. This type of coupling is referred to as ($L$-$S$) or Russell-Saunders coupling and occurs when orbit-orbit (**l**, **l**) and spin-spin (**s**, **s**) interactions are stronger than spin-orbit (**s**, **l**) interactions. Since a given **L**, **S**, and **J** may result from more than one arrangement of the individual **l**'s, **m**'s, and **s**'s of the orbitals, the complete wave function will in general consist of a *sum* of Slater determinants and thus is indeed very complicated to compute.

In calculating transition probabilities, it is generally assumed that only one orbital is different between the initial and final state, i.e., only one electron undergoes the transition. In this case, the matrix element $r_{ij}$ can be split into factors, one coming from the initial and final radial wave functions and another depending upon the angular and spin functions. It is customary, therefore, to write the expression for the line strength in the form

$$S(n', L', S', J'; n, L, S, J) = a_0{}^2 e^2 \sigma^2(n', l'; n, l) \mathscr{S}(\mathscr{M}) \mathscr{S}(\mathscr{L}) \tag{4-176}$$

Here

$$\sigma^2 = \frac{1}{4l_{\max}^2 - 1} \left( \int_0^\infty P_{n'l'} P_{nl}\, r\, dr \right)^2 \tag{4-177}$$

where $l_{\max} = \max(l' l)$. The factor $\mathscr{S}(\mathscr{M})$ is the strength of the multiplet, depending on $nLS$ and $n'L'S'$, the factor $\mathscr{S}(\mathscr{L})$ is the strength of the line within the multiplet, depending on $L'$, $S'$, $J'$ and $L$, $S$, $J$. Extensive tables of $\mathscr{S}(\mathscr{M})$ and $\mathscr{S}(\mathscr{L})$ have been given by Goldberg (*Ap. J.*, **82**, 1, 1935, and *Ap. J.*, **84**, 11, 1936) and reproduced by Aller (Ref. 4, Chap. 8 and the appendix). General formulae have been given by Rohrlich (*Ap. J.*, **129**, 941, 1959, and *Ap. J.*, **129**, 449, 1959). By far the most difficult part of the calculation is the determination of $\sigma^2$. Serious complications also occur when there are deviations from $L$-$S$ coupling.

(b) The Coulomb Approximation. Because of the labor involved in obtaining values for $\sigma^2$ from Hartree-Fock calculations, it is highly desirable to have an approximate method that can be applied easily. Such an approach has been developed by Bates and Damgaard (*Phil. Trans. Roy. Soc. London*, **242A**, 101, 1949), who pointed out that often the largest contribution to the radial integral comes at large values of $r$, where the electron moves in a very nearly coulomb potential. In this case, the integral can be approximated using hydrogenic wave functions, provided that the principal quantum numbers are chosen to give the observed energy of the levels. If $Z^*$ is the charge in the asymptotic potential, then appropriate effective quantum numbers can be defined as $n_l^* = Z^*/\varepsilon_{nl}^{1/2}$ where $\varepsilon_{nl}$ is the energy level below the continuum expressed in Rydbergs. Bates and Damgaard then show that we may write

$$\sigma(n_{l-1}^*, l-1; n_l^*, l) = \frac{1}{Z^*}\mathscr{F}(n_l^*, l)\mathscr{I}(n_{l-1}^*, n_l^*, l) \tag{4-178}$$

The functions $\mathscr{F}$ and $\mathscr{I}$ are tabulated for a wide range of arguments in the above reference. An extensive tabulation of coulomb approximation $f$-values is given by Griem (Ref. 14, pp. 363–441). An extension of this theory has been given by Lawrence (*Ap. J.*, **147**, 293, 1967). Because of the simplicity of this method, it has been applied widely in astrophysical analyses.

(c) Experimental Methods. In many cases the coulomb approximation is inaccurate while a more elaborate quantum mechanical calculation simply is too complicated to be carried out. In these cases, experiment is often the only way $f$-values can be determined. Brief summaries of the experimental methods have been given by Aller (Ref. 4, pp. 300–310), Griem (Ref. 14, Chap. 15), and K. H. Böhm (Ref. 13, pp. 146–49). A large number of measures exist in the literature and recently have been summarized for the lighter elements by Wiese, Smith, and Glennon (Ref. 37). A complete bibliography of work on atomic transition probabilities has been compiled by Glennon and Wiese (Ref. 12).

## 4-3. The Einstein-Milne Relations for the Continuum

A generalization of the Einstein relations to bound-free absorptions was developed by Milne (*Phil. Mag.*, **47**, 209, 1924) in a paper of considerable interest and importance. Let $p_\nu$ be the probability of photoionization by a photon in the frequency range $(\nu, \nu + d\nu)$ so that the *number* of photoionizations is $p_\nu I_\nu \, d\nu \, dt$. In the inverse process, we may have recaptures that occur spontaneously or are induced by radiation. Let $F(v)\,dv$ be the spontaneous recapture probability (per ion) for electrons with velocity $(v, v + dv)$ and $G(v)\,dv$ be the induced recapture probability. Also, let $N_0$ be the number of neutrals, $N_1$ the number of ions, and $N_e$ the number of electrons contained within a unit volume. Assume that the electrons have a Maxwellian velocity distribution so that the number with velocities between $(v, v + dv)$ is

$$N_e(v)\,dv = N_e 4\pi v^2 \left(\frac{m}{2\pi kT}\right)^{3/2} \exp(-mv^2/2kT)\,dv \tag{4-179}$$

Then in terms of the probabilities postulated above, the number of recaptures in unit time in velocity range $(v, v + dv)$ is

$$N(v) = N_1 N_e[F(v) + I_\nu G(v)]\left(\frac{m}{2\pi kT}\right)^{3/2} 4\pi v^3 \exp(-mv^2/2kT)\,dv \tag{4-180}$$

In *thermodynamic equilibrium* the number of recaptures must balance the number of photoionizations. Moreover, in equilibrium, $I_\nu$ must equal the Planck function so that we may write

$$N_0\, p_\nu\, B_\nu\, d\nu = N_1 N_e[F(v) + B_\nu\, G(v)]\left(\frac{m}{2\pi kT}\right)^{3/2} 4\pi m v^3 \exp(-mv^2/2kT)dv \tag{4-181}$$

Now the energy $h\nu$ of the photon is connected to the kinetic energy $\frac{1}{2}mv^2$ of the electron by the relation

$$h\nu = \chi_I + \tfrac{1}{2}mv^2 \tag{4-182}$$

so that $h\, d\nu = mv\, dv$. Thus

$$p_\nu\, B_\nu = \frac{N_1 N_e}{N_0}\left(\frac{m}{2\pi kT}\right)^{3/2} e^{\chi_I/kT} e^{-h\nu/kT}[F(v) + B_\nu\, G(v)]\,\frac{4\pi h v^2}{m} \tag{4-183}$$

But in thermodynamic equilibrium the Saha equation is also valid, and we have [equation (3-16)]

$$\frac{N_1 N_e}{N_0} = \frac{2U_1(T)}{U_0(T)}\left(\frac{2\pi m kT}{h^2}\right)^{3/2} e^{-\chi_I/kT}$$

Substituting into equation (4-138), we therefore obtain

$$p_\nu\, B_\nu = \frac{2U_1(T)}{U_0(T)}\,\frac{4\pi m^2 v^2}{h^2}\, e^{-h\nu/kT}[F(v) + B_\nu\, G(v)] \tag{4-184}$$

or

$$B_\nu = \left[\frac{F(v)}{G(v)}\right]\left\{\left[\frac{p_\nu\, U_0(T)h^2}{2U_1(T)4\pi m^2 v^2 G(v)}\right]e^{h\nu/kT} - 1\right\}^{-1} \tag{4-185}$$

Equation (4-185) will reduce to the correct expression for the Planck function only if we may make the identifications

$$\frac{F(v)}{G(v)} = \frac{2h\nu^3}{c^2} \tag{4-186}$$

and

$$p_\nu = \frac{2U_1(T)}{U_0(T)}\,\frac{4\pi m^2 v^2}{h^2}\, G(v) = \frac{U_1(T)}{U_0(T)}\,\frac{4\pi m^2 v^2 c^2}{h^3 \nu^3}\, F(v) \tag{4-187}$$

Again we must realize that although we have assumed the state of thermal equilibrium in deriving the above relations, the quantities $p_\nu$, $F(v)$, and $G(v)$ can really depend upon atomic properties only so that the expressions given above must therefore be independent of this assumption and be valid in general. One result of interest is that since $p_\nu$ remains finite at the threshold, where $h\nu \to \chi_I$ and $v \to 0$, we must have, from equation (4-187),

$$\lim_{v\to 0} F(v) = \frac{\text{Const}}{v^2} \tag{4-188}$$

To relate these quantities to the more customary macroscopic absorption coefficients, we write the equation of transfer assuming photoionizations and recombinations to one atomic level only (the generalization to a multilevel or multi-atom case is trivial since each term adds linearly, and our conclusions will still apply to the sum). Then we have

$$-\frac{\mu}{\rho}\frac{dI_\nu}{dz} = \frac{N_0\,p_\nu\,h\nu}{\rho} I_\nu - \frac{N_1 N_e h^2 \nu}{\rho m} 4\pi v^2 \exp\left(\frac{-mv^2}{2kT}\right)\left(\frac{m}{2\pi kT}\right)^{3/2} \times [F(v) + I_\nu G(v)] \tag{4-189}$$

To obtain the transfer equation in the usual form, we clearly must take

$$\kappa_\nu = \frac{N_0\,p_\nu\,h\nu}{\rho}\left[1 - \frac{N_1 N_e 4\pi h}{N_0 m}\left(\frac{m}{2\pi kT}\right)^{3/2} \exp\left(\frac{-mv^2}{2kT}\right)\frac{v^2 G(v)}{p_\nu}\right] \tag{4-190}$$

the second term representing the correction for stimulated emission. But substituting the Saha equation and using the Milne relation between $p_\nu$ and $G(v)$ as stated in equation (4-187), we find that the last term in brackets becomes

$$\left[\frac{2U_1(T)}{U_0(T)}\left(\frac{2\pi mkT}{h^2}\right)^{3/2} e^{-\chi_I/kT}\right]\left[\frac{4\pi h}{m}\left(\frac{m}{2\pi kT}\right)^{3/2} \exp\left(\frac{mv^2}{2kT}\right)\right] \times \left[\frac{U_0(T)}{2U_1(T)}\frac{h^2}{4\pi m^2}\right] = e^{-h\nu/kT} \tag{4-191}$$

so that in LTE, the net opacity, corrected for stimulated emissions, is related to the direct absorptivity by

$$\kappa_\nu = \frac{N_0\,p_\nu\,h\nu}{\rho}(1 - e^{-h\nu/kT}) \tag{4-192}$$

This expression, as mentioned above, still applies when summed over more than one level or more than one atomic species. Thus in LTE, we recover the same stimulated emission factor for continua as was obtained in LTE for bound-bound transitions [equation (4-19)]. In the general non-LTE case this result no longer applies [see equation (4-17)]. We also see that the usual energy-absorption cross-section $\alpha_\nu$ is related to the photoionization probability $p_\nu$ by the relation $\alpha_\nu = p_\nu h\nu$, similar to equation (4-5). Furthermore, by a completely analogous argument, we may show that the source term in equation

(4-190) becomes

$$\frac{N_1 N_e h^2 v}{\rho m} 4\pi v^2 e^{-mv^2/2kT} \left(\frac{m}{2\pi kT}\right)^{3/2} F(v)$$

$$= \frac{N_0\, p_\nu\, h\nu}{\rho} \frac{2h\nu^3}{c^2} e^{-h\nu/kT} = \left(\frac{N_0\, p_\nu\, h\nu}{\rho}\right)(1 - e^{-h\nu/kT}) B_\nu(T) \qquad (4\text{-}193)$$

Thus, in LTE, we recover the Kirchhoff-Planck emission law [equation (1-21)]:

$$j_\nu = \kappa_\nu B_\nu(T)$$

provided we use the opacity (summed over all possible contributors at frequency $\nu$) *corrected for stimulated emission*, as given by equation (4-192). The transfer equation then reduces to its standard form for LTE,

$$-\frac{\mu}{\rho}\frac{dI_\nu}{dz} = \kappa_\nu(I_\nu - B_\nu)$$

Milne's analysis thus allows us to make a complete correspondence between the formalism for continuum processes (in LTE) and that for bound-bound processes presented in Section 4-1. It also allows a rigorous derivation of the LTE stimulated emission correction factor for the continuum and therefore rounds out the discussion given in Sections 1-2 and 1-4. We shall employ the results extensively in Chapter 6. Let us now turn our attention to the calculation of opacities per gram of stellar material. Since care must be taken with stimulated emissions in general, the formulae we will give will *not* be corrected for stimulated emissions but will be direct absorption coefficients.

## 4-4. Continuum Absorption Cross-Sections and the Opacity of Stellar Material

The cross-section for bound-free absorptions can be calculated with quantum mechanics by essentially the same methods used above in the bound-bound case. If we consider absorption from a bound state $n$ to the continuum in a frequency interval $\Delta\nu$, we may consider the continuum made up of $\Delta k$ states within the interval $\Delta\nu$ so that by analogy to equation (4-101), we write

$$\alpha_\nu = \frac{8\pi^2 e^2}{3\hbar^2 c} h\nu\, |\langle k|\mathbf{r}|n\rangle|^2 \frac{\Delta k}{\Delta\nu} \qquad (4\text{-}194)$$

Alternatively, we could consider each continuum state to have an effective oscillator strength $f_k$, and if we take a narrow enough band $\Delta k$, these can be set to some mean value $\bar{f}$. Then we could write

$$\alpha_\nu = \frac{\pi e^2}{mc} \bar{f} \frac{\Delta k}{\Delta\nu} \qquad (4\text{-}195)$$

We will not give the details of the calculation of the free state wave functions. At large distances from the atom these wave functions go over simple spherical waves. The only detailed considerations we will give here are for hydrogen, for which we have an analytical formula for the oscillator strength, which we will generalize and use in equation (4-195).

## HYDROGEN BOUND-FREE AND FREE-FREE ABSORPTION

A direct way of obtaining bound-free and free-free absorption cross-sections for hydrogen was suggested by Menzel and Pekeris (*M.N.*, **96**, 77, 1935). They introduced the formalism of representing bound states by real (integer) quantum numbers whose energies relative to the continuum are given by

$$E_n = -\frac{\mathcal{R}}{n^2} = \chi_n - \chi_{\mathrm{I}} \tag{4-196}$$

so that

$$h\nu_{n'n} = \mathcal{R}\left(\frac{1}{n'^2} - \frac{1}{n^2}\right) \tag{4-197}$$

and free states by imaginary quantum numbers $ik$ such that

$$h\nu_{n'k} = \mathcal{R}\left(\frac{1}{n'^2} + \frac{1}{k^2}\right) = \frac{\mathcal{R}}{n'^2} + \frac{1}{2}mv^2 \tag{4-198}$$

We note that $k \to \infty$ at the ionization limit and becomes small high in the continuum. The expression for the oscillator strength for bound levels, namely,

$$f_{n'n} = \frac{32}{3n'^2}\left|\frac{[(n-n')/(n+n')]^{2n+2n'}\,\Delta(n',n)}{n^2n'^2(n-n')\left(\dfrac{1}{n'^2} - \dfrac{1}{n^2}\right)^3}\right| \tag{4-199}$$

(where the absolute value sign will imply the modulus of a complex number) can then be rewritten as

$$f_{n'k} = \frac{32}{3n'^2}\,\frac{\exp[-4k\tan^{-1}(n'/k)]\,|\Delta(n',ik)|}{k^3n'^3\left(\dfrac{1}{n'^2} + \dfrac{1}{k^2}\right)^{7/2}(1 - e^{-2\pi k})} \tag{4-200}$$

or, alternatively,

$$f_{n'k} = \frac{32}{3\pi\sqrt{3}}\,\frac{1}{n'^5k^3}\left(\frac{1}{n'^2} + \frac{1}{k^2}\right)^{-3} g_{\mathrm{II}}(n',k) \tag{4-201}$$

where

$$g_{\mathrm{II}}(n', k) \equiv \frac{\pi\sqrt{3}\,kn' \exp[-4k \tan^{-1}(n'/k)]\,|\Delta(n', ik)|}{(k^2 + n'^2)^{1/2}(1 - e^{-2\pi k})} \tag{4-202}$$

Approximation formulae for $g_{\mathrm{II}}$ are given by Menzel and Pekeris in the limits $k/n' \gg 1$, i.e., near threshold, and $n'/k \gg 1$, i.e., very large photon energies. An extensive numerical tabulation of $g_{\mathrm{II}}$ is given by Karzas and Latter (*Ap. J. Supp. No. 55*, **6**, 167, 1961).

An expression for the absorption coefficient can now be derived from equation (4-195). From equation (4-198) we have

$$\frac{dk}{d\nu} = -\frac{hk^3}{2\mathscr{R}} \tag{4-203}$$

so substitution into equation (4-195) yields

$$\alpha_\nu = \frac{\pi e^2}{mc}\left(\frac{hk^3}{2\mathscr{R}}\right)\left(\frac{32}{3\pi\sqrt{3}}\right)\frac{1}{k^3 n'^5}\frac{g_{\mathrm{II}}(n', k)}{(h\nu/\mathscr{R})^3} \tag{4-204}$$

or

$$\alpha_\nu = \frac{64\pi^4}{3\sqrt{3}}\frac{me^{10}}{ch^6}\frac{1}{n'^5\nu^3}\,g_{\mathrm{II}}(n', \nu) = \mathscr{K}\,\frac{g_{\mathrm{II}}(n', \nu)}{n'^5\nu^3} \tag{4-205}$$

where $\mathscr{K} = 2.815 \times 10^{29}$. From the above, we see that the bound-free absorption from level $n$ commences abruptly at frequency

$$\nu_n = (\chi_{\mathrm{I}} - \chi_n)/h = \mathscr{R}/hn^2 \tag{4-206}$$

and falls off at higher frequencies as $\nu^{-3}$ (neglecting the variation of the Gaunt factor), in agreement with the results of Kramers and Milne. By analogy with bound-bound transitions, a total oscillator strength for the continuum can be defined as

$$\begin{aligned} f_c \equiv \frac{mc}{\pi e^2}\int_0^\infty \alpha_\nu\, d\nu &= \frac{16}{3\pi\sqrt{3}}\left(\frac{\mathscr{R}}{h}\right)^2 \frac{\bar{g}_{\mathrm{II}}}{n^5}\int_{\nu_n}^\infty \frac{d\nu}{\nu^3} \\ &= \frac{8}{3\pi\sqrt{3}}\frac{\bar{g}_{\mathrm{II}}}{n} \end{aligned} \tag{4-207}$$

where $\bar{g}_{\mathrm{II}}$ is an appropriate mean Gaunt factor. Values of $f_c$ are listed in Table 4-1, both with $\bar{g}_{\mathrm{II}}$ taken to be unity and with the proper value included. It is interesting to see that even the entire integrated continuum is not as strong as a very strong line transition (for which $f$ may approach unity).

The opacity per gram of stellar material can now be computed. Basically, we simply multiply the cross-section for level $n$ by the total number of

TABLE 4-1. Continuum Oscillator Strengths for Hydrogen

| | $f_c$ | |
|---|---|---|
| $n$ | $\bar{g}_{II} \equiv 1$ | $\bar{g}_{II}$ *included* |
| 1 | 0.490 | 0.436 |
| 2 | 0.245 | 0.238 |

atoms, per gram of stellar material, in that level, and sum together the contributions of all levels that can absorb at frequency $\nu$. In particular, if we assume LTE, then

$$N_n = N_0 \frac{g_n e^{-\chi_n/kT}}{U_0(T)} = N_0 \frac{2n^2 e^{-\chi_n/kT}}{U_0(T)} \tag{4-208}$$

where $N_0$ is the number of neutral hydrogen atoms *per gram of stellar material.* Now the number of H-atoms per gram of hydrogen is $1/m_H$ while number of grams of hydrogen per gram of stellar material equals $1/\mu'$ [defined by equation (3-53)]. Further, if we let $f_H$ be the fraction of all hydrogen that is neutral (computed from the Saha equation), then clearly

$$N_0(H) = \frac{f_H}{\mu' m_H} \tag{4-209}$$

Now, at a given frequency $\nu$, we sum over all levels such that $\nu \geq \nu_n$, i.e., over all $n \geq n^*$ where $n^*$ is the smallest integer that satisfies the relation

$$n^* \geq (\mathcal{R}/h\nu)^{1/2} \tag{4-210}$$

Thus, combining equations (4-205), (4-208), (4-209), and (4-210), we find the mass absorption coefficient

$$a_\nu(b-f) = \sum_{n=n*}^{\infty} \alpha_\nu(n) N_n = \left[\frac{2f_H}{\mu' m_H U_H(T)}\right] \frac{\mathcal{K}}{\nu^3} \sum_{n*}^{\infty} \frac{g_{II}(n,\nu) e^{-\chi_n/kT}}{n^3} \tag{4-211}$$

or if we define, as is customary,

$$u_n \equiv (\chi_I - \chi_n)/kT \tag{4-212}$$

we can write

$$a_\nu(b-f) = \left[\frac{2f_H}{\mu' m_H U_H(T)}\right] \frac{\mathcal{K}}{\nu^3} e^{-u_I} \sum_{n*}^{\infty} \frac{e^{u_n} g_{II}(n,\nu)}{n^3} \tag{4-213}$$

In practice we cannot extend the summation over an infinite number of levels. Two approaches are commonly taken. First, we might suppose that, due to perturbations, all bound levels starting at some maximum $\bar{n} = n_{max} + 1$

are unbound. We might then sum over all levels up through $n_{\max}$, and lower the effective ionization energy to $\chi_{\mathrm{I}} - \mathscr{R}/\bar{n}^2$. An alternative approach is to sum over a finite number of levels and integrate over the remainder. Thus, symbolically, we write

$$\sum_{n*}^{\infty} = \sum_{n*}^{n_{\max}} + \int_{\bar{n}}^{\infty} \tag{4-214}$$

In the integration procedure we set the Gaunt factors for the higher levels to unity and write

$$\sum_{\bar{n}}^{\infty} \frac{e^{u_n}}{n^3} \approx -\frac{1}{2}\int_{\bar{n}}^{\infty} \exp(\mathscr{R}/n^2 kT)\, d\left(\frac{1}{n^2}\right)$$

$$= \frac{1}{2u_1}(e^{\bar{u}} - 1) \tag{4-215}$$

Then we may write, finally,

$$a_\nu(b-f) = \left[\frac{2f_{\mathrm{H}}}{\mu' m_{\mathrm{H}} U_{\mathrm{H}}(T)}\right] \frac{\mathscr{K}}{\nu^3} e^{-u_1} \left[\sum_{n*}^{n_{\max}} \frac{e^{u_n} g_{\mathrm{II}}(n, \nu)}{n^3} + \frac{(e^{\bar{u}} - 1)}{2u_1}\right] \tag{4-216}$$

The bound-free opacity of hydrogen calculated in this way has a jagged character, as shown in Figure 4-1. Except for the very hottest stars, most of the hydrogen is in the ground state, and the absorption edge at $\lambda$ 912 Å is usually extremely strong. For 912 Å $\leq \lambda \leq$ 3647 Å, absorptions from the ground state no longer occur, and the dominant opacity source is photoionizations from the $n = 2$ level (Balmer continuum). Similarly, for 3647 Å $\leq \lambda \leq$ 8206 Å, the dominant continuum is from $n = 3$ (Paschen continuum) and so on. Actually, this situation is idealized in that there exists a series of lines converging upon each series limit. Near the limit, these transitions contribute strongly to the opacity and indeed merge smoothly into the continuum. In much work on stellar atmospheres, the lines have been ignored, simply to keep the problem tractable, or have been treated only very schematically. Some of the more recent work has begun to account for the overlappping lines, but still only with the simplifying assumption of LTE. The presence of lines alters both the distribution of the emergent flux with frequency and the run of temperature with depth. Collectively these effects are called *line blanketing*; we will discuss them in more detail in Chapter 14.

Let us now consider the free-free opacity of hydrogen. By analogy with our previous work we introduce imaginary quantum numbers for both the initial and final states, say, *ik* and *il*, such that if $v$ is the initial velocity of the electron, and $\nu$ is the frequency of the absorbed photon, then

$$\frac{\mathscr{R}}{k^2} = \frac{1}{2} m v^2 \tag{4-217}$$

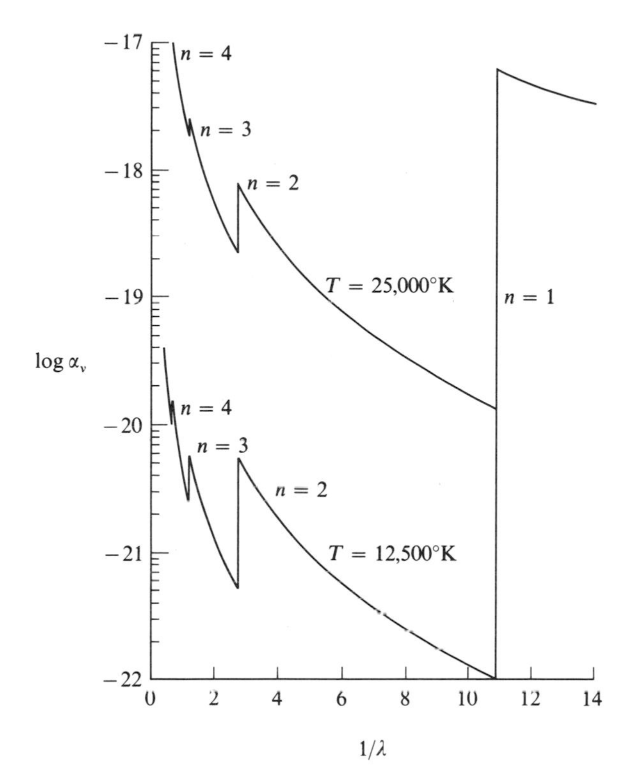


FIG. 4-1. Opacity due to hydrogen at $T = 12{,}500°$K and $T = 25{,}000°$K; photoionization edges are labeled with quantum number of state from which they arise. Ordinate gives sum of bound-free and free-free opacity in cm$^1$/atom; abscissa gives $1/\lambda$, where $\lambda$ is in microns.

and

$$\frac{\mathscr{R}}{k^2} + h\nu = \frac{\mathscr{R}}{l^2} \tag{4-218}$$

We assume that absorptions take place from a band of states $dk$ into a band of states $(dl/d\nu)\,d\nu$ so that using equation (4-195), we identify

$$\alpha(\nu, v) = \frac{\pi e^2}{mc} f_{kl}\, dk\, \frac{dl}{d\nu} \tag{4-219}$$

as the absorption coefficient per ion and per electron moving with velocity $v$. Now by analogy with equations (4-168) and (4-169), we may write

$$f_{kl} = \frac{64}{3\pi\sqrt{3}} \frac{1}{g_k} \left(\frac{1}{k^2} - \frac{1}{l^2}\right)^{-3} \frac{g_{\mathrm{III}}(k, l)}{k^3 l^3} \tag{4-220}$$

Now

$$\frac{1}{k^2} - \frac{1}{l^2} = \frac{h\nu}{\mathscr{R}}$$

while $g_k$ is the statistical weight of a free electron and is

$$g_k = \frac{8\pi m^3 v^2\, dv}{h^3} = \frac{16\pi \mathscr{R} m^2 v}{h^3 k^3}\, dk \tag{4-221}$$

where we have made use of equation (4-217). Thus

$$\alpha(\nu, v) = \frac{\pi e^2}{mc} \frac{64}{3\pi\sqrt{3}} \frac{h^3 k^3}{16\pi \mathscr{R} m^2 v} \left(\frac{\mathscr{R}}{h\nu}\right)^3 \frac{g_{\rm III}(\nu, v)}{k^3 l^3} \frac{dl}{d\nu}$$

$$= \frac{4\mathscr{R}^2 e^2}{3\pi\sqrt{3} m^3 c} \frac{g_{\rm III}(\nu, v)}{v l^3 \nu^3} \frac{dl}{d\nu} \tag{4-222}$$

But for $k$ (or $v$) fixed, we have, from equation (4-218),

$$\frac{dl}{d\nu} = \frac{h l^3}{2\mathscr{R}}$$

and thus

$$\alpha(\nu, v) = \frac{2\mathscr{R} h e^2}{3\pi\sqrt{3} m^3 c} \frac{g_{\rm III}(\nu, v)}{\nu^3 v} \tag{4-223}$$

We must now sum over all electron velocities. The electrons may be assumed to have a Maxwellian velocity distribution

$$N_e(v)\, dv = N_e \left(\frac{m}{2\pi kT}\right)^{3/2} 4\pi v^2 e^{-mv^2/2kT}\, dv \tag{4-224}$$

so multiplying equation (4-223) by the electron and ion density and integrating over velocity, we have

$$a_\nu = N_i N_e \frac{4\mathscr{R} h e^2}{3\pi\sqrt{3} m^3 c} \left(\frac{m}{2\pi kT}\right)^{1/2} \frac{1}{\nu^3} \int_0^\infty g_{\rm III}(\nu, v) \exp(-mv^2/2kT) \left(\frac{mv}{kT}\right) dv \tag{4-225}$$

Defining $\bar{g}_{\rm III}(\nu, T)$ as the thermal average of the Gaunt factor, as written in the integral in equation (4-225), and making use of Saha's equation to eliminate the product $N_i N_e$ in terms of $N_0$, we obtain

$$a_\nu = \frac{16\mathscr{R}^2 e^2}{3\sqrt{3} h^2 mc} \frac{N_0}{U_{\rm H}(T)} \frac{e^{-u_1}}{u_1} \frac{\bar{g}_{\rm III}(\nu, T)}{\nu^3}$$

$$= \mathscr{K} N_0 \left[\frac{2}{U_{\rm H}(T)}\right] \frac{e^{-u_1}}{2u_1} \frac{\bar{g}_{\rm III}(\nu, T)}{\nu^3} \tag{4-226}$$

Finally, substituting from equation (4-209), we obtain

$$a_\nu(f-f) = \left[\frac{2f_H}{\mu' m_H U_H(T)}\right] \mathscr{K} \frac{\bar{g}_{III}(\nu, T)}{\nu^3} \frac{e^{-u_1}}{2u_1}$$

Adding this expression to the bound-free opacity as given by equation (4-216), we find that the *total* opacity of hydrogen becomes

$$a_\nu = \frac{2f_H}{\mu' m_H U_H(T)} \frac{\mathscr{K}}{\nu^3} e^{-u_1} \left\{ \sum_{n*}^{n_{max}} \frac{e^{u_n} g_{II}(n, \nu)}{n^3} + \frac{1}{2u_1} [e^{\bar{u}} - 1 + \bar{g}_{III}(\nu, T)] \right\} \tag{4-227}$$

Approximation formulae for $\bar{g}_{III}$ have been given by Menzel and Pekeris (*M. N.*, **96**, 77, 1935), and extensive tables by Karzas and Latter (*Ap. J. Supp. No. 55*, **6**, 167, 1961), and by Berger (*Ap. J.*, **124**, 550, 1956). The free-free opacity plays an ever more important role at longer wavelengths because of the decreasing number of bound-free edges that can contribute. Also, at high temperatures the free-free opacity grows relative to the bound-free since the $1/u_1$ term grows larger as $T$ increases. Physically, this is reasonable since as the temperature increases, the number of atoms in bound states steadily decreases relative to the number of ions and electrons.

In addition to atomic hydrogen, there are many other stellar opacity sources that have been identified and whose contributions can be calculated. Let us now turn our attention to a few of the most important of these.

### NEGATIVE HYDROGEN ION ABSORPTION

Hydrogen, because of its large polarizability, can form a negative ion consisting of a proton and two electrons. This ion has a single bound state with a binding energy of 0.754 eV. Because of its low binding energy, $H^-$ does not exist at very high temperatures but is prevalent mainly in the atmospheres of solar-type stars and cooler stars. The absorption cross-section of $H^-$ is large, and although only a small fraction of the hydrogen exists in this form, the opacity due to $H^-$ is the dominant one for the cooler stellar atmospheres. The negative hydrogen ion can interact with the radiation field via both bound-free and free-free-absorptions, i.e.,

$$H^- + h\nu \rightleftarrows H + e(v) \tag{4-228}$$

where $\frac{1}{2}mv^2 = h\nu - 0.75$ eV, and

$$H + e(v) + h\nu \rightleftarrows H + e(v') \tag{4-229}$$

where $\frac{1}{2}mv'^2 = h\nu + \frac{1}{2}mv^2$. The bound-free absorption has its threshold at about 16,500 Å, corresponding to the detachment energy. It reaches a maximum of about $4 \times 10^{-17}$ cm$^2$ at 8500 Å and decreases toward shorter wavelengths. The summed opacity has a minimum at around 16,000 Å, which is important because it implies that emergent fluxes for cool stars could be large near this minimum. Actually, other opacity sources act to wash out the sharp minimum, but, on the whole, the opacity is smallest for cool stars near this wavelength.

The determination of the cross-sections for the two processes mentioned above is difficult and has been approached both theoretically and experimentally. Very elaborate wave functions are required to give the desired accuracy. Pioneer calculations that gave fairly good values were carried out by Chandrasekhar and Breen (*Ap. J.*, **104**, 430, 1946). These were shown to be in accord with empirically deduced values for the absorption coefficient in the sun and led to the firm identification of $H^-$ as the major opacity source in the solar atmosphere (see Chapter 2). More accurate values have been obtained by Geltman [*Ap. J.*, **136**, 935, 1962 (bound-free), and *Ap. J.*, **141**, 376, 1965 (free-free)]; these are in good agreement with experimental values.

In model atmosphere calculations, it is convenient to convert the absorption coefficient $\alpha_\nu(H^-)$ per $H^-$ ion to an absorption per neutral H atom by multiplying by $n(H^-)/n(H)$ using a Saha equation in the form

$$\frac{n(H^-)}{n(H)} = p_e\,\varphi(T) \tag{4-230}$$

Then in LTE the mass absorption coefficient is simply

$$a_\nu = \left(\frac{f_H}{\mu' m_H}\right) p_e[\alpha_\nu(H^-)\varphi(T)] \tag{4-231}$$

A very useful approximation formula for the term in brackets has been given by Gingerich (Ref. 15, p. 17).

The fact that $a_\nu(H^-)$ is proportional to the electron pressure implies that it will be a more important opacity source in dwarfs than in giants where low surface gravities imply low pressures. Also, in cool stars that are very metal-poor, there will be fewer free electrons donated by the metals, and hence $a_\nu(H^-)$ will decrease. Thus, there may be appreciable differences in the atmospheric structure among these types of stars.

### He I and He II absorption

Since helium is the most abundant element next to hydrogen, it can make appreciable contribution to the opacity. Indeed, in stars where the helium abundance is anomalous and approaches or exceeds that of hydrogen, helium can dominate the opacity. Helium is observed in stellar spectra both in its neutral and singly ionized state. Because the ionization potential of neutral helium is 24.58 eV, it persists at temperatures where hydrogen is already strongly ionized. The absorption edge from the ground state lies at 504 Å and thus dominates the far ultraviolet spectrum of a star.

The excited states of helium fall into two groups, singlets and triplets, and each $(n, l, s)$ state has a different ionization energy. Roughly speaking, the ionization energies lie close to those of hydrogen states with the same $n$; thus helium contributes a multiplicity of absorption edges near each hydrogen edge. Because the excitation energy of even the lowest excited state is so large (19.72 eV), helium adds to the opacity in the visible regions of stellar spectra only for fairly hot stars (B stars). Moreover, because this excitation enregy lies close to the ionization energy, helium is already appreciably ionized when the excited states begin to contribute to the opacity.

Since helium is a three-body system, exact wave functions cannot be written down as in the case of hydrogen, and approximations must be used in the calculation of opacities. Accurate wave functions exist only for the ground state. For excited states only very approximate results exist.

For the ground state, an accurate Hartree-Fock wave function and its absorption cross-section have been given by Stewart and Webb [*Proc. Roy. Soc.* (London), **82**, 532, 1963]; these results are in good agreement with experiment. Wave functions and photoionization cross-sections for the $2^1S$ and $2^3S$ levels have been calculated by Huang (*Ap. J.*, **108**, 354, 1948). For the $2^1P$ and $2^3P$ states, approximate cross-sections have been calculated by Goldberg (*Ap. J.*, **90**, 414, 1939). Thus for these first five states we may write, in LTE,

$$a_\nu = N(\mathrm{He}) \sum_{i*} \frac{a_\nu(i) g_i e^{-\chi_i/kT}}{U_{\mathrm{He}}(T)} \tag{4-232}$$

where $N$ (He) is the number of neutral helium atoms per gm and where again we sum only over those levels for which $(\chi_I - \chi_i) \leq h\nu$. The statistical weights and ionization energies for the individual $(n, L, S)$ states of helium are given in Table 4-2. Convenient numerical interpolation formulae for the cross-

TABLE 4-2. Photoionization Properties of the $n = 1$ and $n = 2$ States of Helium I

| *State* | $\chi_i$ eV | $g_i$ | $\lambda$ (Å) *at edge* |
|---|---|---|---|
| $1^1$S | 0 | 1 | 504.3 |
| $2^3$S | 19.72 | 3 | 2601.0 |
| $2^1$S | 20.51 | 1 | 3122.0 |
| $2^3$P | 20.86 | 9 | 3422.0 |
| $2^1$P | 21.11 | 3 | 3680.0 |

sections $\alpha_\nu(i)$ in equation (4-232) have been given by Gingerich (Ref. 15, p. 17). For higher states, one can argue that the remaining $1s$ electron shields the nucleus so effectively that the excited electron is essentially in an excited state around a single nuclear charge. The wave function should, therefore, be nearly hydrogenic. Thus, to at least a rough approximation, we can write for these states

$$a_\nu = N(\mathrm{He}) \frac{\mathscr{K}}{\nu^3} \sum_{n*} \frac{4n^2 e^{-\chi_n/kT}}{n_{\mathrm{eff}}^5 U_{\mathrm{He}}(T)} \tag{4-233}$$

where

$$n_{\mathrm{eff}}^2 = \chi_{\mathrm{I}}(\mathrm{H})/[\chi_{\mathrm{I}}(\mathrm{He}) - \chi_n] \tag{4-234}$$

The statistical weight of state $n$ is now $4n^2$ instead of $2n^2$ because in hydrogen each state is a doublet while in helium the states are singlets and triplets. In a similar way, we may argue that the free-free contribution is also essentially hydrogenic and may be written

$$a_\nu = N(\mathrm{He}) \left[\frac{2U_{\mathrm{He}^+}(T)}{U_{\mathrm{He}}(T)}\right] \frac{\mathscr{K}}{\nu^3} \frac{\exp[-\chi_{\mathrm{I}}(\mathrm{He})/kT]}{[2\chi_{\mathrm{I}}(\mathrm{H})/kT]} \tag{4-235}$$

In equations (4-232), (4-233), and (4-235) the number of neutral helium atoms per gram of stellar material is

$$N(\mathrm{He}) = \frac{Y f_{\mathrm{He}}}{\mu' m_{\mathrm{H}}} \tag{4-236}$$

where $Y = N(\mathrm{He})/N(\mathrm{H})$, and $f_{\mathrm{He}}$ is the fraction of all helium that is neutral.

At very high temperatures, characteristic of the O stars for example, both hydrogen and helium are strongly ionized, and singly ionized helium becomes an important opacity source. This ion consists of a single electron and a nucleus with charge $Z = 2$. Thus the atom is hydrogenic, and hydrogenic cross-sections again apply. Simple theory shows the energies in a hydrogenic ion scale as $Z^2$; therefore the ionization energy for He II is 54.4 eV, and similarly, the frequencies of the ionization edges are scaled up by a factor of four. When one works out the cross-section for hydrogenic ions, one finds the same formula as for hydrogen but scaled by a factor of $Z^4$. Thus the opacity of singly ionized helium is simply

$$a_\nu = \left[\frac{2Y f_{\mathrm{He}^+}}{\mu' m_{\mathrm{H}} U_{\mathrm{He}^+}(T)}\right] \frac{16\mathscr{K}}{\nu^3} e^{-u_1} \times \left\{ \sum_{n*}^{n_{\max}} \frac{e^{u_n} g_{\mathrm{II}}(n, \nu)}{n^3} + \frac{1}{2u_1} [e^{\bar{u}} - 1 + \bar{g}_{\mathrm{III}}(\nu, T)] \right\} \tag{4-237}$$

where now $u_n \equiv 4\chi_{\mathrm{I}}(\mathrm{H})/n^2 kT$, $f_{\mathrm{He}^+}$ is the fraction of all helium that is singly ionized, and the Gaunt factors must allow for the factor of four scaling in frequency. Because of the strong temperature sensitivity of the He II opacity, it normally can be neglected except for stars of class B0 and earlier, although it can contribute in the deeper layers of somewhat cooler stars.

## ABSORPTION BY OTHER IONS OF HYDROGEN AND HELIUM

Hydrogen occurs in two other forms that can contribute significantly to the opacity in stellar atmospheres, namely $H_2^+$ and $H_2^-$. The former ion consists of two protons and one electron. Absorption cross-sections for this ion have been calculated by Bates (*M. N.*, **112**, 40, 1952) and by Buckingham, Reid, and Spence (*M. N.*, **112**, 382, 1952). Since the number density of $H_2^+$ is proportional to $N_H \cdot N_{H^+}$, an important contribution is made to the opacity only for that temperature-pressure range where both neutral and ionized hydrogen atoms are abundant, i.e., the hydrogen is about half-ionized. This range is characteristic of the A stars, and $H_2^+$ makes about a 10% contribution at wavelengths in the visible part of the spectrum for such stars. The absorption coefficient of $H_2^+$ actually peaks near 1100 Å, but since the absorption from $n = 2$ of hydrogen is so much stronger, $H_2^+$ is important only on the redward side of the Balmer jump.

The negative molecular ion $H_2^-$ exists only at relatively cool temperatures, characteristic of the M stars, and makes a significant contribution to the total opacity at long wavelengths. Also, the $H_2^-$ continuum tends to fill in the opacity minimum of $H^-$ at 1.6 microns. A calculation of the cross-section for the free-free opacity has been given by Somerville (*Ap. J.*, **139**, 192, 1964), who also shows that the bound-free opacity is negligible.

Helium can also exist in the form of a negative ion and again give rise to a free-free opacity in cool stars. Cross-sections for this process have been given by Somerville (*Ap. J.*, **141**, 811, 1965) and by John (*Ap. J.*, **149**, 449, 1967).

## ABSORPTION BY HEAVIER ELEMENTS

Although hydrogen and helium in their various forms dominate the opacity, at certain wavelengths absorption from heavier elements can also be important. The most comprehensive study of all these sources has been made by members of the Kiel group and the Los Alamos group. A set of graphs showing the run of opacity with wavelength has been published by Vitense (*Z. Ap.*, **28**, 81, 1951; see also Ref. 36, pp. 182–96). This work made use of the hydrogenic approximation for the heavy elements. An extension of this study using improved cross-sections has been carried out by Bode (Ref. 7), who gives both graphs and tables of the opacity coefficient. The Los Alamos data (Cox, Stewart, and Eilers, *Ap. J. Supp. No. 94*, **11**, 1, 1965) pertains mainly to the stellar interior, but some of the results are relevant to stellar atmospheres. A comprehensive review of this work has been given by Cox (Ref. 5, p. 195). An important approach to the calculation of photoionization cross-sections for nonhydrogenic atoms is the *quantum defect method* of Burgess and Seaton (*M. N.*, **120**, 121, 1960; see also Peach, *Mem. R. A. S.*, **71**, 1, 1967), which has been applied extensively to a number of astrophysically important atoms and ions. As work on compiling accurate wave functions for atoms progresses, we can expect ever-improved values to be obtained for opacities.

### ABSORPTION BY NEGATIVE IONS OF HEAVIER ELEMENTS

The possible astrophysical importance of absorption by the ions $OH^-$, $CN^-$, $C_2{}^-$, $O^-$, and $C^-$ has been considered by Branscomb and Pagel (*M. N.*, **118**, 258, 1958), who concluded the most important sources were $C^-$ and $O^-$. Experimental work on these ions has been summarized by Branscomb (Ref. 6, Chap. 4). Recently, Vardya (*Mem. R. A. S.*, **71**, 249, 1967) has pointed out that other ions such as $Cl^-$, $CN^-$, $SH^-$, and $H_2O^-$ occur fairly abundantly in late-type stars, and it is possible that these ions may be of significance in the atmospheric structure of such stars.

### EFFECTS OF BOUND-BOUND TRANSITIONS

Bound-bound transitions can affect the opacity in stellar atmospheres quite significantly. For example, in solar-type stars, there are literally thousands of lines distributed over the entire spectrum but most heavily concentrated in the ultraviolet. In such a case, one recognizes the impossibility of treating all the lines in detail and adopts a statistical treatment, as will be described in Chapter 14. For later stars (late K and M), both observations from balloons (Woolf, M. Schwarzschild, and Rose, *Ap. J.*, **140**, 833, 1964) and theory (Auman, *Ap. J. Supp. No. 127*, **14**, 171, 1967) have shown that absorption by the molecular bands of $H_2O$ dominates the infrared spectrum where the bulk of the flux emerges. For the A stars, the main contributors to the bound-bound opacity are the hydrogen Balmer lines, though metal lines still play an important role in the late A stars. For the B stars, important contributions to the opacity are made by the hydrogen Lyman lines and, even more importantly, lines from numerous highly ionized atoms.

Only a few calculations of stellar atmospheres have been made including the effects of spectrum lines; these will be discussed in Chapter 14.

## 4-5. Continuum Scattering Cross-Sections

As mentioned in Chapter 1, radiation in the continuum may be scattered as well as absorbed. As emphasized in our earlier discussion, photons are destroyed in absorption processes, the energy being contributed at least partly to the thermal content of the gas. In a scattering process, the photon is not destroyed but is merely redistributed in angle and frequency. Let us consider here the two most important continuum scattering sources.

### THOMSON SCATTERING

The scattering of light by free electrons is referred to as *Thomson scattering*. The classical formula for this process may be obtained directly from equation (4-65) by noting that for an unbound electron, one would expect both the resonant frequency $\omega_0$ and the damping parameter $\gamma$ to be zero so that

$$\sigma(\omega) = \frac{8\pi e^4 \omega^4}{3m^2 c^4} \frac{1}{\omega^4} = \frac{8\pi e^4}{3m^2 c^4} \equiv \sigma_T \tag{4-238}$$

This cross-section has been verified, in the limit of low photon energies, by quantum mechanical calculations; at high photon energies one should employ the *Klein-Nishina* formula (Ref. 17, Sec. 22) though it is never necessary to do so in stellar atmospheres calculations. Numerically, $\sigma_T = 6.65 \times 10^{-25}$ cm$^2$.

Note from equation (4-238) that the cross-section is frequency-independent so that this process is equally efficient throughout the entire spectrum. In the derivation of equation (4-65) we averaged over angle and thus suppressed the angular dependence of the scattering coefficient. The correct angular dependence is given by the dipole phase function [equation (1-30)]

$$g(\mathbf{n}', \mathbf{n}) = \tfrac{3}{4}(1 + \cos^2 \Phi)$$

where $\cos \Phi = \mathbf{n}' \cdot \mathbf{n}$, $\mathbf{n}'$ is the direction along the incident beam, and $\mathbf{n}$ the direction along the scattered beam. In many applications, this angular dependence can be ignored and the process considered to be isotropic. We have also neglected the frequency redistribution caused by Doppler shifts in the laboratory frame; we shall return to this question again in Chapter 10. In the continuum, the effects of frequency redistribution may usually be neglected and the scattering taken to be coherent; near a spectral line, it may be necessary to account for the frequency redistribution.

To obtain the macroscopic scattering coefficient per gram of stellar material, we must multiply $\sigma_T$ by the number of free electrons per gram, namely,

$$\frac{N_e}{\rho} = \frac{p_e}{\mu m_{\rm H}(p_g - p_e)} \tag{4-239}$$

where we have made use of equation (3-50). Thus

$$\sigma_e = \frac{N_e \sigma_T}{\rho} = \frac{\sigma_T p_e}{\mu m_H (p_g - p_e)} \tag{4-240}$$

## RAYLEIGH SCATTERING

The term *Rayleigh scattering* refers to the scattering of radiation by bound systems, such as atoms or molecules, at frequencies much lower than characteristic transition frequencies of the system. Again using equation (4-65), this may be described classically by representing real transitions of the system with equivalent classical oscillators of appropriate strengths $f_{ij}$ and resonant frequencies $\omega_{ij}$ equal to the actual transition frequency. Then, for $\omega \ll \omega_{ij}$, equation (4-65) simplifies to yield

$$\sigma(\omega) = f_{ij}\left(\frac{8\pi e^4}{3m^2c^4}\right)\frac{1}{\left(\dfrac{\omega_{ij}^2}{\omega^2} - 1\right)} = \frac{\sigma_T f_{ij}}{\left(\dfrac{\omega_{ij}^2}{\omega^2} - 1\right)} \tag{4-241}$$

Far from the resonant frequency, $\sigma(\omega)$ varies as $\omega^4$ or $\lambda^{-4}$, which leads to a strong color dependence of the scattered radiation; a well-known example of this is the blue color of the sky, resulting from sunlight scattered by molecules of air.

Rayleigh scattering can be important in the atmospheres of stars of moderate temperature (spectral types G and K). Here most of the hydrogen is neutral and in the ground state. The resonant frequencies corresponding to the Lyman transitions ($1 \rightarrow n$) lie far in the ultraviolet, and visible photons will interact with these transitions by the mechanism of Rayleigh scattering. Summing over all transitions and multiplying by the number density hydrogen in the ground state, we obtain a macroscopic scattering coefficient

$$\sigma_\nu = \left[\frac{2f_H}{\mu' m_H U_H(T)}\right]\sigma_T \sum_n f_{1n}\left(\frac{\nu_{1n}^2}{\nu^2} - 1\right)^{-1} \tag{4-242}$$

where the symbols in the brackets have the same meaning as in equation (4-227).

Rayleigh scattering by neutral hydrogen can dominate the opacity at relatively low temperatures and high frequencies (see the graphs of Vitense and Bode, cited above). Moreover, in stars with low metal abundances, the number of free electrons is greatly reduced, and accordingly the opacity of $H^-$ is greatly diminished so that the importance of Rayleigh scattering is greatly enhanced in such stars.

Molecular hydrogen, $H_2$, may scatter radiation in a completely analogous fashion. The cross-section per molecule (Dalgarno and Williams, *Ap. J.*, **136**,

690, 1962) is comparable with the cross-section per atom for atomic hydrogen. At low temperatures, $H_2$ is much more prevalent than atomic H, and the molecular Rayleigh scattering dominates. We should note in passing that in the scattering process there is no analogue of the stimulated emission that occurs in absorption processes. Thus, there is no correction factor for stimulated emission such as appears in equations (4-192) and (4-193). Rather, the scattering coefficient appearing in the transfer equation, e.g., in equation (1-43), is given directly by the appropriate sum over equations such as (4-240) and (4-242).

## 4-6. Non-LTE Effects upon Opacities

In the formulae for opacities per gram of stellar material given in the preceding sections, we have made use of LTE occupation numbers. In the general case, these formulae must be modified to use the occupation numbers resulting from a self-consistent solution of the statistical equilibrium and radiative transfer equations. In this sense, it is more general to use formulae of the form

$$a_\nu = \sum_n \alpha_\nu(n) N_n$$

that appeared, for example, as an intermediate step in equation (4-211). We may retain the basic form of the equations we have written if we employ the formalism [equation 3-55)]

$$b_n \equiv \frac{N_n}{N_n^*}$$

Then for hydrogen, for example, we could write

$$a_\nu = \left[\frac{2f_H}{\mu' m_H U_H(T)}\right] \frac{\mathcal{K} e^{-u_1}}{\nu^3} \times \left\{ \sum_{n*}^{n_{max}} \frac{b_n e^{u_n} g_{II}(n, \nu)}{n^3} + \frac{1}{2u_1} [e^{\bar{u}} - 1 + g_{III}(\nu, T)] \right\} \tag{4-243}$$

Note that $f_H$ and $U_H(T)$ now signify the non-LTE values of these quantities. We also see that the term involving free-free absorption is unaltered since it is a strict continuum-interaction process and hence is automatically in LTE relative to the kinetic temperature of the electrons. If we now define the departure coefficient

$$d_n \equiv b_n - 1 \tag{4-244}$$

which is clearly a measure of the fractional deviation from LTE, we may isolate the non-LTE effects in a single term by writing

$$a_\nu = a_\nu{}^* + \left[\frac{2 f_{\rm H} \cdots}{\mu' m_{\rm H} U_{\rm H}(T)}\right] \frac{\cdots}{\nu^3} \sum_{n*}^{\cdots} \frac{d_n e^{\cdots} g_{\rm II}(n, \nu)}{n^3} = a_\nu{}^* + \delta a_\nu \qquad (4\text{-}245)$$

a form that we will find convenient in some of the discussion in Chapter 7. Here $a_\nu$* denotes the opacity that would be computed with LTE formula (though using non-LTE ionization fractions and partition functions) and $\delta a_\nu$ describes the change due to non-LTE effects. Similar formulae can be written for other atoms.

Having obtained the formulae required to calculate the opacity, let us now turn our attention to the specification of the statistical equilibrium equations.

# 5 | The Equations of Statistical Equilibrium

One of the most fundamental features of the physical situation in stellar atmospheres is that the statistical equilibrium equations governing the occupation numbers of bound states of atoms depend upon and indeed are dominated by the radiation field. This fact introduces the essential difficulty of the calculation since the radiation field, in turn, depends upon the occupation numbers via the absorptivity and emissivity of the stellar material. Thus, what is required is a completely selfconsistent solution of both the radiative transfer and statistical equilibrium equations. It is for this reason that we have repeatedly stated above that the a priori assumption of LTE can be regarded *only* as a computational convenience and that this hypothesis must be tested by an analysis in which one actually solves self-consistently for the radiation field and occupation numbers. If, in any particular case, the occupation numbers obtained in the general analysis happen to agree with those predicted by LTE, then one may legitimately use the LTE assumption; generally, of course, such agreement is not expected.

The mathematical methods used to obtain this self-consistent solution will be described completely in Chapters 7, 12, 13, and 14. A fairly wide variety of techniques may be used, depending upon the particular physical situation under study. In this chapter we shall derive the statistical equilibrium equations and consider their implications in certain limiting cases.

## 5-1. The Microscopic Requirements of LTE

Before we give the actual equations of statistical equilibrium, it is worthwhile to discuss, in a qualitative way, the implications, on a microscopic level, of the assumption of LTE. This assumption is essentially equivalent to the statements that: (a) the electron and ion velocity distributions are Maxwellian; (b) the ionization equilibrium is given by Saha's equation; and (c) the excitation equilibrium is given by the Boltzmann distribution. An interesting commentary on these three points has been given by R. H. Bohm (Ref. 13, Chap. 3). We will summarize and discuss his analysis here.

### THE ELECTRON VELOCITY DISTRIBUTION

In stellar atmospheres, the free electrons are produced by photoionization and collisional ionizations. The inverse processes are radiative recombination and three-body collisions, which remove electrons from the continuum and restore them to bound states. During its stay in the continuum, an electron may undergo elastic collisions with other electrons and inelastic collisions with atoms and ions, resulting in excitations or ionizations of bound electrons. The elastic collisions redistribute energy among the colliding particles and tend to lead to an equilibrium partitioning—hence a Maxwellian velocity distribution. If a Maxwellian velocity distribution is in fact attained, it can be taken as a reference standard, and the local temperature can be defined as the kinetic temperature of the electrons. On the other hand, the inelastic collisions tend to disturb the achievement of a Maxwellian velocity distribution because they involve electrons only in certain velocity ranges and tend systematically to shift electrons from these ranges to much lower velocities. Whether or not the Maxwellian velocity distribution is disturbed depends upon how fast the thermalization by elastic collisions proceeds compared with the inelastic collisions; if it occurs very much more rapidly, then we can expect the velocity distribution to recover its Maxwellian character quickly.

The rate at which thermalization occurs can be measured in terms of the *relaxation time* of the system. If we consider a group of particles interacting with themselves, we find for the relaxation time

$$t_c = \frac{m^{1/2}(3kT)^{3/2}}{17.9\,N_e e^4 Z^4 \ln(D/p_0)}\ \text{sec} \tag{5-1}$$

(see, e.g., the derivation by Spitzer, Ref. 32, Chap. 5) where $D$ is the *Debye length* (to be discussed further in Chapter 9)

$$D = \left(\frac{kT}{4\pi e^2 N_e}\right)^{1/2} \tag{5-2}$$

and $p_0$ is the impact parameter for a collision resulting in a $90°$ deflection (see Figure 5-1)

$$p_0 = \frac{e^2}{mv^2} \tag{5-3}$$

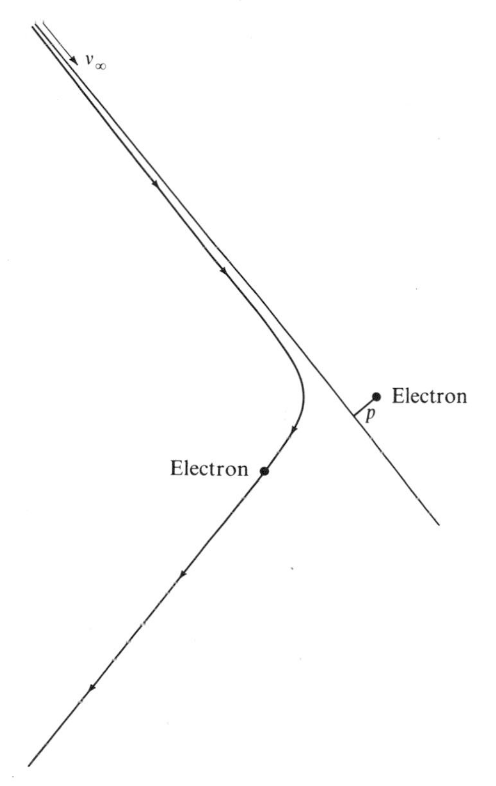


FIG. 5-1. The impact parameter $p$ is defined as the perpendicular distance to the tangent of the incoming particle path at infinity.

The relaxation time must be compared with the rate of recombinations and the rate of inelastic collisions. Consider first the recombinations. If $\sigma$ is the average cross-section for recombination, then the mean time between recombinations is

$$t_r = \frac{1}{N\sigma\langle v\rangle} = \frac{1}{N\sigma}\left(\frac{m}{3kT}\right)^{1/2} \text{ sec} \tag{5-4}$$

where $N$ is the density of the particles with which recombination occurs. Two types of recombination of astrophysical importance are: (a) recombination with H to form $H^-$, and (b) recombination with $H^+$ to form H. At $T \sim 6000$°K (a typical solar temperature), the cross-section for forming $H^-$ is about $3 \times 10^{-22}$ cm$^2$, and the ratio $N_e/N_H$ is about $10^{-4}$. At $T \sim 10{,}000$°K, the cross-section for forming H is about $6 \times 10^{-21}$ cm$^2$ and $N_e/N_{H^+} \sim 1$. Now taking the ratio $t_r/t_c$, we have

$$\frac{t_r}{t_c} = \frac{\text{number of elastic collisions}}{\text{number of recombinations}} = \frac{17.9\, e^4 \ln(D/p_0)}{(3kT)^2\sigma}\frac{N_e}{N} \tag{5-5}$$

Substituting the estimates given above, we find that if we consider the first mechanism, we find $t_r/t_c \sim 10^5$ while the second mechanism yields $t_r/t_c \sim 10^7$. Thus under representative conditions in stellar atmospheres, an electron in the continuum will undergo an *enormous* number of elastic scatterings compared with recombinations and will have an adequate opportunity to establish thermal equilibrium.

Let us now consider the inelastic collisions. The most abundant element in the atmosphere is hydrogen, and electron-hydrogen collisions occur frequently. However, the first excited state of hydrogen lies some 10 eV above the ground state while a typical thermal energy of the electrons is about 1 eV. Thus, only $3 \times 10^{-5}$ of the electrons have enough energy to induce the excitation. Of these, only a fraction will be effective. In fact, using typical cross-sections (see below), we find that (at 10,000°K) the rate of inelastic excitations is of the same order as the recombination rate, so that it is again small compared with the rate of elastic collisions. The same arguments hold for the excitation of helium, only even more strongly because the first excited state lies at even higher energy. We must also consider collisions with other elements. Three groups of elements are important: (a) the alkalis, which have large cross-sections but low abundances ($10^{-6}$); (b) iron atoms, having a large number of low-lying levels, moderate cross-sections, and larger abundances ($10^{-4}$); and (c) carbon, oxygen, and nitrogen, which have small cross-sections but large abundance ($10^{-3}$). Most of the levels involved in categories (b) and (c) are metastable so that the great majority of collisional excitations are subsequently canceled by a collisional de-excitation. If we ignore this situation we greatly overestimate the number of inelastic collisions. Taking the various factors into account and ignoring compensating de-excitation, K. H. Böhm estimates that elastic collisions still outnumber inelastic by at least a factor of $10^3$, and hence again we conclude a Maxwellian velocity distribution will be established.

Thus there is little doubt that the electron velocity distribution is Maxwellian. Since this defines the local $T_e$, it is clear that processes involving only the interaction of particles in the continuum (e.g., free-free absorptions and emissions, and radiative recombinations) can be regarded as occurring at the LTE rate calculated with the local $N_e$ and $T_e$.

One final question to be answered is whether the atoms and ions in the medium also have a Maxwellian velocity distribution and whether the corresponding kinetic temperature agrees with that of the electrons. This question has been examined by Bhatnagar, Krook, Menzel, and Thomas (*Vistas Astron.*, **1**, 296, 1955), who consider the interaction of electrons, ions, atoms, and radiation in a pure hydrogen atmosphere. They examined the relative rates of energy exchange between atoms (and ions) and electrons and the rates of energy losses by the electrons in inelastic collisions and to radiation, while demanding a steady state solution. They found that if $N_e > 10^{10}$ (a condition met in most of the atmosphere), then $|T_k - T_e| \lesssim 10^{-3} T_e$ on the range $5 \times 10^3 < T_e < 10^5$, where $T_k$ is the temperature of the atoms and ions. Thus, it appears safe to conclude that a unique kinetic temperature applies to all of the particles.

### THE IONIZATION EQUILIBRIUM

The basic question we wish to examine here is, What are the relative rates of radiative and collisional ionizations? It will suffice to obtain only an order-of-magnitude estimate.

The direct rate of photoionization per atom is simply

$$R_{i\kappa} = 4\pi \int_{\nu_0}^{\infty} \frac{\alpha_\nu J_\nu \, d\nu}{h\nu}$$

where $\alpha_\nu$ is the photoionization cross-section. In LTE this would equal

$$R_{i\kappa}^* = 4\pi \int_{\nu_0}^{\infty} \frac{\alpha_\nu B_\nu \, d\nu}{h\nu}$$

If we want only a rough estimate, we may use the latter. For simplicity we may write [see equation (4-207)]

$$\alpha_\nu = \frac{2\pi e^2}{mc} f_c \frac{{\nu_0}^2}{\nu^3} \tag{5-6}$$

where $f_c$ is the total oscillator strength for the transition from the bound state to the continuum. Then the radiative rate of ionization is approximately

$$R = \frac{8\pi^2 e^2 {\nu_0}^2}{mc} f_c \int_{\nu_0}^{\infty} \frac{2h\nu^3}{c^2 h\nu^4} (e^{h\nu/kT} - 1)^{-1} \, d\nu$$

$$\approx \frac{16\pi^2 e^2 {\nu_0}^2}{mc^3} f_c E_1(h\nu_0/kT) \tag{5-7}$$

where the last equality holds only for $h\nu_0/kT \gg 1$. This rate is to be compared with the rate of collisional ionizations. If $\sigma(v)$ represents the collisional ionization cross-section for electrons of velocity $v$, then the rate of ionization by collisions is

$$C = N_e \int_{v_0}^{\infty} f(v)\sigma(v)v \, dv \tag{5-8}$$

To obtain a result, to order of magnitude only, K. H. Böhm makes use of the semiclassical Thomson formula

$$\sigma(v) = 3f \frac{\pi e^2}{E} \left( \frac{1}{h\nu_0} - \frac{1}{E} \right) \tag{5-9}$$

where $E$ is the kinetic energy of the electron (more accurate cross-sections will be discussed later). By integration and comparison with the radiative rate, one finds

$$\frac{R}{C} = \frac{4\sqrt{2}\,\pi^{3/2}\nu_0^{\,2}(kT)^{3/2}}{3m^{1/2}e^2c^3N_e}\left(\frac{h\nu_0}{kT}\right)\frac{E_1(h\nu_0/kT)}{E_2(h\nu_0/kT)}$$

$$= 8.9 \times 10^{-23}\,\frac{\nu_0^{\,2}T^{3/2}}{N_e}\,x_0\,\frac{E_1(x_0)}{E_2(x_0)} \tag{5-10}$$

where $x_0 = h\nu_0/kT$. K. H. Böhm has carried out numerical computations with this formula, which are shown graphically in his article (Ref. 13, Chap. 3). He considers representative cases of levels with an ionization potential of 8 eV and of 1 eV for conditions that are characteristic of the outer layers ($\tau \sim 0.05$) of the sun and an O9 star (10 Lac). In particular, for the sun he adopts $N_e \approx 3 \times 10^{12}$ and $T \approx 5 \times 10^{3}$°K, while for 10 Lac he adopts $N_e \approx 3 \times 10^{14}$ and $T \approx 3.2 \times 10^{4}$°K, and finds the values of $R/C$ that are listed in Table 5-1. It is clear that for the low-lying levels, the radiative rates

TABLE 5-1. Ratio of Radiative to Collisional Ionization Rates

| Star | $\chi_i = 8$ eV | $\chi_i = 1$ eV |
|---|---|---|
| Sun | $\sim 10^3$ | $\sim 2.0$ |
| 10 Lac | $\sim 20$ | $\sim 0.2$ |

*Source:* From data by K. H. Böhm, in J. L. Greenstein, ed., *Stellar Atmospheres*, Chicago: University of Chicago Press, 1960, pp. 102–3; by permission.

dominate; indeed only at very high temperatures and densities, and for very high-lying levels, will the collision rate dominate. This is an essential point, for if collisions dominate, we would have been assured of LTE. Since the radiation field dominates, we can argue rigorously that LTE will be obtained *only* if the radiation field is $B_\nu$ (though even if it is not, LTE might nearly be attained "accidentally").

K. H. Böhm suggests that an estimate of how serious the deviations from LTE are can be obtained by comparing the value

$$4\pi\int_0^\infty \frac{\kappa_\nu J_\nu\,d\nu}{h\nu} \qquad \text{with} \qquad 4\pi\int_0^\infty \frac{\kappa_\nu B_\nu\,d\nu}{h\nu}$$

(where $\kappa_\nu$ is the opacity from all overlapping continua) as obtained from model atmosphere calculations. The basic idea is that since radiative ionizations dominate collisional ones, we may estimate how seriously the ionization equilibrium departs from the Saha equation by comparing the actual ionization rate with its LTE value. If the rates are equal, the statement is made that

LTE is self-consistent. There are, however, several possible criticisms that can be made of this approach. First, it is clear that the *integrated* rates contain information from *all* continua and may be subject to intricate cancellations and compensations, i.e., it is not at all clear what a specific deviation between these integrals implies for a *given* level. Second, if the assumption of LTE was used to compute the model, then $J_\nu$ will necessarily approach $B_\nu$ strongly at depth since equation (1-81) shows that for $\tau_\nu \gg 1$,

$$J_\nu(\tau_\nu) = \Lambda_{\tau_\nu}(B_\nu) = B_\nu + O(e^{-\tau_\nu})$$

Thus the two integrals will automatically become equal when $\tau_\nu \gtrsim 1$ at all frequencies. For example, Böhm states that for the *Fe* I-*Fe* II equilibrium in a model solar atmosphere, the two rates have the ratio 2.88 at $\bar{\tau} = 0.01$, 1.30 at $\bar{\tau} = 0.05$, and essentially unity at $\bar{\tau} = 0.1$. From this situation one might conclude that the Saha equation is valid below $\bar{\tau} = 0.1$. There is, however, an important point to be considered here. In actual fact, $B_\nu$ is *not* the correct source function. Indeed, as we shall show later, an essential feature of non-LTE transfer problems is that $S_\nu$ can in general differ significantly from $B_\nu$ over a *large* range of optical depths (we cannot yet prove this point here, so we shall merely assert it). Strictly speaking, $J_\nu$ must be absolutely consistent with $S_\nu$. A possible approach (though an inefficient one in general; see Chapter 7) is to substitute the computed $J_\nu$ into the statistical equilibrium equations, recompute $S_\nu$, and iterate until strict consistency is obtained. Only if this is done is the estimate of $J_\nu$ meaningful. If one merely carries out a *single* iteration, one inevitably obtains an answer very close to LTE. The closeness is misleading, however, since further iteration will in almost all cases show a continuing, progressive departure from LTE, as information about the existence of the boundary penetrates to deeper layers via the iteration procedure. In short, the disagreement between the two rates may really be much larger and extend far deeper than the first iteration indicates. This fact was not realized in much of the early work, and erroneous conclusions were drawn. We shall study this crucially important point in detail in Chapters 7 and 12; further discussion has been given by Thomas (Ref. 33, pp. 141–47).

In reality, one can safely say only the following: Since radiative rates dominate over the collisional, we must in general expect that LTE will *not* be valid and from the outset carry out a solution of both the statistical equilibrium and transfer equations. Only when a self-consistent solution is obtained is it possible to decide in which regions LTE actually occurs. To base estimates upon a single iteration starting from LTE models is worthless.

### THE EXCITATION EQUILIBRIUM

As in the case of ionizations, the question we ask again here is whether collisional or radiative processes dominate. The radiative excitation rate is given by $4\pi B_{ij} J_\nu$, which, for purposes of estimation, we can approximate by $4\pi B_{ij} B_\nu$. A collision cross-section of the same form as that mentioned above can be adopted, and one finds the ratio of the rate of collisions to the rate of photoexcitation to be

$$\frac{C}{R} = 1.2 \times 10^{15} p_e \theta^{1/2} \lambda^4 g(h\nu/kT) \tag{5-11}$$

where $\theta = 5040/T$, and $g(x) = (e^x - 1)E_2(x)$ (see Ref. 38, p. 121). Calculating this ratio for typical conditions in the outer layers of the sun, and of 10 Lac, as mentioned above, we obtain the results listed in Table 5-2. We again see

TABLE 5-2. Ratio of Collisional to Radiative Excitation Rates

| *Star* | $\lambda$ (Å) = 3000 | 4000 | 5000 | 6000 | 7000 | 8000 | 9000 |
|---|---|---|---|---|---|---|---|
| Sun | 0.003 | 0.007 | 0.017 | 0.035 | 0.061 | 0.099 | 0.15 |
| 10 Lac | 0.19 | 0.44 | 0.85 | 1.4 | 2.2 | 3.1 | 4.2 |

*Source:* K. H. Böhm, in J. L. Greenstein, ed., *Stellar Atmospheres*, Chicago: University of Chicago Press, 1960, p. 105; by permission.

that the radiative rates dominate (except in the red and infrared of 10 Lac, which is of small comfort since all the important transitions are in the ultraviolet). In the outermost layers, where $J_\nu \neq B_\nu$, the populations of the levels are therefore being determined by processes that do not occur at the LTE rate. Again we are forced to conclude that we must simply solve the statistical equilibrium and the transfer equations in a self-consistent manner.

## 5-2. The Rate Equations

Let us now consider the equations by which we calculate the actual occupation numbers of the atoms. The basic procedure is to consider from a microscopic point of view the rates of all processes by which an atom leaves a certain state $i$ to some other state $j$ (either bound or free) and return rates of processes from all states $j$ back to state $i$. Then the rate of change of the population of state $i$ is simply

$$\frac{dN_i}{dt} = \sum_{j \neq i} N_j P_{ji} - N_i \sum_{j \neq i} P_{ij} \tag{5-12}$$

One such equation exists for each state. In general the rates $P_{ij}$ and $P_{ji}$ will contain both radiative and collisional terms, i.e., we may write

$$P_{ij} = R_{ij} + C_{ij} \tag{5-13}$$

where $R_{ij}$ and $C_{ij}$ are the radiative and collisional rates respectively, (per atom) from level $i$ to level $j$. In general, both free and bound states will consist of sublevels (in the former case because the energies of free particles are unquantized and in the latter because of perturbations of the eigenstates by various interactions; see Chapter 9). Thus the rate written in equation (5-13) will involve integrals over a range of particle or photon energies and contain appropriate cross-sections or profiles.

If we now require the atmosphere to be in a *steady state*, then we have simply $dN_i/dt = 0$ so that

$$\sum_{j \neq i} N_j P_{ji} - N_i \sum_{j \neq i} P_{ij} = 0 \tag{5-14}$$

These equations may be written more concisely in terms of net rates. We may define a *net radiative rate*

$$\mathscr{R}_{ij} \equiv N_i R_{ij} - N_j R_{ji} \tag{5-15}$$

and a *net collision rate*

$$\mathscr{C}_{ij} \equiv N_i C_{ij} - N_j C_{ji} \tag{5-16}$$

which simply measure the excess of the number of transitions out of level $i$ over those into level $i$. Then, steady state requires

$$\sum_{j \neq i} \mathscr{R}_{ij} + \sum_{j \neq i} \mathscr{C}_{ij} + \mathscr{R}_{i\kappa} + \mathscr{C}_{i\kappa} = 0 \tag{5-17}$$

where the subscript $\kappa$ denotes the continuum. Let us now consider the detailed form of the various rates entering the equations. In what follows, we shall assume that only *one* ionization stage is possible (thus referring particularly to hydrogen) to simplify the analysis; when several ionization states co-exist, the equations are similar but in general involve $b$-factors for the various continua.

## RADIATIVE RATES

(a) Bound-Bound Transitions. Consider transitions from a bound level $i$ to a higher bound level $j$; for simplicity assume that the emission profile is identical to the absorption profile $\varphi_\nu$. The radiative rates can be expressed directly in terms of the Einstein coefficients. Thus, the rate of direct absorption is

$$N_i(\nu) R_{ij} \frac{d\omega}{4\pi} = N_i B_{ij} I_\nu \varphi_\nu \frac{d\omega}{4\pi} \tag{5-18}$$

Integrating over all frequencies and angles, we have the total absorption rate

$$N_i R_{ij} = N_i B_{ij} \int \varphi_\nu J_\nu \, d\nu = b_i N_i^* B_{ij} \int \varphi_\nu J_\nu \, d\nu \tag{5-19}$$

The total rate of stimulated emissions from level $j$ to level $i$ is

$$N_j B_{ji} \int \varphi_\nu J_\nu \, d\nu = b_j N_j^* B_{ji} \int \varphi_\nu J_\nu \, d\nu \tag{5-20}$$

(here we have explicitly assumed that the absorption and emission profiles are identical, i.e., complete redistribution; see Chapter 10). The rate of spontaneous emissions from level $j$ to $i$ is simply

$$N_j A_{ji} = b_j N_j^* A_{ji} \tag{5-21}$$

so that the total rate of radiative transition from level $j$ to level $i$ is

$$N_j R_{ji} = b_j N_j^* \left( A_{ji} + B_{ji} \int \varphi_\nu J_\nu \, d\nu \right) \tag{5-22}$$

Thus the net rate of radiative bound-bound transitions from level $i$ to level $j$ is

$$\begin{aligned} \mathscr{R}_{ij} &= (N_i B_{ij} - N_j B_{ji}) \int \varphi_\nu J_\nu \, d\nu - N_j A_{ji} \\ &= N_i^* B_{ij} \left( b_i - b_j \frac{N_j^* B_{ji}}{N_i^* B_{ij}} \right) \int \varphi_\nu J_\nu \, d\nu - N_j A_{ji} \end{aligned} \tag{5-23}$$

But

$$\frac{N_j^*}{N_i^*} = \frac{g_j}{g_i} e^{-h\nu_{ij}/kT}$$

and

$$g_j B_{ji} = g_i B_{ij}$$

so we may also write

$$\mathscr{R}_{ij} = N_i^* B_{ij}(b_i - b_j e^{-h\nu_{ij}/kT}) \int \varphi_\nu J_\nu \, d\nu - N_j A_{ji} \tag{5-24}$$

Let us now consider the spontaneous emission term in further detail to demonstrate the use of a *detailed balancing* argument. In LTE, spontaneous emissions must just balance absorptions—corrected for stimulated emissions. Thus, in LTE we would expect from detailed balancing that

$$N_j^* A_{ji} = N_i^* B_{ij} B_\nu (1 - e^{-h\nu_{ij}/kT}) \tag{5-25}$$

Let us verify this result by an alternative direct calculation. We recall from equations (4-2) and (4-11) that

$$A_{ji} = \frac{2h\nu_{ij}^3}{c^2} B_{ji} = \frac{2h\nu_{ij}^3}{c^2} \frac{g_i B_{ij}}{g_j}$$

Then we may rewrite the spontaneous emission rate as

$$N_j A_{ji} = N_j \frac{2h\nu_{ij}^3}{c^2} \frac{g_i B_{ij}}{g_j} = b_j \left( N_j^* \frac{g_i}{g_j} \right) \frac{2h\nu_{ij}^3}{c^2} B_{ij}$$

$$= b_j N_i^* e^{-h\nu_{ij}/kT} \frac{2h\nu_{ij}^3}{c^2} B_{ij} \tag{5-26}$$

But

$$B_\nu(1 - e^{-h\nu_{ij}/kT}) = \frac{2h\nu_{ij}^3}{c^2} e^{-h\nu_{ij}/kT}$$

so clearly equation (5-26) reduces to

$$N_j A_{ji} = b_j N_i^* B_{ij} B_\nu(1 - e^{-h\nu_{ij}/kT}) \tag{5-27}$$

confirming the detailed balancing argument. We will frequently use detailed balancing arguments because of their convenience, with the understanding that a careful direct analysis will always lead to the same result. Thus from equations (5-24) and (5-27), the alternative form for $\mathcal{R}_{ij}$ is

$$\mathcal{R}_{ij} = N_i^* \left[ B_{ij}(b_i - b_j e^{-h\nu_{ij}/kT}) \int \varphi_\nu J_\nu \, d\nu - b_j B_{ij}(1 - e^{-h\nu_{ij}/kT}) B_\nu \right] \tag{5-28}$$

Often it is convenient to replace the Einstein coefficients by line absorption cross-sections. If we write $l_{ij} = (\pi e^2/mc) f_{ij}$ as the integrated energy absorption coefficient per atom, then

$$l_{ij} \varphi_\nu = \frac{h\nu_{ij} B_{ij}}{4\pi} \varphi_\nu \tag{5-29}$$

so that we may write equation (5-28) in the equivalent form

$$\mathcal{R}_{ij} = \frac{N_i^* 4\pi l_{ij}}{h\nu_{ij}} \left[ (b_i - b_j e^{-h\nu_{ij}/kT}) \int \varphi_\nu J_\nu \, d\nu - b_j(1 - e^{-h\nu_{ij}/kT}) B_\nu \right] \tag{5-30}$$

This form is useful because of its formal resemblance to the corresponding expression in the continuum.

All of the expressions given above are for the net rate of excitation from a *lower* level $i$ to an *upper* level $j$. If we wish $\mathcal{R}_{ij}$ to give the net rate of de-excitation from level $i$ (now the upper level) to lower level $j$, then we have

$$\mathcal{R}_{ij} = -\mathcal{R}_{ji} \tag{5-31}$$

But $\mathcal{R}_{ji}$ can be written immediately from the above expressions for net excitation rates merely by interchanging $i$ and $j$. Thus, equivalent forms for $\mathcal{R}_{ij}$

when $E_i > E_j$ are

$$\begin{aligned}
\mathscr{R}_{ij} &= -N_j^* B_{ji}\left[(b_j - b_i e^{-h\nu_{ij}/kT}) \int \varphi_\nu J_\nu \, d\nu - b_i B_\nu(1 - e^{-h\nu_{ij}/kT})\right] \\
&= -N_i^* \left(\frac{g_j e^{h\nu_{ij}/kT}}{g_i}\right) B_{ji} \\
&\quad \times \left[(b_j - b_i e^{-h\nu_{ij}/kT}) \int \varphi_\nu J_\nu \, d\nu - b_i B_\nu(1 - e^{-h\nu_{ij}/kT})\right] \\
&= -N_i^* \left(\frac{g_j e^{h\nu_{ij}/kT}}{g_i}\right) \frac{4\pi l_{ji}}{h\nu_{ij}} \\
&\quad \times \left[(b_j - b_i e^{-h\nu_{ij}/kT}) \int \varphi_\nu J_\nu \, d\nu - b_i B_\nu(1 - e^{-h\nu_{ij}/kT})\right] \qquad (5\text{-}32)
\end{aligned}$$

We can see clearly from equation (5-32) that the net rate of de-excitation per atom in the upper state is simply $(g_j/g_i) \exp(h\nu_{ij}/kT)$ times the net excitation rate per atom in the lower state.

An alternative form of the net radiative bound-bound rate first introduced by Thomas is very useful. Here we write the rate explicitly in terms of the *upper* level (de-excitation rate). Thus

$$\begin{aligned}
\mathscr{R}_{ji} &= N_j A_{ji} + N_j B_{ji} \int \varphi_\nu J_\nu \, d\nu - N_i B_{ij} \int \varphi_\nu J_\nu \, d\nu \\
&= N_j A_{ji}\left[1 + \left(\frac{B_{ji} - N_i B_{ij}/N_j}{A_{ji}}\right) \int \varphi_\nu J_\nu \, d\nu\right] \qquad (5\text{-}33)
\end{aligned}$$

We may now recall equation (4-15) for the line source function $S_{ij}$, namely,

$$S_{ij} = \frac{A_{ji}}{(N_i B_{ij}/N_j) - B_{ji}}$$

Then, substitution into equation (5-33) yields simply

$$\mathscr{R}_{ji} = N_j A_{ji}\left(1 - \frac{\int \varphi_\nu J_\nu \, d\nu}{S_{ij}}\right) \equiv N_j A_{ji} Z_{ji} \qquad (5\text{-}34)$$

The term $Z_{ji}$ is called the *net radiative bracket* of the line transition *i* to *j*. We shall employ this notation in further developments in Chapters 12, 13, and 14. Besides being a very concise notational device, the net radiative bracket has certain advantages. One of these is that because the bracket involves only a ratio of $\int \varphi_\nu J_\nu \, d\nu$ and $S_l$, it is often true that the bracket can be computed to a much higher order of accuracy in an iterative procedure than either $S_l$ or $\int \varphi_\nu J_\nu \, d\nu$ themselves. Under favorable conditions, this may significantly enhance the convergence of the solution in multilevel problems.

(b) Bound-Free Transitions. Let us now calculate the net radiative rate from bound level $i$ to the continuum $\kappa$. Let $\alpha_\nu$ be the photoionization cross-section of an atom in state $k$, First, the number of direct photoionizations is simply

$$N_i R_{i\kappa} = 4\pi N_i \int_{\nu_0}^{\infty} \frac{\alpha_\nu J_\nu \, d\nu}{h\nu} = 4\pi N_i^* b_i \int_{\nu_0}^{\infty} \frac{\alpha_\nu J_\nu \, d\nu}{h\nu} \tag{5-35}$$

The induced emission rate can be written by direct analogy to the bound-bound case, equation (5-20), as

$$4\pi N_i^* b_\kappa \int_{\nu_0}^{\infty} \frac{\alpha_\nu J_\nu}{h\nu} e^{-h\nu/kT} \, d\nu = 4\pi N_i^* \int_{\nu_0}^{\infty} \frac{\alpha_\nu J_\nu}{h\nu} e^{-h\nu/kT} \, d\nu \tag{5-36}$$

where we have set $b_\kappa - 1$ since by definition the (single) continuum is in LTE. (If several ionization states exist, one would in general need to introduce appropriate $b$-factors describing these states.) Finally, the spontaneous rate can be obtained either by direct analogy with the bound-bound rate, equation (5-26), as

$$4\pi N_i^* b_\kappa \int_{\nu_0}^{\infty} \frac{2h\nu^3}{c^2} e^{-h\nu/kT} \frac{\alpha_\nu}{h\nu} \, d\nu = 4\pi N_i^* \int_{\nu_0}^{\infty} \frac{\alpha_\nu B_\nu}{h\nu} (1 - e^{-h\nu/kT}) \, d\nu \tag{5-37}$$

or the same result can be obtained from detailed balancing arguments, realizing that the spontaneous captures (which are in LTE) must equal the LTE rate of photoionizations corrected for stimulated emission. Combining equations (5-36) and (5-37), we find the total recombination rate is

$$N_i^* R_{\kappa i} = 4\pi N_i^* \left\{ \int_{\nu_0}^{\infty} \frac{\alpha_\nu}{h\nu} [B_\nu (1 - e^{-h\nu/kT}) + J_\nu e^{-h\nu/kT}] \, d\nu \right\} \tag{5-38}$$

Thus, the net radiative bound-free rate is

$$\mathscr{R}_{i\kappa} = 4\pi N_i^* \left\{ b_i \int_{\nu_0}^{\infty} \frac{\alpha_\nu J_\nu}{h\nu} \, d\nu - \int_{\nu_0}^{\infty} \frac{\alpha_\nu}{h\nu} [B_\nu (1 - e^{-h\nu/kT}) + J_\nu e^{-h\nu/kT}] \, d\nu \right\} \tag{5-39}$$

If we define $d_i \equiv b_i - 1$, then we may also write

$$\mathscr{R}_{i\kappa} = 4\pi N_i^* \left\{ d_i \int_{\nu_0}^{\infty} \frac{\alpha_\nu J_\nu}{h\nu} \, d\nu + \int_{\nu_0}^{\infty} \frac{\alpha_\nu}{h\nu} (J_\nu - B_\nu)(1 - e^{-h\nu/kT}) \, d\nu \right\} \tag{5-40}$$

### COLLISION RATES

In addition to radiative excitations and ionizations, the same processes may occur collisionally. The material in stellar atmospheres is a plasma consisting of atoms, ions, and electrons, among which a wide variety of collision processes may occur. Since the primary concern of this book is early-type stars, we may confine our attention to collisions of an atom with ions and electrons and neglect atom-atom collisions, since the former are more effective by virtue of the long-range nature of coulomb interactions. Moreover, since the total collision frequency is proportional to the velocity of the colliding particles and in thermal equilibrium the ratio of electron to ion velocities will be of the order $(m_p/m_e)^{1/2} \approx 43$, we normally need to consider only collisions with electrons. If we denote the collisional excitation cross-section for an electron of velocity $v$ as $\sigma_{ij}(v)$, then the number of collisional excitations from level $i$ to level $j$ is

$$N_i C_{ij} = N_i N_e \int_{v_0}^{\infty} \sigma_{ij}(v) f(v) v \, dv = b_i N_i^* N_e \int_{v_0}^{\infty} \sigma_{ij}(v) f(v) v \, dv \tag{5-41}$$

Similarly, the de-excitation rate can be written in terms of a cross-section $\sigma_{ji}(v)$, and we have

$$N_j C_{ji} = N_j N_e \int_0^{\infty} \sigma_{ji}(v) f(v) v \, dv = b_j N_j^* N_e \int_0^{\infty} \sigma_{ji}(v) f(v) v \, dv \tag{5-42}$$

However, we can eliminate $\sigma_{ji}$ in terms of $\sigma_{ij}$ since by detailed balancing we must have

$$N_i^* N_e \int_{v_0}^{\infty} \sigma_{ij}(v) f(v) v \, dv = N_j^* N_e \int_0^{\infty} \sigma_{ji}(v) f(v) v \, dv \tag{5-43}$$

Thus, the net collision rate from level $i$ to level $j$ is

$$\mathscr{C}_{ij} = (b_i - b_j) N_i^* N_e \int_{v_0}^{\infty} \sigma_{ij}(v) f(v) v \, dv = (b_i - b_j) N_i^* N_e \Omega_{ij}(T) \tag{5-44}$$

where we have defined

$$\Omega_{ij}(T) \equiv \int_{v_0}^{\infty} \sigma_{ij}(v) f(v) v \, dv \tag{5-45}$$

Strictly speaking, the above definition of $\Omega$ applies only to excitations (i.e., $E_i < E_j$). For de-excitation we may write the rate in the same form but with the understanding that when $E_i > E_j$, we actually use the relation

$$\Omega_{ij}(T) = \frac{g_i}{g_j} \exp(h\nu_{ij}/kT) \Omega_{ji}(T) \tag{5-46}$$

which follows from the detailed balancing result that

$$N_i^* N_e \Omega_{ij}(T) = N_j^* N_e \Omega_{ji}(T) \tag{5-47}$$

If we again write $d_i = b_i - 1$, clearly

$$\mathscr{C}_{ij} = (d_i - d_j) N_i^* N_e \Omega_{ij}(T) \tag{5-48}$$

In a completely analogous way, if we write the cross-section for collisional ionization as $\sigma_{i\kappa}(v)$ and define

$$\Omega_{i\kappa}(T) = \int_{v_0}^{\infty} \sigma_{i\kappa}(v) f(v) v \, dv \tag{5-49}$$

then the net rate of collisional ionization is

$$\mathscr{C}_{i\kappa} = (b_i - 1) N_i^* N_e \Omega_{i\kappa}(T) = d_i N_i^* N_e \Omega_{i\kappa}(T) \tag{5-50}$$

where we have made use of the fact that $b_\kappa \equiv 1$.

Let us now examine the form of $\Omega(T)$ in a bit more detail. Normally, cross-sections $\sigma(v)$ are measured in units of $\pi a_0^2$ where $a_0$ is the Bohr radius, i.e., we write $\sigma_{ij} = \pi a_0^2 Q_{ij}$. The velocity distribution is assumed to be Maxwellian, so

$$v f(v) \, dv = 4\pi v^3 \left(\frac{m}{2\pi kT}\right)^{3/2} \exp\left(-\frac{mv^2}{2kT}\right) dv \tag{5-51}$$

Since cross-sections are more commonly tabulated in terms of the energy $E$ of the exciting particle, it is convenient to convert to energy units. Writing $E = mv^2/2$ and converting equation (5-51), we have

$$v f(v) \, dv = \left(\frac{8}{m\pi}\right)^{1/2} (kT)^{-3/2} E \exp(-E/kT) \, dE \tag{5-52}$$

Then substitution into equation (5-45) yields

$$\Omega_{ij}(T) = \pi a_0^2 \left(\frac{8}{m\pi}\right)^{1/2} (kT)^{-3/2} \int_{E_0}^{\infty} Q_{ij}(E) \exp(-E/kT) E \, dE \tag{5-53}$$

where $E_0$ is the threshold energy of the excitation. Now let $u = E/kT$; then

$$\Omega_{ij}(T) = \pi a_0^2 \left(\frac{8k}{m\pi}\right)^{1/2} T^{1/2} \int_{u_0}^{\infty} Q_{ij}(ukT) u e^{-u} \, du \tag{5-54}$$

The coefficient has the value

$$C_0 = \pi a_0^2 \left(\frac{8k}{m\pi}\right)^{1/2} = 5.465 \times 10^{-11}$$

Now let $x = u - u_0 = u - E_0/kT$; then we may write finally

$$\Omega_{ij}(T) = C_0 T^{1/2} \exp(-E_0/kT) \Gamma_{ij}(T) \tag{5-55}$$

where

$$\Gamma_{ij}(T) \equiv \int_0^{\infty} Q_{ij}(E_0 + xkT) e^{-x} (x + u_0) \, dx \tag{5-56}$$

The advantage of this method of writing the collision rate is that $\Gamma(T)$ is a relatively insensitive function of temperature, and the principal sensitivity to $T$ has been factored out in the term $T^{1/2} \exp(-E_0/kT)$.

The main problem in application is to obtain reliable values for the cross-section $Q_{ij}$. Considerable (and ever increasing) effort is being devoted to this task, and good progress has been made in the past few years. These cross-sections are important not only in astrophysics but also in plasma physics, high-speed aerodynamics, and many other areas; as a result, a large demand exists for them. Both experiment and theory may be used to obtain values for $Q(E)$. For hydrogen and helium, the experiments have yielded very reliable values for collisional excitation and ionization from the ground state while for excitations and ionizations from excited levels, one must generally rely upon theory. An excellent review of both theoretical and experimental methods of determining collision cross-sections has been given by Moiseiwitsch and Smith (*Rev. Mod. Phys.*, **40**, 238, 1968). A comprehensive bibliography of electron-collision cross-section data has been published by Kieffer (Ref. 22). A critical summary of cross-sections useful in astrophysical calculation has recently been given by Sampson (*Ap. J.*, **155**, 575, 1969).

## COMPLETE RATE EQUATIONS

Collecting together the results from above, we can now write the complete statistical equilibrium equation for level $i$ as

$$\begin{aligned}
b_i\Bigg[4\pi \int_{\nu_0}^{\infty} \frac{\alpha_\nu J_\nu}{h\nu}\, d\nu &+ \sum_{j>i} B_{ij} \int \varphi_\nu J_\nu \, d\nu + \sum_{j<i} \frac{g_j B_{ji}}{g_i} \\
&\times \left(\frac{2h\nu_{ij}^3}{c^2} + \int \varphi_\nu J_\nu \, d\nu\right) + N_e \sum_{j\neq i} \Omega_{ij} + N_e \Omega_{i\kappa}\Bigg] \\
&- \sum_{j<i} b_j \left(N_e \Omega_{ij} + \frac{g_j B_{ji}}{g_i} e^{h\nu_{ij}/kT} \int \varphi_\nu J_\nu \, d\nu\right) \\
&- \sum_{j>i} b_j \left[N_e \Omega_{ij} + B_{ij} e^{-h\nu_{ij}/kT} \left(\frac{2h\nu_{ij}^3}{c^2} + \int \varphi_\nu J_\nu \, d\nu\right)\right] \\
= 4\pi \Bigg[\int_{\nu_0}^{\infty} \frac{\alpha_\nu B_\nu}{h\nu} &(1 - e^{-h\nu/kT})\, d\nu + \int_{\nu_0}^{\infty} \frac{\alpha_\nu J_\nu}{h\nu} e^{-h\nu/kT}\, d\nu\Bigg] + N_e \Omega_{i\kappa}
\end{aligned} \tag{5-57}$$

Note that we have explicitly canceled out the population density $N_i^*$ from all terms. An equation of this form exists for each bound state of the atom. If the radiation field is known, then these equations are simply a set of equations linear in the $b$'s, which may be solved by the usual techniques of linear algebra. If one prefers to use line opacities instead of Einstein coefficients, then the equations become

$$
\begin{aligned}
b_i \Bigg[ 4\pi \int_{\nu_0}^{\infty} \frac{\alpha_\nu J_\nu}{h\nu}\, d\nu &+ 4\pi \sum_{j>i} \frac{l_{ij}}{h\nu_{ij}} \int \varphi_\nu J_\nu \, d\nu + 4\pi \sum_{j<i} \frac{g_j l_{ji}}{g_i h\nu_{ij}} \\
&\times \left( \frac{2h\nu_{ij}^3}{c^2} + \int \varphi_\nu J_\nu \, d\nu \right) + N_e \left( \sum_{j \neq i} \Omega_{ij} + \Omega_{i\kappa} \right) \Bigg] \\
&- \sum_{j<i} b_j \left[ N_e \Omega_{ij} + \frac{4\pi g_j l_{ji}}{g_i h\nu_{ij}} \exp(h\nu_{ij}/kT) \int \varphi_\nu J_\nu \, d\nu \right] \\
&- \sum_{j>i} b_j \left[ N_e \Omega_{ij} + \frac{4\pi l_{ij}}{h\nu_{ij}} \exp(-h\nu_{ij}/kT) \left( \frac{2h\nu_{ij}^3}{c^2} + \int \varphi_\nu J_\nu \, d\nu \right) \right] \\
&= 4\pi \left[ \int_{\nu_0}^{\infty} \frac{\alpha_\nu B_\nu}{h\nu} (1 - e^{-h\nu/kT})\, d\nu + \int_{\nu_0}^{\infty} \frac{\alpha_\nu J_\nu}{h\nu} e^{-h\nu/kT}\, d\nu \right] + N_e \Omega_{i\kappa}
\end{aligned}
\tag{5-58}
$$

The bound-bound rates appearing in these equations can be written much more concisely in terms of net radiative brackets. In this form we find

$$
\begin{aligned}
b_i \Bigg[ 4\pi \int_{\nu_0}^{\infty} \frac{\alpha_\nu J_\nu}{h\nu}\, d\nu &+ \sum_{j<i} A_{ij} Z_{ij} + N_e \left( \sum_{j \neq i} \Omega_{ij} + \Omega_{i\kappa} \right) \Bigg] \\
&- \sum_{j<i} b_j N_e \Omega_{ij} - \sum_{j>i} b_j \left[ N_e \Omega_{ij} + \frac{g_j}{g_i} A_{ji} Z_{ji} \exp(-h\nu_{ij}/kT) \right] \\
&= 4\pi \left[ \int_{\nu_0}^{\infty} \frac{\alpha_\nu B_\nu}{h\nu} (1 - e^{-h\nu/kT})\, d\nu + \int_{\nu_0}^{\infty} \frac{\alpha_\nu J_\nu}{h\nu} e^{-h\nu/kT}\, d\nu \right] + N_e \Omega_{i\kappa}
\end{aligned}
\tag{5-59}
$$

Note however that equation (5-59) still involves as many unknown $b$-factors as either equation (5-57) or equation (5-58), some of them merely being hidden implicitly within the net radiative brackets.

Each of these forms has certain advantages and can be employed interchangeably as convenience dictates since all of them contain precisely the same physical information. All of the above forms of the statistical equilibrium equations can be solved immediately if the radiation field is known. As described earlier, however, the radiation field will depend in general upon the occupation numbers so that both the transfer equation and the statistical equilibrium equations must be solved either by iteration or by a simultaneous solution. We will discuss methods of treating this coupled problem in Chapters 7, 12, 13, and 14.

## 5-3. Solution in Limiting Cases

Before we leave the question of the formulation of the statistical equilibrium equations, it is very instructive to consider their solution in certain limiting cases. First, consider an imaginary atom consisting of a single-bound level and a continuum. For such an atom, equations (5-57) through (5-59) reduce simply to

$$b_i\left(4\pi \int_{\nu_0}^{\infty} \frac{\alpha_\nu J_\nu}{h\nu}\, d\nu + N_e \Omega_{i\kappa}\right)$$

$$= 4\pi \int_{\nu_0}^{\infty} \frac{\alpha_\nu}{h\nu} [B_\nu(1 - e^{-h\nu/kT}) + J_\nu e^{-h\nu/kT}]\, d\nu + N_e \Omega_{i\kappa} \tag{5-60}$$

or

$$b_i = \frac{4\pi \int_{\nu_0}^{\infty} \frac{\alpha_\nu}{h\nu} [B_\nu(1 - e^{-h\nu/kT}) + J_\nu e^{-h\nu/kT}]\, d\nu + N_e \Omega_{i\kappa}}{4\pi \int_{\nu_0}^{\infty} \frac{\alpha_\nu J_\nu}{h\nu}\, d\nu + N_e \Omega_{i\kappa}} \tag{5-61}$$

From equation (5-61) we may observe several interesting results. First, in the limit of very high densities, so high that the collision rates are much larger than the radiative rates, we see that

$$\lim_{N_e \to \infty} b_i = \lim_{N_e \to \infty} \frac{N_e \Omega_{i\kappa}}{N_e \Omega_{i\kappa}} = 1 \tag{5-62}$$

i.e., in the limit of large densities, we recover LTE populations, as would be expected. Second, we note that at very great depth in the atmosphere we expect that $J_\nu \to B_\nu$. Then

$$\lim_{\tau_\nu \to \infty} b_i = \frac{4\pi \int_{\nu_0}^{\infty} \frac{\alpha_\nu B_\nu}{h\nu}\, d\nu + N_e \Omega_{i\kappa}}{4\pi \int_{\nu_0}^{\infty} \frac{\alpha_\nu B_\nu}{h\nu}\, d\nu + N_e \Omega_{i\kappa}} = 1 \tag{5-63}$$

so that when the radiation field is perfectly Planckian, we again recover LTE as expected. Two comments are necessary here, however: (a) If we are so deep in the atmosphere that *all* transitions (of a real atom) are extremely opaque, then a result of the form of equation (5-63) will apply for each transition and the atom as a whole is in LTE. If, however, *any* of the transitions are transparent, then we will not in general obtain complete LTE (unless the densities are sufficiently high). (b) The question of how opaque is "extremely opaque" is left unanswered. As we have indicated earlier in this chapter, it is *not* sufficient merely to have $\tau_\nu \gtrsim 1$; in general, very large optical depths are required. We shall give precise estimates in Chapters 7 and 12. Finally, consider equation (5-61) when stimulated emissions can be neglected, i.e., $h\nu/kT \gg 1$. This will be a good approximation for transitions whose binding energy is much larger than the mean thermal energy in the gas. Then in the extreme low-density situation, we have

$$\lim_{N_e \to 0} b_i = \lim_{N_e \to 0} \frac{4\pi \int_{\nu_0}^{\infty} \frac{\alpha_\nu B_\nu}{h\nu}\, d\nu + N_e \Omega_{i\kappa}}{4\pi \int_{\nu_0}^{\infty} \frac{\alpha_\nu J_\nu}{h\nu}\, d\nu + N_e \Omega_{i\kappa}} = \frac{\int_{\nu_0}^{\infty} \frac{\alpha_\nu B_\nu}{h\nu}\, d\nu}{\int_{\nu_0}^{\infty} \frac{\alpha_\nu J_\nu}{h\nu}\, d\nu} \tag{5-64}$$

In complete accordance with what we would expect physically, we see that when the recombination rate exceeds the photoionization rate, the level is overpopulated ($b_i > 1$), and when photoionization exceeds recombination, the level is underpopulated.

Let us now consider a bound-bound transition and consider the behavior of the net radiative bracket. Again at sufficiently great depth $S_\nu \to B_\nu$ and $J_\nu \to B_\nu$. Since the line profile function is sharply peaked, the variation of $B_\nu$ over the line can be neglected, so $\int \varphi_\nu J_\nu \, d\nu \to B_\nu$, and thus $Z_{ji} \to 0$. When $Z_{ji} = 0$ or, at least, is negligible compared with any other net rate from levels $i$ or $j$, we say that the bound-bound transition is in *radiative detailed balance*. At the surface, $Z_{ji}$ is nonzero. In LTE, $S_\nu$ again is $B_\nu$ while $\int \varphi_\nu J_\nu \, d\nu$ will be dominated by the line center value of $J_\nu$, which in an opaque line will be approximately $\frac{1}{2}B_\nu$. Thus in LTE, $Z_{ji}$ at the surface approaches 1/2, i.e., only 1/2 of the downward transitions are compensated by upward radiative transitions. On the other hand, for a line with a source function dominated by scattering, $S_\nu \to J_\nu$ quite closely and $Z_{ji}$ may be almost zero even at the surface (see Chapter 12).

We have now assembled all of the basic physical information required to calculate model atmospheres, at least for the continuum. Let us turn our attention first to LTE continuum models and then to non-LTE continuum models.

# 6 | LTE Model Atmospheres for the Continuum

We have now assembled all of the data required to carry out the program outlined in Chapter 1 for the computation of model atmospheres that will describe the formation of the continuum. Let us now consider the actual methods used in such computations, making the simplifying assumption of LTE; non-LTE models will be discussed in the next chapter.

## 6-1. The Equation of Hydrostatic Equilibrium

As mentioned earlier, we will assume that the atmosphere is in hydrostatic equilibrium so that the *total* pressure gradient is given by

$$\frac{dp}{dz} = -\rho g \tag{6-1}$$

or

$$\frac{dp}{d\tau} = \frac{g}{(\kappa_\nu + \sigma_\nu)_{\text{std}}} \tag{6-2}$$

where $\kappa$ and $\sigma$ are the mass absorption (corrected for stimulated emission) and scattering coefficients, respectively. These are to be either evaluated at some standard wavelength or chosen to be some mean value (Rosseland, Planck, etc.). The total pressure gradient consists of two parts: the actual gas pressure gradient, and the gradient of the pressure exerted by radiation upon the absorbing material in the atmosphere. Because of their very different nature, it is useful to distinguish between these two gradients and to write

$$\frac{dp_g}{d\tau} = \frac{g}{(\kappa_\nu + \sigma_\nu)_{\text{std}}} - \frac{dp_R}{d\tau} \tag{6-3}$$

Recalling equation (1-35), we see that we can write

$$\frac{dp_R}{d\tau} = \frac{\pi}{c} \frac{\int_0^\infty (\kappa_\nu + \sigma_\nu) F_\nu \, d\nu}{(\kappa_\nu + \sigma_\nu)_{\text{std}}} \tag{6-4}$$

Since the fluxes are initially unknown, one must devise some estimate of $dp_R/d\tau$ for the first integration; one might use the value $\sigma T_{\text{eff}}^4/c$, which would be valid if $(\kappa_\nu + \sigma_\nu)_{\text{std}}$ were the flux-weighted mean. Once a first model has been constructed, a revised value of $dp_R/d\tau$ can be obtained for use in subsequent calculations. Normally the $dp_R/d\tau$ term is important only when $g$ is small, $(\kappa_\nu + \sigma_\nu)_{\text{std}}$ is very large, or $T_{\text{eff}}$ is high. In these cases the radiation pressure gradient can substantially reduce the pressure gradient by partly overcoming the force of gravity. Indeed, if $dp_R/d\tau$ becomes sufficiently large or $g/(\kappa_\nu + \sigma_\nu)_{\text{std}}$ sufficiently small, the radiation pressure may dominate and lead to a negative gas pressure gradient. The conditions under which this occurs have been examined by Underhill (*M. N.*, **109**, 563, 1949). Generally speaking, the $dp_R/d\tau$ term will be much less than $g/(\kappa_\nu + \sigma_\nu)_{\text{std}}$ for most cases of interest, and usually the estimates converge very rapidly.

Since the pressure in the atmosphere runs over several orders of magnitude, it is convenient to use logarithmic units and to integrate

$$\frac{d \log p_g}{d\tau} = \frac{1}{p_g \ln 10} \left[ \frac{g}{(\kappa_\nu + \sigma_\nu)_{\text{std}}} - \frac{dp_R}{d\tau} \right] \tag{6-5}$$

STARTING VALUES

To carry through the integration of this equation, we assume that we know the run of $T(\tau)$. We now need to find a starting value. For high-temperature stars this is easily obtained since in the very outer layers the hydrogen is strongly ionized (since the pressure is low) so that $\kappa \ll \sigma_e = \sigma_T/\mu' m_H$. For $\tau_0 \ll 1$, we can assume therefore that $\sigma$ is nearly constant, and we thus obtain

$$p_g(\tau_0) = \tau_0 \left[ \frac{g\mu' m_H}{\sigma_T} - \left. \frac{dp_R}{d\tau} \right|_{\tau_0} \right] \tag{6-6}$$

for the starting value. For very cool stars where $H^-$ is the dominant opacity source, $\kappa$ is proportional to $p_e$ while $p_g = Ap_e$, where $A$ equals the hydrogen/metal ratio. Also, at low temperatures the term in $dp_R/d\tau$ is usually negligible; thus we may write

$$A \frac{dp_e}{d\tau} = \frac{g}{\kappa_0 p_e} \tag{6-7}$$

or

$$p_e = \left(\frac{2g\tau_0}{A\kappa_0}\right)^{1/2} \tag{6-8}$$

which implies

$$p_g(\tau_0) = \left(\frac{2Ag\tau_0}{\kappa_0}\right)^{1/2} \tag{6-9}$$

We have made some slightly restrictive assumptions in deriving equations (6-6) and (6-9). More general methods may be employed if desired (see, e.g., Ref. 3, p. 25, and Ref. 15, p. 56 ff.).

## CONTINUING THE INTEGRATION

We now have $T$, $p_g$, $p_e$, and $\kappa$ at $\tau = \tau_0$ and wish to take an integration step to $\tau_1 = \tau_0 + \Delta\tau$. To do this we could write

$$\Delta \log p_g = (\log p_g)_1 - (\log p_g)_0 = \left.\frac{d \log p_g}{d\tau}\right|_{\tau_0} \Delta\tau \tag{6-10}$$

This would give us $\log p_g$ at $\tau_1$, and this, with knowledge of $T$, yields $p_e$ and $(\kappa_\nu + \sigma_\nu)_{\text{std}}$ and hence $(d \log p_g/d\tau)_1$. If desired, an improved estimate of $(\log p_g)_1$ can be obtained by now writing

$$\Delta \log p_g = \frac{1}{2}\left(\left.\frac{d \log p_g}{d\tau}\right|_{\tau_0} + \left.\frac{d \log p_g}{d\tau}\right|_{\tau_0}\right) \Delta\tau \tag{6-11}$$

This procedure may be iterated until the value of $(\log p_g)_1$ used to compute the derivative is consistent with the value of $(\log p_g)_1$ yielded by the integration. In this way we can proceed step by step to calculate $p_g(\tau)$.

Actually the integration scheme outlined above is quite crude, and in practice one makes use of more accurate and elaborate numerical integration formulae. Thus, if $f_i$ represents the derivative at the $i$th point and if $y_i$ represents the dependent variable at the $i$th point and $\Delta x$ is the step size of the independent variable, then the value of $y_{i+1}$ can be estimated as

$$y_{i+1}^p = y_i + \frac{\Delta x}{24}(55f_i - 59f_{i-1} + 37f_{i-2} - 9f_{i-3}) \tag{6-12}$$

This "predicted" value of $y_{i+1}$ may be used to calculate the derivative $f_{i+1}^p = f(x_{i+1}, y_{i+1}^p)$ and a "corrected" value of higher accuracy can be obtained from the formula

$$y_{i+1}^c = y_i + \frac{\Delta x}{24}(9f_{i+1}^p + 19f_i - 5f_{i-1} + f_{i+2}) \tag{6-13}$$

Again, the procedure can be iterated until $y_{i+1}^c$ agrees with the value of $y$ used to compute $f_{i+1}$ to some desired accuracy. A wide variety of alternative formulae may also be employed (see Ref. 15, p. 56 ff.).

Having found the run of $p_g$ and $p_e$ with depth, we can now compute all other physical variables, such as, $\rho(\tau)$, $f_{\mathrm{H}}(\tau)$, $f_{\mathrm{H}^+}(\tau)$, $\kappa_\nu(\tau)$, $\sigma_\nu(\tau)$, and so on, since by the assumption of LTE, these quantities depend only upon $T(\tau)$ and $p_e(\tau)$. Having $\kappa_\nu$ and $\sigma_\nu$, we may now construct the optical depth scale at each frequency by calculating the integral

$$\tau_\nu(\tau) = \int_0^\tau \frac{\kappa_\nu(t) + \sigma_\nu(t)}{[\kappa_\nu(t) + \sigma_\nu(t)]_{\mathrm{std}}}\, dt \tag{6-14}$$

using, for example, Simpson's rule.

We may also determine the actual physical depth in the atmosphere by integrating

$$z(b) - z(a) = -\int_{p_g(a)}^{p_g(b)} \left[\rho g - \rho(\kappa_\nu + \sigma_\nu)_{\mathrm{std}} \frac{dp_R}{d\tau}\right]^{-1} dp_g \tag{6-15}$$

In short, an assumed $T(\tau)$ allows us to obtain the variation with depth of all the usual thermodynamic properties of the atmosphere.

## 6-2. Solution of the Transfer Equation

Having obtained opacities and optical depths at each frequency, we may now solve the equation of transfer. For simplicity, let us consider here just the case of a source function with an LTE thermal emission component plus an isotropic coherent scattering component so that from equation (1-43) we have

$$S_\nu = \frac{\kappa_\nu}{\kappa_\nu + \sigma_\nu} B_\nu + \frac{\sigma_\nu}{\kappa_\nu + \sigma_\nu} J_\nu \equiv (1 - \rho_\nu)B_\nu + \rho_\nu J_\nu$$

and from equation (1-40) we write

$$\mu \frac{dI_\nu}{d\tau_\nu} = I_\nu - S_\nu$$

Here, as was shown in Section 4-3, $\kappa_\nu$ denotes the sum over all mass absorption coefficients, as given in Chapter 4, that contribute at frequency $\nu$, corrected by the LTE stimulated emission factor $(1 - e^{-h\nu/kT})$. Note that by the nature of the physical process, there is *no* correction factor for the scattering term $\sigma_\nu$. The solution of this pair of equations may be obtained by several techniques. In this section we shall consider the properties of some of the more important methods that have been developed.

## INTEGRAL EQUATION METHODS—ITERATION

The solution of the transfer equation can be obtained directly from equations (1-62) and (1-63) and may be written

$$J_\nu(\tau_\nu) = \Lambda_{\tau_\nu}[S_\nu(t_\nu)] = \Lambda_{\tau_\nu}[B_\nu] + \Lambda_{\tau_\nu}[\rho_\nu(J_\nu - B_\nu)] \tag{6-16}$$

Thus $J_\nu(\tau_\nu)$ is the solution of an *integral equation.* This solution may be effected in several different ways; one of the most common methods is to solve the equation by iteration. Since we know that in LTE $J_\nu \to B_\nu$ as $\tau_\nu \to \infty$, it is expedient to write

$$(J_\nu - B_\nu) = \bar{B}_\nu - B_\nu + \Lambda_{\tau_\nu}[\rho_\nu(J_\nu - B_\nu)] \tag{6-17}$$

where $\bar{B}_\nu(\tau_\nu) \equiv \Lambda_{\tau_\nu}[B_\nu]$. Now if $\rho_\nu$ were everywhere zero, the value of $(J_\nu - B_\nu)$ would be simply $(\bar{B}_\nu - B_\nu)$. When $\rho_\nu$ is not zero, we may regard this as a first approximation for $(J_\nu - B_\nu)$ and calculate an improved estimate as

$$\begin{aligned} (J_\nu - B_\nu)^{(1)} &= (\bar{B}_\nu - B_\nu) + \Lambda_{\tau_\nu}[\rho_\nu(J_\nu - B_\nu)^{(0)}] \\ &= (\bar{B}_\nu - B_\nu) + \Lambda_{\tau_\nu}[\rho_\nu(\bar{B}_\nu - B_\nu)] \\ &= (\bar{B}_\nu - B_\nu) + \Delta^{(1)} \end{aligned} \tag{6-18}$$

This, in turn, may be substituted back into the $\Lambda$-operator to derive a new estimate

$$\begin{aligned} (J_\nu - B_\nu)^{(2)} &= (\bar{B}_\nu - B_\nu) + \Lambda_{\tau_\nu}[\rho_\nu(J_\nu - B_\nu)^{(1)}] \\ &= (\bar{B}_\nu - B_\nu) + \Delta^{(1)} + \Lambda_{\tau_\nu}[\rho_\nu\Delta^{(1)}] \\ &= (\bar{B}_\nu - B_\nu) + \Delta^{(1)} + \Delta^{(2)} \end{aligned} \tag{6-19}$$

In this way we may proceed iteratively to obtain

$$(J_\nu - B_\nu)^{(i)} = (\bar{B}_\nu - B_\nu) + \sum_{j=1}^{i} \Delta^{(j)} \tag{6-20}$$

where

$$\Delta^{(j)} \equiv \Lambda_{\tau_\nu}[\rho_\nu \Delta^{(j-1)}]$$

and $\Delta^{(1)}$ is defined above. The usual criterion of convergence is that

$$\left| \frac{2\Delta^{(j)}}{(J_\nu - B_\nu)^{(j)} + (J_\nu - B_\nu)^{(j-1)}} \right| \le \varepsilon \tag{6-21}$$

over the entire range of optical depths; $\varepsilon$ may be chosen as small as the accuracy of the quadrature formula for the $\Lambda$-operator and the limits decided upon for computing time will permit. Typically, we might choose $\varepsilon \sim 10^{-3}$ to $10^{-4}$. Methods for enhancing the convergence of the iteration cycle have been suggested (see, e.g., Ref. 3, pp. 29–30).

It should be stresseed that satisfaction of the criterion given in equation (6-21) does not necessarily guarantee that the final value of $(J_\nu - B_\nu)$ is accurate to the tolerance $\varepsilon$. This is because the sequence $\{(J_\nu - B_\nu)^{(i)}\}$ usually approaches its final value *monotically*. Therefore, while the fractional difference between successive iterations may be $\varepsilon$, there is no guarantee that, say, $1/\varepsilon$ more iterations may not actually be required. This is an extremely important point that must be understood completely. To this end, let us consider the following simplified analysis.

Suppose that the Planck function distribution can be represented with sufficient accuracy by a linear expansion

$$B_\nu(\tau_\nu) = a_\nu + b_\nu \tau_\nu \tag{6-22}$$

and suppose, in addition, that the parameter $\rho_\nu = \sigma_\nu/(\kappa_\nu + \sigma_\nu)$ is constant with depth. Now from the zero-order moment of the transfer equation we have

$$\frac{dH_\nu}{d\tau_\nu} = J_\nu - S_\nu = (1 - \rho_\nu)(J_\nu - B_\nu) = \lambda_\nu(J_\nu - B_\nu) \tag{6-23}$$

where we have substituted equation (1-43) for $S_\nu$. Similarly, the first-order moment yields

$$\frac{dK_\nu}{d\tau_\nu} = H_\nu \tag{6-24}$$

If we now make the Eddington approximation that $J_\nu = 3K_\nu$, then we may substitute equation (6-24) into equation (6-23) to obtain

$$\frac{1}{3}\frac{d^2J_\nu}{d\tau_\nu^{\,2}} = \lambda_\nu(J_\nu - B_\nu) \tag{6-25}$$

and since we have assumed equation (6-22) is valid, we have

$$\frac{1}{3}\frac{d^2(J_\nu - B_\nu)}{d\tau_\nu^{\,2}} = \lambda_\nu(J_\nu - B_\nu) \tag{6-26}$$

The solution of equation (6-26) may be written directly as

$$J_\nu - B_\nu = \alpha_\nu \exp(-\sqrt{3\lambda_\nu}\,\tau_\nu) + \beta_\nu \exp(\sqrt{3\lambda_\nu}\,\tau_\nu) \tag{6-27}$$

and since we demand $J_\nu \to B_\nu$ as $\tau_\nu \to \infty$, we must have $\beta_\nu \equiv 0$ so that

$$J_\nu(\tau_\nu) = a_\nu + b_\nu \tau_\nu + \alpha_\nu \exp(-\sqrt{3\lambda_\nu}\,\tau_\nu) \tag{6-28}$$

To evaluate $\alpha_\nu$ we make use of the boundary condition $J_\nu(0) = \sqrt{3}\,H_\nu(0)$ suggested by the gray solution [equation (2-106)]. Thus, from equation (6-24) we have

$$H_\nu = \frac{1}{3}\frac{dJ_\nu}{d\tau_\nu} = \frac{1}{3}[b_\nu - \alpha_\nu\sqrt{3\lambda_\nu}\exp(-\sqrt{3\lambda_\nu}\,\tau_\nu)] \tag{6-29}$$

so that applying the boundary condition, we may write

$$J_\nu(0) = a_\nu + \alpha_\nu = \sqrt{3}\,H_\nu(0) = \frac{1}{\sqrt{3}}(b_\nu - \alpha_\nu\sqrt{3\lambda_\nu}) \tag{6-30}$$

from which we obtain

$$\alpha_\nu = \frac{b_\nu - \sqrt{3}\,a_\nu}{\sqrt{3}(1 + \sqrt{\lambda_\nu})} \tag{6-31}$$

and finally

$$J_\nu(\tau_\nu) = a_\nu + b_\nu\tau_\nu + \left[\frac{b_\nu - \sqrt{3}\,a_\nu}{\sqrt{3}(1 + \sqrt{\lambda_\nu})}\right]\exp(-\sqrt{3\lambda_\nu}\,\tau_\nu) \tag{6-32}$$

We may now obtain some crucial insight into the problem by examination of equation (6-32). To begin with, we note that $J_\nu(\tau_\nu) \rightarrow B_\nu(\tau_\nu)$ only at depths $\tau_\nu \gtrsim 1/\sqrt{\lambda_\nu}$. When $J_\nu$ has approached $B_\nu$ arbitrarily closely, we say that the solution has *thermalized*; we may thus refer to the depth $1/\sqrt{\lambda_\nu}$ as a *thermalization depth*. We shall find this concept to be extremely useful and will refer to it again in Chapters 7, 11, and 12 in other contexts. We may obtain an intuitive understanding of the thermalization depth from the following physical argument.

The parameter $\lambda_\nu = \kappa_\nu/(\kappa_\nu + \sigma_\nu)$ is simply a measure of the probability that in any given absorption and emission the photon is destroyed and converted into thermal energy: Between each absorption and emission, the photon random-walks over an optical depth $\Delta\tau$. If the probability of destruction is $\lambda_\nu$, then we expect the photon to be absorbed and re-emitted of the order $n = 1/\lambda_\nu$ times before it thermalizes. Thus, the total optical thickness through which the photon can progress by a random-walk before it thermalizes is simply $n^{1/2}\Delta\tau = \Delta\tau/\sqrt{\lambda_\nu} \approx 1/\sqrt{\lambda_\nu}$ since we expect the mean path length $\Delta\tau$ to be of order unity. We thus see that when scattering terms dominate the solution, the mean intensity may depart from the thermal source term over large optical depths in the atmosphere (we have previously alluded to this fact in Chapter 5). Moreover, the departure may be large. In particular, suppose that in equation (6-22), $b_\nu = 0$, so $B_\nu(\tau_\nu) \equiv a_\nu$. Then from equation (6-32) we find that

$$J_\nu(0) = \frac{\sqrt{\lambda_\nu}}{1 + \sqrt{\lambda_\nu}} a_\nu = \frac{\sqrt{\lambda_\nu}}{1 + \sqrt{\lambda_\nu}} B_\nu \tag{6-33}$$

Thus, when $\lambda_\nu$ is very small, $J_\nu$ may be very much smaller than $B_\nu$ over very large depths in the atmosphere.

These facts show clearly that when the scattering terms dominate, the iteration method of solving equation (6-16) will fail when we start with the estimate $J_\nu \approx \bar{B}_\nu$. The reason is that each Λ-iteration will propagate information about the departure of $J_\nu$ from $\bar{B}_\nu$ only over a depth of order $\Delta\tau \approx 1$ (recall the discussion in Section 2-3). Thus, we may have to perform of order $1/\sqrt{\lambda_\nu}$ iterations to obtain the correct solution, which clearly is prohibitive when $\lambda_\nu \ll 1$. In this case, one must employ one of the alternative methods described below.

Let us mention some of the physical circumstances where we may find $\lambda_\nu \ll 1$. First, in the atmospheres of very hot stars, essentially all the atoms are ionized, and the principal opacity source is electron scattering, with $\rho_\nu$ essentially unity throughout the entire atmosphere. Indeed, values of $\lambda_\nu$ of order $10^{-3}$ or even $10^{-4}$ are not unusual. In these cases the iteration procedure simply fails. Another situation of this kind occurs in the atmospheres of very cool stars. Here all of the hydrogen may be essentially neutral to fairly great depths in the atmosphere and become excited and ionized only at depth. Then in the ultraviolet, the opacity can be dominated by Rayleigh scattering (both by H and $H_2$), and $\rho_\nu$ is essentially unity to great depth. At some point the hydrogen rather abruptly begins to become excited and ionized, and $\rho_\nu$ suddenly drops. While the initial estimate $J_\nu = \bar{B}_\nu$ is satisfactory for $\tau_\nu > \tau^*$ beyond the point where this sudden drop occurs, it is completely unreliable for $\tau_\nu < \tau^*$. If $\tau^* \gg 1$ (as is often the case), the iteration procedure is again doomed to failure.

### INTEGRAL EQUATION METHODS—DIRECT SOLUTION BY LINEAR EQUATIONS

As we have seen above, there are situations of astrophysical interest where the solution of equation (6-16) by iteration fails. An alternative approach is to approximate the run of the source function by an interpolation formula that is valid between a discrete set of depth points. Then, the integral operator can be replaced by a sum, and the integral equation can be reduced to a set of linear algebraic equations, which may then be solved by standard numerical methods. This approach has been employed by a number of authors, e.g., Avrett (Ref. 16, p. 63) and Gebbie (*M. N.*, **135**, 181, 1967). We shall describe here the particular formulation given by Kurucz (*Ap. J.*, **156**, 235, 1969).

Suppose that we regard the integration in the Λ-operator as the sum of integrals taken over $N$ discrete intervals and write (dropping the subscript $\nu$ for brevity)

$$J_l = J(\tau_l) = \frac{1}{2} \sum_{j=1}^{N} \int_{\tau_j}^{\tau_{j+1}} S(t) E_1 |t - \tau_1| \, dt \tag{6-34}$$

Now assume that on the interval $(\tau_j, \tau_{j+1})$, $S(t)$ can be represented by a quadratic interpolation formula of the form

$$S(t) = \sum_{k=1}^{3} t^{k-1} \sum_{i=1}^{N} C_{jki} S_i \tag{6-35}$$

where the $C_{jki}$'s are simply interpolation coefficients involving only the $\tau_i$'s. One might take these coefficients to be, for example, the average of the coefficients in the forward-parabolic and backward-parabolic formulae (see the Kurucz reference for details). Inserting equation (6-35) into equation (6-34), we see that the $j$th term of the sum may be written

$$\begin{aligned} J_{lj} &= \frac{1}{2} \int_{\tau_j}^{\tau_{j+1}} dt \, E_1 |t - \tau_l| \sum_{k=1}^{3} t^{k-1} \sum_{i=1}^{N} C_{jki} S_i \\ &= \sum_{k=1}^{3} \eta_{ljk} \sum_{i=1}^{N} C_{jki} S_i \end{aligned} \tag{6-36}$$

where

$$\eta_{ljk} \equiv \frac{1}{2} \int_{\tau_j}^{\tau_{j+1}} t^{k-1} E_1 |t - \tau_l| \, dt \tag{6-37}$$

This integral may be evaluated analytically to yield

$$\begin{aligned} \eta_{ljk} = \tfrac{1}{2} \big| &(\tau_j^{k-1} E_2 |\tau_l - \tau_j| - \tau_{j+1}^{k-1} E_2 |\tau_l - \tau_{j+1}|) \\ &+ (k-1)(\tau_j^{k-2} E_3 |\tau_l - \tau_j| - \tau_{j+1}^{k-2} E_3 |\tau_l - \tau_{j+1}|) \\ &+ (k-1)(k-2)(\tau_j^{k-3} E_4 |\tau_l - \tau_j| - \tau_{j+1}^{k-3} E_4 |\tau_l - \tau_{j+1}|) \big| \end{aligned} \tag{6-38}$$

Thus we may write finally

$$J_l = \sum_{j=1}^{N} \sum_{k=1}^{3} \eta_{ljk} \sum_{i=1}^{N} C_{jki} S_i = \sum_{i=1}^{N} \Lambda_{li} S_i \tag{6-39}$$

where

$$\Lambda_{li} \equiv \sum_{j=1}^{N} \sum_{k=1}^{3} \eta_{ljk} C_{jki} \tag{6-40}$$

Then we may write, for the problem under consideration,

$$S_l = (1 - \rho_l) B_l + \rho_l J_l = (1 - \rho_l) B_l + \rho_l \sum_{i=1}^{N} \Lambda_{li} S_i \tag{6-41}$$

or in the matrix notation

$$\mathbf{S} = (\mathbf{I} - \boldsymbol{\rho})\mathbf{B} + \boldsymbol{\rho}\boldsymbol{\Lambda}\mathbf{S} \tag{6-42}$$

where **S** and **B** represent $N$-component vectors

$$\mathbf{S} = (S_1, \ldots, S_N) \tag{6-43}$$

and

$$\mathbf{B} = (B_1, \ldots, B_N) \tag{6-44}$$

while $\boldsymbol{\rho}$ is the diagonal matrix with elements $(\boldsymbol{\rho})_{ii} = \rho_i$, and $\boldsymbol{\Lambda}$ is the matrix defined in equation (6-40). We may now manipulate equation (6-42) to obtain simply

$$(\mathbf{I} - \boldsymbol{\rho}\boldsymbol{\Lambda})\mathbf{S} = (\mathbf{I} - \boldsymbol{\rho})\mathbf{B} \tag{6-45}$$

so that

$$\mathbf{S} = (\mathbf{I} - \boldsymbol{\rho}\boldsymbol{\Lambda})^{-1}(\mathbf{I} - \boldsymbol{\rho})\mathbf{B} \tag{6-46}$$

The advantage of this scheme is that for a prechosen set $\{\tau_i\}$ the $\boldsymbol{\Lambda}$-matrix may be computed once and for all. The indicated matrix inverse may be computed in a number of standard ways; according to Kurucz, the Gauss-Seidel iteration scheme is found to be the most efficient. Having obtained **S**, we may compute finally

$$\mathbf{J} = \boldsymbol{\Lambda}\mathbf{S} \tag{6-47}$$

Since the opacity varies from frequency to frequency, we will find that in an actual model, the depth scale $\tau_\nu$ at any given frequency will not in general match the prechosen set. Thus an interpolation procedure is necessary; the method of performing this interpolation has been described by Kurucz, and the reader should refer to his paper for details.

The advantage of this method is that unlike the iteration procedure, it constitutes a direct solution of the integral equation and is, therefore, invulnerable to the difficulties encountered in the iteration method.

### DIFFERENTIAL EQUATION METHODS— THE EIGENVALUE APPROACH

The above methods solve the transfer equation as an integral equation. An alternative would be to treat the transfer equation as a differential equation, though the complication of the angular integral for the mean intensity must now be considered. The method that naturally suggests itself is to replace the angular integration by a quadrature sum and to integrate numerically the system of equations

$$\mu_i \frac{dI_i}{d\tau} = I_i + \frac{\rho}{2} \sum_{j=-n}^{n} a_j I_j + (1 - \rho)B \tag{6-48}$$

(for the case we have been considering). The essential difficulty that now appears is in the application of the boundary conditions. The boundary conditions, as described in Chapter 1, are split into two groups:

$$I_{-i}(0) = 0$$

and

$$I_{+i}(\tau_{max}) = g(\mu_i)$$

where $\tau_{max}$ refers to the deepest point considered in the semi-infinite atmosphere. If we wanted to start the integration at $\tau = 0$ and proceed step by step inward, we would not know what values to use for $I_i(0)$; similarly at $\tau_{max}$ we would be lacking values for $I_{-i}(\tau_{max})$. Thus we have what amounts to an eigenvalue problem of order $n$. For example, we might guess at a set of values for $I_{-i}(\tau_{max})$ and use these to integrate toward the surface. When we reached the surface, we would in general find $I_{-i}(0) \neq 0$. In principle, we could then adjust the values of $I_{-i}(\tau_{max})$ to force $I_{-i}(0) = 0$. In practice, this method is fundamentally strongly unstable and can work only if $\tau_{max}$ is not very large. We can see this as follows. As we know from the gray case, a discretization of the above kind leads to a set of exponential solutions of the form $\exp(\pm k\tau)$ where the $k$'s are of order $1/\mu$; in the case where the coefficients (such as $\rho_\nu$) are variable, this is no longer the exact solution, but it is still true that the solution has an exponential character, perhaps $f(\tau) \times \exp(\pm k\tau)$, where $f$ is a weak function of $\tau$. In the semi-infinite case, we desire to suppress the ascending exponentials. In the gray solution this can be done explicitly since we have the analytical form to work with. In the nongray variable coefficient case, the solution is known only numerically and, in general, contains both the ascending and descending exponentials. Only for *exactly* the right choice of boundary conditions will the coefficients of the ascending exponentials be zero. For any other values the terms in $\exp(k\tau)$ will remain; these terms are called *parasites*. Now clearly the parasites increase at a rate of order $\exp(2k\tau)$ relative to the true solution. Thus, if the boundary conditions are wrong by fractional error $\varepsilon$, the parasite will be of order $\varepsilon \exp(2k\tau_{max})$ relative to the true solution at the other boundary. Obviously, this means that unless our initial choice is *very* good ($\varepsilon \ll 1$), the parasite will completely swamp the true solution, which will then be lost. Now since $\exp(2k\tau_{max}) \sim 10^{k\tau_{max}}$, we see that if we have $n$ significant figures in the computer, we must have $n \sim k\tau_{max}$ if we are to retain any vestige of the true solution compared with the parasite. This shows that if many quadrature points are chosen (in which case some $\mu \ll 1$ and some $k \gg 1$), and if $\tau_{max}$ is large (even of order 10), we will lose the solution. In practice $\tau_{max}$ may be $10^4$ or larger—which shows the hopelessness of this approach.

Thus, it is clear that a completely different method will be required. We will consider two such methods below, each of which accounts explicitly for the two-point nature of the boundary conditions (i.e., separation into two groups, one at the upper and one at the lower boundaries) and thus prevents the parasitic solution from creeping into the calculation. Before we discuss these methods, however, it is worthwhile to give an example illustrating the points mentioned above.

For simplicity, suppose the radiation field can be broken into an outgoing ($I_+$) and incoming ($I_-$) radiation field, and take $\mu_\pm = \pm 1/2$. Then

$$\frac{1}{2}\frac{dI_+}{d\tau} = I_+ - B \tag{6-49}$$

and

$$-\frac{1}{2}\frac{dI_-}{d\tau} = I_- - B \tag{6-50}$$

where we have assumed an LTE source function. For simplicity assume that $B$ is constant. Then by subtraction of equations (6-49) and (6-50) we have

$$\frac{1}{2}\frac{d(I_+ + I_-)}{d\tau} = I_+ - I_- \tag{6-51}$$

and by addition,

$$\frac{1}{2}\frac{d(I_+ - I_-)}{d\tau} = I_+ + I_- - 2B \tag{6-52}$$

Let

$$J = \tfrac{1}{2}(I_+ + I_-) \tag{6-53}$$

Then substitution of equations (6-51) and (6-53) into equation (6-52) gives

$$\frac{1}{4}\frac{d^2J}{d\tau^2} = J - B \tag{6-54}$$

The general solution for $J$ is then

$$J = ae^{2\tau} + be^{-2\tau} + B \tag{6-55}$$

while

$$I_+ - I_- = \frac{dJ}{d\tau} = 2ae^{2\tau} - 2be^{-2\tau} \tag{6-56}$$

so that

$$I_+ = 2ae^{2\tau} + B \tag{6-57}$$

and

$$I_- = 2be^{-2\tau} + B \tag{6-58}$$

The boundary conditions we wish to satisfy are that

$$I_+(\tau_{max}) = B \tag{6-59}$$

and

$$I_-(0) = 0 \tag{6-60}$$

These imply, from equations (6-57) and (6-58), that

$$a = 0$$

and

$$b = -\frac{B}{2}$$

Thus, the correct solutions are

$$I_+(\tau) = B \tag{6-61}$$

$$I_-(\tau) = B(1 - e^{-2\tau}) \tag{6-62}$$

and

$$J(\tau) = B(1 - \tfrac{1}{2}e^{-2\tau}) \tag{6-63}$$

Now suppose instead we had wanted to integrate the equations numerically, starting at $\tau = \tau_{max}$ and progressing toward the surface. We must then make a guess at the value of $I_-$ at the lower boundary; a reasonable guess would be

$$I_-(\tau_{max}) = B \tag{6-64}$$

By comparison with the true solution given by equation (6-62), we are making an error

$$\varepsilon = Be^{-2\tau_{max}} \tag{6-65}$$

by using equation (6-64), and one might imagine that with $\tau_{max}$ large, this error is negligibly small. However, as we stated above, *any* error will introduce a parasitic solution, and, as we will see, in this case the consequences are serious. Equations (6-59) and (6-64) imply from equations (6-57) and (6-58) that $a = 0$ so that $I_+(\tau) \equiv B$ and also $b = 0$ so that $I_-(\tau) \equiv B$. We have, in short, a completely erroneous surface value of $I_-(\tau)$. By comparison with the true solution, we see that this run of $I_-(\tau)$ contains a parasite given by

$$Be^{-2\tau} = \varepsilon \exp[2(\tau_{max} - \tau)] \tag{6-66}$$

Thus from a very small error at large depth, the parasite grows sufficiently to dominate the nature of the solution at the surface. If we were to consider more complex quadrature formulae, the solution would be more complicated, but the qualitative result remains unchanged. Thus we see that it is critical to account for the correct boundary conditions from the outset.

### DIFFERENTIAL EQUATION METHODS—TWO-POINT BOUNDARY CONDITION APPROACHES

Let us now describe two very powerful methods of solving transfer equations that not only eliminate the difficulties described above but, in addition, are capable of solving transfer problems of very great generality. For simplicity, we will write general forms of the equations and then use the source function given by equation (1-43) as a specific example. We will write out detailed forms in other cases when they arise.

(a) The Riccati Transformation. This approach for solving transfer problems was introduced by Rybicki (Ref. 16, p. 149). We again write the transfer equation in differential equation form; for example,

$$\mu \frac{dI}{d\tau} = I - \rho J - (1 - \rho)B \tag{6-67}$$

and introduce a quadrature sum for the integral over $\mu$. Denote outgoing streams as $I_\alpha^{\,1}$, $(\alpha = 1, \ldots, n)$, and incoming streams as $I_\alpha^{\,2}$, and let the quadrature weights $w_\alpha$ be chosen such that

$$J = \frac{1}{2}\int_{-1}^{1} I(\mu)\, d\mu \approx \sum_\alpha w_\alpha(I_\alpha^{\,1} + I_\alpha^{\,2}) \tag{6-68}$$

Then we may replace equation (6-67) by a set of equations of the form

$$\mu_\alpha \frac{dI_\alpha^{\,1}}{d\tau} = I_\alpha^{\,1} - \rho \sum_\alpha w_\alpha(I_\alpha^{\,1} + I_\alpha^{\,2}) - (1 - \rho)B \tag{6-69}$$

and

$$-\mu_\alpha \frac{dI_\alpha^{\,2}}{d\tau} = I_\alpha^{\,2} - \rho \sum_\alpha w_\alpha(I_\alpha^{\,1} + I_\alpha^{\,2}) - (1 - \rho)B \tag{6-70}$$

More generally, if we write the equations in vector form and characterize the outgoing radiation field by a set of numbers $\mathbf{f}^1$, related to $I^1$, and the incoming field by $\mathbf{f}^2$, related to $I^2$, we may consider equations of the form

$$\frac{d\mathbf{f}^1}{d\tau} = \mathbf{\Gamma}^{11}\mathbf{f}^1 + \mathbf{\Gamma}^{12}\mathbf{f}^2 + \mathbf{h}^1 \tag{6-71}$$

and

$$-\frac{d\mathbf{f}^2}{d\tau} = \mathbf{\Gamma}^{21}\mathbf{f}^1 + \mathbf{\Gamma}^{22}\mathbf{f}^2 + \mathbf{h}^2 \tag{6-72}$$

where the $\mathbf{\Gamma}$'s are matrices coupling the various components, and the $\mathbf{h}$'s represent inhomogeneous (thermal) terms. The boundary conditions can be stated formally as

$$\mathbf{f}^1(T) = \mathbf{E}^1 \tag{6-73}$$

and

$$\mathbf{f}^2(0) = \mathbf{E}^2 \tag{6-74}$$

It should be emphasized that equations (6-71) and (6-72) are much more general than equations (6-69) and (6-70), which we regard merely as one simple example. The coupling matrices $\mathbf{\Gamma}^{ij}$ may, in general, involve sums over angle *and* frequencies and thus allow for both types of redistribution.

We now wish to rewrite the equations so that starting at $\tau = T$, we deal only with quantities whose boundary conditions are known there, i.e., we must somehow eliminate $\mathbf{f}^2$. Rybicki suggested that this might be done by defining an auxiliary vector $\boldsymbol{\psi}$ and a matrix $\mathbf{R}$ such that

$$\mathbf{f}^1 = \boldsymbol{\psi} + \mathbf{R}\mathbf{f}^2 \tag{6-75}$$

Then substitution into equations (6-71) and (6-72) yields

$$\frac{d\boldsymbol{\psi}}{d\tau} + \mathbf{R}\frac{d\mathbf{f}^2}{d\tau} + \mathbf{f}^2\frac{d\mathbf{R}}{d\tau} = \mathbf{\Gamma}^{11}\boldsymbol{\psi} + \mathbf{\Gamma}^{11}\mathbf{R}\mathbf{f}^2 + \mathbf{\Gamma}^{12}\mathbf{f}^2 + \mathbf{h}^1 \tag{6-76}$$

and

$$-\mathbf{R}\frac{d\mathbf{f}^2}{d\tau} = \mathbf{R}\mathbf{\Gamma}^{21}\boldsymbol{\psi} + \mathbf{R}\mathbf{\Gamma}^{21}\mathbf{R}\mathbf{f}^2 + \mathbf{R}\mathbf{\Gamma}^{22}\mathbf{f}^2 + \mathbf{R}\mathbf{h}^2 \tag{6-77}$$

By addition of equations (6-76) and (6-77), we have

$$\frac{d\boldsymbol{\psi}}{d\tau} = (\mathbf{\Gamma}^{11} + \mathbf{R}\mathbf{\Gamma}^{21})\boldsymbol{\psi} + (\mathbf{h}^1 + \mathbf{R}\mathbf{h}^2) + \left(\mathbf{\Gamma}^{11}\mathbf{R} + \mathbf{R}\mathbf{\Gamma}^{21}\mathbf{R} + \mathbf{R}\mathbf{\Gamma}^{22} + \mathbf{\Gamma}^{12} - \frac{d\mathbf{R}}{d\tau}\right)\mathbf{f}^2 \tag{6-78}$$

Now we can eliminate the term involving $\mathbf{f}^2$ if we *define* $R$ to be the solution of the matrix differential equation

$$\frac{d\mathbf{R}}{d\tau} = \mathbf{\Gamma}^{11}\mathbf{R} + \mathbf{R}\mathbf{\Gamma}^{21}\mathbf{R} + \mathbf{R}\mathbf{\Gamma}^{22} + \mathbf{\Gamma}^{12} \tag{6-79}$$

This equation is then to be integrated simultaneously with the equation obtained from equations (6-78) and (6-79):

$$\frac{d\boldsymbol{\psi}}{d\tau} = (\mathbf{\Gamma}^{11} + \mathbf{R}\mathbf{\Gamma}^{21})\boldsymbol{\psi} + (\mathbf{h}^1 + \mathbf{R}\mathbf{h}^2) \tag{6-80}$$

starting with the values

$$\boldsymbol{\psi}(T) = \mathbf{E}^1 \tag{6-81}$$

and

$$\mathbf{R}(T) = 0 \tag{6-82}$$

When the integration is complete, we have at our disposal values of **R** and $\boldsymbol{\psi}$ on the range $0 \leq \tau \leq T$. We may now integrate equation (6-77) in the form

$$-\frac{d\mathbf{f}^2}{d\tau} = (\mathbf{\Gamma}^{21}\mathbf{R} + \mathbf{\Gamma}^{22})\mathbf{f}^2 + (\mathbf{\Gamma}^{21}\boldsymbol{\psi} + \mathbf{h}^2) \tag{6-83}$$

from 0 to $T$, starting with

$$\mathbf{f}^2(0) = \mathbf{E}^2 \tag{6-84}$$

Finally, we complete the solution by calculating

$$\mathbf{f}^1 = \boldsymbol{\psi} + \mathbf{R}\mathbf{f}^2 \tag{6-85}$$

We thus see how the method reduces the *two-point boundary-value problem* to *two* simple *initial-value problems*, thereby eliminating the difficulty associated with the eigenvalue approach. The equations for $\boldsymbol{\psi}$, **R**, and $\mathbf{f}^2$ are well chosen in that they do not tend to show exponential increases but rather exponential decreases since they are of the form $y^1 = -ay + b$ (recalling that the $\boldsymbol{\psi}$ and **R** equations are integrated with a *negative* step).

A simple physical interpretation can be suggested for the various quantities introduced in the equations. Thus we can regard $\boldsymbol{\psi}$ as that part of the outgoing radiation field due to sources only in deeper layers. The matrix **R** can be regarded as a *reflection matrix* so that the term $\mathbf{R}\mathbf{f}^2$ can be interpreted as incoming radiation that is reflected back in the outgoing direction. One can argue that since a well-defined physical meaning can be given to each term, one can reasonably expect them to be well defined and well behaved mathematically.

This method is capable of treating accurately the situations described above where the simple iteration procedure fails (as well as much more general cases). In practice, it turns out that considerably fewer computations are required if they are organized in terms of auxiliary quantities showing certain desirable symmetry properties. The requisite transformations are displayed in detail by Hummer and Rybicki (Ref. 3, p. 53, and *Ap. J.*, **150**, 607, 1967) and will not be discussed here. On the whole, this method may be regarded as one of the more powerful tools in our arsenal for solving transfer equations.

(b) The Difference-Equation Method. Let us now consider an alternative approach to solving the transfer equation which again treats explicitly the two-point nature of the boundary values. This method was developed for transfer equations by P. Feautrier (*C. R. Acad. Sci. Paris*, **258**, 3189, 1964) and is also a very general and flexible technique. As before, the transfer equation is

$$\mu \frac{dI_\nu}{d\tau} = \chi_\nu(I_\nu - S_\nu) \tag{6-86}$$

where $\chi_\nu$ is defined as the ratio of the opacity at frequency $\nu$ to the standard opacity used to specify $\tau$. In the case of coherent scattering, we set $\chi_\nu \equiv 1$ since only one frequency enters the problem. In noncoherent cases where several frequencies are involved, $\chi_\nu$ must be carried along. In the above equation we write $S_\nu$ with the understanding that in practice one substitutes an expression for $S_\nu$ explicitly involving terms in intensities and thermal sources; for example, for the coherent scattering case we would write explicitly

$$S_\nu = (1 - \rho_\nu)J_\nu + \rho_\nu B_\nu \tag{6-87}$$

Again we introduce quadrature sums for any angular (frequency) integrations appearing in $S_\nu$, for example,

$$S_\nu = \frac{1}{2}(1 - \rho_\nu) \sum_{j=-n}^{n} a_j I_\nu(\mu_j) + \rho_\nu B_\nu \tag{6-88}$$

Now *define*

$$P_\nu(\tau, \mu) \equiv \tfrac{1}{2}[I_\nu(\tau, \mu) + I_\nu(\tau, -\mu)] \tag{6-89}$$

and

$$R_\nu(\tau, \mu) \equiv \tfrac{1}{2}[I_\nu(\tau, \mu) - I_\nu(\tau, -\mu)] \tag{6-90}$$

which have a mean intensity-like and a flux-like character, respectively. Then by considering the transfer equation at $\pm\mu$, we have

$$\mu \frac{dI_\nu(\tau, \mu)}{d\tau} = \chi_\nu[I_\nu(\tau, \mu) - S_\nu] \tag{6-91}$$

and

$$-\mu \frac{dI_\nu(\tau, -\mu)}{d\tau} = \chi_\nu[I_\nu(\tau, -\mu) - S_\nu] \tag{6-92}$$

so that by addition

$$\mu \frac{dR_\nu(\tau, \mu)}{d\tau} = \chi_\nu[P_\nu(\tau, \mu) - S_\nu] \tag{6-93}$$

and by subtraction

$$\mu \frac{dP_\nu(\tau, \mu)}{d\tau} = \chi_\nu R_\nu(\tau, \mu) \tag{6-94}$$

Substituting equation (6-94) into equation (6-93), we can eliminate $R_\nu$ and obtain

$$\frac{\mu^2}{\chi_\nu} \frac{d}{d\tau}\left[\frac{1}{\chi_\nu} \frac{dP_\nu(\tau, \mu)}{d\tau}\right] = P_\nu(\tau, \mu) - S_\nu \tag{6-95}$$

This transfer equation is to be solved subject to the boundary conditions

$$I_\nu(0, -\mu) = 0 \tag{6-96}$$

and

$$I_\nu(T, \mu) = I_+ \tag{6-97}$$

Now at, $\tau = 0$, equation (6-96) implies

$$R_\nu(0, \mu) \equiv P_\nu(0, \mu) \tag{6-98}$$

and therefore equation (6-94) becomes simply

$$\mu \left. \frac{dP_\nu}{d\tau} \right|_{\tau_0} = \chi_\nu P_\nu \tag{6-99}$$

At $\tau = T$ we can write

$$\begin{aligned} R_\nu(T, \mu) &= \tfrac{1}{2}[I_\nu(T, \mu) - I_\nu(T, -\mu)] \\ &= I_\nu(T, \mu) - \tfrac{1}{2}[I_\nu(T, \mu) + I_\nu(T, -\mu)] \\ &= I_+ - P_\nu(T, \mu) \end{aligned} \tag{6-100}$$

Thus, from equation (6-94) the boundary condition at $\tau = T$ can be written

$$\mu \left. \frac{dP}{d\tau} \right|_T = \chi_\nu(I_+ - P_\nu) \tag{6-101}$$

In practice, one might adopt $I_+ = B_\nu$ at the lower boundary. A better estimate of $I_+$ results from assuming that near $\tau = T$,

$$B_\nu(\tau) = B_\nu(\tau - T) + \left. \frac{dB_\nu}{d\tau} \right|_T (\tau - T) \tag{6-102}$$

Then we may write

$$\begin{aligned} I_+(\mu) &= \int_0^\infty \left[ B_\nu(\tau = T) + \frac{dB_\nu}{d\tau} x \right] e^{-x/\mu} \frac{dx}{\mu} \\ &= B_\nu(\tau = T) + \mu \left. \frac{dB_\nu}{d\tau} \right|_T \end{aligned} \tag{6-103}$$

We now convert the differential equation of transfer to a *difference equation* by introducing a discrete set of depth points $\{\tau_i\}$, $(i = 1, \ldots, N)$, angle points $\{\mu_j\}$, $(j = 1, \ldots, M)$, and (if relevant) frequency points $\{\nu_k\}$, $(k = 1, \ldots, K)$. We replace integrals by sums and derivatives by differences. Thus, we write

$$\left( \frac{dX}{d\tau_\nu} \right)_{i+1/2} = \frac{X(\tau_{i+1}) - X(\tau_i)}{\frac{1}{2}[\chi_\nu(\tau_{i+1}) + \chi_\nu(\tau_i)](\tau_{i+1} - \tau_i)} = \frac{(\Delta X)_{i+1/2}}{(\Delta \tau_\nu)_{i+1/2}} \tag{6-104}$$

and

$$\left(\frac{d^2X}{d\tau_\nu^2}\right)_i = \frac{\left(\frac{dX}{d\tau_\nu}\right)_{i+1/2} - \left(\frac{dX}{d\tau_\nu}\right)_{i-1/2}}{\frac{1}{2}[(\Delta\tau_\nu)_{i+1/2} + (\Delta\tau_\nu)_{i-1/2}]} \tag{6-105}$$

Then the transfer equation becomes

$$\frac{8\mu_j^2\left\{\frac{P_\nu(\tau_{i+1}, \mu_j) - P_\nu(\tau_i, \mu_j)}{[\chi_\nu(\tau_{i+1}) + \chi_\nu(\tau_i)](\tau_{i+1} - \tau_i)} - \frac{P_\nu(\tau_i, \mu_j) - P_\nu(\tau_{i-1}, \mu_j)}{[\chi_\nu(\tau_i) + \chi_\nu(\tau_{i-1})](\tau_i - \tau_{i-1})}\right\}}{[\chi_\nu(\tau_{i+1}) + \chi_\nu(\tau_i)](\tau_{i+1} - \tau_i) + [\chi_\nu(\tau_i) + \chi_\nu(\tau_{i-1})](\tau_i - \tau_{i-1})}$$

$$= P_\nu(\tau_i, \mu_j) - S_\nu(\tau_i) \tag{6-106}$$

There are $M$ such equations (or $M \times K$ if frequencies enter) at each of the $N - 2$ depth points, excluding the boundaries. If we write a vector $\mathbf{P}_i$ of dimension $M$ (or $M \times K$) such that

$$(\mathbf{P}_i)_l = P_{\nu_k}(\tau_i, \mu_j) \tag{6-107}$$

where $l = j$ or $l = j + M\ (k - 1)$ (whichever pertains), then the above difference equation is clearly of the form

$$-\mathbf{A}_i\mathbf{P}_{i-1} + \mathbf{B}_i\mathbf{P}_i - \mathbf{C}_i\mathbf{P}_{i+1} = \mathbf{Q}_i \tag{6-108}$$

where the matrices **A**, **B**, and **C** and the vector **Q** will, in general, depend upon the form of $S$, as described below. The upper boundary condition, equation (6-99), becomes

$$\mu_j \frac{P_\nu(\tau_2, \mu_j) - P_\nu(\tau_1, \mu_j)}{(\tau_2 - \tau_1)} = \chi_\nu(\tau_1)P_\nu(\tau_1, \mu_j) \tag{6-109}$$

which is of the form $\mathbf{B}_1\mathbf{P}_1 - \mathbf{C}_1\mathbf{P}_2 = 0$, so $\mathbf{A}_1 \equiv 0$, and $\mathbf{Q}_1 \equiv 0$.

Similarly, the lower boundary condition, equation (6-101), is

$$\mu_j \frac{P_\nu(\tau_N, \mu_j) - P_\nu(\tau_{N-1}, \mu_j)}{(\tau_N - \tau_{N-1})} = \chi_\nu(\tau_N)[I_+(\mu_j) - P_\nu(\tau_N, \mu_j)] \tag{6-110}$$

which is of the form $-\mathbf{A}_N\mathbf{P}_{N-1} + \mathbf{B}_N\mathbf{P}_N = \mathbf{Q}_N$ so that $\mathbf{C}_N \equiv 0$. Now observe that from the upper boundary condition we may write

$$\mathbf{P}_1 = \mathbf{B}_1^{-1}\mathbf{C}_1\mathbf{P}_2 = \mathbf{D}_1\mathbf{P}_2 \tag{6-111}$$

We can thus eliminate $\mathbf{P}_1$ terms of $\mathbf{P}_2$ and write equation (6-108) with $i = 2$ in the form

$$(\mathbf{B}_2 - \mathbf{A}_2\mathbf{D}_1)\mathbf{P}_2 - \mathbf{C}_2\mathbf{P}_3 = \mathbf{Q}_2 \tag{6-112}$$

This in turn suggests that we write

$$\mathbf{P}_2 = (\mathbf{B}_2 - \mathbf{A}_2\mathbf{D}_1)^{-1}\mathbf{C}_2\mathbf{P}_3 + (\mathbf{B}_2 - \mathbf{A}_2\mathbf{D}_1)^{-1}\mathbf{Q}_2$$
$$\equiv \mathbf{D}_2\mathbf{P}_3 + \boldsymbol{\psi}_2 \tag{6-113}$$

By successive substitutions we can perform successive eliminations of the above type at each point in the atmosphere by writing

$$\mathbf{P}_i = \mathbf{D}_i\mathbf{P}_{i+1} + \boldsymbol{\psi}_i \tag{6-114}$$

where

$$\mathbf{D}_i = (\mathbf{B}_i - \mathbf{A}_i\mathbf{D}_{i-1})^{-1} \tag{6-115}$$

and

$$\boldsymbol{\psi}_i = (\mathbf{B}_i - \mathbf{A}_i\mathbf{D}_{i-1})^{-1}(\mathbf{Q}_i + \mathbf{A}_i\boldsymbol{\psi}_{i-1}) \tag{6-116}$$

for $i = 1, \ldots, N$. Note that these definitions and the facts that $\mathbf{A}_1 = 0$, $\mathbf{C}_N = 0$, and $\mathbf{Q}_1 = 0$ imply that $\mathbf{D}_1 = \mathbf{B}_1^{-1}\mathbf{C}_1$ and $\boldsymbol{\psi}_1 = 0$ from the upper boundary condition, and $\mathbf{D}_N = 0$ and $\mathbf{P}_N = \boldsymbol{\psi}_N$ from the lower. Having obtained $\mathbf{P}_N$, we then determine the radiation field at all other points by successive back-substitutions into the elimination scheme:

$$\mathbf{P}_i = \mathbf{D}_i\mathbf{P}_{i+1} + \boldsymbol{\psi}_i \tag{6-117}$$

Finally, we can compute the actual mean intensity at each point as

$$J_\nu(\tau_i) = \sum_{j=1}^{M} a_j P_\nu(\tau_i, \mu_j) \tag{6-118}$$

Although this method differs almost completely in form from the Riccati transformation described above, it accomplishes the same goals by providing a *general* method that accounts explicitly for the two-point nature of the boundary conditions. In practice, the computation proceeds quite efficiently. The most difficult (and time-consuming) aspect is the need to calculate $N$ inverses of $L \times L$-matrices, where $L = M \times K$, and to store the **D**-matrices and $\boldsymbol{\psi}$-vectors. The accuracy of the solution depends both upon the accuracy of the quadrature formulae employed and upon the choice of the depth points. If the depth points are too widely spaced, the difference approximation of the derivative may become inaccurate; if the points are too closely spaced, roundoff errors may become troublesome. In practice, however, the latitude between these extremes is large, and usually no trouble in constructing a depth scale is encountered. It is also worth noting that the difference-equation method is very stable. Finally, we might note that the first-order upper boundary condition is not as precise as sometimes desired; a second-order form of much higher accuracy has been developed by Auer (*Ap. J. Letters*, **150**, L53, 1967).

Before leaving the difference equation method, it is quite instructive to write out explicitly the **A**-, **B**-, and **C**-matrices, and the vector **Q** in the simple case where the source function is given by equation (6-88). In this case

$$\mu^2 \frac{d^2P(\tau, \mu)}{d\tau^2} = P(\tau, \mu) - \rho \sum_{j=1}^{M} a_j P(\tau, \mu_j) - (1 - \rho)B_\nu \tag{6-119}$$

Here we have set $\chi \equiv 1$ since the scattering is coherent, and only one frequency enters. Introducing the difference approximation discussed above, we find that $\mathbf{A}_i$ is a diagonal matrix with elements

$$(\mathbf{A}_i)_{jj} = 2\mu_j^2[(\tau_{i+1} - \tau_{i-1})(\tau_i - \tau_{i-1})]^{-1} \tag{6-120}$$

$\mathbf{C}_i$ is a diagonal matrix with elements

$$(\mathbf{C}_i)_{jj} = 2\mu_j^2[(\tau_{i+1} - \tau_{i-1})(\tau_{i+1} - \tau_i)]^{-1} \tag{6-121}$$

$\mathbf{B}_i$ is a square matrix with diagonal elements

$$(\mathbf{B}_i)_{jj} = (\mathbf{A}_i)_{jj} + (\mathbf{C}_i)_{jj} + 1 - \rho_i a_j \tag{6-122}$$

and off-diagonal elements

$$(\mathbf{B}_i)_{jk} = -\rho_i a_k \tag{6-123}$$

while $\mathbf{Q}_i$ is a vector with elements

$$(\mathbf{Q}_i)_j = (1 - \rho_i)B_\nu(\tau_i) \tag{6-124}$$

At the upper boundary (using the first-order equations) $\mathbf{A}_i$ and $\mathbf{Q}_1$ are zero, $\mathbf{B}_1$ is the diagonal matrix

$$(\mathbf{B}_1)_{jj} = 1 + \mu_j(\tau_2 - \tau_1)^{-1} \tag{6-125}$$

and $\mathbf{C}_1$ is the diagonal matrix

$$(\mathbf{C}_1)_{jj} = \mu_j(\tau_2 - \tau_1)^{-1} \tag{6-126}$$

At the lower boundary $\mathbf{C}_N = 0$ and $\mathbf{A}_N$ is the diagonal matrix

$$(\mathbf{A}_N)_{jj} = \mu_j(\tau_N - \tau_{N-1})^{-1} \tag{6-127}$$

and $\mathbf{B}_N$ is the diagonal matrix

$$(\mathbf{B}_N)_{jj} = 1 + \mu_j(\tau_N - \tau_{N-1})^{-1} \tag{6-128}$$

and $\mathbf{Q}_N$ is the vector

$$(\mathbf{Q}_N)_j = B_\nu(\tau_N) + \mu_j \left.\frac{dB_\nu}{d\tau}\right|_{\tau_N} \tag{6-129}$$

The elimination and back-substitution follow exactly the outline given above. We observe from the above expressions that the matrices **A** and **C** are simply difference operators, the matrix **B** contains the coupling of the $i$th angle (and, in general, frequency) point to all other angle (and frequency) points, and the inhomogeneous vector **Q** contains the thermal source term. In the case of coherent scattering, the matrix size $(M \times M)$ is usually so small [say, $(4 \times 4)$ to $(6 \times 6)$ depending upon the accuracy desired] that the computation is very speedy indeed and normally is much faster than the interation procedure even for those cases where the iterations converge. Moreover, the method yields the answer with the same ease even for cases where the iteration procedure fails entirely. We now have at our command two quite general and flexible methods, which can be applied in numerous other contexts.

### CALCULATION OF THE FLUX

Having obtained $J_\nu(\tau_\nu)$ and hence $S_\nu(\tau_\nu)$, we may calculate the flux at each frequency and depth point from the $\Phi$-operator, namely [equation (1–64)],

$$F_\nu(\tau_\nu) = 2\left[\int_{\tau_\nu}^{\infty} S_\nu(t_\nu)E_2(t_\nu - \tau_\nu)\,dt_\nu - \int_0^{\tau_\nu} S_\nu(t_\nu)E_2(\tau_\nu - t_\nu)\,dt_\nu\right]$$

Numerical formulae have been developed by Gingerich (*Ap. J.*, **138**, 576, 1963), Cayrel (*Ann. d'Ap.*, **23**, 235, 1961), Reiz (Ref. 9, p. 67), and Norton (Ref. 3, p. 33). Reiz's results are well suited for hand calculation but lack the accuracy obtained by the others. Norton's formula offers very good accuracy with relatively little computation.

Having obtained the monochromatic flux, we can now check that the integrated flux at each point in the atmosphere achieves the desired value, namely [equation (1–191)],

$$\int_0^{\infty} F_\nu\,d\nu = \frac{\sigma T_{\text{eff}}^4}{\pi}$$

If it does not, we must adjust the temperature distribution, as discussed in the next section. If we are satisfied with the accuracy of the model, we can compare the emergent fluxes with observations of stars to examine points of agreement and disagreement and to see if the properties of the stellar atmosphere can be inferred; we will discuss such comparisons in Section 6-4.

## 6-3. Temperature-Correction Procedures

Having outlined the procedures used to construct an LTE model atmosphere, we must now discuss the methods of enforcing the requirement of radiative equilibrium.

THE LAMBDA-ITERATION PROCEDURE

We have shown previously that the requirement that the derivative of the integrated flux be zero leads to equation (1-92):

$$\int_0^\infty \kappa_\nu B_\nu \, d\nu = \int_0^\infty \kappa_\nu J_\nu \, d\nu$$

with scattering terms canceling out. Suppose now that we have computed a model using some temperature distribution $T_0(\tau)$ and have derived mean intensities $J_\nu$. Then suppose we find that the above requirement is not satisfied, i.e.,

$$\int_0^\infty \kappa_\nu B_\nu(T_0) \, d\nu \neq \int_0^\infty \kappa_\nu J_\nu \, d\nu$$

We now assume that the run of temperature $T(\tau)$ which *does* satisfy the radiative equilibrium condition can be written as $T(\tau) = T_0(\tau) + \Delta T(\tau)$, and we request that $\Delta T$ be such that

$$\int_0^\infty \kappa_\nu B_\nu(T_0 + \Delta T) \, d\nu = \int_0^\infty \kappa_\nu J_\nu \, d\nu \tag{6-130}$$

To first order,

$$B_\nu(T_0 + \Delta T) = B_\nu(T_0) + \left.\frac{\partial B_\nu}{\partial T}\right|_{T_0} \Delta T$$

so that substituting into equation (6-130), we write

$$\int_0^\infty \kappa_\nu B_\nu(T_0) \, d\nu + \Delta T \int_0^\infty \kappa_\nu \frac{\partial B_\nu}{\partial T} \, d\nu = \int_0^\infty \kappa_\nu J_\nu \, d\nu \tag{6-131}$$

or

$$\Delta T = \frac{\int_0^\infty \kappa_\nu [J_\nu - B_\nu(T_0)] \, d\nu}{\int_0^\infty \kappa_\nu \frac{\partial B_\nu}{\partial T} \, d\nu} \tag{6-132}$$

Because we are employing the current values $J_\nu(\tau) = \Lambda_\tau[B_\nu(T_0)]$, this procedure is called the *lambda-iteration procedure*. When one carries through the calculation and recomputes a model with the new temperature distribution, he will, in general, find some improvement in satisfying equation (1-92). The procedure, however, suffers from several very severe drawbacks.

(a) In the first place, since we are writing $J_\nu = \Lambda[B_\nu]$, we must automatically find $J_\nu \to B_\nu + 0(e^{-\tau_\nu})$ at depth, and hence the temperature correction goes strongly to zero at depth, even though we may actually have a completely incorrect solution there. We have seen precisely this same trouble before in the use of the lambda operator to improve the gray solution. At depth the improvement is "infinitely slow."

(b) A related difficulty occurs if the opacity is very much higher in certain frequency regions than in others. Then again since $J_\nu \rightarrow B_\nu$ for these frequencies, these bands make no contribution to the integral in the numerator, while contributing strongly to the denominator, which again decreases $\Delta T$ and slows convergence. Another way of seeing both these problems is to recall that the range of the lambda operator is only of order $\Delta\tau \sim 1$. If the atmosphere departs seriously from radiative equilibrium over very large depth, then a correspondingly large (usually prohibitively large) number of iterations will be required.

(c) Since the condition of radiative equilibrium as expressed above is essentially a condition on the flux *derivative*, we have no way of specifying the actual value of the flux to which the solution converges, i.e., the lambda-iteration procedure yields a zero temperature correction if the flux is constant, even if this flux is not the desired value. This difficulty can be overcome only by introducing information about the deviation of the computed flux from some desired value; we shall employ this kind of information in the procedures discussed below.

(d) Finally, it must be realized that the lambda-iteration procedure could not really be expected to be an effective method inasmuch as it is not actually a proper solution of the integral equation, i.e., the $\Delta T$ introduced at some point $\tau$ will have an influence at all other points $\tau'$ in the atmosphere because $J_\nu(\tau')$ is really $\Lambda_{\tau'}[B_\nu(\tau) + \Delta B_\nu(\tau)]$. This effect is totally ignored in the lambda-iteration procedure, and spurious $\Delta T$ values result. A method of meeting this criticism will be described below.

### THE UNSÖLD-LUCY PROCEDURE

In view of our criticisms of the lambda-iteration procedure, it is clear that it is desirable to develop an approach which makes use of information about errors in the flux as well as in the flux derivative. A way of doing this was first suggested by Unsöld for the gray case and subsequently generalized by Lucy for the nongray case. Let $F(\tau)$ be the flux actually obtained in the model calculation, and let $F^*$ be the desired flux. Write $\Delta F(\tau) = F(\tau) - F^*$. We wish to use the run of $\Delta F(\tau)$ to compute a correction $\Delta B$ to the source function such that $B(\tau) - \Delta B(\tau)$ will yield the correct flux $F^*$. Here $B = \sigma T^4/\pi$, and $\Delta B = 4\sigma T^3 \Delta T/\pi$ so that we are, in effect, calculating a correction to the temperature distribution. Unsöld proceeds as follows. The transfer equation and its first two moments are

$$\mu \frac{dI}{d\tau} = I - B \tag{6-133}$$

$$\frac{dH}{d\tau} = J - B \tag{6-134}$$

and

$$\frac{dK}{d\tau} = H \tag{6-135}$$

Since $B$ is not the correct source function, the computed value of $H$ will be constant. We can, however, still write

$$K(\tau) = \int_0^\tau H(\tau')\,d\tau' + C' = \frac{1}{4}\int_0^\tau F(\tau')\,d\tau' + C' \tag{6-136}$$

and if we make the Eddington approximation $J(\tau) = 3K(\tau)$, then

$$J(\tau) \approx \frac{3}{4}\int_0^\tau F(\tau')\,d\tau' + C \tag{6-137}$$

If we now make the further approximation $J(0) = 2H(0) = \frac{1}{2}F(0)$, then we may evaluate $C$ and obtain

$$J(\tau) \approx \frac{3}{4}\int_0^\tau F(\tau')\,d\tau' + \frac{1}{2}F(0) \tag{6-138}$$

Now from equation (6-134) we have

$$B(\tau) = J(\tau) - \frac{1}{4}\frac{dF}{d\tau}$$

Then substituting equation (6-138), we find

$$B(\tau) \approx \frac{3}{4}\int_0^\tau F(\tau')\,d\tau' + \frac{1}{2}F(0) - \frac{1}{4}\frac{dF}{d\tau} \tag{6-139}$$

Equation (6-139) cannot be exact because of the approximations used to derive it, but it can be used with sufficient accuracy to compute perturbations. Thus, demanding a $\Delta B(\tau)$ that yields the correct flux, we write

$$B(\tau) - \Delta B(\tau) \approx \frac{3}{4}\int_0^\infty F^*\,d\tau' + \frac{1}{2}F^* \tag{6-140}$$

and subtracting, we have

$$\Delta B(\tau) = \frac{3}{4}\int_0^\tau \Delta F(\tau')\,d\tau' + \frac{1}{2}\Delta F(0) - \frac{1}{4}\frac{d(\Delta F)}{d\tau} \tag{6-141}$$

which gives the desired temperature correction. In practice this method is much more effective than the lambda-iteration procedure, as we shall show numerically below. It should be realized that although the perturbation equation written above is still based upon approximations, this does not mean that successive applications of this equation to revise the temperature distribution will lead to an incorrect result. The reason is that the quantities $\Delta F$

and $dF/d\tau$ are computed by exact formulae and are independent of any approximations in the correction procedure. The approximations in the temperature correction can, at worst, merely slow the *rate* of convergence but will not lead to an erroneous result.

Let us now consider Lucy's generalization of this procedure (Ref. 15, p. 93). We now have

$$-\frac{\mu}{\rho}\frac{dI_\nu}{dz} = (\kappa_\nu + \sigma_\nu)I_\nu - \kappa_\nu B_\nu - \sigma_\nu J_\nu \tag{6-142}$$

which implies

$$-\frac{1}{4}\frac{1}{\rho}\frac{dF_\nu}{dz} = \kappa_\nu J_\nu - \kappa_\nu B_\nu \tag{6-143}$$

and

$$-\frac{1}{\rho}\frac{dK_\nu}{dz} = \frac{1}{4}(\kappa_\nu + \sigma_\nu)H_\nu \tag{6-144}$$

Integrating over frequency and introducing the Planck, absorption, and flux means [equations (2-30) and (2-31)]

$$\kappa_P B = \int_0^\infty \kappa_\nu B_\nu \, d\nu$$

$$\kappa_J J = \int_0^\infty \kappa_\nu J_\nu \, d\nu$$

and [equation (2-25)]

$$\kappa_F F = \int_0^\infty \kappa_\nu F_\nu \, d\nu$$

and choosing the Planck-mean depth scale $d\tau = -\kappa_P \rho \, dz$, we have

$$\frac{1}{4}\frac{dF}{d\tau} = \frac{\kappa_J}{\kappa_P} J - B \tag{6-145}$$

and

$$\frac{dK}{d\tau} = \frac{1}{4}\frac{\kappa_F}{\kappa_P} F \tag{6-146}$$

Now proceeding just as before, again making use of the Eddington approximation, we find

$$K(\tau) = \frac{1}{4}\int_0^\tau \left(\frac{\kappa_F}{\kappa_P}\right) F(\tau') \, d\tau' + C' \tag{6-147}$$

so that

$$J(\tau) \approx \frac{3}{4}\int_0^\tau \left(\frac{\kappa_F}{\kappa_P}\right) F(\tau')\,d\tau' + \frac{1}{2}F(0) \tag{6-148}$$

and

$$B(\tau) \approx \left(\frac{\kappa_J}{\kappa_P}\right)\left[\frac{3}{4}\int_0^\tau \left(\frac{\kappa_F}{\kappa_P}\right) F(\tau')\,d\tau' + \frac{1}{2}F(0)\right] - \frac{1}{4}\frac{dF}{d\tau} \tag{6-149}$$

Then treating equation (6-149) as a perturbation equation, we have

$$\Delta B(\tau) = \left(\frac{\kappa_J}{\kappa_P}\right)\left[\frac{3}{4}\int_0^\tau \left(\frac{\kappa_F}{\kappa_P}\right) \Delta F(\tau')\,d\tau' + \frac{1}{2}\Delta F(0)\right] - \frac{1}{4}\frac{d\Delta F}{d\tau} \tag{6-150}$$

Two two factors $(\kappa_J/\kappa_P)$ and $(\kappa_F/\kappa_P)$ account properly for the different ways of weighting the radiation field and, in practice, make the generalization much more accurate than Unsöld's original approach in nongray problems. An important point to note in equation (6-141) is that at $\tau = 0$, only the second and third terms determine $\Delta B$. The term $-\frac{1}{4}\,dF/d\tau$ is the same as appears in the lambda-iteration procedure. However, the term $\frac{1}{2}\,\Delta F(0)$ prevents the method from reducing completely to the lambda procedure at the surface and, in particular, prevents the procedure from stabilizing upon an unwanted, though constant, flux. At depth in the atmosphere, the terms involving

$$\frac{3}{4}\int_0^\tau \Delta F(\tau')\,d\tau'$$

dominate and provide a temperature correction which leads to an improved solution long after the lambda-iteration term has gone to zero.

Let us now consider a specific example to illustrate this method. Suppose we are dealing with the gray case and that we start with an approximate solution of the form

$$B_0(\tau) = \tfrac{3}{4}F(\tau + c) \tag{6-151}$$

We know that the lambda-iteration procedure leads to a new solution

$$B(\tau) = \frac{3}{4}F\left[(\tau + c) + \frac{1}{2}E_3(\tau) - \frac{c}{2}E_2(\tau)\right] \tag{6-152}$$

so that we would have

$$\Delta B = \frac{3}{8}F\left[E_3(\tau) - \frac{c}{2}E_2(\tau)\right] \tag{6-153}$$

As noted before, this solution goes to zero as $\tau \rightarrow \infty$, so no correction is made at depth.

Now the flux from $B_0(\tau)$ is easily calculated by an application of the $\Phi$-operator to yield

$$F(\tau) = F + \tfrac{3}{2}F[cE_3(\tau) - E_4(\tau)] \tag{6-154}$$

i.e.,

$$\Delta F(\tau) = \tfrac{3}{2}F[cE_3(\tau) - E_4(\tau)] \tag{6-155}$$

This implies

$$\frac{\Delta F(0)}{2} = \frac{3}{4}F\left[\frac{c}{2} - \frac{1}{3}\right] \tag{6-156}$$

and

$$-\frac{1}{4}\frac{d(\Delta F)}{d\tau} = \frac{3}{8}F[cE_2(\tau) - E_3(\tau)] \tag{6-157}$$

while

$$\frac{3}{4}\int_0^\tau \Delta F(\tau')\,d\tau' = \frac{9}{8}F\left[\frac{c}{3} - cE_4(\tau) - \frac{1}{4} + E_5(\tau)\right] \tag{6-158}$$

Thus, the change in source function predicted by Unsöld's method is

$$\Delta B = \frac{3}{4}F\left[c - \frac{17}{24} + \frac{c}{2}E_2(\tau) - \frac{3}{2}cE_4(\tau) - \frac{1}{2}E_3(\tau) + \frac{3}{2}E_5(\tau)\right] \tag{6-159}$$

and the improved source function is $B(\tau) = B_0(\tau) - \Delta B(\tau)$, or

$$B(\tau) = \frac{3}{4}F\left[\tau + \frac{17}{24} + \frac{3}{2}cE_4(\tau) - \frac{c}{2}E_2(\tau) + \frac{1}{2}E_3(\tau) - \frac{3}{2}E_5(\tau)\right] \tag{6-160}$$

Here we see that unlike the lambda-iterated solution, we obtain a new value of $q(\infty)$ *independent of the original choice of c*. This value is $q(\infty) = 17/24 = 0.708$, which is in close agreement with the exact result 0.710. Thus we see quite clearly the important effects of the terms involving the flux error. A perhaps even more remarkable result is the value of $q(0)$. Since $E_2(0) = 1$ and $E_4(0) = 1/3$, the terms involving $c$ drop out, and we obtain $q(0) = (17/24) + (1/2)E_3(0) - (3/2)E_5(0) = 7/12 = 0.583$, again independent of the original choice of $c$ and in good agreement with the exact result $q(0) = 1/\sqrt{3} = 0.577$. In contrast, the lambda operator yields $q(0) = (c/2) + 1/4$. This value agrees with that given by Unsöld's method when $c = 2/3$ (the Eddington value) but in general will be far less accurate depending upon the choice of $c$.

### THE AVRETT-KROOK PROCEDURE

A rather ingenious and quite powerful method of satisfying the requirement of radiative equilibrium was developed by Avrett and Krook (*Ap. J.*,

**137**, 874, 1963). Their method has been employed to calculate models in a wide variety of physical situations and has enjoyed very good success. Moreover, it is in a form that allows easy generalization to cases where only part of the flux is transported by radiation. The method makes no particular assumption about the magnitudes of the deviations from grayness and has been effective in extremely nongray situations.

The unique feature of this method is that it permits a two-dimensional adjustment of the temperature distribution, i.e., instead of calculating a temperature correction $\Delta T$ at a fixed point $\tau$, adjustments are allowed both in the dependent variable $T$ *and* independent variable $\tau$, and the total correction is a combination of the two.

For simplicity, let us discuss first the case where the source function is entirely thermal so that $S_\nu = B_\nu$. The transfer equation is

$$\mu \frac{dI_\nu}{d\tau_\nu} = I_\nu - B_\nu \tag{6-161}$$

or

$$\mu \frac{dI_\nu}{d\tau} = \chi_\nu(I_\nu - B_\nu) \tag{6-162}$$

where $\chi_\nu \equiv \kappa_\nu/\kappa$, $\kappa$ being the opacity at some standard wavelength. Now suppose the present optical depth scale is denoted by $t$ and the present temperature scale is called $T_0(t)$. Let the final optical depth scale be $\tau$ and the desired temperature distribution be $T(\tau)$. We suppose that the present and desired values are related in terms of small perturbations $\tau_1$ and $T_1$, such that

$$T = T_0 + \lambda T_1 + \cdots \tag{6-163}$$

and

$$\tau = t + \lambda\tau_1 + \cdots \tag{6-164}$$

where $\lambda$ is merely a separation parameter which is used to keep track of the order of the perturbation and which will ultimately be set equal to unity. Now $T(\tau)$ is the distribution which satisfies equation (6-162) exactly, subject to the condition of radiative equilibrium. Since $T(\tau)$ differs from $T_0(t)$, the *present* radiation field will not meet these requirements. We assume now that we can expand the desired properties in terms of present values (denoted by superscript 0) and small perturbations (denoted by superscript 1) as follows:

$$I_\nu = I_\nu^{\,0} + \lambda I_\nu^{\,1} + \cdots \tag{6-165}$$

$$\chi_\nu(\tau) = \chi_\nu(t) + \lambda\chi_\nu'(t)\tau_1 + \cdots \tag{6-166}$$

and

$$B_\nu(T) = B_\nu(T_0) + \lambda\frac{\partial B_\nu}{\partial T}T_1 + \cdots \tag{6-167}$$

where the prime denotes the derivative $d/dt$. If we substitute these expansions into the transfer equation

$$\mu \frac{dI_\nu}{dt} = \left(\frac{d\tau}{dt}\right)\chi_\nu(I_\nu - B_\nu) \tag{6-168}$$

we obtain

$$\mu\left(\frac{dI_\nu^{\,0}}{dt} + \lambda\frac{dI_\nu^{\,1}}{dt} + \cdots\right) = (1 + \lambda\tau_1' + \cdots)[\chi_\nu(t) + \lambda\chi_\nu'(t)\tau_1 + \cdots]$$
$$\times \left[I_\nu^{\,0} + \lambda I_\nu^{\,1} + \cdots - B_\nu(T_0) - \lambda\frac{\partial B_\nu}{\partial T}T_1 - \cdots\right] \tag{6-169}$$

We now collect together terms of equal orders in $\lambda$, and setting $\lambda$ equal to unity, we obtain the zero-order equation

$$\mu\frac{dI_\nu^{\,0}}{dt} = \chi_\nu[I_\nu^{\,0} - B_\nu(T_0)] \tag{6-170}$$

which is the equation we have actually solved to determine the present radiation field, and a first-order equation

$$\mu\frac{dI_\nu^{\,1}}{dt} = \chi_\nu\left(I_\nu^{\,1} - T_1\frac{\partial B_\nu}{\partial T}\right) + (\tau_1'\chi_\nu + \tau_1\chi_\nu')\left(I_\nu^{\,0} - \frac{\partial B_\nu}{\partial T}T_1\right) \tag{6-171}$$

which we will utilize to calculate the perturbations $\tau_1$ and $T_1$. To obtain two equations with which to determine the two unknown perturbations, Avrett and Krook took the zero and first moments of the first-order equation. Let us define

$$J_\nu^{\,0} \equiv \frac{1}{2}\int_{-1}^{1} I_\nu^{\,0}\,d\mu \qquad J^0 \equiv \int_0^\infty J_\nu^{\,0}\,d\nu$$

$$H_\nu^{\,0} \equiv \frac{1}{2}\int_{-1}^{1} I_\nu^{\,0}\mu\,d\mu \qquad H^0 \equiv \int_0^\infty H_\nu^{\,0}\,d\nu$$

and, by analogy, define

$$J_\nu^{\,1} \equiv \frac{1}{2}\int_{-1}^{1} I_\nu^{\,1}\,d\mu \qquad J^1 \equiv \int_0^\infty J_\nu^{\,1}\,d\nu$$

$$H_\nu^{\,1} \equiv \frac{1}{2}\int_{-1}^{1} I_\nu^{\,1}\mu\,d\mu \qquad H^1 \equiv \int_0^\infty H_\nu^{\,1}\,d\nu$$

$$K_\nu^{\,1} \equiv \frac{1}{2}\int_{-1}^{1} I_\nu^{\,1}\mu^2\,d\mu \qquad K^1 \equiv \int_0^\infty K_\nu^{\,1}\,d\nu$$

Taking the moments of equation (6-171), we find

$$\frac{dH_\nu^{\,1}}{dt} = \chi_\nu\left(J_\nu^{\,1} - \frac{\partial B_\nu}{\partial T} T_1\right) + (\tau_1{}'\chi_\nu + \tau_1\chi_\nu{}')[J_\nu^{\,0} - B_\nu(T_0)] \tag{6-172}$$

and

$$\frac{dK_\nu^{\,1}}{dt} = \chi_\nu H_\nu^{\,1} + (\tau_1{}'\chi_\nu + \tau_1\chi_\nu{}')H_\nu^{\,0} \tag{6-173}$$

The perturbations $T_1$ and $\tau_1$ are to be determined in such a way that if $\mathscr{H} = \sigma T_{\text{eff}}^4/4\pi$ is the desired flux, then

$$H^0(t) + H^1(t) = \mathscr{H} \tag{6-174}$$

Now we make further progress, we must introduce some relation among the moments $J^1$, $H^1$, and $K^1$. To do this, we may employ the Eddington approximations in the form

$$K_\nu^{\,1} = \tfrac{1}{3}J_\nu^{\,1} \tag{6-175}$$

and

$$J_\nu^{\,1}(0) = \sqrt{3}H_\nu^{\,1}(0) \tag{6-176}$$

A very reasonable estimate of $H_\nu^{\,1}(0)$ can be written as

$$H_\nu^{\,1}(0) = \left[\frac{H^1(0)}{H^0(0)}\right]H_\nu^{\,0}(0) \tag{6-177}$$

In addition, we make the *arbitrary* assumption

$$\frac{dJ_\nu^{\,1}}{dt} = 0 \tag{6-178}$$

which implies

$$J_\nu^{\,1}(t) \equiv J_\nu^{\,1}(0) = \sqrt{3}H_\nu^{\,1}(0) = 3K_\nu^{\,1}(t) \tag{6-179}$$

Then equation (6-173) may be written

$$\chi_\nu H_\nu^{\,1} + (\tau_1{}'\chi_\nu + \tau_1\chi_\nu{}')H_\nu^{\,0} = 0 \tag{6-180}$$

or

$$\tau_1{}'H_\nu^{\,0} + \tau_1 \frac{H_\nu^{\,0}\chi_\nu{}'}{\chi_\nu} = -H_\nu^{\,1} \tag{6-181}$$

Integrating over frequency, we obtain

$$\tau_1{}'H^0 + \tau_1 \int_0^\infty \frac{H_\nu^{\,0}\chi_\nu{}'}{\chi_\nu}\,d\nu = -H^1 \tag{6-182}$$

or, finally,

$$\tau_1' + \tau_1 \int_0^\infty \frac{H_\nu^0 \chi_\nu'}{H^0 \chi_\nu} \, d\nu = \left(1 - \frac{\mathscr{H}}{H^0}\right) \tag{6-183}$$

This is a simple linear, first-order differential equation, which may be integrated straightaway for $\tau_1$, subject to the initial value $\tau_1(0) = 0$. Note that all of the assumptions above, including the arbitrary condition on $dJ_\nu^1/dt$, will in some (unknown) way influence $\tau_1$. However, since $\tau_1$ will enter the $T_1$ equation, one might hope that compensating effects will be made in $T_1$. In any case, we again realize that these approximations can influence only the *rate* of convergence, not the solution itself.

To obtain an equation for $T_1$, we now consider equation (6-171), and integrating over frequency, we have

$$\frac{dH^1}{dt} = \int_0^\infty \chi_\nu J_\nu^1 \, d\nu - T_1 \int_0^\infty \chi_\nu \frac{\partial B_\nu}{\partial T} \, d\nu + \tau_1 \int_0^\infty \chi_\nu'[J_\nu^0 - B_\nu(T_0)] \, d\nu$$
$$+ \tau_1' \int_0^\infty \chi_\nu [J_\nu^0 - B_\nu(T_0)] \, d\nu \tag{6-184}$$

Now observing that

$$\frac{dH^1}{dt} = \frac{d(\mathscr{H} - H^0)}{dt} = -\frac{dH^0}{dt} = -\int_0^\infty \chi_\nu [J_\nu - B_\nu(T_0)] \, d\nu \tag{6-185}$$

and

$$\int_0^\infty \chi_\nu J_\nu^1 \, d\nu = \sqrt{3} \int_0^\infty \chi_\nu H_\nu^1(0) \, d\nu = \sqrt{3} \left[\frac{\mathscr{H}}{H^0(0)} - 1\right] \int_0^\infty \chi_\nu H_\nu^0(0) \, d\nu \tag{6-186}$$

we obtain

$$T_1(t) = \left\{(1 + \tau_1') \int_0^\infty \chi_\nu [J_\nu^0 - B_\nu(T_0)] \, d\nu \right.$$
$$+ \sqrt{3} \left(\frac{\mathscr{H}}{H^0(0)} - 1\right) \int_0^\infty H_\nu^0(0) \chi_\nu \, d\nu$$
$$\left. + \tau_1 \int_0^\infty \chi_\nu'[J_\nu^0 - B_\nu(T_0)] \, d\nu \right\} \bigg/ \int_0^\infty \chi_\nu \frac{\partial B_\nu}{\partial T} \, d\nu \tag{6-187}$$

Since we already have values for $\tau_1$ and $\tau_1'$ from equation (6-183), we may substitute them into equation (6-187) and obtain directly $T_1(t)$. Thus, we obtain the values $T(\tau) = T_0(t) + T_1(t)$ at depth points $\tau = t + \tau_1(t)$. Since we normally desire the temperatures at some standard depth points, we must finally interpolate the perturbed temperature scale against the perturbed depth scale to find the new temperature distribution at these standard points.

Let us now consider the case in which the source function includes a scattering term so that [equation (1-43)]

$$S_\nu = (1 - \rho_\nu)B_\nu + \rho_\nu J_\nu$$

where

$$\rho_\nu = \frac{\sigma_\nu}{\kappa_\nu + \sigma_\nu}$$

Then we proceed as before, with the additional perturbation equation

$$\rho_\nu(\tau) = \rho_\nu(t) + \lambda \rho_\nu'(t)\tau_1 + \cdots \tag{6-188}$$

We neglect terms of the form $\partial \rho_\nu / \partial T$ since they are small compared with $\rho_\nu'$. Carrying through the analysis, we recover equation (6-183) for $\tau_1$ in unaltered form while the correction $T_1$ becomes

$$T_1 = \left\{(1 + \tau_1') \int_0^\infty \chi_\nu (1 - \rho_\nu)[J_\nu^0 - B_\nu(T_0)]\, d\nu \right.$$
$$+ \sqrt{3}\left(\frac{\mathscr{H}}{H^0(0)} - 1\right) \int_0^\infty \chi_\nu H_\nu^0(0)(1 - \rho_\nu)\, d\nu$$
$$\left. + \tau_1 \int_0^\infty [\chi_\nu'(1 - \rho_\nu) - \chi_\nu \rho_\nu'] [J_\nu^0 - B_\nu(T_0)]\, d\nu \right\} \bigg/ \int_0^\infty \chi_\nu (1 - \rho_\nu) \frac{\partial B_\nu}{\partial T}\, d\nu \tag{6-189}$$

This shows that the $T_1$ correction depends only upon thermal absorption and emission terms, as might be expected.

The Avrett-Krook procedure has been employed by several astronomers to construct models under a wide variety of physical conditions. Normally the results converge swiftly, and the final solution satisfies the radiative equilibrium condition to good accuracy. If care is taken in performing the various integrations required, one can obtain flux constancy of the order $\pm 0.1\%$ and accuracy of $|d \ln F / d\tau|$ of the order of $0.1\%$. Computational experience has shown that the $\tau_1$ correction leads to an effective revision of the temperature scale at depth while the $T_1$ correction is most important near the surface. As in the case of Unsöld's procedure, the $T_1$ correction contains terms depending upon the flux error at the boundary as well as a lambda-iteration term. The term involving $\tau_1$ in the $T_1$ correction is usually quite small compared with the others (and is equal to zero at the very boundary).

It should be emphasized that at the boundary $\tau = 0$, the flux is, for all practical purposes, determined by the temperature distribution at greater depths. Thus, the flux error is actually not the dominant quantity near the surface, but instead the flux derivative is. This is an important point, for when large opacity ratios occur (perhaps due to strong spectral lines or the Lyman continuum), convergence of the lambda operator (which contributes heavily to both the Unsöld and Avrett-Krook procedures near the surface) may be very slow; in these cases a more powerful technique is required.

### THE BÖHM-VITENSE PROCEDURE

One of the criticisms made against the lambda-iteration procedure is that it is simply not a correct solution of the integral equation

$$\int_0^\infty \kappa_\nu B_\nu(T + \Delta T)\, d\nu = \int_0^\infty \kappa_\nu \Lambda_\tau[B_\nu(T + \Delta T)]\, d\nu \tag{6-190}$$

This point was emphasized by Böhm-Vitense, who suggested that the solution should be carried out directly (Ref. 15, p. 99). We again assume that we may write $B_\nu(T) = B_\nu(T_0) + \Delta B_\nu$ and that the temperature change can be written as $B = B_0 + \Delta B$, where $B = \sigma T^4/\pi$, and $B_0 = \sigma T_0{}^4/\pi$. The relation between $\Delta B_\nu$ and $\Delta B$ is assumed to be of the form

$$\Delta B_\nu = \frac{dB_\nu}{dT}\frac{dT}{dB}\Delta B = \frac{dB_\nu}{d\ln T}\frac{\Delta \ln B}{4} \tag{6-191}$$

Then the condition of radiative equilibrium reads

$$\int_0^\infty \kappa_\nu B_\nu(T_0)\, d\nu + \frac{\Delta \ln B}{4}\int_0^\infty \kappa_\nu \frac{dB_\nu}{d\ln T}\, d\nu$$
$$= \int_0^\infty \kappa_\nu J_\nu{}^0\, d\nu + \frac{1}{4}\int_0^\infty \kappa_\nu \Lambda_{\tau_\nu}\left[\Delta \ln B \frac{dB_\nu}{d\ln T}\right] d\nu \tag{6-192}$$

or

$$\int_0^\infty \kappa_\nu \left[\frac{1}{2}\int_0^\infty \Delta \ln B \frac{dB_\nu}{d\ln T} E_1|t_\nu - \tau_\nu|\, dt_\nu - \Delta \ln B \frac{dB_\nu}{d\ln T}\right] d\nu$$
$$= 4\int_0^\infty \kappa_\nu[B_\nu(T_0) - J_\nu{}^0]\, d\nu \tag{6-193}$$

The term on the right-hand side is considered known from the present solution. The left-hand side is to be used to determine the run of $\Delta B(\tau)$, where $\tau$ is the standard optical depth. To solve this equation, we replace $\Delta \ln B(\tau)$ by some functional relation $\sum c_j f_j(\tau)$. Since we have a numerical solution in which the various physical quantities are specified at a finite set of depth points $\{\tau_i\}$, we actually want only $\Delta_i \equiv \Delta \ln B(\tau_i)$. We might therefore assume that between points $\tau_i$ and $\tau_{i+1}$, $\Delta(\tau)$ is simply given by the straight-line segment

$$\Delta(\tau) = \Delta(\tau_i)\frac{(\tau_{i+1} - \tau)}{(\tau_{i+1} - \tau_i)} + \Delta(\tau_{i+1})\frac{(\tau - \tau_i)}{(\tau_{i+1} - \tau_i)} \tag{6-194}$$

and that a similar linear formula is valid at each frequency [i.e., equation (6-194) still holds when we replace $\tau$ by $\tau_\nu$, $\tau_i$ by $\tau_\nu(i)$, etc.]. Then the integration against $E_1$ could be carried out and analytically at each frequency at depth point $i$ and the integral over $\kappa_\nu$ carried out numerically. We obtain, finally, a linear set of equations of the form

$$\sum_j A_{ij}\Delta_j = 4\int_0^\infty \kappa_\nu(\tau_i)[B_\nu(\tau_i) - J_\nu{}^0(\tau_i)]\,d\nu = b_i \tag{6-195}$$

These equations may then be solved by standard methods to yield $\Delta \ln B(\tau)$.

Since this approach represents the actual solution of the integral equation, the temperature correction obtained should be much superior to that given by the lambda-iteration procedure. The discussion given by Böhm-Vitense shows, in particular, how the correction is superior to the lambda procedure at depth. For simplicity, assume that $\Delta \ln B$ can be replaced by some mean value $\overline{\Delta \ln B}$ and $dB_\nu/dT$ by some mean $\overline{dB_\nu/dT}$ so that

$$\begin{aligned} 4\int_0^\infty \kappa_\nu[B_\nu(T_0) - J_\nu{}^0]\,d\nu &\approx \overline{\Delta \ln B}\int_0^\infty \kappa_\nu \overline{\frac{dB_\nu}{d\ln T}}\left[\frac{1}{2}\int_0^\infty E_1|t_\nu - \tau_\nu|\,dt_\nu - 1\right]d\nu \\ &\approx -\frac{1}{2}\overline{\Delta \ln B}\,\overline{E_2(\tau_\nu)}\int_0^\infty \kappa_\nu \overline{\frac{dB_\nu}{d\ln T}}\,d\nu \end{aligned} \tag{6-196}$$

Then

$$\overline{\Delta \ln B}\,\overline{\tfrac{1}{2}E_2(\tau_\nu)} = 4\int_0^\infty \kappa_\nu[J_\nu{}^0 - B_\nu(T_0)]\,d\nu \Big/ \int_0^\infty \kappa_\nu \overline{\frac{dB_\nu}{d\ln T}}\,d\nu \tag{6-197}$$

In contrast, the lambda operation yields

$$\Delta \ln B = 4\int_0^\infty \kappa_\nu[J_\nu{}^0 - B_\nu(T_0)]\,d\nu \Big/ \int_0^\infty \kappa_\nu \frac{dB_\nu}{d\ln T}\,d\nu \tag{6-198}$$

which is smaller than the value predicted by equation (6-197) be a factor of $\overline{E_2(\tau_\nu)/2}$. This shows that the lambda operation yields far too small a change at depth (as we already know) and offers promise that the integral-equation approach will do much better.

### A DIFFERENCE-EQUATION PROCEDURE

The Böhm-Vitense procedure avoids, in principle, most of the difficulties associated with the lambda-iteration procedure both at the surface and at depth. A difficulty associated with it is that it is somewhat time-consuming to carry out the required integrations; also the desired value of the flux does not appear explicitly in the equations, and this is a disadvantage. In most ways, it is more straightforward to solve differential equations, which are also often easier to generalize (say, to non-LTE cases). A procedure that makes use of the two-point boundary method described previously has been developed by Auer and Mihalas (*Ap. J.*, **151**, 311, 1968). Let us write (for the simple thermal case)

$$\mu \frac{dI_\nu}{d\tau} = \chi_\nu(I_\nu - B_\nu{}^*) \tag{6-199}$$

with the radiative equilibrium condition

$$\int_0^\infty \chi_\nu(J_\nu - B_\nu{}^*)\, d\nu = 0 \tag{6-200}$$

In difference-equation form, we write

$$\mu_i{}^2 \frac{d^2 P_i}{d\tau_i{}^2} = P_i - B_i{}^* \tag{6-201}$$

and

$$\sum a_j \chi_j (P_j - B_j{}^*) = 0 \tag{6-202}$$

where $i$ denotes a specific angle *and* frequency choice; sums over $j$ therefore imply sums over both angle and frequency. As before [equation (6-89)], we have defined

$$P_j \equiv \tfrac{1}{2}[I(\tau, \mu_j, \nu_j) + I(\tau, -\mu_j, \nu_j)]$$

Now if the temperature structure of the atmsophere were precisely that which produced radiative equilibrium, then the solutions of equation (6-199) or equation (6-201) would automatically satisfy equation (6-200) or equation (6-201). Usually, however, this particular temperature distribution, and hence $B_i{}^*$, is unknown. If we suppose that the desired temperature distribution differs from the present distribution $T(\tau)$ by some amount $\Delta T(\tau)$, we can then write, to first order,

$$B_i{}^* = B_i + \frac{\partial B_i}{\partial T} \Delta T \tag{6-203}$$

Then

$$\Delta T = \frac{\sum a_j \chi_j (P_j - B_j)}{\sum a_j \chi_j \dfrac{dB_j}{dT}} \tag{6-204}$$

which can be used to eliminate $\Delta T$ from equation (6-201), yielding

$$\mu_i^2 \frac{d^2 P_i}{d\tau_i^2} = P_i - B_i - \frac{\partial B_i}{\partial T} \frac{\sum a_j \chi_j (P_j - B_j)}{\sum a_j \chi_j \dfrac{dB_j}{dT}} \tag{6-205}$$

as the transfer equation to be solved. Note that this equation is formally identical to equation (6-108) so that the solution proceeds exactly as before. Only the details of the **A**-, **B**-, and **C**-matrices have been altered. In particular, the **B**-matrix now contains terms coupling the radiation field at any given frequency to that at *all* other frequencies. In this form, we are essentially solving the transfer equation *subject to a constraint* of radiative equilibrium. Having obtained the solution of equation (6-205), one can later compute the run of $\Delta T(\tau)$ from equation (6-204) and iterate if necessary.

It is important to realize that this approach is *not* in any sense a lambda iteration but is equivalent to a direct solution of the integral equation. A lambda iteration results only if the last term of equation (6-205) is suppressed. The essential point is that although we have introduced a local perturbation $\Delta T(\tau)$, we have incorporated it into the transfer equation in such a way as to include explicitly from the outset the coupling to all other points in the atmosphere via the mean intensity. Thus, convergence will be global. In fact, for the case where the $\chi$'s are independent of temperature, this approach would, in principle, give the correct solution in a single step, but, in practice, it is limited by the accuracy of the linearization in equation (6-203).

The solution obtained is, of course, the solution of the *difference* equations. Because of discretizations both in angle and in depth, this solution may differ slightly from the solution of the differential equation. On the other hand, convergence is not inhibited by large opacity ratios since these influence only the coefficients of a linear algebraic set of equations, and insofar as the correct inverses are obtained for this set, we still obtain the correct solution. This clearly is an important advantage.

The use of correct boundary conditions is quite important. At the upper boundary one requires, as usual, that

$$I(0, -\mu, \nu) = 0$$

At the lower boundary we have

$$\mu_i \frac{dP_i}{d\tau_i} = I_{+i} - P_i \tag{6-206}$$

where

$$I_{+i} = B_i + \frac{\partial B_i}{\partial T} \Delta T + \mu \frac{\partial B_i}{\partial \tau_i} \tag{6-207}$$

If we assume that $B = \int_0^\infty B_\nu \, d\nu$ is linearly expandable on *some* optical depth scale, we can write

$$\frac{\partial B_i}{\partial \tau_i} = \left(\frac{\partial B_i}{\partial T}\right)\left(\frac{dT}{dB}\right)\left(\frac{dB}{d\tau}\right)\left(\frac{d\tau}{d\tau_i}\right) = \frac{\pi}{4\sigma T^3 \chi_i} \frac{\partial B_i}{\partial T} Z \tag{6-208}$$

so that

$$I_{+i} = B_i + \frac{\partial B_i}{\partial T} \Delta T + \frac{\pi \mu_i}{4\sigma T^3 \chi_i} \frac{\partial B_i}{\partial T} Z \tag{6-209}$$

In equation (6-209), we can eliminate $\Delta T$ via equation (6-204), and $Z$ via the requirement that

$$\mathscr{H} = \frac{\sigma T_{\text{eff}}^4}{4\pi} = \sum a_j \mu_j (I_{+j} - P_j) \tag{6-210}$$

which introduces explicitly the desired total flux. Thus, using equations (6-204), (6-209), and (6-210), equation (6-206) becomes

$$\mu_i \frac{dP_i}{d\tau_i} = B_i - P_i + \frac{\partial B_i}{\partial T}\left(\sum a_j \chi_j P_j - E_4\right)/E_1$$
$$+ \frac{\mu_i}{\chi_i}\frac{\partial B_i}{\partial T}\left[\mathscr{H} + \sum a_j \mu_j P_j - E_5 + \frac{E_2}{E_1}\left(E_4 - \sum a_j \chi_j P_j\right)\right] \Big/ E_3 \tag{6-211}$$

where

$$E_1 = \sum a_j \chi_j \frac{\partial B_j}{\partial T} \tag{6-212}$$

$$E_2 \equiv \sum a_j \mu_j \frac{\partial B_j}{\partial T} \tag{6-213}$$

$$E_3 \equiv \sum \frac{a_j \mu_j^2}{\chi_j} \frac{\partial B_j}{\partial T} \tag{6-214}$$

$$E_4 \equiv \sum a_j \chi_j B_j \tag{6-215}$$

and

$$E_5 \equiv \sum a_j \mu_j B_j \tag{6-216}$$

The explicit introduction of the flux into equation (6-211) assures that the solution will converge to the desired value.

The principal limitation of this method is that the matrix sizes become large, and from restrictions in computing time or machine size, it may not be easy to carry through the computation in practice. On the other hand, this is basically a technical problem which may be overcome by more advanced computing machines. Moreover, the method can easily be generalized to more complex forms for the source function. We shall demonstrate this in Chapter 7.

## 6-4. Results of LTE Model Atmosphere Calculations for Early-Type Stars

The greatest number of models has been computed for early-type stars (F0 and earlier), and we will confine our attention here to this group. Two large groups of models for normal early-type stars have been published by Mihalas (*Ap. J. Supp. No. 92*, **9**, 321, 1965, and *Ap. J. Supp. No. 114*, **13**, 1, 1966) and by Strom and Avrett (*Ap. J. Supp. No. 103*, **12**, 1, 1965). A series of model atmospheres for white dwarfs has been given by Matsushima and Terashita (*Ap. J.*, **156**, 203, 1969). Models for yellow supergiants (5400°K $\leq T_{\text{eff}} \leq$ 6600°K) have been published by Parsons (*Ap. J. Supp. No. 159*, **18**, 127, 1969). Numerous other models have been published by several other authors; a fairly complete list has been given by Pecker (*Ann. Rev. Astron. Ap.*, **3**, 135, 1965) and need not be repeated here since the results in the above references are fairly typical. These calculations all assume LTE and enforce the radiative equilibrium condition with the Avrett-Krook procedure. The main results from these calculations are emergent fluxes—which can be compared with observation—and the run of the physical variables with depth.

### THE EMERGENT ENERGY DISTRIBUTION

The emergent flux distribution from models can be compared directly with observations with narrow-band photoelectric scanners. Normally, the observed fluxes are expressed in energy units, corrected for the sensitivity of the instrument but containing an arbitrary multiplying factor (since actual absolute energies are difficult to measure at low light levels). These observed energy distributions are usually calibrated in terms of the star $\alpha$ Lyrae (Vega), whose energy distribution has, in turn, been compared with the radiation from standard lamps and blackbody furnaces. The details of this procedure would take us too far afield here. [See, e.g., the discussions by Code (Ref. 13, Chap. 2) and by Oke (*Ap. J.*, **131**, 358, 1960, and *Ap. J.*, **140**, 189, 1964) for a description of methods and results.] The calibration of $\alpha$ Lyrae is, therefore, fundamental to our knowledge of stellar energy distributions, and it is important that the calibration be as accurate as possible since it serves as a strong test of the accuracy of the stellar atmospheres models.

In comparing theory with observation in early-type stars, there are a few outstanding features in the flux distribution that are normally considered. The first of these is the slope of the energy distribution in the *Paschen continuum* (3650 Å $\leq \lambda \leq$ 8206 Å). This region of the spectrum derives its name from the fact that the dominant opacity source in early-type stars on this wavelength range is bound-free absorptions from the $n = 3$ level of hydrogen. Two other important features are the *Balmer jump*,

$$D_B = 2.5 \log[F_\nu(\lambda 3650^+)/F_\nu(\lambda 3650^-)] \tag{6-217}$$

and the *Paschen jump*,

$$D_P = 2.5 \log[F_\nu(\lambda 8206^+)/F_\nu(\lambda 8206^-)] \tag{6-218}$$

These parameters give a measure of the decrease in flux as one compares wavelengths at which there are no absorptions from the $n = 2$ level (in the case of the Balmer jump) to those that lie in the $n = 2$ photoionization continuum. The much higher opacity on the short wavelength sides of these jumps allows us to see only the outer cooler layers, from which the emitted flux is much lower. Thus, there are relatively sudden drops in the flux across these boundaries (actually the drop is not completely sharp because of overlapping spectrum lines near the photoionization edges).

Until very recently, the confrontation between theory and observation has not been entirely satisfactory. For example, in a comparison of the energy distribution of Vega with (hydrogen-line blanketed) models, Mihalas found the best fit was only fairly good and was compatible with the Balmer jump measured by Bahner (*Ap. J.*, **138**, 1314, 1963) but inconsistent with the older measures of Chalonge and Divan (*Ann. d'Ap.*, **15**, 201, 1952) by about 0.06 magnitudes (6%). The effective temperature derived from Vega was 9600°K. More recently, Hayes has recalibrated Vega by photoelectric scanner comparisons with standard lamps. He obtains a significantly different energy distribution, which leads to an effective temperature of 9940°K. Hayes's measured energy distribution agrees very closely with the prediction of models at this temperature (see Woolf, Kuhi, and Hayes, *Ap. J.*, **152**, 871, 1968).

If one feels the calibration of Vega is sufficiently reliable, he may then compare other stars with model atmosphere calculations. With the older calibration of Vega, the agreement between theory and observation was not too good. Generally, it was found that the temperature derived from the slope of the continuum was inconsistent with that derived from the Balmer jump. In particular, when a fit was made to the slope of the Paschen continuum, the computed Balmer jump was invariably found to be too large (see Figure 6-1). Three possibilities were suggested to account for this disagreement: (a) errors in the calibration of Vega; (b) inadequacies of the models (e.g., neglect of non-LTE effects); and (c) influence of interstellar reddening. Let us examine these in turn.

The effects of interstellar reddening would decrease the slope of the Paschen continuum while leaving the Balmer jump unaltered. In principle, this change could explain the discrepancy between observation and theory,

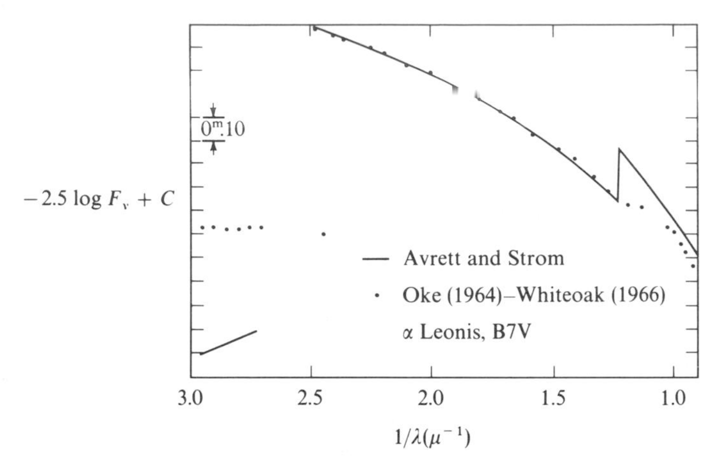


FIG. 6-1. Comparison of the energy distribution of α Leo, using the old calibration, with the model atmosphere that best fits the Paschen continuum and with $T_{eff} = 10{,}000°K$ and log $g = 4$. Note that the derived effective temperature is much too low and the computed Balmer jump is much too large. (From D. Hayes, Ph.D. thesis, University of California, Los Angeles, 1968; by permission.)

but in practice it can be excluded because many of the stars are known to be too nearby to be affected by reddening to the degree that would be required.

The second possibility, of non-LTE effects, will be discussed in the next chapter, but we can state briefly here that the effects turn out to be negligible for main-sequence stars. Thus, by exclusion, one might argue that errors existed in the earlier calibration work. This argument is supported by the newer calibration of Hayes, which, as mentioned above, leads to very good agreement with models. For example, the same star shown in Figure 6-1 is plotted again in Figure 6-2 with the new calibration; the improvement is striking. We emphasize that while the present results are quite satisfactory, the calibration is not yet definitive, and further observational work is urgently needed on this fundamental problem.

Accurate flux distributions are not known for large numbers of stars, and for a given star, one may have to rely upon less observational information. For many stars, Balmer jumps are available from the work of Chalonge and his associates (*Ann. d'Ap.*, **15**, 201, 1952, and references cited therein). This parameter is useful because model computations show that it is very sensitive to $T_{eff}$ above about 10,000°K and insensitive to gravity at these temperatures (see Figure 6-3). Thus, one can estimate $T_{eff}$ from the Balmer jump, but he must remember that the observational material was obtained photographically and may contain important calibration errors. At temperatures below about 8000°K the Balmer jump becomes very sensitive to the gravity, and if the

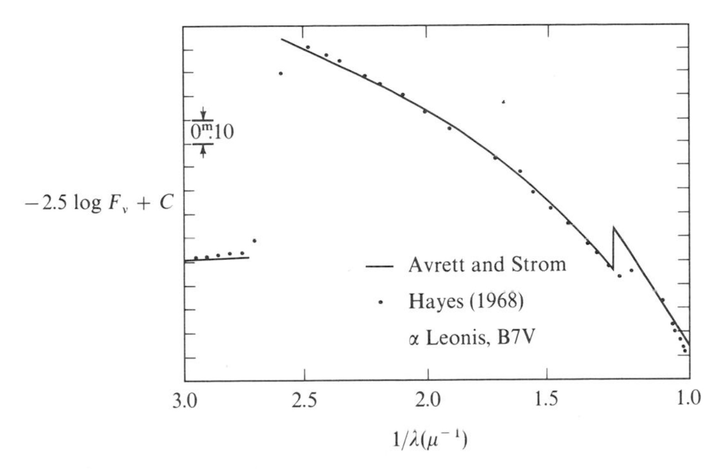


FIG. 6-2. Comparison of the energy distribution of α Leo, using the new calibration by Hayes, with the model atmosphere that best fits the Paschen continuum, namely, with $T_{\text{eff}} = 13{,}000°\text{K}$ and $\log g = 4$. Note that computed and observed Balmer jumps are consistent. (From D. Hayes, Ph.D. thesis, University of California, Los Angeles, 1968; by permission.)

effective temperature can be estimated by some independent means, the Balmer jump serves as a gravity indicator. One advantage of the Balmer jump is that it is unaffected by interstellar reddening, though this must be weighed against the difficulties of measurement and calibration mentioned above.

For a still larger sample of stars, the most reliable observational data are colors measured with broad-band photometers. These may be measured with high precision and can be corrected for the effects of interstellar reddening. One problem that arises in comparing observed colors with theory is the presence of spectral lines which block the flux in some windows and redistribute it into others. Many theoretical models have ignored the presence of lines and, therefore, yield essentially unusable computed colors, except at very high temperatures where the lines are weak. Models allowing for the effects of hydrogen lines have been computed and can be compared with hot stars, in which metal lines are weak, or with stars of very low metal abundance. Again the agreement between theory and observation is satisfactory (see Figure 6-4).

We have sketched here only a few of the more obvious kinds of checks that can be made between theory and observation. Actually, a substantial amount of literature on this subject exists, and we could not hope to do it justice in a few sentences. We can only remark that the agreement between observation and theory in the continuum seems now to be fairly good but that very high accuracy in the observations is desirable if we hope to be able to discriminate physically important effects, such as deviations from LTE. Thus,

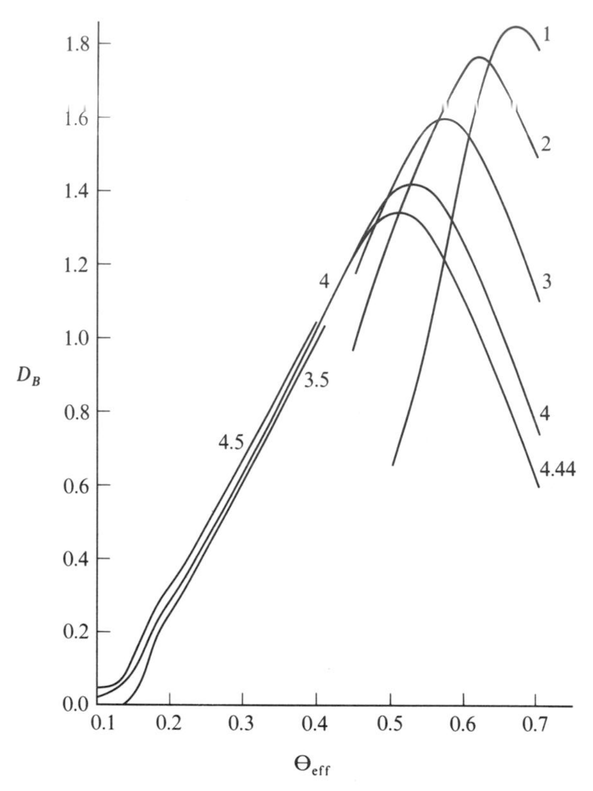


FIG. 6-3. Balmer jumps computed from LTE model atmospheres, as a function of effective temperature and gravity. Ordinate: Balmer jump in magnitude units; abscissa: $\theta_{\text{eff}} = 5040/T_{\text{eff}}$. Curves are labeled with log $g$.

in a sense, the future lies completely in the hands of the observer, who must find whether or not the theory works and thereby force the theoretician to improve his calculations where necessary.

## THE TEMPERATURE STRUCTURE OF THE ATMOSPHERE

In addition to emergent fluxes, the references mentioned above give results for the run of the physical variables $T(\tau)$, $P_g(\tau)$, $P_e(\tau)$, $(\kappa + \sigma)_{\text{std}}$, and $\tau_\nu(\tau)$, $S_\nu(\tau)$, $\kappa_\nu/(\kappa + \sigma)_{\text{std}}$, and $\sigma_\nu/(\kappa + \sigma)_{\text{std}}$ at selected wavelengths. These data can be used in other kinds an analyses, such as the calculation of spectral line strengths and profiles.

Among these results we will consider here the temperature distribution and, in particular, the ratio $T_0/T_{\text{eff}}$, which can be compared with the gray value of 0.811. Figure 6-5 shows typical values of this ratio found

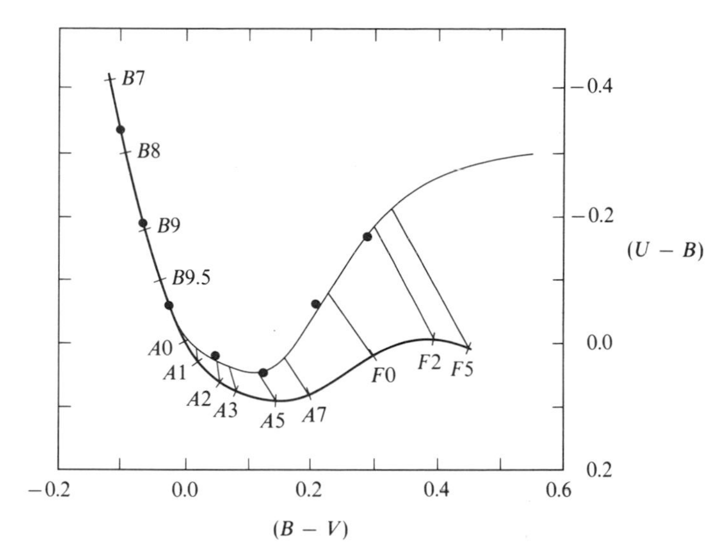


FIG. 6-4. Comparison of computed colors with observed colors. *Heavy curve*: observed colors of main sequence stars; *light curve*: observed colors of metal-poor stars; and *dots*: computed colors from hydrogen line-blanketed models.

from model atmosphere calculations; we see that at the cool end, $T_0/T_{\text{eff}}$ is nearly the gray value. This is not surprising, inasmuch as the opacity of $\text{H}^-$, which is dominant here, is only weakly frequency-dependent and may be considered as practically gray. As we proceed to higher temperatures, the boundary temperature drops, which is an indication of the effects of non-grayness in the atmosphere (the Balmer jump). At $\theta_{\text{eff}} = 0.23$, $(\theta = 5040/T)$, the curve shows a sharp break due to the effects of the Lyman continuum, which was not included in the calculation at lower effective temperatures. Qualitatively, it is easy to see that the presence of a large opacity jump, such as the Lyman jump, will cause the boundary temperature to drop. Suppose the opacity rises from $\kappa$ to $\eta\kappa$ at $\nu = \nu_0$. Then radiative equilibrium demands

$$\kappa \int_0^{\nu_0} B_\nu(T_0)\, d\nu + \eta\kappa \int_{\nu_0}^{\infty} B_\nu(T_0)\, d\nu = \kappa \int_0^{\nu_0} J_\nu\, d\nu + \eta\kappa \int_{\nu_0}^{\infty} J_\nu\, d\nu \tag{6-219}$$

If $\eta = 1$, we find (roughly) that $J_\nu \approx \frac{1}{2}B_\nu(\tau \approx 1)$, so integrating over frequency implies $T_0{}^4 \approx \frac{1}{2}T_{\text{eff}}^4$, as obtained earlier for the gray problem. If, now, $\eta \gg 1$, then for $\nu \geq \nu_0$, $J_\nu \approx \frac{1}{2}B_\nu(T_0)$, not $\frac{1}{2}B_\nu(\tau \approx 1)$, so

$$\int_0^{\infty} B_\nu(T_0)\, d\nu = \int_0^{\infty} J_\nu\, d\nu - \frac{(\eta - 1)}{2} \int_{\nu_0}^{\infty} B_\nu(T_0)\, d\nu \tag{6-220}$$

If the last term were zero, we would recover the $T_0$ found above. Since it is negative, clearly $T_0$ will now be lower than the gray value. As we proceed to

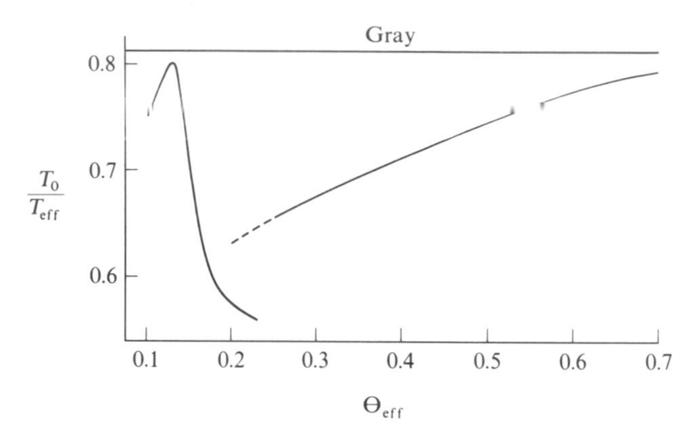


FIG. 6-5. Ratio of boundary temperature $T_0$ to effective temperature $T_{eff}$ as a function of $T_{eff}$. Break at 21,000°K is due to inclusion of Lyman continuum in high-temperature models. Upper line gives value of $T_0/T_{eff}$ for a gray atmosphere.

still higher values of $T_{eff}$, hydrogen becomes more strongly ionized and the effect of the Lyman jump diminishes. Indeed, at very high temperatures, this effect—also the fact that now the maximum of the radiation has shifted beyond the Lyman jump—allows $T_0$ to rise to its gray value again. At still higher temperatures, there is an indication that $T_0$ drops again, now due to radiation encountering the opacity jump at the $\lambda$504 Å of He I.

It is possible to construct a simple theory that yields a semiquantitative estimate of the effects of an opacity jump. The form of the equations is identical to that used first by Chandrasekhar (*M. N.*, **96**, 21, 1935) and later by Münch (*Ap. J.*, **104**, 87, 1946) in the picket fence model of line blanketing. We will return to this model in its original context in Chapter 14. Let us suppose that we can divide the frequency range into two bands, with $\kappa_\nu \equiv \kappa$ for $0 \leq \nu \leq \nu_0$ (band 1) and $\kappa_\nu \equiv \eta\kappa$ for $\nu_0 \leq \nu \leq \infty$ (band 2). Then in these bands the transfer equations are:

$$\mu \frac{dI_\nu^{(1)}}{d\tau} = I_\nu^{(1)} - B_\nu^{(1)} \tag{6-221}$$

and

$$\frac{\mu}{\eta} \frac{dI_\nu^{(2)}}{d\tau} = I_\nu^{(2)} - B_\nu^{(2)} \tag{6-222}$$

If we integrate these equations over frequency and write

$$\int_0^{\nu_0} B_\nu \, d\nu = w_1 B$$

and

$$\int_{\nu_0}^{\infty} B_\nu \, d\nu = w_2 B$$

we have

$$\mu \frac{dI^{(1)}}{d\tau} = I^{(1)} - w_1 B \tag{6-223}$$

and

$$\frac{\mu}{\eta} \frac{dI^{(2)}}{d\tau} = I^{(2)} - w_2 B \tag{6-224}$$

For simplicity we now *assume* $w_1$ and $w_2$ are constant with depth; we might take for these constant values the numbers appropriate at $T = T_{\text{eff}}$. As pointed out by Münch (Ref. 13, p. 38) this approximation is very crude, but we will employ it because it greatly simplifies the analysis. In the picket fence model the interpretation of $w_1$ and $w_2$ is different, as will be explained in Chapter 14.

Now the requirement of radiative equilibrium demands [equation (1-92)]:

$$\int_0^{\infty} \kappa_\nu J_\nu \, d\nu = \int_0^{\infty} \kappa_\nu B_\nu \, d\nu$$

or

$$\kappa[J^{(1)} + \eta J^{(2)}] = \kappa[w_1 B + w_2 \eta B] \tag{6-225}$$

so that

$$B = \frac{J^{(1)} + \eta J^{(2)}}{w_1 + w_2 \eta} \tag{6-226}$$

and therefore

$$\mu \frac{dI^{(1)}}{d\tau} = I^{(1)} - \frac{w_1}{w_1 + w_2 \eta} [J^{(1)} + \eta J^{(2)}] \tag{6-227}$$

and

$$\frac{\mu}{\eta} \frac{dI^{(2)}}{d\tau} = I^{(2)} - \frac{w_2}{w_1 + w_2 \eta} [J^{(1)} + \eta J^{(2)}] \tag{6-228}$$

Now following the discrete-ordinate approach, we write

$$J^{(\alpha)} = \frac{1}{2} \sum_{j=-n}^{n} a_j I_j^{(\alpha)} \tag{6-229}$$

Also for brevity write $\eta_1 \equiv 1$ and $\eta_2 \equiv \eta$ so that the above equations can be written

$$\frac{\mu_i}{\eta_\alpha}\frac{dI_i^{(\alpha)}}{d\tau} = I_i^{(\alpha)} - \frac{w_\alpha}{\sum_\beta w_\beta \eta_\beta}\frac{1}{2}\sum_\beta \eta_\beta \sum_{j=-n}^{n} a_j I_j^{(\beta)} \qquad \begin{matrix}(\alpha = 1, 2)\\(i = \pm 1, \ldots, \pm n)\end{matrix} \tag{6-230}$$

If we now assume a solution of the form

$$I_i^{(\alpha)} = \frac{Cw_\alpha e^{-k\tau}}{1 + k\mu_i/\eta_\alpha} \tag{6-231}$$

we find that the $k$'s must satisfy the characteristic equation

$$w_1 + \eta w_2 = w_1 \sum_{j=1}^{n} \frac{a_j}{1 - k^2\mu_j^{\,2}} + w_2 \eta \sum_{j=1}^{n} \frac{a_j}{1 - k^2\mu_j^{\,2}/\eta^2} \tag{6-232}$$

This equation yields $2n - 1$ nonzero roots for $k^2$ and hence $4n - 2$ values for $k$ of the form $\pm k_i$. In addition, $k^2 = 0$ is a root of the characteristic equation. A particular solution for the zero root must be of the form

$$I_i^{(\alpha)} = bw_\alpha(\tau + q_i) \tag{6-233}$$

which leads to

$$I_i^{(\alpha)} = bw_\alpha(\tau + Q + \mu_i/\eta_\alpha) \tag{6-234}$$

by direct substitution into the transfer equation. Thus, the general solutions of the form

$$I_i^{(\alpha)} = w_\alpha b\left(\tau + Q + \frac{\mu_i}{\eta_\alpha} + \sum_{\beta=1}^{2n-1} \frac{L_\beta e^{-k_\beta \tau}}{1 + k_\beta \mu_i/\eta_\alpha} + \sum_{\beta=1}^{2n-1} \frac{L_{-\beta} e^{k_\beta \tau}}{1 - k_\beta \mu_i/\eta_\alpha}\right) \qquad \begin{matrix}(\alpha = 1, 2)\\(i = \pm 1, \ldots, +n)\end{matrix} \tag{6-235}$$

If the solution is to remain bounded as $\tau \to \infty$, we must demand $L_{-\beta} \equiv 0$ for all values of $\beta$. Requiring that

$$F = 2\sum_{\alpha=1}^{2} \sum_{j=-n}^{n} a_j \mu_j I_j^{(\alpha)}$$

leads to

$$b = \frac{\frac{3}{4}F}{\sum_{\alpha=1}^{2} w_\alpha \eta_\alpha^{-1}} \tag{6-236}$$

The constant $Q$ and $L_\beta$'s are determined from the surface boundary conditions $I_i^{(\alpha)} \equiv 0$ which imply

$$Q + \sum_{\beta=1}^{2n-1} \frac{L_\beta}{1 - k_\beta \mu_i/\eta_\alpha} = \frac{\mu_i}{\eta_\alpha} \qquad \begin{matrix}(\alpha = 1, 2)\\(i = 1, \ldots, n)\end{matrix} \tag{6-237}$$

There are $2n$ unknowns and $2n$ equations. Thus, finally,

$$I_i^{(\alpha)} = \frac{\frac{3}{4}Fw_\alpha}{\sum w_\alpha \eta_\alpha^{-1}} \left( \tau + \frac{\mu_i}{\eta_\alpha} + Q + \sum_{\beta=1}^{2n-1} \frac{L_\beta e^{-k_\beta \tau}}{1 + k_\beta \mu_i/\eta_\alpha} \right) \tag{6-238}$$

$$J^{(\alpha)} \equiv \frac{1}{2} \sum_{j=-n}^{n} a_j I_j^{(\alpha)} = \frac{\frac{3}{4}Fw_\alpha}{\sum w_\alpha \eta_\alpha^{-1}} \left( \tau + Q + \sum_{\beta=1}^{2n-1} L_\beta e^{-k_\beta \tau} \sum_{j=1}^{n} \frac{a_j}{1 - k_\beta^2 \mu_j^2/\eta_\alpha^2} \right) \tag{6-239}$$

$$\begin{aligned} F^{(\alpha)} &\equiv 2 \sum_{j=-n}^{n} a_j \mu_j I_j^{(\alpha)} \\ &= \frac{Fw_\alpha}{\sum w_\alpha \eta_\alpha^{-1}} \left( \frac{1}{\eta_\alpha} - 3\eta_\alpha \sum_{\beta=1}^{2n-1} L_\beta e^{-k_\beta \tau} \frac{k_\beta}{\eta_\alpha} \sum_{j=1}^{n} \frac{a_j \mu_j^2}{1 - k_\beta^2 \mu_j^2/\eta_\alpha^2} \right) \end{aligned} \tag{6-240}$$

and, by equation (6-226),

$$B(\tau) = \frac{\frac{3}{4}F}{\sum w_\alpha \eta_\alpha^{-1}} \left( \tau + Q + \sum_{\beta=1}^{2n-1} L_\beta e^{-k_\beta \tau} \right) \tag{6-241}$$

As in the case of the discrete-ordinate representation of the gray problem, it is possible to eliminate the constants analytically. If we define the function

$$S(\mu) = \sum_{\beta=1}^{2n-1} \frac{L_\beta}{1 - \mu k_\beta} + Q - \mu \tag{6-242}$$

we may show, by an analysis similar to the gray-atmosphere case, that

$$S(\mu) = k_1 \ldots k_{2n-1} \frac{\prod_{\alpha=1}^{2} \prod_{i=1}^{n} \left( \frac{\mu_i}{\eta_\alpha} - \mu \right)}{\prod_{\alpha=1}^{2n-1} (1 - \mu k_\alpha)} \tag{6-243}$$

which implies

$$S(0) = \frac{k_1 \ldots k_{2n-1} \mu_1^2 \ldots \mu_n^2}{\eta^n} \tag{6-244}$$

On the other hand, by an analysis of the characteristic equation we may show

$$k_1 \ldots k_{2n-1} \mu_1^2 \ldots \mu_n^2 \eta^{-n} = \left[ \frac{w_1 + w_2/\eta}{3(w_1 + w_2 \eta)} \right]^{1/2} \tag{6-245}$$

which, by substitution into equation (6-244), yields

$$S(0) = \frac{1}{\sqrt{3}} \left[ \frac{w_1 + w_2/\eta}{w_1 + w_2 \eta} \right]^{1/2} \tag{6-246}$$

But if we compare the expression for $S(0)$ and $B(0)$ as given by equation (6-241), we find

$$B(0) = \frac{\frac{3}{4}FS(0)}{(w_1 + w_2/\eta)} \tag{6-247}$$

so that

$$B(0) = \frac{\sqrt{3}}{4} F[(w_1 + w_2 \eta)(w_1 + w_2/\eta)]^{-1/2} \tag{6-248}$$

Now we note that

$$\bar{\kappa}_P \equiv \int \frac{\kappa_\nu B_\nu}{B} \, d\nu = \frac{\kappa}{B}(w_1 B + w_2 \eta B) = \kappa(w_1 + w_2 \eta) \tag{6-249}$$

while

$$\frac{1}{\bar{\kappa}_R} \equiv \frac{\int \frac{1}{\kappa_\nu} \frac{\partial B_\nu}{\partial T} \, d\nu}{(dB/dT)} = \left(\frac{dB}{dT}\right)^{-1} \left(\frac{w_1}{\kappa} \frac{dB}{dT} + \frac{w_2}{\eta\kappa} \frac{dB}{dT}\right)$$

or

$$\bar{\kappa}_R = \kappa(w_1 + w_2/\eta)^{-1} \tag{6-250}$$

so that

$$\frac{\bar{\kappa}_R}{\bar{\kappa}_P} = [(w_1 + w_2 \eta)(w_1 + w_2/\eta)]^{-1} \tag{6-251}$$

Thus we may rewrite equation (6-248) as

$$B(0) = \frac{\sqrt{3}}{4} F \left(\frac{\bar{\kappa}_R}{\bar{\kappa}_P}\right)^{1/2} \tag{6-252}$$

or

$$\frac{T_0}{T_{\text{eff}}} = \left(\frac{\sqrt{3}}{4}\right)^{1/4} \left(\frac{\bar{\kappa}_R}{\bar{\kappa}_P}\right)^{1/8} \tag{6-253}$$

This expression clearly gives the correct result in the gray case. Let us now use this result to estimate the effect of the Lyman continuum upon the boundary temperature of a stellar atmosphere. We assume that the Rosseland mean opacity—because it is a *reciprocal* mean—is essentially unaltered by the Lyman continuum, and then we calculate the Planck mean with and without the Lyman continuum. We then use these values to estimate $T_0$ with and without the Lyman continuum. To calculate the Planck mean, we include hydrogen opacity only, and from the results of Chapter 4, we use an absorption coefficient of the form

$$\kappa_\nu = \frac{C}{u^3}(1 - e^{-u})\left[\sum_{u>u_n} \frac{2u_1 \exp(u_1/n^2)}{n^3} + 1\right]$$

Also the Planck function is of the form

$$B_\nu = \frac{C' u^3 e^{-u}}{(1 - e^{-u})}$$

Here $\mu = h\nu/kT_{\text{eff}}$, $u_1 = \chi_I/kT_{\text{eff}}$, etc. We extend the integral over the frequency from 0 to $u_0$, where $u_0 = u_1$ when the Lyman continuum is excluded, and $u_0 = \infty$ when it is included. Then we obtain

$$\begin{aligned} \bar{\kappa}_P(u_0) &= C \int_0^{u_0} e^{-u}\left[\sum_{u>u_n} \frac{2u_1 \exp(u_1/n^2)}{n^3} + 1\right] du \\ &= C\left\{1 - e^{-u_0} + \sum_n \frac{2u_1}{n^3}\left[1 - \exp\left(\frac{u_1}{n^2} - u_0\right)\right]\right\} \end{aligned} \tag{6-254}$$

Now for $\theta_{\text{eff}} = 0.23$, $u_1 = 2.302 \times 0.23 \times 13.6 = 7.2$. If $u_0 = \infty$, the term $\exp(u_1/n^2 - u_0)$ is simply zero. If $u_0 = u_1$, then the exponential can be neglected unless $n = 1$. Therefore,

$$\bar{\kappa}_P(\infty) = C\left(1 + 2u_1 \sum_{n=1}^{\infty} \frac{1}{n^3}\right) = C(1 + 2.4u_1) \tag{6-255}$$

while

$$\bar{\kappa}_P(u_1) = C\left(1 + 2u_1 \sum_{n=2}^{\infty} \frac{1}{n^3}\right) = C(1 + 0.4u_1) \tag{6-256}$$

Thus

$$\frac{\bar{\kappa}_P\,(\text{with Lyman continuum})}{\bar{\kappa}_P\,(\text{without Lyman continuum})} = \frac{1 + 2.4 \times 7.2}{1 + 0.4 \times 7.2} = 4.68$$

which implies

$$\frac{T_0\,(\text{with Lyman continuum})}{T_0\,(\text{without Lyman continuum})} = \left(\frac{1}{4.68}\right)^{1/8} = 0.825$$

In Figure 6-5, we see that if we extrapolate the results *without* Lyman continuum to $\theta_{\text{eff}} = 0.23$, we find $T_0/T_{\text{eff}} = 0.647$. On the other hand, by including the Lyman continuum we find $T_0/T_{\text{eff}} = 0.560$. Thus

$$\frac{T_0\,(\text{with Lyman continuum})}{T_0\,(\text{without Lyman continuum})} = 0.865$$

This is in excellent agreement with the value predicted by the simple theory developed above, considering all of the approximations that were made. We emphasize that this steep drop occurs only at very shallow optical depths where the Lyman continuum becomes transparent. At depths of even $10^{-4}$ in the visible, the Lyman continuum is opaque and the temperatures of models with and without Lyman continuum agree.

## 6-5. Convection and Models for Late-Type Stars

As was discussed in Section 1-5, energy transport in a stellar atmosphere may proceed by radiative transfer or by convection. As we shall see below, the process that prevails at any point in the atmosphere is the one which can transport the flux more efficiently. We should state at the outset that at present time, the theory of convection is not in an entirely satisfactory state. The basic reason is that simple dimensional arguments show immediately that the flow is *turbulent* (see, e.g., the discussion by Biermann, *Z. Ap.*, **25**, 135, 1948) and consists of a complicated hierarchy of "eddies" or "bubbles" moving and interacting in an involved way. We are therefore faced with an extremely complex physical situation, and many difficult questions arise about the correct physical and mathematical formulation of the problem. The equations of hydrodynamics in this situation become hopelessly difficult to solve. As a result, despite the very diligent efforts of a number of investigators, we are still far from a rigorous theory for convection. On the other hand, it has been possible to construct a rather phenomenological theory, which contains a number of fairly basic ingredients of the situation in a plausible (if not entirely precise) form. This theory, the so-called *mixing-length theory*, has been widely used in astronomy and, aside from a few more precise studies of convection in the solar atmosphere, still provides most of our theoretical knowledge of the structure of the envelopes and atmospheres of late-type stars. In this section, we shall outline the broad features of this approach as employed in stellar atmospheres computations.

### THE SCHWARZSCHILD STABILITY CRITERION

Suppose we have an atmosphere in radiative equilibrium. We then ask whether an element of material, when displaced from its original position, experiences forces which tend to move it farther in the direction of its displacement. If so, the atmosphere is unstable against mass motions, and convection will occur; if not, the convective motions will be damped and will die out, and radiative equilibrium will persist. The basic criterion for stability against convection was established by K. Schwarzschild (Ref. 26, p. 25). The basic argument proceeds as follows.

Consider a small element of gas; imagine that its position is perturbed upward by a distance $\Delta r$ in the atmosphere. We suppose that the movement occurs so slowly that the element remains in hydrostatic equilibrium with its surroundings, i.e., the pressure inside the element is exactly equal to that outside. Further, suppose that element does not exchange energy with its surroundings so that the motion is *adiabatic*. Since the pressure drops as the element rises, the gas will expand and the density inside will decrease by an amount

$$(\Delta\rho)_E = \left(\frac{d\rho}{dr}\right)_A \Delta r \tag{6-257}$$

where the subscript $E$ denotes "element" and the subscript $A$ denotes "adiabatic." If, at its new position, the density of the element is less than that of its surroundings, it will experience a bouyancy force and will continue to rise, i.e., we will have *instability* against convective motions if

$$(\Delta\rho)_E = \left(\frac{d\rho}{dr}\right)_A dr < (\Delta\rho) = \left(\frac{d\rho}{dr}\right)_R dr \tag{6-258}$$

where the subscript $R$ denotes "radiative" and the derivative $(d\rho/dr)_R$ describes the density gradient in the surroundings (assumed radiative). Recall here that the derivatives are *negative* since the density decreases outward; then equivalently, we have convection if

$$\left|\frac{d\rho}{dr}\right|_A > \left|\frac{d\rho}{dr}\right|_R \tag{6-259}$$

This equation may be cast into a more convenient form. In the adiabatic element (which we shall momentarily assume is a perfect gas) the equation of state is

$$\ln p = \gamma \ln \rho + C \tag{6-260}$$

so that

$$\left(\frac{d \ln \rho}{dr}\right)_A = \frac{1}{\gamma}\left(\frac{d \ln p}{dr}\right)_A \tag{6-261}$$

In the surrounding material we may write (again assuming a perfect gas)

$$\ln p = \ln \rho + \ln T + C \tag{6-262}$$

so that

$$\left(\frac{d \ln \rho}{dr}\right)_R = \left(\frac{d \ln p}{dr}\right)_R - \left(\frac{d \ln T}{dr}\right)_R \tag{6-263}$$

Substituting into equation (6-259), we find that the condition for instability is

$$\left(\frac{\gamma - 1}{\gamma}\right)\left(\frac{-d \ln p}{dr}\right)_R < \left(\frac{-d \ln T}{dr}\right)_R \tag{6-264}$$

or

$$\left(\frac{d \ln T}{d \ln p}\right)_R > \left(\frac{\gamma - 1}{\gamma}\right) = \left(\frac{d \ln T}{d \ln p}\right)_A \tag{6-265}$$

It is customary to use the notation $\nabla = (d \ln T/d \ln p)$; thus the Schwarzschild instability criterion can be written

$$\nabla_R > \nabla_A \tag{6-266}$$

In actual stellar atmospheres the gas is not perfect because of the effects of radiation pressure and ionization. In this case we write (in the customary notation)

$$\left(\frac{d \ln T}{d \ln p}\right)_A = \frac{\Gamma_2 - 1}{\Gamma_2}$$

with the understanding that in general $\Gamma_2$ will not equal its value for a monatomic perfect gas, namely, $\gamma = c_p/c_v = 5/3$. Convenient formulae for the calculation of $\Gamma_2$ allowing for radiation pressure and ionization have been given by Unsöld (Ref. 36, Sec. 56), Vardya (*M. N.*, **129**, 205, 1965), Krishna-Swamy (*Ap. J.*, **134**, 1017, 1961), and others. The effects mentioned may be of considerable importance. Thus, for a perfect monatomic gas the critical value for $\nabla_R$ is $(2/3)/(5/3) = 0.4$. For pure radiation pressure, $\Gamma_2 = 4/3$; so the critical value of $\nabla_R$ is $(1/3)/(4/3) = 0.25$. In regions where hydrogen is ionizing strongly, $\Gamma_2$ may become as small as 1.1; so the critical $\nabla_R$ may drop to 0.1! This clearly suggests that we may expect convection zones to be associated with regions where ionization of hydrogen (and to a lesser extent of other elements, e.g., helium) occurs. This expectation is further strengthened when we examine the expression for $\nabla_R$. Thus in the limit of the diffusion approximation (see Sections 1-6 and 2-2) we may write

$$-\frac{dT}{dr} = \left(\frac{3\pi}{16\sigma}\right)\frac{\rho \bar{\kappa}_R F}{T^3} \tag{6-267}$$

which implies

$$\nabla_R = \left(\frac{3\pi}{16\sigma}\right)\frac{\bar{\kappa}_R F p}{g T^4} \tag{6-268}$$

Here we see that large values of the opacity require that the radiative gradient must be large in order to drive the flux $F$ through the atmosphere. This fact is of fundamental importance in stellar atmospheres. A large increase in the opacity occurs when hydrogen becomes appreciably excited into its higher states; this happens at just about the same temperatures and pressures where ionization occurs, with its consequent decrease in $\Gamma_2$. Taken together, these two effects imply that the radiative gradient will exceed the adiabatic gradient in the hydrogen convection zone so that convection occurs. The importance of these mechanisms and the existence of extensive hydrogen convection zones was first recognized by Unsöld (*Z. Ap.*, **1**, 138, 1931).

In the earliest-type stars, hydrogen is very strongly ionized throughout the envelope so that the circumstances described above do not occur, and radiative equilibrium prevails. Actually there are very weak, thin convection zones associated with the ionization of He I and He II, but these transport only a tiny fraction of the flux. As we reach the A stars, their hydrogen convection zones begin to develop at shallow depths in the atmosphere ($\tau \approx 0.2$). In the F stars the convection starts somewhat deeper, and becomes much thicker, and by types F3 to F5 will transport essentially *all* of the flux at some point within the zone. As we pass to later and later types, the zone extends ever deeper and convection becomes more efficient until in the latest types the convective envelope is so extensive that it determines the structure of the star as a whole (Limber, *Ap. J.*, **127**, 363, 1958). Also, in the later types, additional destabilizing influences exist, in particular the dissociation of $H_2$ into atomic hydrogen.

## THE MIXING-LENGTH THEORY

The basic physical picture used in the mixing-length theory is that the transport in the unstable layer is effected by turbulent elements or "bubbles" moving upward and downward through some surrounding environment. The upward moving elements have an excess of thermal energy over the surrounding material while downward moving elements have a defect of thermal energy. At the end of some characteristic distance, the *mixing-length*, one envisions that these elements "dissolve" and merge smoothly into the surroundings, delivering any excess energy they possess or absorbing any defect. Thus a direct transport of energy occurs, and the temperature gradient in the material is lowered below that which would obtain if the only transport mechanism were radiation. To characterize this situation, we will introduce the following temperature gradients: $\nabla_R$ is the fictitious radiative gradient that would occur if convection were suppressed; $\nabla_A$ is the gradient for adiabatic processes; $\nabla_E$ is the gradient in individual convective elements; and $\nabla$ is the gradient in the mean surroundings in the final state where both radiation and convection act together to transport the total energy flux. In general we must have

$$\nabla_R \geq \nabla \geq \nabla_E \geq \nabla_A \tag{6-269}$$

Consider now a rising element of material. If $\delta T$ is the temperature difference between the element and its surroundings, then its excess energy content per unit volume is $\rho C_p\,\delta T$. The temperature difference $\delta T$ can be expressed in terms of the difference in gradients between the element and the surroundings. Thus after traveling over a distance $\Delta r$,

$$\delta T = \left[\left(\frac{-dT}{dr}\right) - \left(\frac{-dT}{dr}\right)_E\right] \Delta r \tag{6-270}$$

so that the energy flux transported by elements moving with an average velocity $\bar{v}$ is

$$\pi F_{\text{conv}} = \rho C_p \bar{v}\left[\left(\frac{-dT}{dr}\right) - \left(\frac{-dT}{dr}\right)_E\right] \Delta r \tag{6-271}$$

At a given horizontal level in the atmosphere we will encounter elements distributed over all points on their paths of travel; thus, averaging over all elements, we set $\Delta r = l/2$ where $l$ is the mixing length. To recast the above equation into a more convenient form, we make use of the hydrostatic equation (6-1) to obtain

$$\pi F_{\text{conv}} = \frac{1}{2}\frac{g\rho^2 C_p T \bar{v}}{p}\, l(\nabla - \nabla_E) \tag{6-272}$$

A further simplification results if we introduce the *pressure scale height H* (the distance over which the pressure changes by a factor of $e$ in an isothermal atmosphere), namely,

$$\frac{1}{H} = -\frac{d \ln p}{dr} = \frac{g\rho}{p} \tag{6-273}$$

We then have

$$\pi F_{\text{conv}} = \frac{1}{2}\rho C_p \bar{v} T\left(\frac{l}{H}\right)(\nabla - \nabla_E) \tag{6-274}$$

We must now obtain an expression for the velocity $\bar{v}$. We estimate this velocity by setting the mean kinetic energy of an element equal to the work done upon it by buoyant forces. If $\delta\rho$ is the density difference between the element and its surroundings, the buoyant force is

$$f_b = -g\,\delta\rho \tag{6-275}$$

From the equation of state we can write

$$\ln \rho = \ln p - \ln T + \ln \mu$$

where $\mu$ is now considered to be variable to allow for changes in ionization, etc. Therefore,

$$\frac{d\rho}{\rho} = \frac{dp}{p} - \frac{dT}{T} + \left(\frac{\partial \ln \mu}{\partial \ln T}\right)_p \frac{dT}{T}$$

$$= \frac{dp}{p} - Q\frac{dT}{T}$$

where

$$Q \equiv 1 - \left(\frac{\partial \ln \mu}{\partial \ln T}\right)_p \tag{6-276}$$

Taking differences between the element and its surroundings and noting that for pressure equilibrium $\delta p \equiv 0$, we have simply

$$\frac{\delta\rho}{\rho} = -Q\frac{\delta T}{T}$$

Then by substitution into equation (6-275) and expansion of $\delta T$ in terms of differences in gradients, we have

$$f_b = \frac{gQ\rho}{T}\,\delta T = \frac{gQ\rho}{T}\left[\left(\frac{-dT}{dr}\right) - \left(\frac{-dT}{dr}\right)_E\right]\Delta r \tag{6-277}$$

This force is linear in the displacement distance $\Delta r$; integrating over some total displacement $\Delta$, we obtain the work done

$$W = \int_0^\Delta F(\Delta r)\,d(\Delta r) = \frac{1}{2}\frac{gQ\rho}{T}\left[\left(\frac{-dT}{dr}\right) - \left(\frac{-dT}{dr}\right)_E\right]\Delta^2 \tag{6-278}$$

Averaging over all elements in a given layer, we set $\Delta = l/2$, and introducing the pressure scale height (equation 6-273), we obtain

$$W = \frac{1}{8}(\rho g Q H)(\nabla - \nabla_E)\left(\frac{l}{H}\right)^2 \tag{6-279}$$

Of this work, we estimate that about one-half will end up as the kinetic energy of the element and the other half will be lost to "friction" in pushing aside other turbulent elements. Therefore we set

$$\frac{1}{2}\rho\bar{v}^2 = \frac{1}{2}W = \frac{1}{16}(\rho g Q H)(\nabla - \nabla_E)\left(\frac{l}{H}\right)^2$$

so that

$$\bar{v} = \frac{1}{2\sqrt{2}}(gQH)^{1/2}(\nabla - \nabla_E)^{1/2}\left(\frac{l}{H}\right) \tag{6-280}$$

Substituting this result into equation (6-274), we have, finally,

$$\pi F_{\text{conv}} = \frac{1}{4\sqrt{2}}(gHQ)^{1/2}(\rho C_p T)(\nabla - \nabla_E)^{3/2}\left(\frac{l}{H}\right)^2 \qquad (6\text{-}281)$$

One of the fundamental uncertainties in the theory is the question of how to choose an appropriate value of $l$. The usual prescription is to assume that it is simply some multiple, say 1 or 2, times the local pressure scale height $H$. While there is no fundamental justification for such a choice, it nevertheless has been found to yield reasonable answers in stellar structure computations. In particular, theoretical models of the sun, allowing for evolution, can be fitted to present-day observations of the radius (which is the parameter most sensitive to the mixing length) with a value of $(l/H) \approx 1.5$. A number of other prescriptions have been suggested, such as using the density scale height or a nonlocal scale height estimated by some kind of averaging process over the path length, or using some specified fraction of the thickness of the convective layer. Each of these procedures enjoys certain advantages but at the same time is to a certain extent arbitrary. We will not comment further upon these various alternatives here but will merely note that the uncertainty in the choice of mixing length introduces an important source of uncertainty in the computation of the convective flux.

To complete the theory, we now must consider the overall efficiency of the convection process, and to do this we will follow the suggestions of Unsöld. As an element rises, its temperature exceeds that of its surroundings (which, of course, accounts for its ability to effect a net transport of energy). But since the element is hotter than its surroundings, it will radiate some of its energy out of the "sides" of the element into the surroundings. This energy lost horizontally will diminish the excess energy content of the element and therefore the net energy that the element will yield when it finally dissolves into its surroundings at the end of the mixing length. These considerations suggest that we introduce an efficiency parameter $\gamma$ defined as

$$\gamma = \frac{\text{excess energy content at time of dissolution}}{\text{energy lost by radiation during the lifetime of the element}} \qquad (6\text{-}282)$$

As discussed previously in the derivation of equation (6-274), the energy excess should be proportional to $(\nabla - \nabla_E)$; this represents the energy that will actually be yielded by the dissolved element after losses are allowed for. On the other hand, if the element had moved adiabatically, and had suffered no losses to its surroundings until it finally dissolved, the energy excess would have been proportional to $(\nabla - \nabla_A)$, with the same proportionality factor as before. The energy lost by radiation is proportional to the difference in these to quantities, namely, $(\nabla - \nabla_A) - (\nabla - \nabla_E) = (\nabla_E - \nabla_A)$. Therefore we may write

$$\gamma = \frac{\nabla - \nabla_E}{\nabla_E - \nabla_A} \tag{6-283}$$

Alternatively, we may calculate each of the terms in equation (6-282) directly in terms of local variables. Thus, for an element of volume $V$ and temperature excess $\delta T$, the excess energy content is simply $\rho C_p V\, \delta T$. The rate of energy loss by radiation will depend upon whether the element is optically thick or thin. In the optically thick case, we may use the diffusion approximation (equation 2-29), which yields rate of energy loss per unit area of

$$\frac{16\sigma T^3}{3\rho\bar{\kappa}_R}\frac{dT}{dr} \approx \frac{16\sigma T^3}{3\rho\bar{\kappa}_R}\frac{\delta T}{l}$$

where we assume a temperature fluctuation $\delta T$ occurs over a characteristic element size $l$. If the element has surface area $A$ and lifetime $(l/\bar{v})$, the energy loss over the element's lifetime is approximately

$$\frac{16\sigma T^3}{3\rho\bar{\kappa}_R}\frac{A}{\bar{v}}\,\delta T$$

Combining these expressions, we estimate that for optically thick elements $(\tau_e = \bar{\kappa}_R \rho l \gg 1)$, the convective efficiency is

$$\gamma_{\text{thick}} = \frac{3\rho C_p \bar{v}}{16\sigma T^3}\,\bar{\kappa}_R\, \rho\left(\frac{V}{A}\right)$$

The choice of $(V/A)$ is ambiguous. If the elements were spherical and had radius $l$, then $(V/A) \sim l/3$, so

$$\gamma_{\text{thick}} = \frac{\rho C_p \bar{v}}{16\sigma T^3}\,\tau_e \tag{6-284}$$

Other values for $(V/A)$ have been employed by various investigators; again the choice is largely arbitrary and is another source of uncertainty in the theory. A similar calculation may be made for a transparent element $(\tau_e \ll 1)$. Here the average rate of energy loss will be [recall the discussion leading to equation (1-49)] the volume emissivity $4\pi\, \Delta B \bar{\kappa}_R \rho$ times the volume $V$ times the lifetime $(l/v)$, i.e.,

$$4\pi\left(\frac{4\sigma T^3}{\pi}\right)\frac{\delta T}{2}\,(V\bar{\kappa}_R\,\rho)\left(\frac{l}{\bar{v}}\right)$$

where $(\delta T/2)$ is the average value over the lifetime of the element. Thus we have

$$\gamma_{\text{thin}} = \frac{C_p \bar{v}}{8\sigma T^3 \bar{\kappa}_R\, l} = \frac{\rho C_p \bar{v}}{8\sigma T^3}\frac{1}{\tau_e} \tag{6-285}$$

To interpolate between the two extremes represented by equations (6-284) and (6-285), we adopt a linear relation

$$\gamma = \frac{\rho C_p \bar{v}}{8\sigma T^3} \frac{(1 + \frac{1}{2}\tau_e^2)}{\tau_e} \tag{6-286}$$

so that combining equations (6-283) and (6-286) and substituting equation (6-280), we have

$$\frac{\nabla_E - \nabla_A}{(\nabla - \nabla_E)^{1/2}} = \frac{16\sqrt{2}\,\sigma T^3}{\rho C_p (gQH)^{1/2}(l/H)} \frac{\tau_e}{(1 + \frac{1}{2}\tau_e^2)} \equiv B \tag{6-287}$$

The final requirement we may impose upon the theory is that the correct total flux be transported when both radiation and convection occur, i.e.,

$$\pi F_{\text{rad}} + \pi F_{\text{conv}} = \pi F = \sigma T_{\text{eff}}^4 \tag{6-288}$$

To determine the structure in an atmosphere in which convection occurs, we proceed as follows. First we compute the structure assuming that transport is by radiation only, using the methods described earlier in this chapter. We then check the stability condition at each point; if it is violated, convection occurs, and the true gradient $\nabla$, $\nabla_R \geq \nabla \geq \nabla_A$ must be found by satisfying equation (6-288). If the instability occurs deep enough in the atmosphere that the diffusion approximation is valid, then for a gradient $\nabla$ we may write

$$\frac{F_{\text{rad}}}{F} = \frac{\nabla}{\nabla_R} \tag{6-289}$$

Thus we may rewrite equation (6-288) as

$$\pi F_{\text{conv}} = \sigma T_{\text{eff}}^4 \left(1 - \frac{\nabla}{\nabla_R}\right) \tag{6-290}$$

or, by substitution of equation (6-281), we find

$$A(\nabla - \nabla_E)^{3/2} = \nabla_R - \nabla \tag{6-291}$$

where

$$A \equiv \frac{\nabla_R (l/H)^2 (gQH)^{1/2} \rho C_p T}{4\sqrt{2}\,\sigma T_{\text{eff}}^4} \tag{6-292}$$

Adding and subtracting $(\nabla_E - \nabla_A)$ to the above equation, we may rewrite it as

$$A(\nabla - \nabla_E)^{3/2} + (\nabla - \nabla_E) + (\nabla_E - \nabla_A) = (\nabla_R - \nabla_A) \tag{6-293}$$

and substituting from equation (6-287), we have

$$A(\nabla - \nabla_E)^{3/2} + (\nabla - \nabla_E) + B(\nabla - \nabla_E)^{1/2} = (\nabla_R - \nabla_A) \tag{6-294}$$

This is a cubic equation in $x = (\nabla - \nabla_E)^{1/2}$. The root may be found by standard methods, and thus we have

$$\nabla = \nabla_A + Bx + x^2 \tag{6-295}$$

and

$$\nabla_E = \nabla - x^2 \tag{6-296}$$

Having found $\nabla$, we may integrate to find $T$ as a function of $p$; simultaneously we may integrate the equation of hydrostatic equilibrium to find $\tau(p)$.

Because densities and opacities are always very small in the outermost layers of an atmosphere, radiative transport will be very efficient in these layers while convection will not. The upper layers will, therefore, be in radiative equilibrium, and $\nabla$ will equal $\nabla_R$. As we proceed inward, the opacity will grow larger and strong ionization may begin so that convection may occur. The depth at which convection occurs (if at all) and the detailed effects it has upon the structure of the atmosphere will depend strongly upon the effective temperature and the gravity of the model. The basic consideration is always how efficient convective transport is relative to radiation. In a general way, it will have the largest effects in models with low effective temperatures (at least low enough that the hydrogen is not strongly ionized throughout the atmosphere) and high gravities (which imply large densities and heat capacities, and hence efficient mass transport). When convection is inefficient, the true gradient $\nabla$ will lie close to $\nabla_R$, and a substantial part of the total flux may be carried by radiation. When it is very efficient, $\nabla$ will lie close to $\nabla_A$, and practically all the flux will be transported by convection; for example, in computations of stellar interiors convection is so efficient that one may simply set $\nabla \equiv \nabla_A$ and dispense with the mixing-length theory entirely.

When the convection zone lies sufficiently close to the surface of the atmosphere, the layer will no longer be optically deep at all wavelengths, and the approximation inherent in equation (6-289) will no longer be valid. It is then necessary to calculate $F_{\text{rad}}$ directly from the solution of the nongray transfer equation and to employ some form of iterative temperature-correction procedure. In any such procedure an essential ingredient is to allow for the changes in both $F_{\text{rad}}$ and $F_{\text{conv}}$ introduced by the proposed alteration of the temperature structure. Methods for doing this have been suggested by Mihalas (*Ap. J.*, **141**, 564, 1965), Parsons (*Ap. J. Supp. No. 159*, **18**, 127, 1969), and Auman (*Ap. J.*, **157**, 799, 1969).

A number of radiative-convective models for later-type stars have been constructed; of the rather vast literature we shall mention only a few representative papers. In the papers cited above, Parsons has carried out a very thorough analysis of the atmospheres of middle-type supergiants (including certain refinements of the theory not discussed here) while Auman performed an important study of M stars (dwarfs through supergiants), including the very complex absorption spectrum of $H_2O$. An extensive set of computations over a wide range of effective temperatures and gravities has been carried out by Böhm-Vitense (*Z. Ap.*, **46**, 108, 1958); these models assume gray radiative transport and thus employ equation (6-289). The convection zone in the solar atmosphere has been studied both in the mixing-length approximation (Vitense, *Z. Ap.*, **32**, 135, 1953) and in more detailed hydrodynamical approximations (K. H. Böhm, *Ap. J.*, **137**, 881, 1963, and *Ap. J.*, **138**, 1963).

We shall not dwell here upon the limitations of the mixing length theory since they are quite obvious in the derivation given above; it is clear that the theory is only an heuristic attempt to treat an otherwise unapproachable problem. A significant improvement in our understanding of the physical state of the atmospheres of late-type stars will necessarily be delayed until the theory of convection is better understood; therefore an improved theory is eagerly awaited by many astronomers.

## 6-6. Empirical Solar Models

We have discussed at some length methods for the theoretical determination of the temperature distribution in stellar atmospheres. For one star, the sun, we can also obtain an empirical estimate of the temperature distribution. This is a happy circumstance because for the sun the theoretical estimates are relatively poor since they are not yet able to account accurately for the effects of line blanketing (very important in the sun) or of convection (which becomes a more efficient transport mechanism than radiation at depth in the solar atmosphere).

The temperature distribution in the solar atmosphere can be deduced from limb-darkening data. If, at some specific wavelength, we can see into the solar atmosphere to a physical depth $\Delta z$ at the center of the disc (corresponding to $\tau_\nu \approx 1$), then as we observe toward the limb, we see into depths $\mu\,\Delta z$ since the optical path length scales as $1/\mu$. In this way, the observations near the center of the disk give information about the deeper layers while those near the limb yield information about the outermost layers.

Limb-darkening measures yield directly the ratio

$$\varphi_\nu(\mu) = \frac{I_\nu(0, \mu)}{I_\nu(0, 1)} \tag{6-297}$$

while theory gives (with the assumption of LTE)

$$\begin{aligned} \varphi_\nu(\mu) &= \int_0^\infty \frac{B_\nu[T(\tau_\nu)]}{I_\nu(0, 1)} \exp(-\tau_\nu/\mu)\,\frac{d\tau_\nu}{\mu} \\ &= \int_0^\infty b_\nu(\tau_\nu)\exp(-\tau_\nu/\mu)\,\frac{d\tau_\nu}{\mu} \end{aligned} \tag{6-298}$$

Now, if we assume that $b_\nu$ can be represented by some analytical formula containing expansion coefficients, then $\varphi_\nu$ can be computed in terms of these coefficients. By fitting this calculated function to the observations, one can find the coefficients and hence $b_\nu$. For example, if we assume

$$b_\nu = \sum_k a_k \tau_\nu{}^k \tag{6-299}$$

then we find

$$\varphi_\nu(\mu) = \sum_k a_k k! \mu^k \tag{6-300}$$

while if we assume

$$b_\nu = A_\nu + B_\nu \tau_\nu + C_\nu E_2(\tau_\nu) \tag{6-301}$$

we find

$$\varphi_\nu(\mu) = A_\nu + B_\nu \mu + C_\nu \left[1 - \mu \ln\left(1 + \frac{1}{\mu}\right)\right] \tag{6-302}$$

Numerous observations of limb darkening have been made by several astronomers; representative values are given (e.g., by Pierce, *Ap. J.*, **120**, 221, 1954). By fitting the above forms to the observations, we may determine $b_\nu(\tau_\nu)$ at each of several wavelengths. If, in addition, we have measurements if $I_\nu$ (0, 1) in absolute physical units, we can convert $b_\nu(\tau_\nu)$ to $B_\nu(\tau_\nu)$. Measures of $I_\nu(0, 1)$ have been given, for example, by Labs and Neckel (*Z. Ap.*, **55**, 269, 1962; *Ap. J.*, **135**, 969, 1962; and *Ap. J.*, **138**, 296, 1963). In this way we may infer $T(\tau_\nu)$.

The above procedure is subject to some uncertainties. For example, the values of $I_\nu(0, 1)$ contain uncertainties of a few percent. Moreover, there are fundamental limitations on the information obtainable from observations of limb darkening. This may readily be seen as follows. Suppose we expand $B_\nu(\tau_\nu)$ around some point $\tau_\nu{}^*$ in a Taylor's series

$$B_\nu(\tau_\nu) = B_\nu(\tau_\nu{}^*) + \left.\frac{dB_\nu}{d\tau_\nu}\right|_{\tau_\nu{}^*} (\tau_\nu - \tau_\nu{}^*) + \frac{1}{2} \left.\frac{d^2B_\nu}{d\tau_\nu{}^2}\right|_{\tau_\nu{}^*} (\tau_\nu - \tau_\nu{}^*)^2 + \cdots \tag{6-303}$$

Then

$$I_\nu(0, \mu) = B_\nu(\tau_\nu{}^*) + (\mu - \tau_\nu{}^*) \left.\frac{dB_\nu}{d\tau_\nu}\right|_{\tau_\nu{}^*} + \frac{1}{2} [(\mu - \tau_\nu{}^*)^2 + \mu^2] \left.\frac{d^2B_\nu}{d\tau_\nu{}^2}\right|_{\tau_\nu{}^*} + \cdots \tag{6-304}$$

We wish to choose $\tau_\nu{}^*$ such that the dominant contribution to $I_\nu(0, \mu)$ comes from the term $B_\nu(\tau_\nu{}^*)$; in this sense the point $\tau_\nu{}^*$ is representative of the limb-darkening data at $\mu$. We see that the choice $\tau_\nu{}^* = \mu$ simultaneously eliminates the first derivative and minimizes the coefficient of the second so that $I_\nu(0, \mu) \approx$

$B_\nu(\mu)$. Since $\mu \leq 1$, we see that we may gain information only for $\tau_\nu \lesssim 1$. Since some wavelengths are more transparent than others, the situation is improved a bit, but in any case $T(\bar{\tau})$ is essentially undetermined for $\bar{\tau} \gtrsim 4$. At the opposite extreme, information at small optical depths is given by observations at the extreme limb of the sun ($\mu \approx 0$). These observations are difficult to obtain because of seeing effects (turbulence in the earth's atmosphere), scattered light in the spectrograph, uncertainty in the exact position of the true limb, and curvature of and inhomogeneities in the layers being observed. Even if we are extremely optimistic, seeing problems alone limit the resolution to $\sim 1''$, or about 0.001 solar radii, which corresponds to $\mu = 0.045$. More realistically, the limit of resolution might be $\sim 10''$, corresponding to $\mu \sim 0.15$. Thus, in a practical way, the optical depth range over which we may determine $T(\bar{\tau})$ is limited at most to $0.05 \lesssim \bar{\tau} \lesssim 5$.

In addition to these difficulties, there are problems associated with the actual fitting procedure itself. First of all, it is clear that we cannot hope to derive information about very small-scale fluctuations. In practice, both of the source functions sketched in Figure 6-6 would be indistinguishable from limb-darkening data. Moreover, there are limitations on the number of coefficients that may be obtained. For example, if one fits a three-term formula (terms to $\mu^2$) and a five-term formula (terms to $\mu^4$) to $\varphi_\nu(\mu)$, in each case one can reproduce the data to 0.1 %, but the resulting source functions are very different, the five-term fit yielding large oscillations. A discussion of this problem has been given by K. H. Böhm (*Ap. J.*, **134**, 264, 1961), who shows that for data of observational accuracy of the order of $\pm 1$ %, one is restricted to at most three coefficients. An interesting alternative approach using the so-called *Prony algorithm* has been described by White (*Ap. J.*, **152**, 217, 1968). This method has the property of automatically limiting the number of fitting coefficients to those justified by the accuracy of the data. The situation could be improved somewhat by high-precision, high-resolution data, obtained perhaps from balloons and satellites.

If we confine our attention to various three-term solutions, we find temperature differences among them of the order of 150°K to 250°K, even though

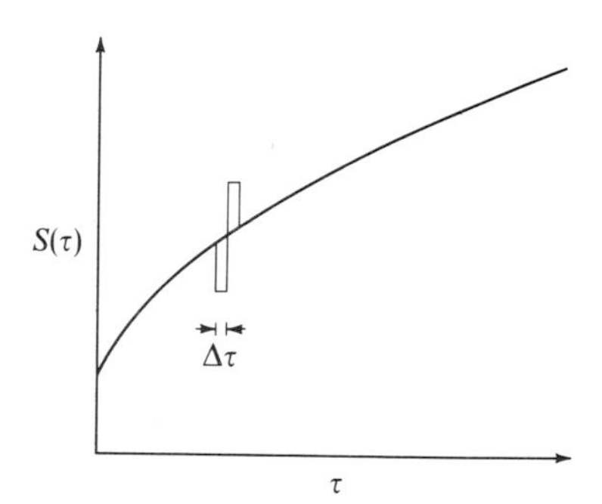


FIG. 6-6. Hypothetical fine structure in run of source function. In practice, for $\Delta T \ll 1$, the fluctuation would not be distinguishable from the smooth curve using the usual inversion procedures.

they all fit the limb-darkening data equally well. Uncertainties of this order can be important in certain contexts, for example, in deriving abundances of elements in the solar atmosphere. In calculations of this kind, we need to know the occupation numbers of the bound states, and uncertainties in temperature introduce uncertainties in these numbers. For example, consider the equilibrium of Fe I, which has a well-represented spectrum in the sun. Iron has an ionization potential of 7.87 eV, which, using typical values of $T$ and $p_e$ from models, implies that most of the iron is in the form of Fe II. Thus, the occupation numbers of state $i$ of Fe I can be written

$$\frac{N_i}{N(\text{FeI})}\frac{N(\text{FeI})}{N(\text{Fe})} \approx \frac{N_i}{N(\text{FeI})}\frac{N(\text{FeI})}{N(\text{FeII})} \propto T^{-5/2}\exp[(\chi_{\text{I}} - \chi_i)/kT]$$
$$= T^{-5/2}10^{\theta(\chi_{\text{I}} - \chi_i)}$$

where $\theta = 5040/T$. Take $T \approx 6000°\text{K}$ or $\theta \approx 5/6$. Then $\Delta T = \pm 200°\text{K}$ implies $\Delta\theta = \mp(1/30)(5/6)$ so that

$$\Delta \log N = -\frac{5}{2}\left(\frac{\Delta T}{T}\right)\Big/\log_e 10 + \Delta\theta(\chi_{\text{I}} - \chi_i) = \frac{\Delta\theta}{\theta}\,[1.1 + \theta(\chi_{\text{I}} - \chi_i)]$$
$$= \mp\frac{1}{30}\left[7.7 - \frac{5}{6}\chi_i\right]$$

The worst uncertainty arises for lines from the ground state, for which $\chi_i = 0$. In this case, $\Delta \log N \approx \mp 0.25$ or $\mp 60\%$ in abundance. In a comprehensive analysis of the solar spectrum by Goldberg, Müller, and Aller (*Ap. J. Supp. No. 45*, **5**, 1, 1960), most of the lines used arise from levels with $\chi_i$ about 0 to 1 eV. Thus, the uncertainties in the model alone may introduce errors of the order of $\pm 50\%$ in the abundance. Allowing for errors in $f$-values, physical assumptions such as the use of LTE, lack of other parameters (e.g., line-broadening constants), etc., one might easily imagine that the accuracy of the determination is no better than about a factor of two. This estimate is more pessimistic than some that appear in the literature but is not unduly so.

If one has established the run of temperature with depth, then further information may be obtained. One important kind of data that can be found are empirical absorption coefficients. We proceed as follows. If we choose a value of $T$, then at each frequency we can calculate $B_\nu$; if we then use the empirical relations for $B_\nu(\tau_\nu)$, we can find the value of $\tau_\nu$ corresponding to this temperature at each wavelength for which we have these relations. Thus, we can plot the curves of constant $T$ in the $(\tau_\nu, \nu)$ plane. Now at a fixed wavelength, we can estimate the derivative $d\tau_\nu/dT$ by graphical differentiation. But $d\tau_\nu/dT$ is simply $\rho\kappa_\nu(dz/dT)$; so at a fixed value of $T$, the ratio $(d\tau_\nu/dT)_1/(d\tau_\nu/dT)_2$ yields simply $(\kappa_\nu)_1/(\kappa_\nu)_2$. In this way we can find the frequency dependence of $\kappa_\nu$ for various temperatures in the atmosphere. A detailed

TABLE 6-1. Pierce-Waddell Model of the Solar Photosphere

| $\log \tau_{5000}$ | $\theta$ | $\log p_g$ | $\log p_e$ |
|---|---|---|---|
| −1.5 | 1.054 | 4.411 | 0.260 |
| −1.4 | 1.041 | 4.466 | 0.324 |
| −1.3 | 1.028 | 4.522 | 0.391 |
| −1.2 | 1.013 | 4.577 | 0.460 |
| −1.1 | 0.9972 | 4.633 | 0.531 |
| −1.0 | 0.9803 | 4.689 | 0.606 |
| −0.9 | 0.9625 | 4.745 | 0.686 |
| −0.8 | 0.9439 | 4.800 | 0.773 |
| −0.7 | 0.9247 | 4.854 | 0.869 |
| −0.6 | 0.9049 | 4.906 | 0.975 |
| −0.5 | 0.8847 | 4.953 | 1.091 |
| −0.4 | 0.8641 | 5.000 | 1.220 |
| −0.3 | 0.8433 | 5.041 | 1.359 |
| −0.2 | 0.8222 | 5.079 | 1.506 |
| −0.1 | 0.8009 | 5.113 | 1.660 |
| 0.0 | 0.7791 | 5.144 | 1.821 |
| 0.1 | 0.7568 | 5.170 | 1.990 |
| 0.2 | 0.7338 | 5.195 | 2.166 |
| 0.3 | 0.7098 | 5.218 | 2.351 |
| 0.4 | 0.6844 | 5.236 | 2.547 |
| 0.5 | 0.6578 | 5.252 | 2.753 |
| 0.6 | 0.6295 | 5.266 | 2.972 |

*Source:* Adapted from A. K. Pierce and J. H. Waddell, *Mem. R. A. S.*, **58**, 89, 1961.

analysis of this kind has been carried out by Pierce and Waddell (*Mem. R. A. S.*, **58**, 89, 1961). They find that the empirically determined absorption coefficient is consistent with the hypothesis that $H^-$ is the dominant opacity source but that the $H^-$ opacity is insufficient to explain all of the opacity in the ultraviolet and infrared. In the ultraviolet, at least part of the discrepancy may be due to neglect of metallic continua and the multitude of weak overlapping metal lines in this region. In the infrared, there may be some errors remaining in the $H^-$ free-free cross-section, and there may be contributions from other sources (e.g., negative ions and molecules).

If we adopt an empirical temperature distribution, we can integrate the equation of hydrostatic equilibrium in the usual way to obtain $p_g(\tau)$, $p_e(\tau)$, etc. Several models of this kind have been constructed by Pierce and Waddell; we reproduce a typical model in Table 6-1. The optical depth scale in this model is measured at $\lambda 5000$ Å; the hydrogen to metals ratio is $10^4$; and the helium to hydrogen ratio is 0.2. To calculate $B_\lambda$ the measurement by Minnaert of $I_\lambda(0, 1) = 4.16 \times 10^{14}$ at $\lambda 5000$ Å has been adopted. More recent attempts have been made to extend the empirical model for the solar atmosphere by using better estimates of the helium abundance (0.10), a revised value of $I_\lambda(0, 1) = 4.05 \times 10^{14}$, and by incorporating observations made from rockets and satellites in the extreme ultraviolet, data obtained during eclipses, and observations at radio wavelengths. These improvements have led to the *Utrecht Reference Model of the Photosphere* (Ref. 37, p. 239) and the *Bilderberg Continuum Atmosphere* (*Solar Phys.*, **3**, 5, 1968). As the data improve and more kinds of information are used, our knowledge of the physical structure of the solar photosphere can be expected to improve. As the theory of stellar atmospheres becomes capable of treating line blanketing and convection more accurately, we may expect improved agreement between the theoretical and empirical models.

# 7 | Non-LTE Model Atmospheres for the Continuum

In this chapter we will describe the methods used to obtain models in which the atomic level populations and the radiation field are computed self-consistently. For the present, we will consider only bound-free transitions; the effects of bound-bound transitions will be discussed in Chapter 14.

To obtain the self-consistent solution we desire, there are two basic approaches that may be followed. In the first approach, we start with an estimated run of occupation numbers (perhaps those predicted by LTE), use these to compute the radiation field, and, finally, use this radiation field in the statistical equilibrium equations to recompute the occupations numbers. This *iteration* procedure could then be repeated again and again until the radiation field and occupation numbers converge to fixed values. While this procedure seems perfectly straightforward in principle, in practice it is fraught with difficulty. These difficulties arise when the opacities of the relevant transitions are large and when the radiation field is only weakly coupled to *local* conditions (e.g., temperatures). When this situation arises, radiative control of level populations extends over large optical path lengths. A typical instance where this condition may occur is when scattering terms are dominant. Under these circumstances, we will find that successive iterations differ only slightly, even when the correct solution bears no resemblance to the current estimates, and that the iterations may in fact *stabilize* on a spurious value without ever converging. Two physical situations in which we encounter such difficulties are in the treatment of the Lyman continuum and of bound-bound transitions. That difficulties occur is not surprising when we recognize that the procedure we have outlined is a lambda-iteration procedure and that the circumstances we have described are precisely those under which a lambda iteration can be expected to fail. On the other hand, if we consider *only* relatively transparent continua and exclude the Lyman continuum and the lines, the iteration procedure is reasonably useful. We will describe this approach further in Section 7-2.

In the second approach, one attempts to overcome the difficulties inherent in the iteration method by incorporating the statistical equilibrium equations directly into the transfer equation, thereby solving both sets of equations *simultaneously*. The details of how this is done will be described in Sections 7-3 and 7-4 and again for lines in Chapters 12 and 13. The resulting transfer equations may be quite complicated because of the appearance of what may be regarded as noncoherent scattering terms, which arise from radiative rates in the statistical equilibrium equations. Thus, one must employ one of the general methods described in Chapter 6 to solve these equations; in this book we shall use the difference-equation method for purposes of illustration. In a general way, this simultaneous approach is very powerful and will lead to the correct solution even when the iteration method fails. In practice, difficulties may still arise because of nonlinearities and complex coupling in the equations and because of the additional constraint of radiative equilibrium. One approach is simply to iterate the solution to account for the nonlinearities. If one *fixes* the temperature distribution, then this iteration scheme will often converge efficiently. If, however, the temperature distribution is unknown, then convergence may not occur. In this case, we must incorporate the radiative equilibrium constraint by a *linearization* procedure, as will be described in Section 7-4. When this is done, one is in essence solving the transfer equation simultaneously with the constraints of statistical equilibrium *and* radiative equilibrium. For models including only bound-free transitions, this provides a very efficient and completely satisfactory computational procedure, when bound-bound transitions are included, however, even this scheme may fail, and one must employ a *complete linearization* method, as will be described in Chapter 13.

Thus we see that there are a variety of techniques of different degrees of generality and reliability; we shall discuss them in turn. Before we do this, however, let us consider the basic physical simplification that will be made in the models described in this chapter.

## 7-1. Detailed Balance in the Lines

Of all the difficulties in non-LTE calculations, those associated with the treatment of the bound-bound transitions (lines) are the greatest. This is because the lines are very opaque, and at the same time the radiation field in them is only weakly coupled to local conditions (e.g., temperatures); we shall show this in detail in Chapter 12. Therefore, the problem mentioned above of specifying the occupation numbers self-consistently with the radiation field over large optical depths becomes very severe. In addition, there is the practical consideration that the inclusion of lines introduces a large number of additional frequency points that are needed to describe the line profiles, and, therefore, the calculation becomes very time-consuming.

For these reasons, we wish to inquire whether it is possible to defer treatment of the lines and to consider only the continua at present. The situation in the continuum is basically simpler (with the exception of the Lyman continuum). In the first place, by the very nature of the processes involved, continuum formation is much more closely coupled to local thermal conditions. Second, the continua are much more transparent, and the self-consistency problem occurs in regions that are not very optically thick (again excepting the Lyman continuum) since densities normally rise fast enough to assure LTE in the continua rather quickly. Moreover, when *all* continua become optically thick, LTE is obtained automatically. For these reasons, the continuum can often be treated successfully by a simple iteration between the transfer and statistical equilibrium equations.

The question we now ask is whether we can, in fact, ever treat the continuum only and ignore the lines. An affirmative answer to this question was given for early-type stars by Kalkofen (Ref. 15, p. 175), who pointed out that the lines are so opaque, that in the region where continuum formation is taking place, the radiative bound-bound rates are in detailed balance, and so the terms $\mathscr{R}_{ij}$ in the statistical equilibrium equations can be set to zero. A detailed analysis of the situation by Kalkofen (*J. Q. S. R. T.*, **6**, 633, 1966) shows that for typical atmospheres with $T_{\text{eff}} \sim 10^4\,^\circ\text{K}$, the assumption of detailed balance is valid for the Lyman and Balmer lines for continuum optical depths $\tau_{5000} \gtrsim 10^{-4}$. Therefore, consideration of the statistical equilibrium equations in the form

$$\mathscr{R}_{i\kappa} + \mathscr{C}_{i\kappa} + \sum_{j \neq i} \mathscr{C}_{ij} = 0 \tag{7-1}$$

leads to correct solutions in the continuum out to points which are already optically thin in the continuum. Thus, even though radiative bound-bound rates will be of importance at points closer to the surface, they will not alter the result for the continuum itself. The only exception to this statement is the Lyman continuum, which is about as opaque as the Balmer lines, so that a proper treatment should allow for the Balmer lines from the outset; we will describe this complete treatment in Chapters 13 and 14.

Physically, the approximation used in equation (7-1) simply recognizes that the photons first "see" the surface in the relatively transparent continuum bands and that free diffusion to the surface in these bands leads to departures from LTE at the greatest geometrical depths in the atmosphere.

In this way, treatment of the simplified continuum problem leads, in some sense, to the correct asymtotic solution at depth and provides a starting point for solutions involving the line terms. Furthermore, from such solutions we may obtain good estimates of the effects of departures from LTE upon continuum fluxes and can use these estimates to interpret the observations.

## 7-2. Solution by Iteration

The first calculation for early-type stars, using the iteration method, was carried out by Kalkofen (Ref. 15, p. 175), who considered a five-level hydrogen atom in a pure hydrogen model atmosphere; in this calculation both the rates $\mathscr{R}_{ij}$ and $\mathscr{C}_{ij}$ were set to zero. In a later calculation, Strom and Kalkofen (*Ap. J.*, **144**, 75, 1966) considered hydrogen-helium model atmospheres in which departures from LTE were allowed for the $n = 2$ and 3 levels of hydrogen (with all other sources in LTE), and again the terms $\mathscr{R}_{ij}$ and $\mathscr{C}_{ij}$ were set to zero (though $\mathscr{C}_{23}$ was included in a schematic way). This stimulating paper led to models with substantial deviations from LTE and predicted that significant changes in the emergent energy distribution occurred. These changes implied an important revision in the relation between $T_{\text{eff}}$ and the Balmer jump $D_B$ and thereby in the effective temperature calibration of early-type stars. While subsequent work has decreased the deviations predicted, this noteworthy paper nevertheless served well to emphasize the possibility of such deviations having important observable consequences and led to a heightened interest in pursuing calculations of this kind. A more complete calculation was then carried out by Kalkofen and Strom (*J. Q. S. R. T.*, **6**, 653, 1966) allowing for departures in $n = 1$, 2, and 3 and again setting the terms $\mathscr{R}_{ij}$ to zero, but now including the terms $\mathscr{C}_{ij}$. This calculation showed somewhat smaller deviations than the earlier result. Subsequently, Mihalas (*Ap. J.*, **149**, 169, 1967) calculated some models with a fifteen-level hydrogen atom of which the first ten levels are allowed to depart from LTE. These results showed deviations in the same directions as those found by Kalkofen and Strom, but they were again smaller. Part of the reason the deviations in this latter calculation are smaller is that there is strong collisional coupling between levels $n$ and $n \pm 1$, which allows an efficient cascade of the LTE condition in the continuum to lower levels via the route $n \to n - 1 \to n - 2 \ldots$, etc. Most of the differences between the calculations are due to the choice of collisional cross-sections employed, those employed by Mihalas being larger and more realistic (see the discussion by Sampson, *Ap. J.*, **155**, 575, 1969). Let us now consider the procedure in a bit more detail.

We showed in Chapter 4 that the absorption coefficient $a_\nu$ can be written as $a_\nu^* + \delta a_\nu$, where the term $\delta a_\nu$ represents the effects of departures from

LTE [see equation (4-245)]. Now the transfer equation is simply

$$\frac{\mu}{\rho}\frac{dI_\nu}{dz} = -I_\nu(a_\nu^* + \delta a_\nu + \sigma_\nu) + a_\nu^* I_\nu e^{-h\nu/kT} + a_\nu^*(1 - e^{-h\nu/kT})B_\nu$$
$$+ \sigma_\nu J_\nu \qquad (7\text{-}2)$$

On the right-hand side, the first term represents the energy removed from the beam. The second term represents the correction for stimulated emission as obtained in equation (5-36). Note that since stimulated emissions take place from the continuum, $b_\kappa$ is unity, and they affect only that part of the absorption arising in $a_\nu^*$. The third represents direct recombinations. Here we have made use of a detailed balancing argument, and recognizing the LTE nature of electrons in the continuum, which implies that the recombination will be an LTE process, we have set the direct recombination rate equal to the LTE rate of absorption, corrected for stimulated emission. The fourth term represents scattering by electrons. We see now that it is useful to define $\kappa_\nu^* = a_\nu^*(1 - e^{-h\nu/kT})$, the usual LTE opacity corrected for stimulated emissions, and we note that the non-LTE change in the opacity is *not* affected by stimulated emissions and therefore introduces a kind of asymmetry into the equation. Thus, the transfer equation may be written

$$\frac{\mu}{\rho}\frac{dI_\nu}{dz} = -(\kappa_\nu^* + \delta\kappa_\nu + \sigma_\nu)I_\nu + \kappa_\nu^* B_\nu + \sigma_\nu J_\nu \qquad (7\text{-}3)$$

or

$$\mu\frac{dI_\nu}{d\tau_\nu} = I_\nu - \xi_\nu B_\nu - \zeta_\nu J_\nu \qquad (7\text{-}4)$$

where

$$\xi_\nu \equiv \frac{\kappa_\nu^*}{\kappa_\nu^* + \delta\kappa_\nu + \sigma_\nu} \qquad (7\text{-}5)$$

and

$$\zeta_\nu \equiv \frac{\sigma_\nu}{\kappa_\nu^* + \delta\kappa_\nu + \sigma_\nu} \qquad (7\text{-}6)$$

where, for consistency of notation, we have replaced $\delta a_\nu$ by $\delta\kappa_\nu$. In analogy with equation (4-18) for bound-bound transitions, we may refer to the source function implied by equations (7-4), (7-5), and (7-6) as the *implicit* form of the continuum source function since the statistical equilibrium equations are implied in the departure coefficients $b_i$ and hence in $\delta\kappa_\nu$. The solution of equation (7-4) may be written

$$J_\nu(\tau_\nu) = \Lambda_{\tau_\nu}[\xi_\nu B_\nu] + \Lambda_{\tau_\nu}[\zeta_\nu J_\nu] \qquad (7\text{-}7)$$

In this solution, we assume the populations, hence $\delta\kappa_\nu$, $\xi_\nu$, and $\zeta_\nu$ are fixed.

If we set the radiative bound-bound rates to zero, the statistical equilibrium equations assume a simple form. Thus setting all terms $Z_{ji}$ in equation (5-59) to zero yields

$$b_i\left[4\pi\int_{\nu_0}^{\infty}\frac{\alpha_\nu J_\nu}{h\nu}\,d\nu + N_e\left(\sum_{j\neq i}\Omega_{ij}+\Omega_{i\kappa}\right)\right] - \sum_{j\neq i} b_j N_e \Omega_{ij}$$
$$= 4\pi\left[\int_{\nu_0}^{\infty}\frac{\alpha_\nu B_\nu}{h\nu}(1-e^{-h\nu/kT})\,d\nu + \int_{\nu_0}^{\infty}\frac{\alpha_\nu J_\nu}{h\nu}e^{-h\nu/kT}\,d\nu\right] + N_e\Omega_{i\kappa} \tag{7-8}$$

or, again introducing the notation $d_i = b_i - 1$,

$$d_i\left[4\pi\int_{\nu_0}^{\infty}\frac{\alpha_\nu J_\nu}{h\nu}\,d\nu + N_e\left(\sum_{j\neq i}^{15}\Omega_{ij}+\Omega_{i\kappa}\right)\right] - \sum_{j\neq i}^{10} d_j N_e \Omega_{ij}$$
$$= 4\pi\int_{\nu_0}^{\infty}\frac{\alpha_\nu}{h\nu}(B_\nu - J_\nu)(1-e^{-h\nu/kT})\,d\nu \tag{7-9}$$

Here we have assumed explicitly a fifteen-level atom with deviations from LTE in the first ten levels. These are linear equations and can be solved by the usual methods for given values of $J_\nu$, $T$, and $N_e$.

In the calculations cited above, the additional assumption was made that the Lyman continuum is in radiative detailed balance as well (since it is about as opaque as the Balmer lines); thus the statistical equilibrium equation for the ground state is somewhat different from equation (7-9), as will be discussed below.

To enforce the constraint of radiative equilibrium, we might employ the Avrett-Krook procedure. If we define

$$\chi_\nu = \frac{(\kappa_\nu^* + \delta\kappa_\nu + \sigma_\nu)}{(\kappa_\nu^* + \delta\kappa_\nu + \sigma_\nu)_{\text{std}}} \tag{7-10}$$

we again obtain equation (6-183):

$$\tau_1' + \tau_1\int_0^{\infty}\frac{H_\nu^0\chi_\nu'}{H^0\chi_\nu}\,d\nu = \left(1-\frac{\mathscr{H}}{H^0}\right)$$

and

$$T_1 = \left\{(1+\tau_1')\frac{dH^0}{dt} + \tau_1\int_0^{\infty}\chi_\nu'[(1-\zeta_\nu)J_\nu^0 - \xi_\nu B_\nu]\,d\nu\right.$$
$$-\tau_1\int_0^{\infty}\chi_\nu(\zeta_\nu' J_\nu + \xi_\nu' B_\nu)\,d\nu$$
$$\left. + \sqrt{3}\left[\frac{\mathscr{H}}{H^0(0)}-1\right]\int_0^{\infty}\chi_\nu(1-\zeta_\nu)H_\nu^0(0)\,d\nu\right\}\Big/\int_0^{\infty}\chi_\nu\xi_\nu\frac{\partial B_\nu}{\partial T}\,d\nu \tag{7-11}$$

Thus, in summary, with a given set of values for the $d_i$'s, we solve equation (7-7), find a new temperature distribution from equations (7-10) and (7-11), revise the populations via equations (7-9), and iterate to consistency.

Let us now consider some results from actual model calculations. In Figure 7-1 we show values of $d_i \times 10^2$ (which then measures the percentage deviations from LTE) for a model with $T_{\text{eff}} = 10{,}000°\text{K}$, $\log g = 4$. Note that the deviations are small at depths representative of continuum formation ($10^{-2}$ to $10^{-1}$ on the $\tau_{4000}$ scale for different continua). At these temperatures the mean intensity in the Paschen and higher continua is smaller than the Planck function; thus these levels are overpopulated. Note that as $n$ grows larger, $d_n$ decreases, as would be expected from the stronger coupling of the upper levels to the continuum via large collision rates. Since the mean intensity in the Balmer continuum exceeds the Planck function, $d_2 < 0$, and the $n = 2$ state is underpopulated. Since the $n = 2$ state is underpopulated and $n = 3$ overpopulated, then at the position of the Balmer jump, we will find $[\kappa_\nu(2+3)/\kappa_\nu(3)] < [\kappa_\nu^*(2+3)/\kappa_\nu^*(3)]$, and hence the non-LTE Balmer jump is smaller than the LTE value. As one proceeds to lower gravities, the densities are of course smaller and the resulting deviations from LTE are larger.

The fact that $d_1$ is shown to be the same as $d_2$ in Figure 7-1 follows from the assumption that the Lyman continuum is in radiative detailed balance.

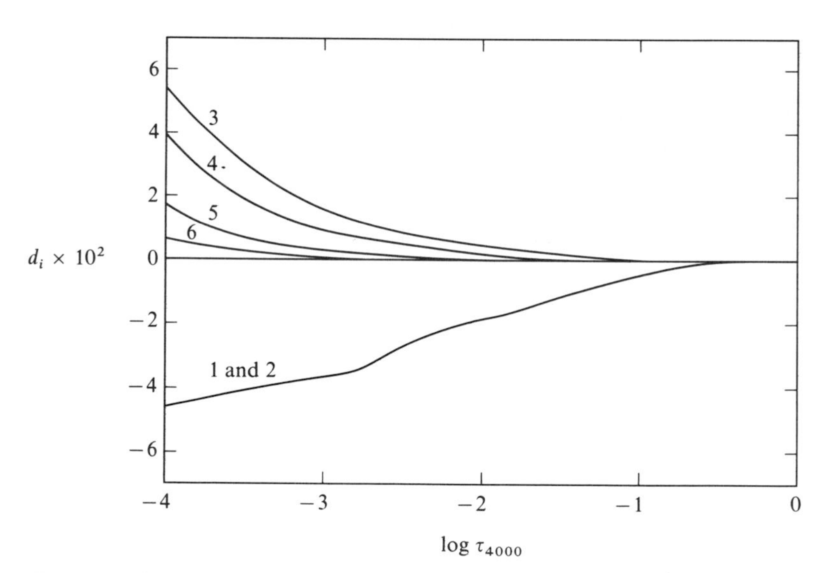


FIG. 7-1. Non-LTE departure coefficients for a model with $T_{\text{eff}} = 10{,}000°\text{K}$ and $\log g = 4$. Ordinate gives $d_i \times 10^2$ for each level of the model atom. Levels 1 and 2 are locked together by the assumption of detailed balance in the Lyman continuum. Note that level 2 is underpopulated while higher levels are overpopulated.

Let us consider this point in a little more detail. At the temperatures occurring in these models, the Lyman continuum is already extremely opaque at $\tau_{4000} \sim 10^{-4}$. In general, $J_\nu = \Lambda(\xi_\nu B_\nu)$; but since the Lyman continuum is so opaque, $\xi_\nu$ is very nearly constant over the range of the lambda operator, and we may write

$$J_\nu \approx \xi_\nu \Lambda(B_\nu) = \xi_\nu B_\nu + O(e^{-\tau_\nu})$$

But

$$\xi_\nu \equiv \frac{\kappa_\nu^*}{\kappa_\nu^* + \delta\kappa_\nu} = \frac{\kappa_\nu^*(1)}{\kappa_\nu^*(1) + d_1\kappa_\nu^*(1)} = \frac{1}{b_1}$$

since the Lyman continuum dominates all the others by orders of magnitude. Then writing

$$R \equiv 4\pi \int_{\nu_0}^{\infty} \frac{\alpha_\nu B_\nu}{h\nu}\, d\nu$$

and neglecting $\exp(-h\nu/kT)$, which is much less than unity at the frequencies and temperatures considered, we have for the statistical equilibrium of the ground state

$$d_1\left[\frac{1}{b_1} R + \sum_{j=2}^{15} N_e\Omega_{1j} + N_e\Omega_{1\kappa}\right] - \sum_{j=2}^{10} d_j N_e \Omega_{1j}$$
$$= R - \frac{1}{b_1} R = \frac{d_1 R}{b_1} \tag{7-12}$$

Thus we see that the radiative terms cancel exactly, and we have

$$d_1\left(\sum_{j=2}^{15} \Omega_{1j} + \Omega_{1\kappa}\right) = \sum_{j=2}^{10} d_j \Omega_{1j} \tag{7-13}$$

Now normally $\Omega_{12} \gg \Omega_{1j}$ for $j > 2$ so that effectively equation (7-13) reduces to $d_1 \Omega_{12} - d_2 \Omega_{12}$ or $d_1 \equiv d_2$. Thus, the behavior of the Balmer level completely drives the ground-state population under these circumstances.

Models of the kind we have described here are useful in calculating Balmer and Paschen jumps that are to be compared with observation. Values obtained by Mihalas (*Ap. J.*, **149**, 169, 1967; *Ap. J.*, **150**, 909, 1967; and *Ap. J.*, **153**, 317, 1968) and by Mihalas and Stone (*Ap. J.*, **151**, 293, 1968) are listed in Table 7-1; the latter calculation allows also for departures from LTE in the bound states of He I and He II (again assuming detailed balance in the lines). Inspection of the results in Table 7-1 shows that the effects of departures from LTE upon $D_B$ and $D_P$ are quite small, except at low gravities and high $T_{eff}$. We note that in the lower range of temperatures, $D_B$ decreases slightly, as mentioned above, while at much higher temperatures, it increases since the $n = 2$ level becomes overpopulated. The Paschen jump is practically

TABLE 7-1. Balmer and Paschen Jumps in LTE and Non-LTE Models

| $T_{eff}$ | $\log g$ | *Balmer jump (mag.)* | | *Paschen jump (mag.)* | |
|---|---|---|---|---|---|
| | | LTE | Non-LTE | LTE | Non-LTE |
| 10,000 | 4 | 1.39 | 1.36 | 0.21 | 0.21 |
| 10,000 | 3 | 1.41 | 1.36 | 0.23 | 0.23 |
| 10,000 | 2 | 1.26 | 1.15 | 0.20 | 0.22 |
| 12,500 | 4 | 1.03 | 1.00 | 0.17 | 0.17 |
| 12,500 | 3 | 0.97 | 0.92 | 0.16 | 0.17 |
| 12,500 | 2 | 0.77 | 0.68 | 0.12 | 0.16 |
| 15,000 | 4 | 0.79 | 0.79 | 0.13 | 0.13 |
| 17,500 | 4 | 0.62 | 0.62 | 0.10 | 0.10 |
| 20,000 | 4 | 0.48 | 0.49 | 0.08 | 0.09 |
| 25,200 | 4 | 0.31 | 0.315 | 0.05 | 0.05 |
| 30,000 | 4 | 0.20 | 0.22 | 0.035 | 0.04 |
| 36,000 | 4 | 0.07 | 0.13 | 0.015 | 0.025 |

unchanged because, as may be seen in Figure 7-1, both $n = 3$ and $n = 4$ are overpopulated by about the same factor so that the opacity ratio at the Paschen edge is unaltered to a high degree of approximation. We shall discuss the comparison of these results with observation in Section 7-5 below.

## 7-3. Formation of the Lyman Continuum

The models described above assume the Lyman continuum is in radiative detailed balance and treat the coupling between the transfer and statistical equilibrium equations by *iteration*. As we proceed to higher temperatures, the Lyman continuum becomes relatively more transparent; at some point, significant transfer occurs and leads to an uncoupling of the $n = 1$ state from the $n = 2$ state. Because the iteration procedure is ineffective for high-opacity transitions, one must approach the problem from the direction of introducing the statistical equilibrium equation directly into the expression for the source function, thereby treating the two sets of equations *simultaneously*. This powerful approach will be used again and again in our later discussions and is well illustrated by the case of the Lyman continuum. For simplicity, assume that the Lyman continuum is so strong that all other opacity sources can be neglected. We may then write the transfer equation as

$$\frac{\mu}{\rho}\frac{dI_\nu}{dz} = -b_1\kappa_0{}^*\varphi_\nu I_\nu + \kappa_0{}^*\varphi_\nu I_\nu e^{-h\nu/kT} + \kappa_0{}^*\varphi_\nu B_\nu(1 - e^{-h\nu/kT}) \tag{7-14}$$

Here $\kappa_0{}^*$ is the LTE opacity at $\nu_0$, the frequency of the head of the Lyman continuum, and $\varphi_\nu$ is the continuum profile function

$$\varphi_\nu \equiv \frac{\kappa(\nu)}{\kappa(\nu_0)} = \frac{g_{II}(\nu)}{g_{II}(\nu_0)}\left(\frac{\nu_0}{\nu}\right)^3 \tag{7-15}$$

Now defining the optical depth scale to be that at frequency $\nu_0$,

$$d\tau = -\rho\kappa_0{}^*(b_1 - e^{-h\nu/kT})\,dz \tag{7-16}$$

we have

$$\mu\frac{dI_\nu}{d\tau} = \varphi_\nu I_\nu - \frac{\varphi_\nu(1 - e^{-h\nu/kT})B_\nu}{(b_1 - e^{-h\nu/kT})} = \varphi_\nu(I_\nu - S_\nu) \tag{7-17}$$

where

$$S_\nu \equiv \frac{(1 - e^{-h\nu/kT})B_\nu}{(b_1 - e^{-h\nu/kT})} \tag{7-18}$$

This is a general expression for the non-LTE source function of a continuum that is not seriously overlapped by other opacity sources and is not restricted to the ground level only. If for simplicity we neglect the stimulated emission factor $\exp(-h\nu/kT)$ and solve for $b_1$ from equation (7-8) for the ground state, we find

$$S_\nu = \frac{B_\nu}{b_1} = \frac{B_\nu\left[4\pi\int_{\nu_0}^{\infty}\frac{\alpha_\nu J_\nu}{h\nu}\,d\nu + N_e\left(\sum_{j=2}^{15}\Omega_{1j} + \Omega_{1\kappa}\right)\right]}{4\pi\int_{\nu_0}^{\infty}\frac{\alpha_\nu B_\nu}{h\nu}\,d\nu + N_e\left(\sum_{j=2}^{15}b_j\Omega_{1j} + \Omega_{1\kappa}\right)} \tag{7-19}$$

This expression can be rewritten into the symbolic form

$$S_\nu = \frac{\tilde{\sigma}4\pi\int_{\nu_0}^{\infty}\frac{\alpha_\nu J_\nu}{h\nu}\,d\nu + \tilde{\varepsilon}B_\nu}{1 + \tilde{\beta} + \tilde{\varepsilon}} = \gamma_\nu\int_{\nu_0}^{\infty}\Phi_\nu J_\nu\,d\nu + \varepsilon_\nu B_\nu \tag{7-20}$$

where

$$\tilde{\sigma} \equiv B_\nu/4\pi\int_{\nu_0}^{\infty}\frac{\alpha_\nu B_\nu}{h\nu}\,d\nu \tag{7-21}$$

$$\tilde{\varepsilon} \equiv N_e\left(\sum_{j=2}^{15}\Omega_{1j} + \Omega_{1\kappa}\right)\Big/4\pi\int_{\nu_0}^{\infty}\frac{\alpha_\nu B_\nu}{h\nu}\,d\nu \tag{7-22}$$

$$\tilde{\beta} \equiv N_e\sum_{j=2}^{15}d_j\Omega_{1j}\Big/4\pi\int_{\nu_0}^{\infty}\frac{\alpha_\nu B_\nu}{h\nu}\,d\nu \tag{7-23}$$

and

$$\Phi_\nu = \frac{4\pi\alpha_\nu}{h\nu} \tag{7-24}$$

The corresponding definitions of $\gamma_\nu$ and $\varepsilon_\nu$ appearing in equation (7-20) are obvious. Thus, the transfer equation to be solved in the Lyman continuum becomes

$$\mu \frac{dI_\nu}{d\tau} = \varphi_\nu\left(I_\nu - \gamma_\nu \int_{\nu_0}^{\infty} \Phi_\nu J_\nu \, d\nu - \varepsilon_\nu B_\nu\right) \tag{7-25}$$

In this form we see very clearly that the source function contains a kind of *noncoherent scattering term* (in that intensities at all frequencies in the entire continuum are coupled) and a *thermal term* (i.e., a term that does not contain the radiation field of the transition under consideration). We shall refer to this type of an expression for a source function as the *explicit* form since scattering and thermal terms appear explicitly. The coupling parameters $\gamma_\nu$ and $\varepsilon_\nu$ are specified in a completely natural way by the statistical equilibrium equations. Note that these coupling parameters contain terms related to the radiation field in *other* transitions inasmuch as the departures $d_j$ that appear in the term $\tilde{\beta}$ depend upon the radiation intensities in the relevant transitions.

To solve the above equation, we may use the difference-equation method. It is instructive to sketch a few of the details involved. As before, we rewrite the transfer equation as a second-order equation, as given in equation (6-95). We then introduce a set of depth points $\{\tau_i\}$, $(i = 1, \ldots, N)$; angle points $\{\mu_j\}$, $(j = 1, \ldots, M)$; and, in addition, frequency points $\{\nu_k\}$, $(k = 1, \ldots, K)$. We also replace integrals over frequency by sums:

$$4\pi \int_{\nu_0}^{\infty} \frac{\alpha_\nu}{h\nu} f(\nu)\, d\nu \rightarrow \sum_{k=1}^{K} c_k f(\nu_k) \tag{7-26}$$

Then we may write

$$\begin{aligned} \frac{2\mu_j^2}{(\tau_{i+1} - \tau_i)} & \left[\frac{P(\tau_{i+1}, \mu_j, \nu_k) - P(\tau_i, \mu_j, \nu_k)}{\tau_{i+1} - \tau_i} - \frac{P(\tau_i, \mu_j, \nu_k) - P(\tau_{i-1}, \mu_j, \nu_k)}{\tau_i - \tau_{i-1}}\right] \\ &= \varphi^2(\nu_k)\Bigg[P(\tau_i, \mu_j, \nu_k) - \gamma(\tau_i, \nu_k) \\ &\quad \times \sum_{j'=1}^{M} \sum_{k'=1}^{K} a_{j'} c_{k'} P(\tau_i, \mu_{j'}, \nu_{k'}) - \varepsilon(\tau_i, \nu_k) B(\tau_i, \nu_k)\Bigg] \end{aligned} \tag{7-27}$$

where we have made use of the fact that $\varphi_\nu$ is independent of depth. As before, we can reduce this equation to the standard form

$$-\mathbf{A}_i \mathbf{P}_{i-1} + \mathbf{B}_i \mathbf{P}_i - \mathbf{C}_i \mathbf{P}_{i+1} = \mathbf{Q}_i \tag{7-28}$$

where now the vectors are of dimension $L = M \times K$ and the matrices are $(L \times L)$. The matrices **A** and **C** are the same as written previously for the coherent scattering case. The matrix **B** now has diagonal elements

$$(\mathbf{B}_i)_{ll} = (\mathbf{A}_i)_{ll} + (\mathbf{C}_i)_{ll} + \varphi^2(\nu_k)[1 - \gamma(\tau_i, \nu_k)a_j c_k] \tag{7-29}$$

where $l = j + M(k - 1)$, and off-diagonal elements

$$(\mathbf{B}_i)_{lm} = -\varphi^2(\nu_k)\gamma(\tau_i, \nu_k)a_{j'} c_{k'} \tag{7-30}$$

where $m = j' + M(k' + 1)$. The vector $\mathbf{Q}_i$ is simply

$$(\mathbf{Q}_i)_l = \varphi^2(\nu_k)\varepsilon(\tau_i, \nu_k)B(\tau_i, \nu_k) \tag{7-31}$$

Boundary conditions similar to those written previously can be found, and the solution proceeds in the standard way. As before, we recover the mean intensity by the sum

$$J(\tau_i, \nu_k) = \sum_{j=1}^{M} a_j P(\tau_i, \mu_j, \nu_k) \tag{7-32}$$

the source function by computing

$$S(\tau_i, \nu_k) = \gamma(\tau_i, \nu_k) \sum_{k'-1}^{K} c_{k'} J(\tau_i, \nu_{k'}) + \varepsilon(\tau_i, \nu_k)B(\tau_i, \nu_k) \tag{7-33}$$

and, using equation (1-64), the flux by direct quadrature.

If we solve the transfer equation as written above, in principle we have satisfied simultaneously the transfer and statistical equilibrium equations, which is greatly superior to an iterative approach. It should be recognized, however, that some iteration will still be required in the solution. In particular, the temperature structure $T(\tau)$, the density structure $N(\tau)$, and the run of the relevant physical parameters will, in general, not be specified a priori on the $\tau$ scale at the head of the Lyman continuum but on some arbitrary scale, say, $\tau_{4000}$. Thus one must use some estimate of $b_1$ to construct the Lyman continuum depth scale, solve the transfer equation, redetermine $b_1$, and iterate until convergence is obtained. Because the scattering term is introduced *explicitly* into the transfer equation, the correct solution is obtained for a given run of source-sink terms over the *entire* range of optical depths in a single step, and the slow convergence properties of the lambda-iteration scheme are avoided.

Calculations of the kind described above have been carried out for a few models by Mihalas (*Ap. J.*, **150**, 909, 1967). Results for $d_1$ and $d_2$ for one of these models is shown in Figure 7-2. We see that the ground state is strongly overpopulated while the $n = 2$ state is underpopulated as before. Only when the Lyman continuum becomes opaque does the ground-state population become strongly coupled to $n = 2$.

These results are only schematic, however, since the Avrett-Krook method was used as a temperature-correction procedure, and unfortunately, because of its lambda-iteration character near the surface, this procedure failed to converge. The temperature structure obtained in the calculation cited above was close to the LTE temperature distribution; as we shall see in the next section, the actual situation is quite different.

Some additional insight into the nature of the problem may be gained following a simplified analysis given by Dietz and House (*Ap. J.*, **141**, 1393, 1965). We consider only an atom consisting of a ground state and continuum. Then from equation (7-20) we may write the source function as

$$S_\nu = \frac{4\pi\tilde{\sigma}\displaystyle\int_{\nu_0}^{\infty}\frac{\alpha_\nu J_\nu}{h\nu}\,d\nu + \tilde{\varepsilon}B_\nu}{1+\tilde{\varepsilon}} \tag{7-34}$$

where $\tilde{\sigma}$ has the same meaning as in equation (7-21), and now

$$\tilde{\varepsilon} = N_e\Omega_{1\kappa}/4\pi\int_{\nu_0}^{\infty}\frac{\alpha_\nu B_\nu}{h\nu}\,d\nu \tag{7-35}$$

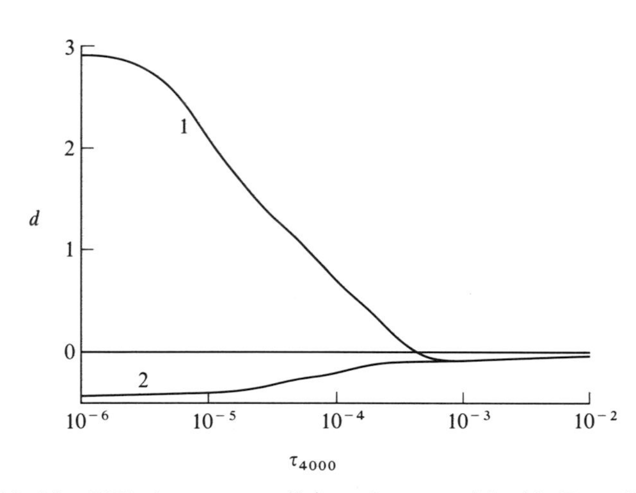


FIG. 7-2. Non-LTE departure coefficients for a model with $T_{\rm eff} = 15{,}000$°K and $\log g = 4$, allowing for transfer in Lyman continuum. Note overpopulation of ground state. These results are somewhat schematic because the temperature-correction procedure failed to converge even though the transfer equation was solved properly.

For simplicity, we consider a constant-temperature, constant-density atmosphere so that $\tilde{\sigma}$ and $\tilde{\varepsilon}$ are constant. We further write

$$J_\nu = (1 - \delta_\nu)B_\nu \tag{7-36}$$

where $0 \leq \delta_\nu \leq 1$. Now, if we make the Eddington approximation $J_\nu = 3K_\nu$, and, to simplify the problem as much as possible, if we choose some constant value $\delta_\nu \equiv \delta$ and ignore the profile function $\varphi_\nu$ by working at the head of the Lyman continuum only, then we find

$$S_\nu = \frac{1 - \delta + \tilde{\varepsilon}}{1 + \tilde{\varepsilon}} B_\nu \tag{7-37}$$

so that the Eddington transfer equation

$$\frac{1}{3}\frac{d^2 J_\nu}{d\tau^2} = J_\nu - S_\nu \tag{7-38}$$

reduces to

$$\frac{d^2\delta}{d\tau^2} = \frac{3\tilde{\varepsilon}}{1 + \tilde{\varepsilon}}\delta = \eta\delta \tag{7-39}$$

The appropriate boundary condition at the surface is that

$$\frac{dJ}{d\tau} = 3H(0) = \sqrt{3}\,J(0) \tag{7-40}$$

or

$$\left.\frac{d\delta}{d\tau}\right|_0 = -\sqrt{3}(1 - \delta) \tag{7-41}$$

The general solution of equation (7-39) is

$$\delta(\tau) = A\exp(-\sqrt{\eta}\tau) + B\exp(\sqrt{\eta}\tau) \tag{7-42}$$

and by demanding that $\delta \to 0$ as $\tau \to \infty$ and applying equation (7-41), we obtain

$$\delta(\tau) = \frac{\exp(-\sqrt{\eta}\,\tau)}{1 + \sqrt{\eta/3}} \tag{7-43}$$

From this very simplified solution we may observe two very important features that are characteristic of this problem. First, we see that $\delta \to 0$ only over depths of the order of $1/\sqrt{\eta} \sim 1/\sqrt{\tilde{\varepsilon}}$. Thus, the rate at which $J$ thermalized to $B$ is controlled entirely by the parameter $\tilde{\varepsilon}$. Since $\tilde{\varepsilon}$ is generally quite small, we expect significant departures of $J$ from $B$ over *large* optical depths in the continuum. A second important feature of the problem is the value of $b_1$ at the boundary. We find easily that

$$\delta(0) = \frac{1}{1 + [\tilde{\varepsilon}/(1 + \tilde{\varepsilon})]^{1/2}} \approx 1 - \tilde{\varepsilon}^{1/2} \tag{7-44}$$

so that

$$b_1(0) = \frac{B_\nu}{S_\nu(0)} \approx \frac{1}{\tilde{\varepsilon}^{1/2}} \tag{7-45}$$

This implies that in general the departure coefficients may be large at the surface. These two features [which are analogous to those encountered in equations (6-32) and (6-33)] are very characteristic of non-LTE transfer problems, and we will encounter them again and again in our future work.

Dietz and House carry through a more accurate solution of the problem, allowing for the frequency dependence of $\delta_\nu$. Introducing a quadrature sum for the integral, they solve the equation

$$\frac{1}{3}\frac{d^2\delta_i}{d\tau^2} = \varphi_i{}^2\left(\delta_i - C\sum_j w_j\,\delta_j\right) \tag{7-46}$$

where $C$ depends on $\bar{\varepsilon}$. The details of the solution differ from the schematic problem considered here, but the main features remain unchanged.

It is interesting to compare the value of $b_1$ (0) given by equation (7-45) with that obtained for the model shown in Figure 7-2. First, we need to calculate $\bar{\varepsilon}$ at the values $N_e(0)$ and $T(0)$ in the model. In this calculation we alter the definition of $\bar{\varepsilon}$ to include $\Omega_{12}$ as well as $\Omega_{1\kappa}$ (the former rate being about 100 times larger). The value of $\bar{\varepsilon}$ so obtained implies that $b_1(0) \sim 300$, in contrast with $b_1(0) \sim 5$ given by the model. The primary reason for this discrepancy is the strong increase of the density inward, which leads to a strong increase in $\bar{\varepsilon}$. Thus, if we use the value of $\bar{\varepsilon}$ at the surface, it will overestimate the thermalization distance and hence $b_1(0)$. An estimate of the effect of the density gradient can be made by writing

$$\eta = \alpha - \beta e^{-\gamma\tau} \tag{7-47}$$

in which $\alpha$ and $\beta$ can be chosen to cancel strongly at the boundary while $\gamma$ can be chosen to describe the gradient. Then equation (7-39) becomes

$$\frac{d^2\delta}{d\tau^2} = (\alpha - \beta e^{-\gamma\tau})\delta \tag{7-48}$$

Now writing

$$x = \frac{2\beta^{1/2}}{\gamma}\,e^{-\gamma\tau/2} \tag{7-49}$$

we find equation (7-48) reduces to *Bessel's equation*

$$x^2\,\delta'' + x\,\delta' + (x^2 - n^2)\delta = 0 \tag{7-50}$$

where $n = 2\alpha^{1/2}/\gamma$. The boundary condition becomes

$$\frac{\gamma}{2} x_0\,\delta' = \sqrt{3}(1 - \delta) \tag{7-51}$$

at $x_0 = 2\beta^{1/2}/\gamma$. The solution is then

$$\delta = \frac{J_n(x)}{\sqrt{\beta/3}\,J_n'(x_0) + J_n(x_0)} \tag{7-52}$$

where $J_n$ is the ordinary Bessel function of order $n$. From equation (7-52), $\delta_0$ and $b_1$ (0) can be computed for a given choice of $\alpha$, $\beta$, and $\gamma$. For the model under consideration, one finds $\eta$ can be schematized with $\alpha = 0.6$, $\beta = 0.999 \times 0.6$, and $\gamma$ on the range 0.003 to 0.01. This leads to $b_1$ (0) of order 15 to 20, in much better agreement with the results shown in Figure 7-2. The residual missing factor of 3 or 4 can be accounted for by the fact that there is a positive gradient of $B_\nu$ inward and that we have ignored the noncoherent coupling to other frequencies in the continuum. Accounting for both these effects should bring still better agreement with the numerical calculation. The important point here is to realize that density gradients may strongly affect estimates of the thermalization lengths.

## 7-4. Solution by Partial Linearization

The basic technique described in the preceding section allows one, in effect, to solve the transfer equations and statistical equilibrium equations simultaneously for a *given* temperature distribution. If one, in addition, wishes to satisfy the condition of radiative equilibrium, the temperature distribution must also be regarded as unknown. One might attempt to determine the temperature structure by iteration with, say, the Avrett-Krook method, but, as mentioned before, experience has shown that this approach will fail for very opaque transitions. Rather, it is more efficient to incorporate the constraint of radiative equilibrium into the equations at the outset, by means of a *linearization* procedure. Thus, in principle, one is solving the transfer equation simultaneously with *both* the constraints of radiative and statistical equilibrium. As we shall see below, one is able to cancel out the scattering terms, to a large extent, and thus attain a strongly convergent method. For the sake of generality and to avoid repetition, we shall write the equations in such a way as to include overlapping transitions and to allow for lines (although we will defer discussion of the lines until Chapter 13). The particular formulation we shall describe is that given by Auer and Mihalas (*Ap. J.*, **156**, 157, 1969, and *Ap. J.*, **156**, 681, 1969).

We will employ the difference-equation technique so that, recalling equation (6-95), the equation of transfer to be solved is of the form

$$\frac{\mu_i^2}{\chi_i}\frac{\partial}{\partial \tau}\left(\frac{1}{\chi_i}\frac{\partial P_i}{\partial \tau}\right) = P_i - S_i \tag{7-53}$$

where $i$ denotes a specific angle and frequency point, $\tau$ is a standard optical depth scale (arbitrarily chosen), and $\chi_i$ is the ratio of opacity at the frequency under consideration to the standard opacity. For simplicity, we will assume the atmosphere consists of pure hydrogen so that we may write $N_{\text{ion}} = N_e$, and we may define LTE populations as [equation (3-56)]

$$\begin{aligned} N_i^* &= N_e^2 \frac{g_i}{2}\left(\frac{h^2}{2\pi mkT}\right)^{3/2} \exp[(\chi_I - \chi_i)/kT] \\ &= N_e^2 \varphi_i(T) \end{aligned}$$

If we include $k - 1$ bound levels and use the index $k$ to denote the continuum, we may write quite generally (for both lines and continua)

$$\begin{aligned} \chi_\nu = \frac{1}{(a_\nu)_{\text{std}}} \Bigg[ \sum_{i=1}^{k-1} \sum_{j>i}^{k} \alpha_{ij}(\nu)(N_i - g_i N_j/g_j) \\ + N_e^2 \alpha_{kk}(\nu)(1 - e^{-h\nu/kT}) + N_e \sigma_e \Bigg] \end{aligned} \tag{7-54}$$

where $\alpha_{ij}(\nu)$ denotes the energy-absorption cross-section in transition $i \to j$, $(a_\nu)_{\text{std}}$ denotes the standard opacity, $\alpha_{kk}(\nu)$ denotes the free-free cross-section, and where we observe the convention that

$$\frac{g_i N_k}{g_k} \equiv N_i^* e^{-h\nu/kT} \tag{7-55}$$

From its fundamental definition in equation (1-39) and in view of the results of Chapter 4 [particularly equations (4-6), (4-8), (4-10), (4-18), (4-189), (4-192), and (4-193)], we may write a completely general expression for the source function as

$$S_\nu = \frac{\dfrac{2h\nu^3}{c^2}\left[\displaystyle\sum_{i=1}^{k-1}\sum_{j>i}^{k} \alpha_{ij}(\nu) N_j g_i/g_j + N_e^2 \alpha_{kk}(\nu) e^{-h\nu/kT}\right] + N_e \sigma_e J_\nu}{\displaystyle\sum_{i=1}^{k-1}\sum_{j>i}^{k} \alpha_{ij}(\nu)(N_i - g_i N_j/g_j) + N_e^2 \alpha_{kk}(\nu)(1 - e^{-h\nu/kT}) + N_e \sigma_e} \tag{7-56}$$

In particular, if LTE obtains, it is easy to show that equation (7-56) reduces immediately to equation (1-43). This expression may be rewritten in the form

$$S_\nu = \psi_\nu B_\nu(T_e) + \zeta_\nu J_\nu \tag{7-57}$$

where

$$\psi_\nu \equiv \frac{(1 - e^{-h\nu/kT})[\sum_{i=1}^{k-1} \sum_{j<i}^{k} \alpha_{ij}(\nu) N_j N_i^*/N_j^* + N_e^2 \alpha_{kk}(\nu)]}{\sum_{i=1}^{k-1} \sum_{j>i}^{k} \alpha_{ij}(\nu)(N_i - g_i N_j/g_j) + N_e^2 \alpha_{kk}(\nu)(1 - e^{-h\nu/kT}) + N_e \sigma_e} \tag{7-58}$$

and

$$\zeta_\nu \equiv \frac{N_e \sigma_e}{\sum_{i=1}^{k-1} \sum_{j>i}^{k} \alpha_{ij}(\nu)(N_i - g_i N_j/g_j) + N_e^2 \alpha_{kk}(\nu)(1 - e^{-h\nu/kT}) + N_e \sigma_e} \tag{7-59}$$

At any given frequency, a single transition will in general dominate the transfer problem. As we have seen in the preceding sections, it is important to employ the explicit form of the source function for the dominant transition in the solution of the transfer equation. We therefore factor out appropriate terms for each transition and then substitute solutions of the statistical equilibrium equations into these terms. In particular, if the transition from level $i$ to level $j$ dominates at frequency $\nu$, then we define

$$\xi_\nu = \frac{(N_i - g_i N_j/g_j)}{N_j N_i^*/N_j^*} \psi_\nu \tag{7-60}$$

and write the source function as

$$S_\nu = \left(\frac{N_j N_i^*/N_j^*}{N_i - g_i N_j/g_j}\right) \xi_\nu B_\nu(T_e) + \zeta_\nu J_\nu \tag{7-61}$$

(except for that part of the spectrum dominated by free-free transitions, where we do not perform the factoring but merely set $\xi_\nu \equiv \psi_\nu$). Equation (7-61) is to be understood in the following sense: The terms $\xi_\nu$, $\psi_\nu$, $\zeta_\nu$, and $B_\nu$ are computed from *current* estimates of temperatures and densities; the factor in parenthesis is to be replaced by an expression obtained from the statistical equilibrium equations. Note that for continuum transitions we obtain

$$\frac{N_k N_i^*/N_k^*}{N_i - N_k g_k/g_i} = \frac{1}{b_i - e^{-h\nu/kT}} \tag{7-62}$$

where we have made use of equation (7-55) and the fact that $N_k \equiv N_k^*$. Thus equation (7-61) is completely compatible with the form used in the preceding section [e.g., equation (7-18)]. If we now substitute from the statistical equilibrium equations into equation (7-61), we will obtain in general an expression of the form

$$S_\nu = \xi_\nu \left[\gamma_\nu \sum_l w_l \Phi_l P_l + \varepsilon_\nu B_\nu(T_e)\right] + \zeta_\nu J_\nu \tag{7-63}$$

where the $w$'s are quadrature weights over both angle and frequency, and

$$\Phi_l \equiv \frac{4\pi\,\alpha_{ij}(\nu_l)}{h\nu_l} \tag{7-64}$$

and, as indicated, all coefficients are frequency-dependent. The quadrature sum represents an integral only over the frequency range dominated by the transition considered, say, from $\nu_{ij0}$ to $\nu_{ij1}$. Thus, the transfer equation to be solved may be written as

$$\frac{\mu_i^2}{\chi_i}\frac{\partial}{\partial\tau}\left(\frac{1}{\chi_i}\frac{\partial P_i}{\partial\tau}\right) = P_i - \xi_i\left(\gamma_i \sum_l w_l \Phi_l P_l + \varepsilon_i B_i\right) - \zeta_i J_i \tag{7-65}$$

From the statistical equilibrium equation for level $i$ (including bound-bound transitions for later reference) [equation (5-59)]:

$$b_i\left(R_{i\kappa} + \sum_{j<i} A_{ij} Z_{ij} + N_e \sum_{j\neq i} \Omega_{ij} + N_e \Omega_{i\kappa}\right) - \sum_{j<i} b_j N_e \Omega_{ij}$$
$$- \sum_{j>i} b_j \left[A_{ji} Z_{ji} \frac{g_j}{g_i} \exp(-h\nu_{ij}/kT) + N_e \Omega_{ij}\right]$$
$$= R_{\kappa i} + N_e \Omega_{i\kappa}$$

we may obtain the expression in equation (7-62) immediately. Then grouping terms as indicated in equation (7-63), we find

$$\gamma_\nu = \frac{B_\nu(T_e)}{D_\nu} \tag{7-66}$$

$$\varepsilon_\nu = \frac{1}{D_\nu}\left[\int_{\nu_{ij1}}^{\infty} \Phi_\nu J_\nu \, d\nu + \sum_{j<i} A_{ij} Z_{ij} + N_e\left(\sum_{j\neq i} \Omega_{ij} + \Omega_{i\kappa}\right)\right] \tag{7-67}$$

where

$$D_\nu \equiv (R_{\kappa i} - e^{-h\nu/kT} R_{i\kappa}) + \sum_{j\neq i} (b_j - e^{-h\nu/kT}) N_e \Omega_{ij}$$
$$+ N_e \Omega_{i\kappa}(1 - e^{-h\nu/kT}) - e^{-h\nu/kT} \sum_{j<i} A_{ij} Z_{ij}$$
$$+ \sum_{j>i} b_j A_{ji} Z_{ji} \frac{g_j}{g_i} \exp(-h\nu_{ij}/kT) \tag{7-68}$$

These expressions are quite general. Note the integral over the radiation field contained in the definition of $\varepsilon_\nu$. This *overlap integral* arises because the scattering term in equation (7-63) applies *only* to that frequency range dominated by the transition $i \to j$; all other intervals dominated by overlapping transitions (even if they occur in the range $\nu_{ij0}$ to $\nu_{ij1}$, as would be the case for lines lying in the continuum) are excluded from the scattering term and put into $\varepsilon_\nu$. Normally the overlap integrals are small. With the transformation described above, the most important dependences are accounted for; the nonlinearities must be allowed for by iteration.

We must now incorporate into equation (7-65) the constraint of radiative equilibrium, which requires that

$$\int_0^\infty \chi_\nu(J_\nu - S_\nu)\,d\nu = \sum_j w_j\chi_j\left[P_j - \xi_j\left(\gamma_j \sum_{n_{0j}}^{n_{1j}} w_l \Phi_l P_l + \varepsilon_j B_j\right) - \zeta_j J_j\right]$$

$$= 0 \qquad (7\text{-}69)$$

Note that the inner sum extends only over the appropriate frequencies, as described above, so that the range of the inner summation index depends upon the outer; let us denote this range as $n_{0j}$ to $n_{1j}$. Equation (7-69) will be satisfied only when the temperature distribution is precisely that which yields radiative equilibrium. In general, the current values of $T$ and hence of $B_\nu(T)$ will not satisfy this relation. Suppose however that we can write the radiative equilibrium value of $B_\nu$ in terms its current value of $B_\nu{}^*$ and a perturbation so that

$$B_\nu \approx B_\nu{}^* + \frac{\partial B_\nu}{\partial T}\Delta T \qquad (7\text{-}70)$$

Then by substitution into equation (7-69), one finds

$$\Delta T = \sum_j w_j[\chi_j(1 - \zeta_j)P_j/E_1 - \Phi_j P_j E_{3j}] - E_2 \qquad (7\text{-}71)$$

where

$$E_1 = \sum_j w_j \chi_j \xi_j \varepsilon_j \frac{\partial B_j}{\partial T} \qquad (7\text{-}72)$$

$$E_2 = \frac{1}{E_1}\sum_j w_j \chi_j \xi_j \varepsilon_j B_j \qquad (7\text{-}73)$$

and

$$E_{3j} = \frac{1}{E_1}\sum_{n_{0j}}^{n_{1j}} w_l \chi_l \xi_l \gamma_l \qquad (7\text{-}74)$$

If we now replace the Planck function in equation (7-65) by equations (7-70) and (7-71), we obtain finally

$$\frac{\mu_i^2}{\chi_i}\frac{\partial}{\partial \tau}\left(\frac{1}{\chi_i}\frac{\partial P_i}{\partial \tau}\right) = P_i - \xi_i\left(\gamma_i \sum_{n_{0i}}^{n_{1i}} w_j \Phi_j P_j + \varepsilon_i B_i{}^*\right) - \zeta_i J_i$$

$$-\xi_i \varepsilon_i \frac{\partial B_i}{\partial T}\left\{\sum_j w_j[\chi_j(1 - \zeta_j)P_j/E_1 - \Phi_j P_j E_{3j}] - E_2\right\} \qquad (7\text{-}75)$$

There are several important points to be noted concerning equation (7-75). First, let us emphasize again that since this transfer equation implicitly includes the temperature correction in terms of the *new* radiation field (yet to be determined), in essence we have an equation that will satisfy the constraints of both statistical and radiative equilibrium simultaneously, as desired. Second, we see that scattering terms largely cancel out in the coefficients that appear in the transfer equation. Since this cancellation occurs before we solve the equation, we will obtain very good control of the residual thermal terms, which may be quite small compared with the scattering terms. Third, we see that equation (7-75) is again of the standard form (7-28), and the solution proceeds as outlined in Chapter 6. The upper boundary condition is again

$$\frac{\mu_i}{\chi_i}\frac{\partial P_i}{\partial \tau} = P_i \tag{7-76}$$

while by an analysis similar to that leading to equation (6-211), we find the lower boundary condition may be written

$$\frac{\mu_i}{\chi_i}\frac{\partial P_i}{\partial \tau} = P_i - B_i^* + \frac{\partial B_i}{\partial T}\sum_j w_j P_j\left\{\left[E_{3i}\Phi_j - \frac{\chi_j}{E_1}(1-\zeta_j)\right]\left(1 - \frac{\mu_i E_5}{\chi_i E_6}\right) - \frac{\mu_i \mu_j}{\chi_j E_6}\right\} + E_2\frac{\partial B_i}{\partial T} - \frac{\mu_i}{\chi_i}\frac{\partial B_i}{\partial T}\frac{1}{E_6}(\mathscr{H} - E_4 + E_2 E_5) \tag{7-77}$$

where $E_1$, $E_2$, and $E_{3i}$ are defined in equations (7-72) through (7-74), and

$$E_4 \equiv \sum w_j \mu_j B_j^* \tag{7-78}$$

$$E_5 \equiv \sum w_j \mu_j \frac{\partial B_j}{\partial T} \tag{7-79}$$

$$E_6 \equiv \sum \frac{w_j \mu_j}{\chi_j}\frac{\partial B_j}{\partial T} \tag{7-80}$$

and

$$\mathscr{H} = \frac{\sigma T_{\text{eff}}^4}{4\pi} \tag{7-81}$$

As before, the introduction of $\mathscr{H}$ into the lower boundary condition guarantees that the correct total flux is transported.

Computations of the kind described above have been made by Feautrier (*Ann. d'Ap.*, **31**, 257, 1968) and by Auer and Mihalas (*Ap. J.*, **156**, 157, 1969, and *Ap. J.*, **156**, 681, 1969). Feautrier's method differs markedly in appearance from the particular equations written above but in principle is very similar and also has excellent convergence properties. The results of these computations are quite interesting, the most striking effects being the changes in the temperature structure due to non-LTE effects. In the standard LTE models, the temperature is a monotone decreasing function outward. If the effects of the Lyman continuum are ignored (valid for $\tau_{4000} \gtrsim 10^{-4}$), the boundary temperature for a model with $T_{\text{eff}} = 15{,}000°\text{K}$ and $\log g = 4$ is found to be about 10,400°K. When the Lyman continuum is included, the boundary temperature drops to about 9400°K, a value achieved for $\tau_{4000} \lesssim 10^{-5}$. On the other hand, when the non-LTE calculation is carried out, the boundary temperature (including the Lyman continuum) rises to 10,500°K. The temperature at first decreases outward, reaches a minimum at $\tau_{4000} \approx 5 \times 10^{-5}$, and then *increases* outward (see Figure 7-3). The ground state is found to be strongly overpopulated. The non-LTE energy balance in the continuum is substantially different from that in LTE. For example, in LTE the Lyman continuum is a strong cooling continuum while non-LTE effects cause it to become a mild heating continuum. A detailed discussion of these effects can be found in the papers cited above. All of these effects occur at too small an optical depth in the visible to be observable there; indeed, as found before, the Balmer and Paschen jumps of the LTE and non-LTE models agree quite well. Nevertheless, the results are of considerable theoretical interest and have important implications when line formation problems are considered. Simultaneously, the lines can interact with the continua and lead to important changes in the energy balance and, therefore, the boundary temperature, as will be discussed in Chapter 14.

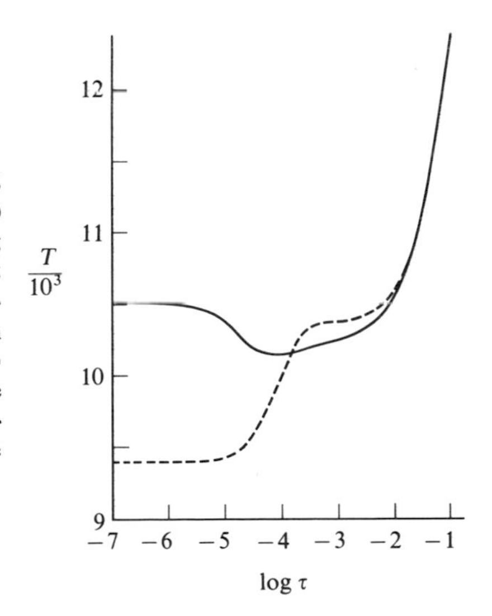


FIG. 7-3. Temperature structure in LTE (dashed curve) and non-LTE (solid curve) atmospheres with $T_{\text{eff}} = 15{,}000°\text{K}$ and $\log g = 4$. In the LTE model, the plateau at $\tau \approx 10^{-3}$ occurs when all continua are completely transparent except for the Lyman continuum, which is opaque; the final drop for $\tau < 3 \times 10^{-4}$ is caused by cooling in the Lyman continuum. In contrast, the non-LTE model shows a temperature rise because of heating in the Lyman continuum.

## 7-5. Comparison with Observation

Let us now consider briefly the comparison of the continuous energy distribution predicted by non-LTE models with observations. As shown in Table 7-1, the effects of departures from LTE upon the emergent energy distribution are small except at low gravities and high $T_{eff}$. Even at low gravities, some observational tests will show no effects at all. For example, in Figure 7-4 we show the values of the Balmer jump $D_B$ versus the flux ratio $F_\nu$ (4000 Å)/$F_\nu$ (6000 Å), a measure of the slope of the Paschen continuum, for LTE and non-LTE models. Points are plotted for models with $T_{eff} =$ 10,000°K and 12,500°K at 3 values of the surface gravity (the points are labeled with log $g$). Values given by LTE models are joined with dotted lines; those given by non-LTE models are joined by solid lines. We see that even though definite changes in $D_B$ and the flux ratio occur, they compensate one another in such a way that at a given gravity there is no clear separation between LTE and non-LTE models. Strom and Kalkofen (*Ap. J.*, **149**, 191, 1967) recognized that a much more sensitive observational parameter is the ratio $\varphi = D_P/D_B$. Since the $n = 2$ level is underpopulated while the $n = 3$ level is overpopulated, $D_B$ should decrease in non-LTE models. Similarly,

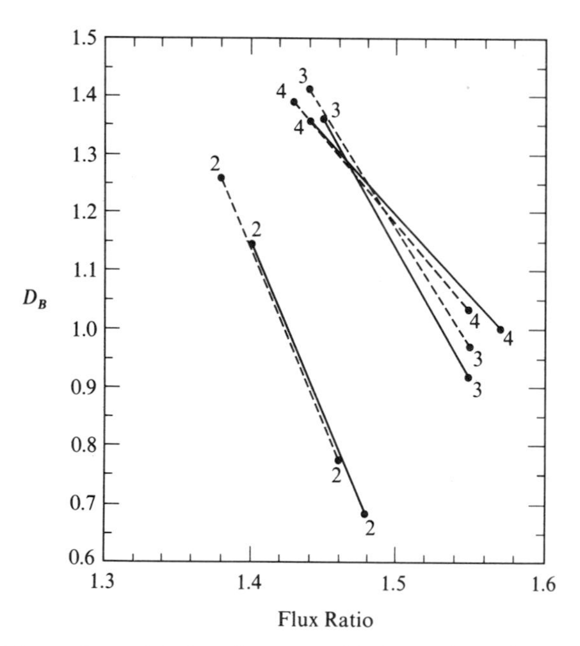


FIG. 7-4. Comparison of Balmer jump $D_B$ with flux ratio $F_\nu(\lambda\,4000)/F_\nu(\lambda\,6000)$ for LTE (solid lines) and non-LTE (dashed lines) models. Models are labeled with the value of log $g$.

level 3 is slightly more overpopulated than level 4 so that $D_P$ should increase slightly (though not much). The ratio $\varphi$ should, therefore, be larger in non-LTE models than in LTE models. The original results of Strom and Kalkofen have been improved upon by Mihalas (*Ap. J.*, **153**, 317, 1968) and by Smith and Strom (*Ap. J.*, **158**, 1161, 1969). Typical results are shown in Figure 7-5. We see here that there is no observable effect for stars near the main sequence while at low enough gravities, the effect is pronounced. Note also that the LTE models all predict values of $\varphi$ near 0.16, independent of gravity, so that larger $\varphi$-values can be explained only by non-LTE effects. As seen in Figure 7-6, these expectations are confirmed completely by the observations of Smith and Strom, which show clearly that the supergiants have much larger values of $\varphi$, at a given $D_B$, than dwarfs and normal giants, as predicted by the non-LTE models. The absolute positions in the observational diagram may not be strictly comparable with those in the theoretical diagram because of difficulties in calibration, but the differential comparison is reliable. It thus appears that the theoretically predicted non-LTE effects in the continuum, though small, can be detected and lead to an improved understanding of the observations of low-gravity stars. At the same time, we see that *near the main sequence*, non-LTE effects in the visible continuum are negligible and that LTE can probably be used as an adequate approximation in the interpretation of the observed continuous energy distribution of A and B stars.

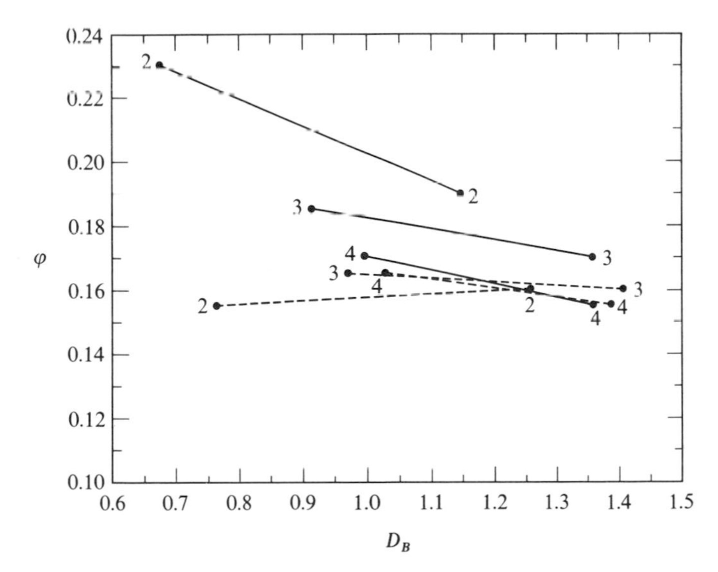


FIG. 7-5. Comparison of ratio $\varphi = D_P/D_B$ with Balmer jump $D_B$ for LTE and non-LTE models.

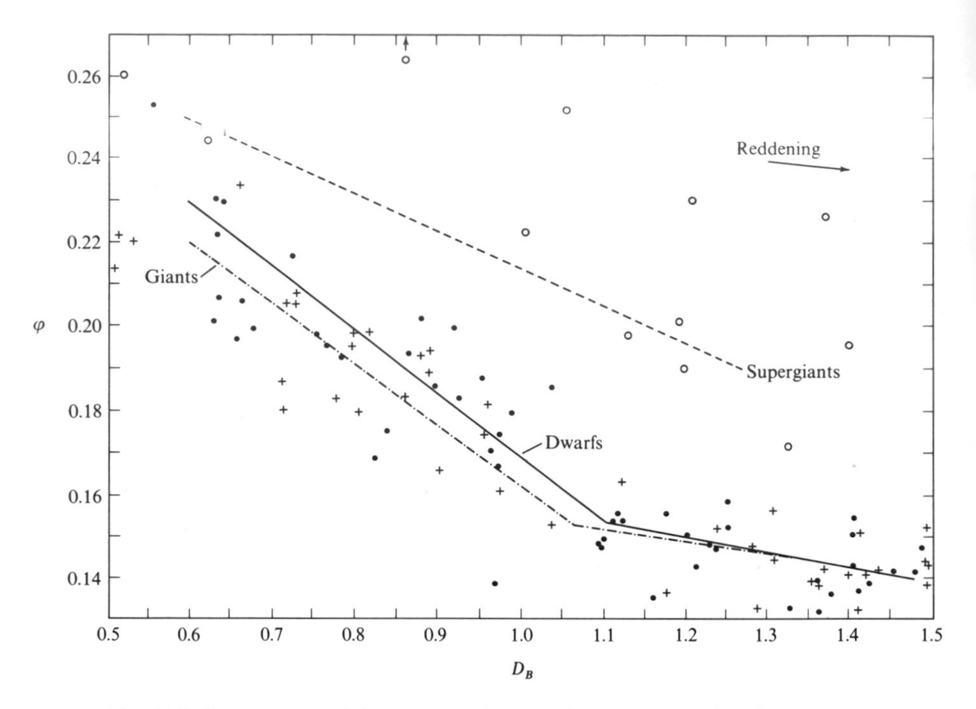


FIG. 7-6. Comparison of observations of supergiants, giants, and main sequence stars. Note that the supergiants have distinctly larger $\varphi$-values, in qualitative agreement with the predictions of theory. (From M. Smith and S. E. Strom, *Ap. J.*, **158**, 1161, 1969; by permission.)

This last statement should, however, not be construed as a blanket endorsement for the indiscriminate use of the LTE assumption outside of the range for which the validation seems proven. In particular, very recent work has shown important non-LTE effects upon $D_B$ for the O stars.

# 8 | The Line Spectrum: An Overview

Superimposed upon the continuum of a star, we observe discrete spectrum lines either in absorption or emission. These lines arise from transitions between bound states of atoms and ions in the star's atmosphere. An extremely wide variety of lines are observed, covering a wide range of atomic and ionic states and leading to very different-looking spectra for different classes of stars. A panoramic view of this variety is best obtained by inspection of actual stellar spectra as shown in particular in References 2 and 28 (see also Ref. 13, Chap. 14, and Ref. 19, Chap. 1). One finds that lines in stellar spectra show enormous ranges in line strength and striking variations in profile. A close examination shows that the spectra may be arranged into a two-dimensional scheme reflecting the effective temperature and luminosity of the star. It would take us too far afield to describe here the details of this procedure—developed to a high state of refinement by Morgan and his collaborators—or the full implications of the results; therefore references cited above should be studied carefully. Suffice it to say, the spectrum lines contain a wealth of information concerning the run of physical variables in the star and therefore provide important diagnostic tools in inferring the state of the atmosphere. Moreover, it is clear that the strength of a given line must contain informaton about, among other things, the abundance in the atmosphere of the star of the element to which it is due. Thus, by a suitable interpretation, the line spectrum offers us the opportunity of performing a quantitative chemical analysis of the material of which stars are formed. This information in turn provides valuable clues when we attempt to construct a coherent picture of the structure and evolution of our Galaxy and the universe.

Therefore, it is of prime importance to develop a theoretical framework that can predict line profiles and that can allow us to infer the desired physical information. A great deal of effort has been devoted by many astronomers to this end, and considerable progress has been made. Actually, much insight into the line formation problem has been gained only recently, and in view of the rapid progress in this field, we should realize that some of the discussion we will give is still subject to change and improvement. Let us summarize here just a few of the basic aspects of the problem, and to orient our later work, let us point out the kinds of information required to effect the solution.

## 8-1. Observational Quantities

A line in a stellar spectrum is most completely characterized by its *profile*, which is merely the observed distribution of energy as a function of frequency. For all stars except the sun, we can observe only the flux integrated over the entire disk of the star. We measure the flux in the line $F_\nu$ relative to that in the continuum $F_c$ and write the profile in terms of the *absorption depth*

$$A_\nu \equiv 1 - \frac{F_\nu}{F_c} \tag{8-1}$$

or the *residual intensity*

$$R_\nu \equiv \frac{F_\nu}{F_c} = 1 - A_\nu \tag{8-2}$$

In the case of the sun, we can observe the frequency distribution of the radiation at each point on the disk. Again, we can describe the profile in terms of the emergent specific intensity, $I_\nu(0, \mu)$, in units of the nearby continuum intensity $I_c(0, \mu)$ and write

$$a_\nu(\mu) \equiv 1 - \frac{I_\nu(0, \mu)}{I_c(0, \mu)} \tag{8-3}$$

or

$$r_\nu(\mu) \equiv \frac{I_\nu(0, \mu)}{I_c(0, \mu)} \tag{8-4}$$

This type of information is extremely valuable since it provides a kind of depth resolution otherwise unavailable and places important constraints on the theory. Often, because of the low light levels involved, it is not possible to measure the stellar spectrum with sufficient resolution to determine the profile in detail, and one then substitutes an integrated line strength—*the equivalent width*—for this more detailed information. The equivalent width is defined as

$$W_\nu \equiv \int_0^\infty A_\nu \, d\nu \tag{8-5}$$

or

$$W_\lambda \equiv \int_0^\infty A_\nu \, d\lambda \tag{8-6}$$

for stellar fluxes and

$$W_\nu(\mu) \equiv \int_0^\infty a_\nu(\mu) \, d\nu \tag{8-7}$$

or

$$W_\lambda(\mu) \equiv \int_0^\infty a_\nu(\mu) \, d\lambda \tag{8-8}$$

in the solar case. The most usual form is $W_\lambda$, expressed in Å (or $m$ Å). The equivalent width clearly is just the width of a perfectly black line with the same area under the profile as the line under study so that it gives a direct measure of the total amount of energy in the continuum influenced by that line.

Ideally, one desires to obtain profiles instead of equivalent widths since they contain far more information. Indeed, there is an infinity of profiles that will produce a given equivalent width. Thus, an interpretation based upon an equivalent width alone can be misleading—the same holds even for profiles! On the other hand, approaches exist that can use equivalent width information from many lines simultaneously, and these can yield important results. We will discuss these various problems in some detail. The actual measurement of the data requires fairly refined and complicated instrumental techniques; we will not discuss these methods here since they lie beyond the scope of this book, but excellent discussions exist elsewhere (see, e.g., Ref. 18, Chaps. 2, 4, and 13).

## 8-2. The Line Formation Problem

As in the case of the continuum, the description of line formation requires the solution of a transfer problem since the radiation emerges more or less efficiently from an entire range of depths, over which the physical properties may vary more or less strongly. Let us merely inquire here into the kinds of information we will need to have in order to approach the problem.

Consider a frequency $\nu$, and suppose we know the continuous opacity

$\kappa_\nu$, the continuum scattering coefficient $\sigma_\nu$, and the line absorption coefficient $l_\nu$ as a function of some standard depth $\tau_{\text{std}}$. Then if, in addition, we knew the run of the source function $S_\nu$, we could immediately calculate

$$F_\nu = 2 \int_0^\infty S_\nu(\tau_\nu) E_2(\tau_\nu)\, d\tau_\nu \tag{8-9}$$

$$F_c = 2 \int_0^\infty S_c(\tau_c) E_2(\tau_c)\, d\tau_c \tag{8-10}$$

and thus equation (8-1),

$$A_\nu = 1 - \frac{F_\nu}{F_c}$$

Here

$$\tau_\nu \equiv \int \frac{(\kappa_\nu + \sigma_\nu + l_\nu)}{(\kappa_\nu + \sigma_\nu)_{\text{std}}}\, d\tau_{\text{std}} \tag{8-11}$$

and

$$\tau_c \equiv \int \frac{(\kappa_\nu + \sigma_\nu)}{(\kappa_\nu + \sigma_\nu)_{\text{std}}}\, d\tau_{\text{std}} \tag{8-12}$$

In practice, of course, it is the source function that must be determined. As we have seen before, in any general (non-LTE) situation the source function and optical depth scale depend explicitly upon the occupation numbers of the particular levels involved. These in turn depend upon the radiation field and hence the source function. Thus we must, as in the case of the continuum, solve the transfer and statistical equilibrium equations self-consistently. Even before we do this, however, some insight into the nature of the problem can be gained by the following phenomenological arguments.

Consider the propagation of a photon in a line with overlapping continuous absorption and scattering. Some photons will be absorbed or scattered by the continuous process. Others will be absorbed in the line. Some of the photons in the line will be scattered in general with a redistribution characterized by $R(\nu', \mathbf{n}'; \nu, \mathbf{n})$. Others may be destroyed by collisional de-excitations or transitions to other levels. Photons may be introduced into the line by collisional excitations or by transitions into the upper level from other levels, with subsequent cascade to the lower level. Thus, the *form* of the transfer equation will, quite generally, be

$$\begin{aligned} \frac{\mu}{\rho}\frac{dI_\nu}{dz} &= -(\kappa_\nu + \sigma_\nu + l_\nu) I_\nu + \kappa_\nu S_c + \sigma_\nu J_\nu \\ &\quad + \tilde{\gamma} l \oint \frac{d\omega}{4\pi} \int_0^\infty I(\nu', \mathbf{n}') R(\nu', \mathbf{n}'; \nu, \mathbf{n})\, d\nu' + \tilde{\varepsilon} l \varphi_\nu B_\nu \end{aligned} \tag{8-13}$$

Here the coupling constants $\tilde{\gamma}$ and $\tilde{\varepsilon}$ describe the fraction of photons scattered and emitted by other processes. In general, the expressions for these quantities may be very complicated and may contain both radiative and collisional rates from the levels giving rise to the line, as well as coupling factors to other levels. To specify $\tilde{\gamma}$ and $\tilde{\varepsilon}$, we will need to consider the relevant statistical equilibrium equations.

At this point we can already see four important ingredients that are necessary to compute line profiles theoretically:

(a) We must be able to calculate the absorption profile $\varphi_\nu$. This will be treated in some detail in Chapter 9.

(b) We must be able to calculate the redistribution function $R(\nu', \mathbf{n}'; \nu, \mathbf{n})$. We will consider this problem in Chapter 10.

(c) We must be able to specify the coupling between photons and the atom via the parameters $\tilde{\gamma}$ and $\tilde{\varepsilon}$. This will be done in Chapters 11, 12, and 13.

(d) We must be able to solve the resulting transfer problem. Here we will refer back to the Riccati transformation or difference-equation methods discussed in Chapter 6, both of which have been used with success. In addition, we will discuss in Chapter 13 the additional formalism needed to include successfully the constraint of radiative equilibrium. In all of these four areas great strides have been made recently. In many ways, however, the greatest improvement in our understanding has come in regard to points (b), (c), and (d). Classically, the question of redistribution was often evaded, and the line scattering treated as coherent; we know today that this is a poor approximation and that, in fact, a better approximation often is the opposite extreme of complete redistribution over the line. In the specification of the parameters $\tilde{\gamma}$ and $\tilde{\varepsilon}$ the classical approach was sketchy and led to some serious misconceptions. More modern treatments have brought to light the importance of a clear understanding of these terms. In the area of actually solving the transfer problems, important advances have been made possible by extensive application of high-speed, large-capacity computers, using the recently developed powerful numerical techniques.

Our approach will be to deal with points (a) and (b) immediately. We then will try to summarize some of the more important classical treatments of the problem of line formation. We caution at the outset that one must realize these treatments are not entirely satisfactory. Nevertheless, they provide useful background, and, moreover, they must be understood if one is to fully appreciate their limitations and the concomitant limitations of the large literature based upon them. We then will turn to more modern treatments of line formation in various levels of physical approximation. Finally, we will discuss the effects of lines upon the structure of stellar atmospheres.

# 9 | The Line Absorption Profile

The profiles of lines in stellar spectra contain information both about the physical conditions and the abundances of chemical elements in the stellar atmosphere. Therefore, they provide extremely valuable diagnostic tools and must be exploited as fully as possible. To be able to carry out such a project, we will need to know how the detailed distribution of opacity with frequency in the line—the *line absorption profile*—depends upon local conditions of temperature, density, etc. We therefore turn to a discussion of the various mechanisms that determine line broadening.

## 9-1. The Natural Damping Profile

*Natural* (or *radiation*) *damping* refers to the line broadening caused by the finite duration of the emitted wavetrain as determined by a simple decay of the radiation process itself. This process is distinguished from broadening due to perturbations of the wavetrain by other atoms or charged particles interacting with the radiator. These latter processes are referred as *pressure broadening* and will be discussed in detail in later sections.

### ENERGY SPECTRA, POWER SPECTRA, AND THE AUTOCORRELATION FUNCTION

Let us now derive some basic relations that we will need in our discussion. Consider a time-varying oscillation of amplitude $f(t)$. We define the Fourier transform $F(\omega)$ to be

$$F(\omega) \equiv \int_{-\infty}^{\infty} f(t)e^{-i\omega t}\, dt \tag{9-1}$$

which satisfies the fundamental reciprocity relation

$$f(t) = \frac{1}{2\pi}\int_{-\infty}^{\infty} F(\omega)e^{i\omega t}\, d\omega \tag{9-2}$$

The quantity

$$E(\omega) \equiv \frac{1}{2\pi} F^*(\omega)F(\omega) \tag{9-3}$$

is called the *energy spectrum* of the oscillator. This designation derives from the fact that

$$\int_{-\infty}^{\infty} E(\omega)\, d\omega = \frac{1}{2\pi}\int_{-\infty}^{\infty} F^*(\omega)F(\omega)\, d\omega = \int_{-\infty}^{\infty} f^*(t)f(t)\, dt \tag{9-4}$$

which may be verified by direct calculation by using equation (9-1). If $f(t)$ were the voltage across a one-ohm resistor, then $f^*(t)f(t)$ would be the instantaneous power delivered to the resistor, and the integral over all time gives the total energy. Thus, $E(\omega)$ is a direct measure of the energy in the wavetrain at frequency $\omega$. In most cases, we will not use the energy spectrum but rather the energy delivered per unit time, i.e., the *power spectrum* $I(\omega)$, which may be defined as

$$I(\omega) \equiv \lim_{T\to\infty} \frac{1}{2\pi T}\left|\int_{-T/2}^{T/2} f(t)e^{-i\omega t}\, dt\right|^2 \tag{9-5}$$

For oscillations of finite duration or with, say, an exponential decay, the power spectrum will of course be zero since on the average over an infinite time interval, the finite total energy emitted yields zero power. In these cases, which are of practical interest to us, we must deal with the energy spectrum itself. We assume here that we observe an *ensemble* of such oscillators, created at a constant rate with random phases; then a finite power will result, with a frequency distribution proportional to the energy spectrum of a single oscillator.

In certain situations we are not able to calculate the power spectrum directly using equation (9-5). It is then valuable to make use of the *autocorrelation function*

$$\Phi(s) \equiv \lim_{T\to\infty} \frac{1}{T}\int_{-T/2}^{T/2} f^*(t)f(t+s)\, dt \tag{9-6}$$

from which the power spectrum may be obtained by the relation

$$I(\omega) = \frac{1}{2\pi} \int_{-\infty}^{\infty} \Phi(s) e^{-i\omega s} \, ds \tag{9-7}$$

as may be verified by direct calculation, using a limiting procedure in the integration over $s$. As we shall see, the autocorrelation function provides a very convenient and powerful tool in the calculation of power spectra.

### THE DAMPED CLASSICAL OSCILLATOR

The simplest picture we can construct of the radiation process is to consider the atom as a classical dipole. As we have seen earlier (Chapter 4), classical electromagnetic theory yields for the equation of motion of such an oscillator

$$m\ddot{x} = -m\omega_0{}^2 x + \frac{2}{3}\frac{e^2}{c^3}\dddot{x} \tag{9-8}$$

The reaction term (third term) is numerically quite small and may be estimated by using the unperturbed solution

$$x = x_0 \, e^{i\omega_0 t} \tag{9-9}$$

so that

$$\dddot{x} = -\omega_0{}^2 \dot{x} - i\frac{2e^2\omega_0{}^3}{3mc^3}\, x = -(\omega_0{}^2 + i\gamma\omega_0)x \tag{9-10}$$

where

$$\gamma \equiv \frac{2}{3}\frac{e^2\omega_0{}^2}{mc^3} \tag{9-11}$$

Then we have, neglecting terms in $\gamma^2$,

$$x = x_0 \, e^{i\omega_0 t} e^{-\gamma t/2} \tag{9-12}$$

which is simply an exponentially damped oscillation. Calculating the Fourier transform, we obtain

$$F(\omega) = x_0 \int_0^{\infty} e^{-i(\omega-\omega_0)t} e^{-\gamma t/2} \, dt = \frac{x_0}{i(\omega - \omega_0) + \gamma/2} \tag{9-13}$$

where we have assumed that the oscillation starts abruptly at time $t = 0$. The energy spectrum is given by

$$E(\omega) = \frac{x_0{}^2}{2\pi} \frac{1}{(\omega - \omega_0)^2 + (\gamma/2)^2} \tag{9-14}$$

Now, as discussed above, when we consider an ensemble of radiators and assume continuous creation with random phases, the power spectrum of the ensemble is proportional to the energy spectrum of an individual oscillator; thus we may write

$$I(\omega) = \frac{C}{(\omega - \omega_0)^2 + (\gamma/2)^2} \tag{9-15}$$

It is customary to normalize $I(\omega)$ such that

$$\int_{-\infty}^{\infty} I(\omega)\, d\omega = 1 \tag{9-16}$$

We then find $C = \gamma/2\pi$ so that

$$I(\omega) = \frac{\gamma/2\pi}{(\omega - \omega_0)^2 + (\gamma/2)^2} \tag{9-17}$$

The profile given by equation (9-17) is referred to as a *Lorentz profile.* Note that when $\Delta\omega = \omega - \omega_0 = \pm\gamma/2$, $I(\omega)$ drops to half its central intensity. Thus $\gamma$ is called the (full) half-intensity width of the line. Classical electromagnetic theory predicts a definite value for $\gamma$ and hence for the half-intensity width of the line. In wavelength units the (full) half-intensity width is

$$\Delta\lambda_c = \frac{2\pi c}{\omega^2}\gamma = \frac{4\pi e^2}{3mc^2} = 1.2 \times 10^{-4}\ \text{Å} \tag{9-18}$$

This width is very much smaller than actually observed either in the laboratory or in stellar spectra. We must therefore consider a more general picture of the radiation process.

## THE QUANTUM MECHANICAL OSCILLATOR

In the preceding section we considered the radiation of a damped classical oscillator. A quantum mechanical analogue can be constructed by assuming that the radiation arises from transitions of an atom from an excited state of finite lifetime to the ground state. Following Wigner and Weisskopf, we assume that the transition is characterized by a spontaneous emission rate $A_{ji}$ and that the probability of finding an atom in the excited state $j$ is given by

$$P_j(t) = \psi_j^* \psi_j e^{-\Gamma t} \tag{9-19}$$

where $\Gamma = A_{ji}$. Then the time development of the wave function of the state must be

$$\psi_j(\mathbf{r}, t)e^{-\Gamma t/2} = u_j(\mathbf{r})e^{-iE_j t/\hbar}e^{-\Gamma t/2} = u_j(\mathbf{r})e^{-i\omega_j t - \Gamma t/2} \tag{9-20}$$

Consistent with the *uncertainty principle*, we may consider that the decaying state $j$ no longer has a perfectly defined energy $E_j$ but rather is a superposition of states with energies spread in some fashion about $E_j$. From the fundamental reciprocity relations of quantum mechanics, the amplitude of the energy distribution is found from the Fourier transform of the time dependence and the probability from the square of the amplitude. It is thus clear that the result will be the same as for a classical oscillator, and by analogous arguments, we write finally

$$I(\omega) = \frac{\Gamma/2\pi}{(\omega - \omega_0)^2 + (\Gamma/2)^2} \tag{9-21}$$

where now $\Gamma$ is interpreted as the reciprocal of the mean lifetime of the upper state.

If the line under consideration arises from a transition between two excited levels, then each will have a characteristic mean lifetime and intrinsic width. In this case we may write for each level

$$\Gamma_U = \sum_{i<U} A_{Ui} \tag{9-22}$$

and

$$\Gamma_L = \sum_{i<L} A_{Li} \tag{9-23}$$

The line profile between these two states will, in general, reflect the width of both states. We assume that the probability distribution of substates relative to the nominal energy of the level is given by a Lorentz profile of the appropriate half-intensity width. Let $\delta = \Gamma/2$ for either level, and let $x = (E - E_0)/\hbar$ be the frequency displacement of a particular substate from the center of the level. If we assume that the probability of ending in a particular substate of the lower level is independent of the substate from which the transition started, then the joint probability of starting in substate $x$ of the upper level and ending in substate $x'$ of the lower level is

$$p(x, x') = \frac{\delta_L \, \delta_U/\pi^2}{(x^2 + \delta_U{}^2)(x'^2 + \delta_L{}^2)} \tag{9-24}$$

If we confine our attention to transitions giving rise to a specific frequency $\omega$, then $x$ and $x'$ must be connected by the relation

$$\omega = \omega_0 + x - x' \tag{9-25}$$

or writing $x_0 = \omega - \omega_0$,

$$x' = x - x_0 \tag{9-26}$$

The total intensity at $\omega$ can be obtained by summing over all upper substates $x$, subject to the condition of equation (9-26). That is,

$$I(\omega) = \int_{-\infty}^{\infty} p(x, x - x_0)\, dx = \frac{\delta_U\, \delta_L}{\pi^2} \int_{-\infty}^{\infty} \frac{dx}{(x^2 + \delta_U{}^2)[(x - x_0)^2 + \delta_L{}^2]} \tag{9-27}$$

This integral may be calculated using the residue theorem. The poles at $z = i\,\delta_U$ and $z = x_0 + i\,\delta_L$ must be accounted for. We then have

$$\begin{aligned} I(\omega) &= \frac{2\pi i \delta_U \delta_L}{\pi^2} \{[2i\delta_U(i\delta_U - x_0 + i\delta_L)(i\delta_U - x_0 - i\delta_L)]^{-1} \\ &\quad + [2i\delta_L(x_0 + i\delta_L + i\delta_U)(x_0 + i\delta_L - i\delta_U)]^{-1}\} \\ &= \frac{2i\delta_L\delta_U}{\pi} \left[\left(\frac{-i}{2\delta_L\delta_U}\right) \frac{(\delta_L + \delta_U)}{x_0{}^2 + (\delta_L + \delta_U)^2}\right] \end{aligned} \tag{9-28}$$

or

$$I(\omega) = \frac{\Gamma/2\pi}{(\omega - \omega_0)^2 + (\Gamma/2)^2} = \frac{1}{\pi} \frac{(\delta_L + \delta_U)}{(\omega - \omega_0)^2 + (\delta_L + \delta_U)^2} \tag{9-29}$$

where now

$$\Gamma = \Gamma_L + \Gamma_U \tag{9-30}$$

Hence, the resulting profile is again Lorentzian but with a half-intensity width equal to sum of the half-intensity widths of each level.

The Lorentz profiles we have calculated are, strictly speaking, emission profiles. If however we assume detailed balancing, we can assume that the absorption profile will have the same form. To convert to actual opacities, recall that we showed in equation (4-68) that

$$\int_{-\infty}^{\infty} \alpha_\nu\, d\nu = \frac{\pi e^2}{mc}.$$

Thus, assuming a general profile of the form of equation (9-29) and converting to ordinary frequency units, we have a absorption coefficient

$$\alpha_\nu = \frac{\pi e^2}{mc} f \frac{\Gamma/4\pi^2}{(\nu - \nu_0)^2 + (\Gamma/4\pi)^2} \tag{9-31}$$

The broadening of lines by radiation damping can be of great importance in astrophysical situations, particularly for strong lines in low-density media. For example, the $L\alpha$ line in interstellar space is broadened by radiation damping. In most cases, however, the line is formed in regions where the density of atoms, ions, and electrons is high enough to perturb the radiating atom, thus leading to pressure broadening of the line, which we will discuss presently. Before we do so, however, let us first consider the effects of motions of the radiating atoms upon the line profile.

## 9-2. The Effects of Doppler Broadening: The Voigt Function

When we observe a line in a stellar atmosphere (or laboratory plasma), we are seeing the combined effects of absorption by all of the atoms in the ensemble. Each of these atoms will have some characteristic velocity along the line of sight measured in the observer's frame, and the absorption profile will be shifted a corresponding amount in frequency. If we can assume that the damping process giving rise to the absorption profile is uncorrelated with the atoms' velocities, then we may superimpose these shifted profiles and obtain the total absorption coefficient of the ensemble by a simple folding procedure.

If the plasma can be characterized by a temperature $T$, then the velocity distribution will be Maxwellian so that if $\xi$ represents the velocity along the line of sight, the probability of finding an atom with velocity on the range $(\xi, \xi + d\xi)$ is

$$W(\xi)\, d\xi = \frac{1}{\sqrt{\pi}} \exp(-\xi^2/\xi_0{}^2)\, \frac{d\xi}{\xi_0} \tag{9-32}$$

The parameter $\xi_0$ is related to the temperature as follows. Let $\langle \xi^2 \rangle^{1/2}$ be the mean velocity in one component. Since each component is statistically independent of the others, the average total velocity $\langle v^2 \rangle = 3\langle \xi^2 \rangle$. But $\langle v^2 \rangle = 3kT/m$. Thus

$$\langle \xi^2 \rangle = kT/m \tag{9-33}$$

But direct calculation yields

$$\langle \xi^2 \rangle = \int_{-\infty}^{\infty} \xi^2 W(\xi)\, d\xi = \frac{1}{2}\xi_0{}^2 \tag{9-34}$$

Therefore,

$$\xi_0 = (2kT/m)^{1/2} = 12.85\left(\frac{T}{10^4 A}\right)^{1/2} \text{km/sec} \tag{9-35}$$

where $A$ is the atomic weight of the atoms under consideration.

Now, if we observe at frequency $\nu$, then an atom with velocity component $\xi$ is absorbing at frequency $\nu - \nu(\xi/c)$ in its own reference frame, and the absorption coefficient for that atom is $\alpha_\nu(\nu - \xi\nu/c)$. Thus, the total absorption coefficient at frequency is given by

$$\alpha_\nu = \int_{-\infty}^{\infty} \alpha(\nu - \xi\nu/c) W(\xi)\, d\xi \tag{9-36}$$

Note further that $\xi_0/c \ll 1$ for astrophysically interesting temperatures. Thus $\Delta\nu/\nu$ will be small, and it is adequate to approximate $\xi\nu/c$ by $\xi\nu_0/c$, where $\nu_0$ is the rest frequency of the line. Then

$$\alpha_\nu = \frac{\sqrt{\pi}\,e^2}{mc} f \frac{1}{\pi} \int_{-\infty}^{\infty} \frac{(\Gamma/4\pi)\exp(-\xi^2/\xi_0^2)}{\left(\nu - \nu_0 - \dfrac{\xi\nu_0}{c}\right)^2 + (\Gamma/4\pi)^2} \frac{d\xi}{\xi_0} \tag{9-37}$$

Let us now define the *Doppler width* of the line as

$$\Delta\nu_0 \equiv \frac{\xi_0\,\nu_0}{c} \tag{9-38}$$

Also the Doppler shift due to velocity $\xi$ is

$$\Delta\nu = \frac{\xi\nu}{c} \tag{9-39}$$

Now define

$$v \equiv \frac{(\nu - \nu_D)}{\Delta\nu_D} \tag{9-40}$$

$$y \equiv \frac{\Delta\nu}{\Delta\nu_D} = \frac{\xi}{\xi_0} \tag{9-41}$$

and

$$a = \frac{\Gamma}{4\pi\Delta\nu_D} \tag{9-42}$$

Then we may rewrite equation (9-37) as

$$\alpha_\nu = \frac{\sqrt{\pi}\,e^2}{mc} f \frac{H(a, v)}{\Delta\nu_D} \tag{9-43}$$

where

$$H(a, v) \equiv \frac{a}{\pi} \int_{-\infty}^{\infty} \frac{\exp(-y^2)\,dy}{(v - y)^2 + a^2} \tag{9-44}$$

is known as the *Voigt function*. As the integral of $H(a, v)$ over all $v$ is $\sqrt{\pi}$, the *normalized Voigt function*

$$U(a, v) = \frac{1}{\sqrt{\pi}} H(a, v) \tag{9-45}$$

is sometimes more convenient. The most extensive tables of $H(a, v)$ and $U(a, v)$ are those of Finn and Mugglestone (*M. N.*, **129**, 221, 1965) and of Hummer (*Mem. R. A. S.*, **70**, 1, 1965), respectively.

Using the well-known expression for the Laplace transformation of the sine function

$$\int_0^\infty e^{-ax} \cos bx \, dx = \frac{a}{a^2 + b^2} \tag{9-46}$$

and the addition rule for cosines, we can write the Voigt function in the form

$$H(a, v) = \frac{1}{\pi} \int_0^\infty e^{-ax} \cos vx \int_{-\infty}^\infty e^{-y^2} \cos xy \, dy \, dx \tag{9-47}$$

The sine transformation of the Gaussian is

$$\int_{-\infty}^\infty e^{-y^2} \cos xy \, dy = \sqrt{\pi} \, e^{-(x/2)^2} \tag{9-48}$$

from which we obtain the desired expression

$$H(a, v) = \frac{1}{\sqrt{\pi}} \int_0^\infty e^{-ax-(x/2)^2} \cos vx \, dx \tag{9-49}$$

When Doppler broadening dominates, the damping constant $a$ is less than unity. In this limit we can usefully expand the Voigt function in powers of $a$ by replacing $\exp(-ax)$ in the preceding equation with its series. Thus we may write

$$H(a, v) = \sum_{n=0}^\infty a^n H_n(v) \tag{9-50}$$

where

$$H_n(v) \equiv \frac{(-1)^n}{\sqrt{\pi} n!} \int_0^\infty e^{-(x/2)^2} x^n \cos vx \, dx \tag{9-51}$$

From equation (9-48) we see immediately that

$$H_0(v) = e^{-v^2} \tag{9-52}$$

and by repeatedly differentiating this expression, we obtain

$$H_2(v) = -\frac{1}{2} \frac{d^2}{dv^2} H_0(v) = (1 - 2v^2)e^{-v^2} \tag{9-53}$$

and

$$H_4(v) = -\frac{1}{12} \frac{d^2}{dv^2} H_2(v) = \left(\frac{1}{2} - 2v^2 + \frac{2}{3} v^4\right) e^{-v^2} \tag{9-54}$$

Higher even-order terms are found in the same way.

The first odd-order term may be integrated by parts to obtain

$$H_1(v) = \frac{-2}{\sqrt{\pi}}\left[1 - v\int_0^\infty e^{-(x/2)^2} \sin vx\, dx\right] \tag{9-55}$$

Utilizing now the sine transformation of the Gaussian

$$\int_0^\infty e^{-t^2} \sin 2vt\, dt = e^{-v^2}\int_0^v e^{t^2}\, dt \equiv F(v) \tag{9-56}$$

we have

$$H_1(v) = \frac{-2}{\sqrt{\pi}}[1 - 2vF(v)] \tag{9-57}$$

$F(v)$ is known as *Dawson's function*. Repeated differentiation of $H_1(v)$ yields the higher-order odd coefficients,

$$H_3(v) = -\frac{1}{6}\frac{d^2}{dv^2}H_1(v) = \frac{-2}{\sqrt{\pi}}\left[\frac{2}{3}(1 - v^2) - 2v\left(1 - \frac{2}{3}v^2\right)F(v)\right] \tag{9-58}$$

Harris (*Ap. J.*, **108**, 112, 1948) has tabulated $H_n(v)$ for $n \leq 4$. Since Dawson's function has the asymptotic expansion

$$F(v) \sim \frac{1}{2v} + \frac{1}{2^2v^3} + \frac{1 \cdot 3}{2^3v^5} + \frac{1 \cdot 3 \cdot 5}{2^4v^7} + \cdots \tag{9-59}$$

as may be seen by introducing a new variable of integration $t' = v - t$ into equation (9-56) and expanding $\exp(-vt')$, we can express the Voigt function for large $v$ in the form

$$H(u, v) \sim \frac{u}{\sqrt{\pi}\, v^2} \tag{9-60}$$

Hence, a schematic representation of the Voigt profile is

$$H(a, v) \sim e^{-v^2} + \frac{a}{\sqrt{\pi}\, v^2} \tag{9-61}$$

where the first term applies in the line core and the second at large values of $v$. The line core is thus seen to be dominated by Doppler broadening while natural broadening dominates the wings. When natural broadening is strong, i.e., for $a \gtrsim 1$, the Voigt function becomes similar to the Lorentz form for all frequencies.

## 9-3. Collision Broadening of Spectrum Lines

Let us now turn to the problem of the broadening of a spectral line by the interaction of the radiating atom with nearby particles. We shall consider two basic pictures of this process. The first theory we shall consider is known as *impact theory*. In this approximation, the radiating atom is assumed to undergo a collision that occurs essentially instantaneously and that interrupts the radiating wavetrain with a sudden phase shift or by inducing a transition. The action of these collisions is thus to cause the radiator to "start" and "stop" in intervals of finite duration; therefore, a Fourier analysis leads to a spread in the frequency of the radiated wavetrain. When averaged over all atoms in the ensemble, the result is a broadened spectral line.

An alternative approach is to consider the atom radiating in the field of an ensemble of perturbers. This ensemble will give rise to some field that will fluctuate about a mean value in a statistical way. At a given value of the field, the energy levels of the radiating atom will be shifted slightly, and, correspondingly, the frequency of the line will be altered. The intensity of the radiation at some specified frequency shift will be proportional to the statistical frequency with which the perturbation of the appropriate field strength occurs.

These two theories represent limiting cases of a more comprehensive theory. We shall consider them first from a semiclassical point of view since the quantum mechanical treatment is somewhat more complicated. The more basic features are revealed by the classical theory, which thus provides excellent background. There are now available several good discussions of collision broadening. We mention in particular Griem's book (Ref. 14), Traving's monograph (Ref. 34), and a review article by Cooper (*Reports Prog. Phys.*, **29**, 35, 1966). We will attempt to summarize the major results of the theory and give appropriate references where details may be found.

## 9-4. The Classical Impact Theory

This theory has its origins in a classical analysis by Lorentz, who considered the atom to be a radiating oscillator, perturbed by nearby passages of particles in such a way that there is a change in phase during the encounter. We assume the collisions occur essentially instantaneously so that the wavetrain suffers an instantaneous phase dislocation. We assume, in addition, that the collisions occur one at a time and occur between the atom and a single perturber. Thus, suppose that $T$ is the time between two successive collisions and that in this interval the radiator emits a monochromatic train

$$f(t) = e^{i\omega_0 t} \tag{9-62}$$

The Fourier transform of this finite wavetrain is

$$F(\omega, T) = \int_0^T e^{i(\omega_0 - \omega)t}\, dt = \frac{\exp[i(\omega - \omega_0)T] - 1}{i(\omega - \omega_0)} \tag{9-63}$$

The energy spectrum of this particular train is given by

$$E(\omega, T) = \frac{1}{2\pi} F^*(\omega, T)F(\omega, T) \tag{9-64}$$

Now in general there will not be a unique time between collisions, but these intervals will be distributed in a probabilistic way. If we assume that the collisions occur in a random-walk process and that the *mean time between collisions* is $\tau$, then the probability that the collision time lies on the range $(T, T + dT)$ is

$$W(T)\,dT = e^{-T/\tau} \frac{dT}{\tau} \tag{9-65}$$

Hence, averaging over all collision times, we have a mean energy spectrum

$$E(\omega) = \langle E(\omega, T)\rangle_T = \frac{1}{2\pi}\int_0^\infty F^*(\omega, T)F(\omega, T)W(T)\,dT \tag{9-66}$$

Computation of this integral with normalization to

$$\int_{-\infty}^{\infty} E(\omega)\,d\omega = 1$$

yields

$$E(\omega) = \frac{1/\pi\tau}{(\omega - \omega_0)^2 + (1/\tau)^2} = \frac{\Gamma/2\pi}{(\omega - \omega_0)^2 + (\Gamma/2)^2} \tag{9-67}$$

We again obtain a Lorentz profile, but now $\Gamma = 2/\tau$. As in the case of a radiation-damped oscillator, we take the profile of the ensemble of randomly phased oscillators created continuously to be proportional to the energy spectrum of a single oscillator (in this case averaged over lifetimes). If both radiation damping, with a width $\Gamma_R$, and collision damping, with a width $\Gamma_C$, occur and are assumed to be *completely uncorrelated*, then the profile resulting from both processes is the convolution of the two Lorentz profiles. By an analysis similar to that leading to equations (9-29) and (9-30), one readily may show that the combined profile is again a Lorentzian, with a total width $\Gamma = \Gamma_R + \Gamma_C$.

The mean collision time may be used to define an effective impact parameter $\rho_0$ such that

$$\frac{1}{\tau} \equiv N\pi\rho_0^2 v \tag{9-68}$$

or

$$\Gamma = 2N\pi\rho_0^2 v \tag{9-69}$$

where $N$ is the perturber density and

$$v = \langle v^2 \rangle^{1/2} = \left[\frac{8kT}{\pi m_H}\left(\frac{1}{A_1} + \frac{1}{A_2}\right)\right]^{1/2} \tag{9-70}$$

The value of $\rho_0$ is fixed by the details of the interaction and must be computed separately. It is important to note that we have obtained a Lorentz profile for the line only because we have assumed that the collisions occur completely randomly. A more detailed theory (considered below) yields a *shift* of the line as well. Let us now consider the calculation of $\rho_0$.

### THE WEISSKOPF APPROXIMATION

To calculate the phase shift due to an encounter, we will assume that: (a) the perturber may be considered as a classical particle; (b) the perturber moves on a straight-line path relative to the atom with impact parameter $\rho$ (see Figure 9-1); (c) the interaction may be described as

$$\Delta\omega(t) = \frac{C_p}{[r(t)]^p} \tag{9-71}$$

and (d) no transitions in the atom are produced by the action of the perturber. The validity of these approximations will be considered again later. The form of the interaction given by equation (9-71) is only approximate but holds over a wide range of distances. The exponent $p$ depends upon the nature of the interaction. Values of astrophysical interest and the interaction they represent are as follows: $p = 2$, *linear Stark effect* (hydrogen + charged particle); $p = 3$, *resonance broadening* (atom $A$ + atom $A$); $p = 4$, *quadratic Stark effect* (non-hydrogenic atom + charged particle); $p = 6$, *van der Waals interaction* (atom $A$ + atom $B$). The value for the interaction constant $C_p$ must be calculated from quantum theory or measured by experiment.

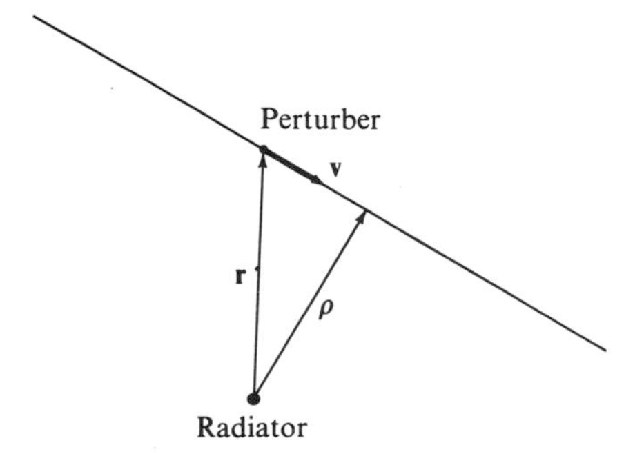


FIG. 9-1. Geometry of interactions between atom and perturber.

The *phase shift* induced by the perturbation is

$$\eta(t) = C_p \int_{-\infty}^{t} \frac{dt'}{[r(t')]^p} = C_p \int_{-\infty}^{t} \frac{dt'}{(\rho^2 + v^2t'^2)^{p/2}} \tag{9-72}$$

The total phase shift, $\eta(\infty)$, may be found by direct calculation and can be written in the form

$$\eta = \frac{C_p \psi_p}{v\rho^{p-1}} \tag{9-73}$$

Values of $\psi_p$ are given in Table 9-1 for various values of $p$. Weisskopf assumed

TABLE 9-1. Total Phase Shifts

| $p$ | $\psi_p$ |
|---|---|
| 2 | $\pi$ |
| 3 | 2 |
| 4 | $\pi/2$ |
| 6 | $3\pi/8$ |

that only those collisions which produced a total phase shift greater than some value $\eta_0$ would be effective in broadening the line. Thus taking

$$\frac{C_p \psi_p}{v\rho_0^{p-1}} = \eta_0 \tag{9-74}$$

we have

$$\rho_0 = \left(\frac{C_p \psi_p}{\eta_0 v}\right)^{1/(p-1)} \tag{9-75}$$

as the effective impact parameter. Weisskopf arbitrarily adopted $\eta_0 = 1$ as the critical phase shift, from which we obtain the *Weisskopf radius*

$$\rho_W = \left(\frac{C_p \psi_p}{v}\right)^{1/(p-1)} \tag{9-76}$$

The corresponding value for the damping constant is

$$\Gamma_W = 2\pi N v \left(\frac{C_p \psi_p}{\eta_0 v}\right)^{2/(p-1)} \tag{9-77}$$

An effective *collision time* $\tau_s$ can be defined such that the mean value of $\Delta\omega$, namely $C_p \rho_0^{-p}$, times $\tau_s$ equals the total phase shift, i.e.,

$$\tau_s C_p \rho_0^{-p} = C_p \psi_p \rho_0^{-p+1} v^{-1} \tag{9-78}$$

or

$$\tau_s = \frac{\psi_p \rho_0}{v} \tag{9-79}$$

In comparison, the mean time between collisions, the *time of flight*, $\tau$, is given by

$$\frac{1}{\tau} = N\pi\rho_0{}^2 v \tag{9-80}$$

so that

$$\frac{\tau_s}{\tau} = N\pi\rho_0{}^3\psi_p \tag{9-81}$$

Now $N = 3/(4\pi r_0{}^3)$ where $r_0$ is the mean interparticle distance; thus

$$\frac{\tau_s}{\tau} = \frac{3}{4}\psi_p\left(\frac{\rho_0}{r_0}\right)^3 \tag{9-82}$$

For the impact theory to be valid, we must assume that only one collision occurs at a time so that $\tau_s \ll \tau$. This implies that $\rho_0 \ll r_0$. Thus the impact theory is valid only if the density of particles is sufficiently low so that the Weisskopf radius is small compared with the interparticle distance.

While the above theory gives results that are of the right order of magnitude, there remain some objections:

(a) The theory does not account for the effects of collisions which produce small phase shifts even though the number of such collisions increase as $\rho^2$.

(b) The choice $\eta_0 = 1$ is arbitrary, and there is no means of determining a prori the correct value of $\eta_0$ to be used.

(c) This theory fails to predict the existence of a line *shift*, as will be found below.

### THE LINDHOLM APPROXIMATION

This impact theory will, to some extent, attempt to meet the objections listed above. We consider the radiator to be an oscillator with instantaneous frequency $\omega(t)$ which may, because of perturbations, differ from the nominal frequency $\omega_0$ by an amount $\Delta\omega(t)$. Thus we write

$$\omega(t) = \omega_0 + \Delta\omega(t) \tag{9-83}$$

and

$$\begin{aligned} f(t) = \exp[i\omega(t)t] &= \exp\left[i\omega_0 t + i\int_0^t \Delta\omega(t')\,dt'\right] \\ &= \exp[i\omega_0 t + \eta(t)] \end{aligned} \tag{9-84}$$

where $\eta(t)$ is the *instantaneous* phase of the oscillator. Now *in principle* we may obtain the power spectrum of the oscillator by the direct calculation

$$I(\omega) = \lim_{T\to\infty} \frac{1}{2\pi T}\left|\int_{-T/2}^{T/} f(t)e^{-i\omega t}\,dt\right|^2 \tag{9-85}$$

In practice, however, we are not actually able to perform this direct calculation since we have no knowledge of the order in which specific collisions occur. These difficulties can be avoided if we calculate instead the autocorrelation function and then use it to calculate the power spectrum. We will make use of the fact that if we were to average phase shifts over a sufficiently long period of time, then collisions at *all* impact parameters would occur with a certain statistical frequency distribution. We therefore replace time averages (on a long enough interval) by averages over the frequency distribution function of impact parameters; this technique invokes the so-called *ergodic hypothesis*. As we shall see, we are able thereby to organize the calculation in a very efficient way.

Let us introduce the function $\varphi(s)$ defined as

$$\varphi(s) = e^{-i\omega_0 s}\Phi(s) \tag{9-86}$$

which eliminates the unperturbed oscillation from the calculation. Then

$$\begin{aligned}\varphi(s) &= \lim_{T\to\infty} \frac{1}{T}\int_{-T/2}^{T/2} e^{-i\omega_0 s}e^{-i[\omega_0 t+\eta(t)]}e^{i[\omega_0(t+s)+\eta(t+s)]}\,dt \\ &= \lim_{T\to\infty} \frac{1}{T}\int_{-T/2}^{T/2} e^{i[\eta(t+s)-\eta(t)]}\,dt\end{aligned} \tag{9-87}$$

In this way $\varphi(s)$ measures the time-averaged value of the additional phase shift occurring in the time interval $s$. For brevity, write

$$\eta(t, s) \equiv \eta(t+s) - \eta(t) \tag{9-88}$$

Then

$$\varphi(s) = \langle \exp[i\eta(t, s)]\rangle_T \tag{9-89}$$

Suppose we consider now $\varphi(s + ds)$. Then

$$\begin{aligned}d\varphi(s) &= \varphi(s+ds) - \varphi(s) = \langle e^{i\eta(t,s+ds)}\rangle_T - \langle e^{i\eta(t,s)}\rangle_T \\ &= \langle e^{i\eta(t,s)}(e^{i\eta'} - 1)\rangle_T\end{aligned} \tag{9-90}$$

where $\eta'$ denotes the *change in phase* which occurs in an additional time $ds$. This change will be due to any impacts which occur in this interval. We now argue that the phase change brought about by these additional impacts is in no way correlated with the current value of the phase since there is no physical reason why it should be. Thus the mean value of the product can be replaced by the product of the means, i.e.,

$$\begin{aligned} d\varphi(s) &= \langle e^{i\eta(t,s)}\rangle_T \langle e^{i\eta'} - 1\rangle_T \\ &= \varphi(s)\langle e^{i\eta'} - 1\rangle_T \end{aligned} \tag{9-91}$$

where we have made use of equation (9-89). If we can calculate the latter average, we will have a differential equation for $\varphi(s)$. Now the above average must include impacts at *all* impact parameters. In forming the average over all time, we note that collisions will occur at all values of $\rho$ with their appropriate statistical frequency. By use of the ergodic hypothesis, we replace the average over time by the appropriate sum over impact parameters. Since we want the change due to impacts in time $ds$, we multiply by the number of impacts occurring in this time interval, namely, $2\pi\rho\, d\rho N v\, ds$. Thus, we write

$$\langle e^{\eta i'} - 1\rangle_T \rightarrow \langle e^{i\eta'(\rho)} - 1\rangle_\rho = 2\pi N v\, ds \int_0^\infty [e^{i\eta(\rho)} - 1]\rho\, d\rho \tag{9-92}$$

Recognizing that the integral in equation (9-92) will have both a real and an imaginary part, we customarily write

$$\langle e^{i\eta'(\rho)} - 1\rangle_\rho = -Nv\, ds(\sigma_R - i\sigma_I) \tag{9-93}$$

where

$$\sigma_R = 2\pi \int_0^\infty [1 - \cos \eta(\rho)]\rho\, d\rho = 4\pi \int_0^\infty \sin^2\left[\frac{\eta(\rho)}{2}\right]\rho\, d\rho \tag{9-94}$$

and

$$\sigma_I = 2\pi \int_0^\infty \sin \eta(\rho)\rho\, d\rho \tag{9-95}$$

We will evaluate these integrals in greater detail later. Combining equations (9-91), (9-92), and (9-93), we have

$$\frac{d\varphi(s)}{\varphi(s)} = -Nv(\sigma_R - i\sigma_I)\, ds \tag{9-96}$$

so that

$$\varphi(s) = \exp[-Nv(\sigma_R - i\sigma_I)s] \tag{9-97}$$

Note that strictly speaking the term in $\sigma_R$ should be considered to be $-Nv\, \sigma_R|s|$. Finally, from equations (9-7), (9-86), and (9-97) we find

$$\begin{aligned} I(\omega) &= \frac{1}{2\pi}\int_{-\infty}^{\infty} \exp[i(\omega_0 + Nv\sigma_I - \omega)s] \exp(-Nv\sigma_R|s|)\, ds \\ &= \frac{1}{\pi} Re[i(\omega_0 - \omega + Nv\sigma_I) - Nv\sigma_R]^{-1} \\ &= \frac{2Nv\sigma_R/\pi}{(\omega - \omega_0 - Nv\sigma_I)^2 + (Nv\sigma_R)^2} \end{aligned} \tag{9-98}$$

and normalizing such that

$$\int_{-\infty}^{\infty} I(\omega)\, d\omega = 1$$

we have

$$I(\omega) = \frac{(Nv\sigma_R/\pi)}{(\omega - \omega_0 - Nv\sigma_I)^2 + (Nv\sigma_R)^2} \tag{9-99}$$

Thus the Lindholm theory predicts *line broadening* with a damping constant

$$\Gamma = 2Nv\sigma_R \tag{9-100}$$

and a *line shift* by an amount

$$\Delta\omega_0 = Nv\sigma_I \tag{9-101}$$

Let us now consider the actual values of $\sigma_R$ and $\sigma_I$ for various values of $p$. Consider first a linear Stark effect with $p = 2$. Because of the long range of the interaction involved, the integrals in this case diverge. In practice, useful results may be obtained simply by introducing a cutoff at a maximum impact parameter or equivalently a minimum phase shift. As will be seen later, maximum impact parameter cutoffs enter the theory in a natural way. If we define $x = 2/\eta$, equation (9-94) can be rewritten

$$\sigma_R = \frac{\pi^3 C_2{}^2}{v^2} \int_0^{2/\eta_{\min}} \sin^2\left(\frac{1}{x}\right) x\, dx$$

$$= \frac{\pi^3 C_2{}^2}{v^2} \left(0.923 - \ln \eta_{\min} + \frac{\eta_{\min}^2}{24} + \cdots\right) \tag{9-102}$$

The line shift in this case is unimportant since under linear Stark effect, the line splits symmetrically and there is no net shift. For values of $p > 2$, the integrals do not diverge and may be calculated directly to yield the results listed in Table 9-2 (see Ref. 34, pp. 14–15, for details of the calculation). The last line of the table gives the value of $\eta_0$ which, when inserted into Weisskopf formula equation (9-77), gives the Lindholm $\Gamma$. As we see, $\eta_0$ is always less than unity so that application of the Weisskopf formula always leads to too small a value of $\Gamma$. The phase shift $\eta_0$ corresponds to a slightly larger impact parameter than the Weisskopf radius.

TABLE 9-2. Results of Lindholm Theory

| $p$ | 3 | 4 | | 6 | |
|---|---|---|---|---|---|
| $\Gamma$ | $2\pi^2 C_3 N$ | 11.37 | $C_4^{2/3} v^{1/3} N$ | 8.08 | $C_6^{2/5} v^{3/5} N$ |
| $\Delta\omega_0$ | | 9.85 | | 2.94 | |
| $\Gamma/\Delta\omega_0$ | | 1.16 | | 2.75 | |
| $\eta_0$ | 0.64 | 0.64 | | 0.61 | |

While the Lindholm theory correctly predicts the existence of line shift and leads to values of the broadening that in principle account for the impacts at all impact parameters, there remain, nevertheless, certain inadequacies of the theory. In particular, as $\rho \to \infty$, the impact time, $\tau_s \sim \rho/v$ must clearly become ever larger and ultimately will exceed $\tau$, the time between impacts. In this case, the collisions can no longer be considered as taking place one at a time, and the theory becomes invalid. Indeed, at sufficiently large distances, the perturber takes so long to finish the encounter, it may in fact be regarded as being essentially at rest. In this case, we must resort to the statistical theory and calculate the strength of the fluctating field due to the ensemble of perturbers. The Lindholm theory also assumes that transitions do not occur (the *adiabatic* assumption) and thus ignores inelastic collisions. In addition, as pointed out above, the impact theory will cease to be valid when the densities rise to the point where there is more than one perturber witin the Weisskopf radius since again more than one effective collision will then occur at a time. Let us now turn our attention to predictions of the statistical theory.

## 9-5. Statistical Broadening Theory

The basic picture in this theory is that the atom finds itself radiating in the statistically fluctuating field of the perturbers. The motions of the perturbers are ignored. This approximation is known as the *quasi-static* approximation. As we shall see later, this approximation is quite good for describing the effects of the relatively slow-moving charged nuclei in the plasma (e.g., protons) upon the radiating atom. In fact, in all specific examples we shall restrict attention to the quasi-static broadening of hydrogen lines by linear Stark-effect interactions with protons, even though this theory applies in other contexts as well

For a given value of the perturbing field, the frequency of the oscillation of the radiator is shifted by some $\Delta\omega$. The intensity of the radiation at this $\Delta\omega$ is assumed to be proportional to the statistical frequency of occurrence of the appropriate field. Thus, the central problem is to determine the probability distribution of the perturbing fields. We can do this to various degrees of approximation.

### THE NEAREST-NEIGHBOR APPROXIMATION

If we assume that the frequency shift due to the presence of a perturber at distance $r$ can be represented by equation (9-71),

$$\Delta\omega = \frac{C_p}{r^p}$$

then we may calculate the frequency spectrum directly in terms of the probability of a perturber being located at this distance, i.e., we write

$$I(\Delta\omega)\,d(\Delta\omega) \propto W(r)\,\frac{dr}{d(\Delta\omega)}\,d(\Delta\omega) \tag{9-103}$$

As a first approximation, we may assume the main effect is due to the strongest perturbation acting at any given instant, i.e., the frequency shift is due to the presence of the *nearest neighbor*. We wish then to calculate the probability that a perturber will be located on the range $(r, r + dr)$ *and* that no perturber will be located at a distance less than $r$. Assume a uniform particle density $N$. Then the probability $W(r)$ that the nearest neighbor lies at distance $r$ is given by

$$W(r)\,dr = \left[1 - \int_0^r W(x)\,dx\right] 4\pi r^2 N\,dr \tag{9-104}$$

where the factor $4\pi r^2 N\,dr$ is simply the (relative) probability of a particle lying in the shell $(r, r + dr)$ while the term in brackets is the probability that no particle lies inside this shell. By differentiation we have directly

$$\frac{d}{dr}\left[\frac{W(r)}{4\pi r^2 N}\right] = -4\pi r^2 N\left[\frac{W(r)}{4\pi r^2 N}\right] \tag{9-105}$$

and by integration

$$W(r) = 4\pi r^2 N \exp(-\tfrac{4}{3}\pi r^3 N) \tag{9-106}$$

Note that as defined $W(r)$ is normalized such that

$$\int_0^\infty W(r)\,dr = 1 \tag{9-107}$$

With this probability distribution, the mean perturber distance is

$$\langle r\rangle = \int_0^\infty W(r) r\,dr = \frac{\Gamma(\frac{4}{3})}{(\frac{4}{3}\pi N)^{1/3}} = 0.554 N^{-1/3} \tag{9-108}$$

By way of comparison, the mean interparticle distance $r_0$ is

$$r_0 = \left(\frac{4\pi}{3}N\right)^{-1/3} = 0.620\,N^{-1/3} \tag{9-109}$$

It is customary to adopt $r_0$ as the *reference distance* which produces the *normal phase shift* $\Delta\omega_0$, where

$$\Delta\omega_0 = \frac{C_p}{r_0{}^p} \tag{9-110}$$

Thus in general we write

$$\left(\frac{\Delta\omega}{\Delta\omega_0}\right)^{1/p} = \frac{r_0}{r} \tag{9-111}$$

Hence, we may rewrite equation (9-106) as

$$\begin{aligned} W(r)\,dr &= 3\left(\frac{r}{r_0}\right)^3 \exp\left[-\left(\frac{r}{r_0}\right)^3\right]\frac{dr}{r_0} \\ &= \exp\left[-\left(\frac{\Delta\omega_0}{\Delta\omega}\right)^{3/p}\right] d\left(\frac{\Delta\omega_0}{\Delta\omega}\right)^{3/p} \end{aligned} \tag{9-112}$$

Therefore

$$I(\Delta\omega)\,d(\Delta\omega) = \frac{3}{p}\left(\frac{\Delta\omega_0}{\Delta\omega}\right)^{(3/p)+1} \exp\left[-\left(\frac{\Delta\omega_0}{\Delta\omega}\right)^{3/p}\right] d\left(\frac{\Delta\omega}{\Delta\omega_0}\right) \tag{9-113}$$

In the case of linear Stark effect, the strength of the perturbing field is

$$F = \frac{e}{r^2} \tag{9-114}$$

It is customary to define a *normal field strength*

$$F_0 = \frac{e}{r_0^{\,2}} = \frac{e}{(\frac{4}{3}\pi N)^{-2/3}} = 2.5985N^{-2/3} \tag{9-115}$$

and to measure $F$ in terms of $F_0$, i.e., we write

$$F = \beta F_0 \tag{9-116}$$

Then nearest-neighbor theory predicts

$$W(\beta)\,d\beta = \tfrac{3}{2}\beta^{-5/2}\exp(-\beta^{-3/2})\,d\beta \tag{9-117}$$

Clearly, as $\beta \to \infty$, $W(\beta) \propto \beta^{-5/2}$, so in the far wings, the statistical theory predicts the absorption coefficient will fall off as $\Delta\omega^{-5/2}$ (for linear Stark effect), in contrast with $\Delta\omega^{-2}$, as predicted by impact theory. We note in passing that the asymptotic expansion for $W(\beta)$, useful in the line wings ($\beta \gg 1$), is

$$W(\beta) \approx \frac{3}{2}\beta^{-5/2}\left(1 - \frac{1}{\beta^{3/2}} + \frac{1}{2\beta^3} - \cdots\right) \tag{9-118}$$

which we can compare with the results of the more precise theory given below.

The basic failing of this approximation is that the profile is really the result of perturbations by *all* particles not just the nearest neighbor. We must therefore construct a more elaborate theory to account for these other perturbers. That the nearest neighbor theory is useful at all is due entirely to the obvious fact that the nearest neighbor does, indeed, exert the strongest perturbation.

HOLTSMARK THEORY

Let us now consider the problem of determining the probability distribution of the net vector field strength at the position of the radiating atom resulting from the superposition of the field vectors of all perturbers. A thorough treatment of this problem has been given by Chandrasekhar (*Rev. Mod. Phys.*, **15**, 1, 1943), whose results we will summarize here.

Let $\boldsymbol{\varphi}_j(\mathbf{r}_j)$ be the field due to a perturber $j$, and let the total field $\boldsymbol{\Phi}$ be given by

$$\boldsymbol{\Phi} = \sum_{j=1}^{n} \boldsymbol{\varphi}_j(\mathbf{r}_j) \tag{9-119}$$

Let the probability that $\mathbf{r}_j$ lies in the range $(\mathbf{r}_j, \mathbf{r}_j + \mathbf{dr}_j)$ be

$$\tau(\mathbf{r}_j)\, dr_j^{(1)}\, dr_j^{(2)}\, dr_j^{(3)} = \tau(\mathbf{r}_j)\, \mathbf{dr}_j \tag{9-120}$$

We wish to determine the probability that $\boldsymbol{\Phi}$ lies in the range

$$\boldsymbol{\Phi}_0 - \tfrac{1}{2}\, \mathbf{d}\boldsymbol{\Phi}_0 \le \boldsymbol{\Phi} \le \boldsymbol{\Phi}_0 + \tfrac{1}{2}\, d\boldsymbol{\Phi}_0 \tag{9-121}$$

for any pre-assigned value of $\boldsymbol{\Phi}_0$. Denote this probability as

$$W(\boldsymbol{\Phi}_0)\, \mathbf{d}\boldsymbol{\Phi}_0 = W(\boldsymbol{\Phi}_0)\, d\Phi_0^{(1)}\, d\Phi_0^{(2)}\, d\Phi_0^{(3)} \tag{9-122}$$

Now $\boldsymbol{\Phi}$ will lie on this prespecified range for certain configurations of the perturbers, and the total probability that this occurs is obtained by summing over all these configurations. Thus

$$W(\boldsymbol{\Phi}_0)\, \mathbf{d}\boldsymbol{\Phi}_0 = \iiint \left[ \prod_{j=1}^{n} \tau(\mathbf{r}_j)\, \mathbf{dr}_j \right] \tag{9-123}$$

where the integral extends over the "appropriate range," which we must now specify. This may be done with the greatest simplicity and generality if we extend the integral over the entire volume containing the perturbers but introduce an astutely chosen factor that becomes zero whenever the intgration range would cause $\boldsymbol{\Phi}$ to lie outside the chosen limits. Thus, we must introduce a factor $\Delta$ such that $\Delta = 1$ when $\boldsymbol{\Phi}$ satisfies equation (9-121) but is zero otherwise. Then we may write

$$W(\boldsymbol{\Phi}_0)\mathbf{d}\boldsymbol{\Phi}_0 = \iiint_{\text{all space}} \Delta \left[ \prod_{j=1}^{n} \tau_j(\mathbf{r}_j)\mathbf{dr}_j \right] \tag{9-124}$$

The function $\Delta$ may be constructed in the following way. We make use of the *discontinuous integral of Dirichlet*, which has the property that

$$\delta_k(\alpha_k, \gamma_k) \equiv \frac{1}{\pi} \int_{-\infty}^{\infty} \frac{\sin \alpha_k \rho_k}{\rho_k} \exp(i\rho_k \gamma_k)\, d\rho_k = 1 \tag{9-125}$$

if $-\alpha_k < \gamma_k < \alpha_k$ and is zero otherwise. For the problem at hand, we use one such Dirichlet integral for each component of the vector field and choose

$$\alpha_k = \tfrac{1}{2}\, d\Phi_0^{(k)} \tag{9-126}$$

and

$$\gamma_k = \Phi_0^{(k)} - \sum_{j=1}^{n} \varphi_j^{(k)}(\mathbf{r}_j) \tag{9-127}$$

Then the desired function $\Delta$ is given by

$$\Delta = \delta_1 \delta_2 \delta_3 = \frac{1}{\pi^3} \iiint_{-\infty}^{\infty} \left[ \prod_{k=1}^{3} \frac{\sin(\frac{1}{2}\rho_k\, d\Phi_0^{(k)})}{\rho_k} \right]$$
$$\times \exp\left[ i\left( \sum_{k=1}^{3} \rho_k \sum_{j=1}^{n} \varphi_j^{(k)} - \sum_{k=1}^{3} \rho_k \Phi_0^{(k)} \right) \right] d\rho_1\, d\rho_2\, d\rho_3 \tag{9-128}$$

Now for $d\Phi_0^{(k)}$ infinitesimal we may write

$$\sin(\tfrac{1}{2}\rho_k\, d\Phi_0^{(k)}) = \tfrac{1}{2}\rho_k\, d\Phi_0^{(k)} \tag{9-129}$$

so that

$$\Delta = \frac{\mathbf{d\Phi}_0}{8\pi^3} \iiint_{-\infty}^{\infty} \exp\left[ i\boldsymbol{\rho} \cdot \left( \sum_{j=1}^{n} \boldsymbol{\varphi}_j(\mathbf{r}_j) - \boldsymbol{\Phi}_0 \right) \right] d\rho_1\, d\rho_2\, d\rho_3 \tag{9-130}$$

The case of greatest interest is where the particles are uniformly distributed over the volume under consideration. Then $\tau = 1/V$, and $V = \frac{4}{3}\pi R^3$, where $R$ is the radius of the sphere containing the $n$ particles. Then we have

$$W(\boldsymbol{\Phi}_0) = \frac{1}{8\pi^3} \iiint_{-\infty}^{\infty} d\rho_1\, d\rho_2\, d\rho_3 \left[ \frac{1}{V^n} \int \cdots \int e^{-i\boldsymbol{\rho}\cdot(\boldsymbol{\Phi}_0 - \sum_{j=1}^{n} \boldsymbol{\varphi}_j)} \mathbf{dr}_1 \cdots \mathbf{dr}_n \right] \tag{9-131}$$

or equivalently

$$W(\boldsymbol{\Phi}_0) = \frac{1}{8\pi^3} \iiint_{-\infty}^{\infty} A(\boldsymbol{\rho}) e^{-i\boldsymbol{\rho}\cdot\boldsymbol{\Phi}_0}\, \mathbf{d\rho} \tag{9-132}$$

where

$$A(\boldsymbol{\rho}) \equiv \left[ \frac{1}{V} \iiint e^{i\boldsymbol{\rho}\cdot\boldsymbol{\varphi}_j(\mathbf{r}_j)}\, \mathbf{dr}_j \right]^n \tag{9-133}$$

As written, $A(\boldsymbol{\rho})$ clearly represents the three-dimensional Fourier transform of $W(\boldsymbol{\Phi}_0)$. Now assume that as $R \to \infty$, $n \to \infty$ in such a way that $N = n/\frac{4}{3}\pi R^3$

is a constant. We may rewrite the expression for $A(\boldsymbol{\rho})$ as

$$A(\boldsymbol{\rho}) = \lim_{R\to\infty} \left[1 - \frac{3}{4\pi R^3} \iiint_{|\mathbf{r}|<R} (1 - e^{-\boldsymbol{\rho}\cdot\boldsymbol{\varphi}(\mathbf{r})})\mathbf{dr}\right]^{(4/3)\pi R^3 N} \tag{9-134}$$

But the definition of $e$, the base of natural logarithms, is

$$e^a \equiv \lim_{y\to\infty} \left(1 + \frac{a}{y}\right)^y \tag{9-135}$$

which implies that we may rewrite equation (9-134)

$$A(\boldsymbol{\rho}) = e^{-ND(\rho)} \tag{9-136}$$

where $D(\rho)$ is given by

$$\begin{aligned} D(\rho) &= \iiint_{-\infty}^{\infty} (1 - e^{i\boldsymbol{\rho}\cdot\boldsymbol{\varphi}})\mathbf{dr} \\ &= \oint d\omega \int_0^\infty (1 - e^{i\boldsymbol{\rho}\cdot\boldsymbol{\varphi}})r^2\, dr \end{aligned} \tag{9-137}$$

Now

$$\varphi = \frac{C_p}{r^p} \tag{9-138}$$

so that

$$r^2\, dr = -\frac{1}{p} C_p^{3/p} \varphi^{-[(p+3)/p]}\, d\varphi \tag{9-139}$$

Also, choosing the polar angle $\alpha$ to be the angle between $\boldsymbol{\varphi}$ and $\boldsymbol{\rho}$ and writing $\mu = \cos\alpha$, we have

$$\begin{aligned} D(\rho) &= \frac{C^{3/p}}{p} \int_0^{2\pi} d\theta \int_{-1}^{1} d\mu \int_0^\infty (1 - e^{i\rho\varphi\mu})\varphi^{-[(p+3)/p]}\, d\varphi \\ &= \frac{2\pi C^{3/p}}{p} \int_{-1}^{1} d\mu \int_0^\infty (1 - \cos\rho\varphi\mu)\varphi^{-[(p+3)/p]}\, d\varphi \end{aligned} \tag{9-140}$$

where we have discarded the sine term which is odd in $\mu$. Carrying out the integration over $\mu$, we obtain

$$D(\rho) = \frac{4\pi C^{3/p}}{p} \int_0^\infty \left(1 - \frac{\sin\rho\varphi}{\rho\varphi}\right)\varphi^{-[(p+3)/p]}\, d\varphi \tag{9-141}$$

Note in passing the singularity that occurs at $\varphi = 0$. If the integral is to converge, we must require that $(3 + p)/p < 3$ or that $p > 3/2$, which includes the cases of physical interest. Now defining $z = \rho\varphi$, we may write finally

$$D(\rho) = \frac{4\pi}{n}(C\rho)^{3/p}\int_0^\infty (z - \sin z)z^{2+(3/p)}\,dz = (\gamma\rho C)^{3/p} \tag{9-142}$$

where $\gamma$ is given by

$$\gamma = \left[\frac{2\pi^2 p}{3(p+3)\Gamma(3/p)\sin(3\pi/2p)}\right]^{p/3} \tag{9-143}$$

Direct substitution leads to $\gamma = 2.6031$ for $p = 2$ (linear Stark effect). Having obtained an expression for $A(\rho)$, we may now compute the distribution function itself. Thus

$$W(\mathbf{\Phi}_0) = \frac{1}{8\pi^3}\int_{-\infty}^{\infty} \exp[-i\boldsymbol{\rho}\cdot\mathbf{\Phi}_0 - N(\gamma C\rho)^{3/p}]\mathbf{d}\boldsymbol{\rho} \tag{9-144}$$

Again, choosing polar coordinates with polar axis along $\mathbf{\Phi}_0$, we have

$$\begin{aligned} W(\mathbf{\Phi}_0) &= \frac{1}{8\pi^3}\int_0^\infty \exp[-N(\gamma C\rho)^{3/p}]\rho^2\,d\rho\int_0^{2\pi} d\theta\int_{-1}^{1} \\ &\quad \times [\cos(\rho\mu\Phi_0) + i\sin(\rho\mu\Phi_0)]\,d\mu \quad . \\ &= \frac{1}{2\pi^2\Phi_0}\int_0^\infty \exp[-N(\gamma C\rho)^{3/p}]\sin(\rho\Phi_0)\rho\,d\rho \end{aligned} \tag{9-145}$$

Let us now define the *normal field* as

$$F_0 \equiv \gamma C N^{p/3} \tag{9-146}$$

For linear Stark effect, $p = 2$, $C = e$, and $\gamma = 2.6031$ so that

$$F_0 = 2.6031\, eN^{2/3} \tag{9-147}$$

In addition, set $x = \rho\Phi_0$. Then we may rewrite equation (9-145) as

$$W(\mathbf{\Phi}_0) = \frac{1}{2\pi^2{\Phi_0}^3}\int_0^\infty \exp\left[-\left(\frac{F_0 x}{\Phi_0}\right)^{3/p}\right]x\sin x\,dx \tag{9-148}$$

Moreover, if we want only the probability distribution for the *magnitude* of $\Phi_0$, we can write

$$W(\Phi_0) = 4\pi{\Phi_0}^2 W(\mathbf{\Phi}_0) \tag{9-149}$$

Finally, let $\beta = \Phi_0/F_0$, i.e., measure the field in units of the normal strength. Then $W(\beta) = F_0\,W(\Phi_0)$, or

$$\begin{aligned} W(\beta) &= \frac{2}{\pi\beta}\int_0^\infty \exp[-(x/\beta)^{3/p}]x\sin x\,dx \\ &= \frac{2\beta}{\pi}\int_0^\infty \exp(-y^{3/p})y\sin\beta y\,dy \end{aligned} \tag{9-150}$$

This integral can be evaluated explicitly for $p = 3/2$ and $p = 3$ but in general must be computed numerically.

If we consider the case of the linear Stark effect ($p = 2$), then one can obtain useful expansions from equation (9-150). In particular, for $\beta$ small,

$$W(\beta) = \frac{4}{3\pi} \sum_{l=0}^{\infty} (-1)^l \Gamma\left(\frac{4l+6}{3}\right) \frac{\beta^{2l+2}}{(2l+1)!} \tag{9-151}$$

so that for $\beta \ll 1$, $W(\beta) \propto \beta^2$. On the other hand, for $\beta \gg 1$, one may obtain an asymptotic expansion

$$W(\beta) = \frac{2}{\pi} \sum_{l=1}^{\infty} \frac{(-1)^{l+1} \Gamma\left(\frac{3l+4}{2}\right) \sin\left(\frac{3\pi l}{4}\right)}{l!\,\beta^{(3l+2)/2}}$$

$$= \frac{1.496}{\beta^{5/2}} \left(1 + \frac{5.107}{\beta^{3/2}} + \frac{14.43}{\beta^3} + \cdots\right) \tag{9-152}$$

Note that the leading term of the asymptotic expansion is essentially just that given by the nearest-neighbor theory, equation (9-118). A convenient tabulation of $W(\beta)$ is given by Chandrasekhar (*Rev. Mod. Phys.*, **15**, 1, 1943).

### THE LINE PROFILE

In many cases of interest and in particular for hydrogen, the levels giving rise to the line under consideration consist of degenerate sublevels. If we assume that each of these sublevels shifts by an amount proportional to the strength of the fluctuating field (*linear* Stark effect) and that no other broadening mechanisms are operative, then the line profile will consist of a number of *Stark components*, arising from transitions between the sublevels of the upper and lower states. Each of these components will have a definite intensity and will be displaced from the line center by a characteristic distance. The line that we observe will be a superposition of these components, weighted by their intensities and the probability of being shifted to the appropriate position. If we denote the fractional intensity of the $k$th component as $I_k$ and characterize its shift in terms of an interaction constant $C_k$, defined such that the shift $\Delta\lambda$ (in Å) of the $k$th component is given by

$$\Delta\lambda = C_k F \tag{9-153}$$

then at a given $\Delta\lambda$, the total line absorption will be

$$I(\Delta\lambda)d(\Delta\lambda) \propto \sum_k I_k W\left(\frac{F}{F_0}\right) \frac{dF}{F_0}$$

$$= \sum_k I_k W\left(\frac{\Delta\lambda}{C_k F_0}\right) \frac{d(\Delta\lambda)}{C_k F_0} \tag{9-154}$$

Defining

$$\alpha \equiv \frac{\Delta\lambda}{F_0} \tag{9-155}$$

we may obtain a normalized profile

$$S(\alpha)\, d\alpha = \sum_k I_k W\left(\frac{\alpha}{C_k}\right) \frac{d\alpha}{C_k} \tag{9-156}$$

which is normalized such that

$$\int_{-\infty}^{\infty} S(\alpha)\, d\alpha = 1$$

To obtain an absorption coefficient $\alpha_\nu$ in terms of this profile function, we may write

$$\alpha_\nu(\Delta\lambda) = \frac{\pi e^2}{mc} fS\left(\frac{\Delta\lambda}{F_0}\right)\left(\frac{10^8 \lambda^2}{cF_0}\right) \tag{9-157}$$

where the factor of $10^8$ enters because $\Delta\lambda$ in equation (9-155) is expressed in Å. Early work on the broadening of hydrogen lines ignored the broadening by electrons and accounted for only the quasi-static broadening by ions. Extensive tables of $S(\alpha)$ for hydrogen lines have been computed under this assumption. Modern work shows, however, that the broadening by electron impacts is nonnegligible and that the complete profile consists of a combination of quasi-static ion broadening and electron impact broadening. We shall discuss this point in detail in Section 9-8, but let us first consider some further aspects of the statistical broadening theory.

### EFFECTS OF INTERACTIONS AMONG PERTURBERS

In the preceding derivation of the probability distribution of the perturbing field, we entirely neglected interactions among the perturbers. In fact, when such interactions are taken into account, the probability that a particle will be found in volume $dV$ is not just $N\,dV$ but depends also upon the electrostatic potential $\varphi$ at the point. We will give here a simple and descriptive treatment of this effect following Ecker (*Z. Ph.*, **140**, 274, 1955; *Z. Ph.*, **140**, 292, 1955; *Z. Ph.*, **148**, 593, 1957; and *Z. Ph.*, **149**, 254, 1957). If the electrostatic potential at a point is greater than zero, electrons will tend to migrate toward the point while ions will tend to migrate away, and vice versa if the potential is less than zero. We may describe this, schematically, by introducing a Boltzmann factor which depends upon $\varphi$. Thus, we write for electrons and ions, respectively,

$$N_e W_e(dV)\,dV = N_e \exp(e\varphi/kT)\,dV \approx N_e\left(1 + \frac{e\varphi}{kT}\right) dV \tag{9-158}$$

and

$$N_i W_i(dV)\,dV = N_i \exp(-Z_i e\varphi/kT)\,dV$$

$$\approx N_i\left(1 - \frac{Z_i e\varphi}{kT}\right) dV \tag{9-159}$$

where we have assumed the electrostatic interaction is much smaller than the mean kinetic energy of the particles. $N_i$ and $N_e$ are the densities of ions and electrons, respectively. Note that since the plasma must be electrically neutral over sufficiently large volumes,

$$N_e = \sum_i N_i Z_i \tag{9-160}$$

Let us now calculate the potential around an ion and account for interactions. To perform the calculation, we make the simplifying assumption that the particles are uniformly smeared out over the volume containing them. Then we may use Poisson's equation to connect the potential and charge density. Thus we write

$$\nabla^2\varphi = -4\pi e\rho \tag{9-161}$$

where

$$e\rho = -eN_e W_e(dV) + e\sum_i Z_i N_i W_i(dV) \tag{9-162}$$

Inserting equations (9-158) and (9-159) for $W_e$ and $W_i$, we find

$$e\rho = e\left[-N_e\left(1 + \frac{e\varphi}{kT}\right) + \sum_i Z_i N_i\left(1 - \frac{Z_i e\varphi}{kT}\right)\right]$$

$$= \frac{-e^2\varphi}{kT}\left(N_e + \sum_i N_i Z_i^2\right) \tag{9-163}$$

accounting for electrical neutrality. Thus we may write

$$\nabla^2\varphi = \frac{\varphi}{D^2} \tag{9-164}$$

where

$$D \equiv \left[\frac{kT}{4\pi e^2\left(N_e + \sum_i Z_i^2 N_i\right)}\right]^{1/2} \tag{9-165}$$

is known as the *Debye length* of the medium. Solving now for $\varphi$, we have

$$\varphi = \frac{A}{r} e^{-r/D} + \frac{B}{r} e^{r/D} \tag{9-166}$$

Demanding that $\varphi \to 0$ as $r \to \infty$, we set $B = 0$. Moreover, in the limit as $r \to 0$, we wish to recover the potential of a single ion which we may regard at the origin; then we set $A = Z_i e$. Thus

$$\varphi = \frac{Z_i e}{r} e^{-r/D} \tag{9-167}$$

From equation (9-167), we see that the Debye length provides a characteristic distance in the plasma, beyond which the potential of a single particle is so strongly altered that the particle no longer effectively interacts (as an individual) with other charges. Physically, this accounts for the fact that a charged particle will tend to polarize the plasma in its vicinity, and the polarization charges shield the particle at points beyond a certain distance.

Before we incorporate shielding into the calculation of $W(\beta)$, it is instructive to examine numerical values. Suppose we have a pure hydrogen plasma; then we may set $Z_i = 1$ and $N_i = N_e$. Then

$$D = 4.8\left(\frac{T}{N_e}\right)^{1/2} \text{ cm} \tag{9-168}$$

Typical values for situations of astrophysical interest are listed in Table 9-3. We see that in stellar atmospheres Debye lengths are small while in the interstellar medium they may become very large.

TABLE 9-3. Debye Lengths in Astrophysical Plasmas

| | | | |
|---|---|---|---|
| Sun | $T = 6 \times 10°K$ | $N_e = 5 \times 10^{13}$ cm$^{-3}$ | $D = 6 \times 10^{-5}$ cm |
| Orion nebula | $T = 10^{4}°K$ | $N_e = 10^4$ | $D = 5$ cm |
| | | $N_e = 10^2$ | $D = 50$ cm |

Now to calculate the effects of shielding upon $W(\beta)$, we will make use of Ecker's simplified approach, in which we assume that

$$\mathbf{\Phi} = \frac{C\mathbf{r}}{r^3} \tag{9-169}$$

for $r \leq D$, and

$$\mathbf{\Phi} \equiv 0 \tag{9-170}$$

for $r > D$. From the development given previously [equation (9-140)], we have

$$D(\rho) = \frac{C^{3/p}}{p} \int d\theta \int d\mu \int (1 - e^{i\boldsymbol{\rho} \cdot \boldsymbol{\varphi}}) \varphi^{-[(p+3)/p]} \, d\varphi$$

If we set $p = 2$, take a cutoff at $r = D$ and set $z = \varphi\rho/c$ then we find

$$D(\rho) = 2\pi\rho^{3/2} \int_{\rho/D^2}^{\infty} \left(1 - \frac{\sin z}{z}\right) z^{-5/2} \, dz = \frac{4\pi D^3}{3} g(y) \tag{9-171}$$

where $y = \rho/D^2$, and

$$g(y) = \frac{3}{2} y^{3/2} \int_{y}^{\infty} \left(1 - \frac{\sin z}{z}\right) z^{-5/2} \, dz \tag{9-172}$$

Then setting $\delta = \frac{4}{3}\pi D^3 N$, which equals the number of perturbers in the Debye sphere, we have

$$A(\rho) = e^{-ND(\rho)} = e^{-\delta g(y)} \tag{9-173}$$

Finally, substituting $A(\rho)$ into the Fourier transform for $W$ and carrying out the integrations, we obtain

$$W(\beta, \delta) = \frac{2}{\pi} \beta \delta^{4/3} \int_0^{\infty} \sin(\delta^{2/3}\beta y) e^{-\delta g(y)} y \, dy \tag{9-174}$$

For large values of $\delta$ one may show that

$$W(\beta, \delta) \to W(\beta, \infty) + \frac{1}{\pi \delta^{1/3} \beta^3} \int_0^{\infty} \sin x \exp\left[-\left(\frac{x}{\beta}\right)^{3/2}\right] x^3 \, dx \tag{9-175}$$

so that for very large numbers of perturbers in the Debye sphere, we recover the Holtsmark distribution, as expected. The connection is shown even more clearly by the asymptotic expansion for large $\beta$, which yields

$$W(\beta, \delta) = \frac{1.496}{\beta^{5/2}} \left[1 + \frac{5.107}{\beta^{3/2}} - \frac{6.12}{\delta^{1/3}\beta^2} + \cdots\right] \tag{9-176}$$

in comparison with equation (9-152):

$$W_H(\beta) = W(\beta, \infty) = \frac{1.496}{\beta^{5/2}} \left[1 + \frac{5.107}{\beta^{3/2}} + \frac{14.43}{\beta^3} + \cdots\right]$$

For small $\delta$, the theory should merge continuously into the single-particle picture, though in practice when $\delta$ is less than about 5, the Debye assumptions break down. A plot of $W(\beta, \delta)$ is shown in Figure 9-2.

More recently, quite thorough discussions of the effects of perturber interactions upon the field at a test point have been given by Mozer and

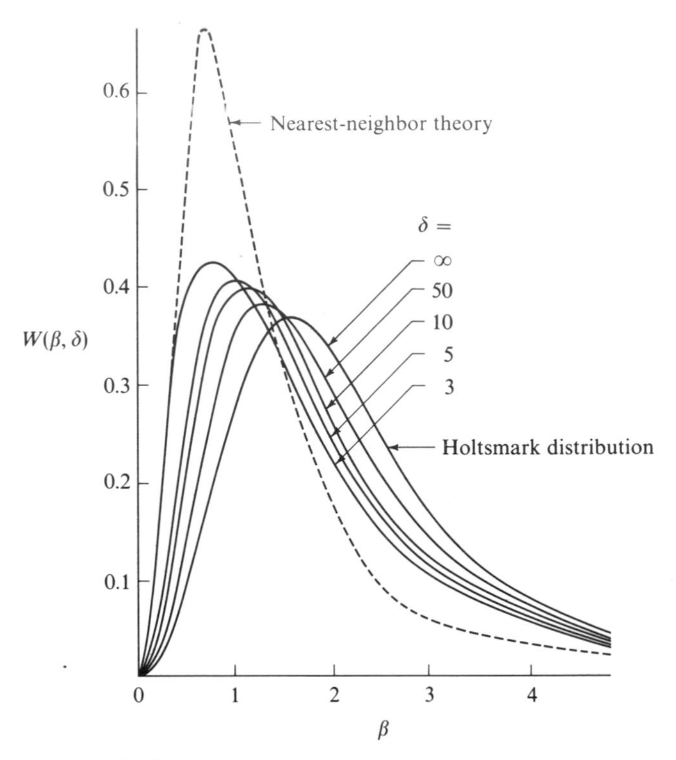


FIG. 9-2. Field distribution at a test point, including shielding effects; $\delta$ is the number of charged particles within the Debye sphere. (From G. Ecker, *Z. Ph.*, **148**, 593, 1957; by permission.)

Baranger (*Phys. Rev.*, **118**, 626, 1960) and by Hooper (*Phys. Rev.*, **165**, 215, 1968). Most modern calculations of line broadening have made use of these improved treatments.

In practice, the effects of shielding are often quite important in laboratory plasmas while in stellar atmospheres the densities are so low that the number of particles in a Debye sphere is large ($\delta \gtrsim 100$), and the deviations from a Holtsmark distribution are not marked.

## LOWERING OF THE IONIZATION POTENTIAL

The presence of nearby charges partly neutralizes the effects of the nuclear charge upon an orbital electron and thereby weakens the potential by which the electron is bound. The reduction in binding energy can be calculated using the Debye potential given in equation (9-167) as

$$\Delta E = \frac{Ze^2}{r} \exp\left(-\frac{r}{D}\right) - \frac{Ze^2}{r} \approx -\frac{Ze^2}{D} \tag{9-177}$$

which is valid for $r \ll D$. Now if we consider an electron which in the unperturbed atom lies at an energy $\Delta\chi$ below the ionization limit, we can argue that this electron will become unbound if $\Delta\chi \lesssim \Delta E$. Thus we may say qualitatively that the ionization potential of the atom is decreased by an amount

$$\Delta\chi = \frac{Ze^2}{D} = 27.2\frac{Za_0}{D}\text{ eV} \tag{9-178}$$

If we assume that all ions are singly charged, then

$$D = \left(\frac{kT}{8\pi e^2 N_e}\right)^{1/2} = \left(\frac{k}{2e\sqrt{2\pi}}\right)\frac{T}{p_e^{1/2}}$$

Inserting numerical constants we find

$$\Delta\chi = 5 \times 10^{-4} Z p_e^{1/2}\theta \text{ eV} \tag{9-179}$$

If the upper levels are hydrogenic, then the maximum bound quantum number is

$$n = \frac{165 Z^{1/2}}{\theta^{1/2} p_e^{1/4}} \tag{9-180}$$

The treatment presented above is naturally quite schematic. A comprehensive discussion of lowering of ionization potentials in a plasma has been given from several theoretical points of view (and documented with experimental data) in Reference 10.

## 9-6. Validity Criteria

Before we discuss the quantum mechanical theory of line broadening, it will be useful to review some questions of the validity of various approximations that we have employed thus far.

### THE TRANSITION BETWEEN IMPACT AND STATISTICAL THEORIES

We have mentioned previously that both the impact theory and the statistical theory are only idealized limiting cases of a complete theory. We pointed out that the basic validity criterion for the impact theory is that the collision time $\tau_s \sim \rho/v$ should be small compared with $\tau$, the time between impacts. This criterion is violated for sufficiently high particle densities (so that several perturbers are inside the Weisskopf radius), sufficiently large impact parameters, and for slow enough moving perturbers.

We wish to show here that, in addition, at sufficiently large frequency separations $\Delta\omega$ from the line core, impact theory ceases to be valid, and statistical theory becomes valid. For simplicity, we consider the impact broadening to be given by the Weisskopf theory. From the general properties of Fourier transforms we expect radiation at a frequency separation $\Delta\omega$ to correspond to characteristic interruption times of order $1/\Delta\omega$. Clearly, for sufficiently large $\Delta\omega$ these times will be less than the characteristic impact time $\rho/v$, and the impact theory fails. Radiation at large $\Delta\omega$ can be produced by collisions with small impact parameters $\rho_s$. During such impacts the oscillator will radiate at frequency $\omega = \omega_0 + \Delta\omega(\rho_s)$, and the intensity of this radiation will be proportional to the statistical frequency of impacts at $\rho_s$, i.e., the statistical broadening picture holds.

While it would be very difficult to construct a theory for this transition region, it is helpful to imagine that the change from one regime to the next is abrupt and occurs at some frequency separation $\Delta\omega_g$. If we assume that the dominant broadening occurs from a typical perturber at $\rho_w$, then we can use the estimate

$$\Delta\omega_g \approx \Delta\omega_w = \left(\frac{v}{\psi_p}\right)^{p/(p-1)} C_p^{-[1/(p-1)]} \tag{9-181}$$

The *boundary frequency* $\Delta\omega_g$ has been discussed in detail by Unsöld (*Vierteljahr. Astron. Ges.*, **78**, 213, 1943) and Holstein (*Phys. Rev.*, **79**, 744, 1950), who derive values that differ from $\Delta\omega_w$ only by factors of order unity. Therefore, we may use $\Delta\omega_g = \Delta\omega_w$ for purposes of rough estimation.

### VALIDITY OF THE CLASSICAL PARTICLE PICTURE OF THE PERTURBER

In all of the above discussion, we have treated the perturber as a classical particle whose position and momentum are precisely defined, moving on a classical orbit (i.e., a straight line past an atom, or an hyperbola past an ion). To simplify the calculation in the quantum mechanical treatment of line broadening, the classical picture of the particle is usually retained. To do so will be valid when the de Broglie wavelength of the perturber is small compared with the impact parameter for those collisions that contribute significantly to the broadening, i.e., we must have

$$\rho \gg \lambdabar = \frac{\hbar}{mv} \tag{9-182}$$

or

$$mv\rho \gg \hbar \tag{9-183}$$

Note that $mv\rho$ is just the angular momentum of the perturber; thus our criterion is equivalent to the requirement that the orbital angular quantum number $l$ be much larger than unity. Under such circumstances we would also expect according to the *correspondence principle* that the classical particle picture is valid.

This validity criterion must always be checked in line-broadening calculations. In most calculations of astrophysical interest, it is found to hold. Some perturbers that violate the condition can always be expected, but the approximation remains useful if they do not contribute strongly to the broadening (see also the discussion by Baranger, Ref. 6, p. 498 ff.).

Another assumption that is made is that the perturber path is fixed. Thus the effect of the perturber upon the atom is accounted for, but the back-reaction of the atom upon the perturber is ignored. This will be valid if the energy gained or lost by the perturber is much less than its typical kinetic energy ($\sim kT$), a condition that is almost always found to be satisfied. A few collisions will always occur in which large energy exchanges take place, but again these will not invalidate the assumption if they do not contribute strongly to the broadening.

A theory that does not make use of the classical path approximation has been developed by Baranger (*Phys. Rev.*, **112**, 855, 1958).

#### THE EFFECTS OF NONADIABATIC IMPACTS

In the classical theory of line broadening one treats the radiator simply as a perturbed dipole. As such, transitions cannot be induced by a perturbation, and no allowance is made in the theory for this possibility. In reality, an atom is a system with a large number of bound states at well-defined energies. Line broadening occurs when these states are shifted relative to one another by the action of perturbers. In addition, collisions with perturbers can cause transitions among the atomic states and thus lead to a violation of the *adiabatic assumption* contained in early theories of line broadening. Actually, these nonadiabatic effects are often very important, particularly for an atom, such as hydrogen, with degenerate levels. Modern quantum mechanical theories of line broadening account fully for transitions, a fact that has led on occasion to a drastic revision of earlier work.

## 9-7. The Quantum Mechanical Calculation of Line Broadening

Very good discussions of the quantum theory of line broadening have been given by Baranger (*Phys. Rev.*, **111**, 481, 1958; *Phys. Rev.*, **111**, 494, 1958; and Ref. 6, Chap. 13), by Cooper (*Rev. Mod. Phys.*, **39**, 167, 1967), and by Griem (Ref. 14, Chap. 4). The development of a satisfactory quantum mechanical theory of line broadening has brought about a major improvement in one of the most important and difficult applications of atomic theory in modern astrophysics.

The main application of this theory that we shall consider is the broadening of hydrogen lines by linear Stark-effect interactions of the radiating atom with both electrons and protons in the plasma. As we will show in more detail below, in situations of astrophysical interest we may always treat the electrons using the impact approximation (except in the very far wings) while the ion broadening always can be treated using the quasi-static approximation (with due allowance for shielding effects). Our procedure will therefore be the following: We will assume that the atom is in a static electric field due to the ions and that the degenerate levels are split into their Stark components (see Section 9-8 for a discussion of the Stark pattern of hydrogen lines) by this field. We will then calculate the broadening effects of the electrons upon these individual components by using the impact approximation. Finally, we will average over the appropriate probability distribution function of the ionic static field strength.

We showed in Chapter 4 that the power radiated by an *isolated* atom in a transition from upper state $j$ to lower state $i$ is as given in equation (4-104):

$$P = \frac{4\omega^4}{3c^3} |\langle i| e\mathbf{r} |j\rangle|^2$$

The total spectrum of the atom summed over all transitions is then

$$P(\omega) = \frac{4\omega^4}{3c^3} \sum_{i,j} \delta(\omega - \omega_{ji}) |\langle i| e\mathbf{r} |j\rangle|^2 \rho_j \tag{9-184}$$

where we have assumed that $\omega^4$ is constant over the line and where $\rho_j$ is the probability of occurrence of the upper state. For an isolated atom in thermodynamic equilibrium.

$$\rho_j = \langle j|\rho|j\rangle = \frac{e^{-E_j/kT}}{U(T)} \tag{9-185}$$

where $U(T)$ is the atomic partition function. If we now wish to calculate the broadening of the lines emitted by an *atom in a plasma,* we consider the radiating system to consist of atom *plus* perturbers and generalize the notion of states $|i\rangle$ and $|j\rangle$ to include perturbers also. The *profile* of the line is written simply as

$$I(\omega) = \sum_{i,j} \rho_j \delta(\omega - \omega_{ji}) |\langle i| e\mathbf{r} |j\rangle|^2 \tag{9-186}$$

Here $\rho_j$ refers to the probability of a particular state of atom *and* pertuber. For a plasma in thermal equilibrium we expect $\rho_j$ to be proportional to

$$e^{-H/kT} = e^{-(H_A + H_P + V)/kT} \tag{9-187}$$

where $H$ is the total Hamiltonian of the complete system, $H_A$ is the Hamiltonian of the atom alone, $H_P$ is the perturber Hamiltonian, and $V$ is the interaction Hamiltonian.

It is easiest to calculate the effects of collisions by using the Fourier transform

$$\varphi(t) = \int_{-\infty}^{\infty} I(\omega)e^{i\omega t}\,d\omega = \sum_{i,j} \rho_j e^{i\omega_{ji}t}|\langle i|\,\mathbf{er}\,|_j\rangle|^2 \tag{9-188}$$

which is completely analogous to the classical autocorrelation function (see, e.g., the discussion by Baranger in Ref. 6, p. 498). As was the case previously, the effects of collisions at a specific impact parameter upon the autocorrelation function can be calculated, and statistical averages over all possible perturber paths can be formed directly. The intensity profile then follows directly from the inverse Fourier transform

$$I(\omega) = \frac{1}{2\pi}\int_{-\infty}^{\infty} \varphi(t)e^{-i\omega t}\,dt = \frac{1}{\pi}\,Re \int_{0}^{\infty} \varphi(t)e^{-i\omega t}\,dt \tag{9-189}$$

where we have made use of the fact that $\varphi(-t) = \varphi^*(t)$. Thus the problem at hand is to obtain an expression for $\varphi(t)$. To do this, we will need to find the change in time of the eigenstates $|i\rangle$ and $|j\rangle$ under the effects of perturbing collisions. To calculate such changes we will use the *time-development* operator $T(t, 0)$. This operator is chosen such that when it operates on the state of the system $|\alpha, 0\rangle$ at time $t = 0$, it results in $|\alpha, t\rangle$, the state of the system at time $t$, i.e.,

$$|\alpha, t\rangle = T(t, 0)\,|\alpha, 0\rangle \tag{9-190}$$

Now since $|\alpha, t\rangle$ still satisfies the Schrödinger equation

$$H\,|\alpha, t\rangle = i\hbar\,\frac{d\,|\alpha, t\rangle}{dt} \tag{9-191}$$

we may derive a Schrödinger equation for $T(t, 0)$ since

$$\begin{aligned} HT(t, 0)\,|\alpha, 0\rangle &= i\hbar\,\frac{d}{dt}\,[T(t, 0)\,|\alpha, 0\rangle] \\ &= i\hbar\,\frac{dT}{dt}\,|\alpha, 0\rangle + i\hbar T(t, 0)\,\frac{d\,|\alpha, 0\rangle}{dt} \end{aligned} \tag{9-192}$$

But $|\alpha, 0\rangle$ is fixed in time; thus the last term in equation (9-192) is zero, and we may write

$$HT(t, 0) = i\hbar\,\frac{dT(t, 0)}{dt} \tag{9-193}$$

A special case occurs when we want the time-development operator of an unperturbed eigenstate; then, as we have shown earlier,

$$|\alpha, t\rangle = e^{-iE_\alpha t/\hbar}\,|\alpha, 0\rangle \tag{9-194}$$

so in this case,

$$T(t, 0) = e^{-iE_a t/\hbar} = e^{-iHt/\hbar} \tag{9-195}$$

where the latter exponential is to be understood as an operator. We may now rewrite equation (9-189) in terms of time-development operators, including perturbations. Thus, writing $e\mathbf{r} = \mathbf{d}$, we have

$$\begin{aligned}\varphi(t) &= \sum_{i,j} \rho_j \exp[i(E_j - E_i)t/\hbar] \, |\langle i | \mathbf{d} | j \rangle|^2 \\ &\sum_{i,j} \rho_j \langle i | \mathbf{d} | j \rangle^* e^{-iE_i t/\hbar} \langle i | \mathbf{d} | j \rangle e^{iE_j t/\hbar} \\ &= \sum_{i,j} \rho_j \langle j | \mathbf{d} | i \rangle e^{-iE_i t/\hbar} \langle i | \mathbf{d} | j \rangle e^{iE_j t/\hbar} \\ &= \sum_{i,j} \rho_j \langle j | \mathbf{d}T | i \rangle \langle i | \mathbf{d}T^\dagger | j \rangle\end{aligned} \tag{9-196}$$

Now in Dirac notation, the expansion rule, relative to a *complete set* of states $|\gamma\rangle$, is

$$\langle \alpha | \beta \rangle = \sum_\gamma \langle \alpha | \gamma \rangle \langle \gamma | \beta \rangle \tag{9-197}$$

so we see that we can rewrite equation (9-196) as

$$\begin{aligned}\varphi(t) &= \sum_j \rho | \langle j | \mathbf{d}T\mathbf{d}T^\dagger | j \rangle \\ &= \text{trace}[\rho \mathbf{d}T\mathbf{d}T^\dagger]\end{aligned} \tag{9-198}$$

The trace is to be carried out over both atomic and perturber states. This expression is quite general; the precise form of $\varphi$ will depend upon the details of $T$.

## THE CLASSICAL PATH APPROXIMATION

Let us now consider the calculation of $T$ in more detail. As mentioned above, we consider the complete system to consist of atom plus perturber. Let the wave function $\psi$ describing this system be the solution of the equation

$$i\hbar \frac{d\psi}{dt} = (H_A + H_P + V)\psi \tag{9-199}$$

where $H_A$ is the Hamiltonian of the atom alone, $H_p$ is the Hamiltonian of the perturber alone, and $V$ is the interaction Hamiltonian between atom and perturber. Note that $H_A$ is independent of perturber coordinates and that $H_P$ is independent of atomic coordinates while $V$ depends upon both. Moreover, since the atom is assumed to be in the ionic static field of strength $F$, we should write $H_A(F)$. For brevity of notation, we will suppress $F$ here, but bear in mind that it is present; we will include it explicitly in our final result.

To make progress, we now assume that the perturber will follow its classical path (straight line for interaction with a neutral atom and a hyperbola for an ion) and write the wave function in separated form, namely,

$$\psi(t) = \alpha(t)\pi(t) \tag{9-200}$$

where $\alpha(t)$ is the atomic wave function, and $\pi(t)$ is the perturber wave function. We emphasize that this form for $\psi(t)$ is only an approximation since strictly speaking the Schrödinger equation is not actually separable (see Ref. 6, p. 512 ff.). We further suppose that the perturber path will be independent of the state of the atom with which the interaction takes place. Then $\pi(t)$ is the solution of the equation

$$i\hbar \frac{d\pi(t)}{dt} = H_P \pi(t) \tag{9-201}$$

normalized such that

$$\int \pi^*(t)\pi(t)\, d\tau_P = 1 \tag{9-202}$$

The time-development operator for the perturber is then simply

$$T_P(t, 0) = e^{-iH_P t/\hbar} \tag{9-203}$$

Consider now the Schrödinger equation for the atom alone. To obtain it, we multiply the Schrödinger equation for the complete system by $\pi^*$ and integrate over perturber coordinates:

$$i\hbar\left[\alpha(t)\int \pi^* \frac{d\pi}{dt}\, d\tau_P + \frac{d\alpha}{dt}\int \pi^* \pi\, d\tau_P\right]$$
$$= \left(H_A \int \pi^* \pi\, d\tau_P + \int \pi^* V \pi\, d\tau_P\right)\alpha(t) + \alpha(t)\int \pi^* H_P \pi\, d\tau_P \tag{9-204}$$

But from equation (9-201) we have

$$i\hbar \int \pi^* \frac{d\pi}{dt}\, d\tau_P = \int \pi^* H_P \pi\, d\tau_P \tag{9-205}$$

Thus, equation (9-204) reduces to

$$i\hbar \frac{d\alpha}{dt} = \left(H_A + \int \pi^* V \pi\, d\tau_P\right)\alpha(t) \tag{9-206}$$

If the wave packets of the perturbers are narrow enough that they may be considered to represent classical particles on classical paths, we *make the identification*

$$\int \pi^* V \pi\, d\tau_P \rightarrow V_{cl}(t) \tag{9-207}$$

where $V_{cl}(t)$ is the classical interaction potential. This is the essence of the *classical path approximation.* Thus, the Schrödinger equation for the time-development operator of the atom becomes

$$i\hbar \frac{dT_A(t, 0)}{dt} = [H_A + V_{cl}(t)]T_A(t, 0) \tag{9-208}$$

and the time-development operator of the complete system is

$$T(t, 0) = T_A(t, 0)T_P(t, 0) = T_A(t, 0)e^{-iH_P t/\hbar} \tag{9-209}$$

Finally, we assume that the probability-density matrix $\rho$ can be written as

$$\rho = \rho_A \rho_P \tag{9-210}$$

where $\rho_A$ refers to atomic states only, and $\rho_P$ refers to perturber states only and is diagonal in the perturber coordinates. We may reasonably do this if the interaction energy $V$ in equation (9-199) is much less than $kT$ so that the back-reaction of the atom upon the perturber can be neglected. Equivalently, we expect this to hold for frequency separations $\Delta\omega$ from line center such that

$$\hbar\, \Delta\omega \ll kT \tag{9-211}$$

When these expressions for $\rho$ and $T$ are inserted into equation (9-198) for $\varphi(t)$, and the separations inherent in equations (9-200), (9-209), and (9-210) are accounted for, we obtain simply

$$\begin{aligned} \varphi(t) &= \text{trace}[\rho\, \mathbf{d}T\, \mathbf{d}T^{\dagger}] \\ &= \text{trace}\{\rho_A\, \mathbf{d}T_A\, \mathbf{d}T_A{}^{\dagger}\} \end{aligned} \tag{9-212}$$

The trace over perturber states has reduced merely to a *thermal average over all perturbers*, which we have denoted by the braces and the trace is now carried out over *atomic states* only.

### THE IMPACT APPROXIMATION

As we mentioned above, we will apply the quantum mechanical calculation to broadening of hydrogen lines by electrons. Since we shall see that this broadening lies in the impact regime, let us now consider this case in more detail.

We assume that both the initial state $a$ and the final state $b$ consist of several substates, denoted by $\alpha$ and $\beta$, respectively. We assume that optical dipole transitions exist only between substates of states $a$ and $b$ but that transitions among the substates of $a$ or $b$ can be ignored. On the other hand, we ignore collision-induced transitions between states $a$ and $b$ and assume that collisions can result only in transitions among substates of $a$ or $b$. Thus, we have terms of the form

$$\langle\alpha|\,\mathbf{d}\,|\alpha'\rangle = 0, \langle\beta|\,\mathbf{d}\,|\beta'\rangle = 0, \quad \text{and} \quad \langle\alpha|\,T_{a,b}\,|\beta\rangle = 0$$

Then writing out the trace in equation (9-212) for $\varphi(t)$, we have

$$\begin{aligned}
\varphi(t) &= \rho_a \sum_{\alpha,\alpha',\beta,\beta'} \{\langle\alpha|\,\mathbf{d}\,|\beta\rangle\langle\beta|\,T_b\,|\beta'\rangle\langle\beta'|\,\mathbf{d}\,|\alpha'\rangle\langle\alpha'|\,T_a^{\dagger}\,|\alpha\rangle\} \\
&= \rho_a \sum_{\alpha,\alpha',\beta,\beta'} \langle\alpha|\,\mathbf{d}\,|\beta\rangle\langle\beta'|\,\mathbf{d}\,|\alpha'\rangle\{\langle\beta|\,T_b\,|\beta'\rangle\langle\alpha'|\,T_a^{\dagger}\,|\alpha\rangle\} \\
&= \rho_a \sum_{\alpha,\alpha',\beta,\beta'} \langle\alpha|\,\mathbf{d}\,|\beta\rangle\langle\beta'|\,\mathbf{d}\,|\alpha'\rangle\{\langle\beta|\,T_b\,|\beta'\rangle\langle\alpha|\,T_a\,|\alpha'\rangle^{*}\} \\
&= \rho_a \sum_{\alpha,\alpha',\beta,\beta'} \langle\alpha|\,\mathbf{d}\,|\beta\rangle\langle\beta'|\,\mathbf{d}\,|\alpha'\rangle\{\langle\alpha|\langle\beta|\,T_b T_a^{*}\,|\alpha'\rangle|\beta'\rangle\} \\
&= \rho_a \sum_{\alpha,\alpha',\beta,\beta'} \langle\alpha|\,\mathbf{d}\,|\beta\rangle\langle\beta'|\,d\,|\alpha'\rangle\langle\alpha|\langle\beta|\,\{T_b T_a^{*}\}\,|\alpha'\rangle|\beta'\rangle
\end{aligned} \tag{9-213}$$

where we have neglected the variation of $\rho_a$ among the upper substates and have noted that only the time-development operators can depend upon statistical averages. It is convenient to replace the complete time-development operator by an operator $U$ defined such that

$$U(t, 0) = e^{iH_A t/\hbar}T(t, 0) \tag{9-214}$$

The operator $U$ is the time-development operator in the *interaction representation.* This definition effectively factors out the oscillatory term for an eigenstate and essentially includes only perturbation terms; indeed for an unperturbed eigenstate $U(t, 0) \equiv 1$. By substitution, we then have

$$\varphi(t) = \rho_a \sum_{\alpha,\alpha',\beta,\beta'} \langle\alpha|\,\mathbf{d}\,|\beta\rangle\langle\beta'|\,\mathbf{d}\,|\alpha'\rangle\langle\alpha|\langle\beta|\,e^{-iH_b t/\hbar}\{U_b U_a^{*}\}e^{iH_a t/\hbar}\,|\alpha'\rangle|\beta'\rangle \tag{9-215}$$

The operator $U(t, 0)$ satisfies the Schrödinger equation (9-208):

$$i\hbar \frac{d}{dt}[e^{-iH_A t/\hbar}U(t, 0)] = [H_A + V_{cl}(t)]e^{-iH_A t/\hbar}U(t, 0)$$

which reduces to

$$i\hbar \frac{dU(t, 0)}{dt} = e^{iH_A t/\hbar}V_{cl}(t)e^{-iH_A t/\hbar}U(t, 0) \equiv V'_{cl}(t)U(t, 0) \tag{9-216}$$

We may solve this equation by iteration, taking as the first approximation $U(t, 0) = 1$. Then we obtain a solution of the form

$$U(t, 0) = 1 + \frac{1}{i\hbar}\int_0^t V'_{cl}(t_1)\,dt_1 + \left(\frac{1}{i\hbar}\right)^2 \int_0^t dt_2\,V'_{cl}(t_2)\int_0^{t_2} dt_1 V'_{cl}(t_1) + \cdots \tag{9-217}$$

We must now consider the calculation of the statistical average $\{U_b U_a^*\}$. We proceed in a way quite analogous to that used in the development of

Lindholm theory. Consider the change of $\{U_b U_a^*\}$ in some time $\Delta t$. Then

$$\Delta\{U_b U_a^*\} = \{U_b(t+\Delta t, 0)U_a^*(t+\Delta t, 0) - U_b(t, 0)U_a^*(t, 0)\}$$
$$= \{[U_b(t+\Delta t, t)U_a^*(t+\Delta t, t) - 1]U_b(t, 0)U_a^*(t, 0)\} \tag{9-218}$$

Here the first term represents the change during the time interval $(t, t+\Delta t)$. This change may be considered to be due to some specific collision. As in the Lindholm theory, we argue that such changes can be assumed to be statistically independent of the current values of $U_b$ and $U_a^*$ so that we can replace the average of the product by the product of averages. We further assume that some time interval $\Delta t$ may be chosen such that the average of

$$[U_b(t+\Delta t, t)U_a^*(t+\Delta t, t) - 1]$$

is small compared with unity and may be calculated by an iterative calculation. By analogy with equation (9-217) for $U(t, 0)$ we have

$$U(t+\Delta t, t) = 1 - \frac{i}{\hbar}\int_t^{t+\Delta t} dt_1 V'_{cl}(t_1) + \left(\frac{i}{\hbar}\right)^2 \int_t^{t+\Delta t} dt_1 V'_{cl}(t_1) \int_t^{t_1} \times dt_2 V'_{cl}(t_2) + \cdots \tag{9-219}$$

Writing $s_1 = t_1 - t$ and $s_2 = t_2 - t$ and recalling the definition in equation (9-216),

$$V'_{cl}(t) = e^{iH_A t/\hbar} V_{cl}(t) e^{-iH_A t/\hbar}$$

we find

$$U(t+\Delta t, t) = 1 - e^{iH_A t/\hbar}\left[\frac{i}{\hbar}\int_0^{\Delta t} ds_1 V'_{cl}(s_1) - \left(\frac{i}{\hbar}\right)^2 \int_0^{\Delta t} ds_1 V'_{cl}(s_1) \int_0^{s} ds_2 V'_{cl}(s_2) + \cdots\right] e^{-iH_A t/\hbar} \tag{9-220}$$

Then by multiplication and grouping of terms, we find

$$\{U_b(t+\Delta t, t)U_a^*(t+\Delta t, t) - 1\}$$
$$= e^{i(H_b - H_a)t/\hbar}\left\{-\frac{i}{\hbar}\int_0^{\Delta t} ds_1 [V_b'(s_1) - V_a^{*\prime}(s_1)] + \left(\frac{i}{\hbar}\right)^2 \left[\int_0^{\Delta t} ds_1 V_b'(s_1) \int_0^{s} ds_2 V_b'(s_2) + \int_0^{\Delta t} ds_1 V_a^{*\prime}(s_1) \int_0^{s} ds_2 V_a^{*\prime}(s_2) - \int_0^{\Delta t} ds_1 V_b'(s_1) \int_0^{s} ds_2 V_a^{*\prime}(s_2)\right] + \cdots\right\} e^{-i(H_b - H_a)t/\hbar} \tag{9-221}$$

Or defining $\Phi_{ab}$ to be $1/\Delta t$ times the expression in braces (a thermal average), we have finally from equations (9-218) and (9-221)

$$\Delta\{U_b U_a^*\} = e^{i(H_b - H_a)t/\hbar}\Phi_{ab}\, e^{-i(H_b - H_a)t/\hbar}\{U_b(t, 0)U_a^*(t, 0)\}\, \Delta t \qquad (9\text{-}222)$$

If $\Delta t$ is sufficiently small, we may regard this as a differential equation for $\{U_b U_a^*\}$, whose solution is

$$\{U_b(t, 0)U_a^*(t, 0)\} = e^{i(H_b - H_a)t/\hbar} \exp[-i(H_b - H_a)t/\hbar + \Phi_{ab}\, t] \qquad (9\text{-}223)$$

Thus substituting into equation (9-215), we obtain

$$\varphi(t) = \rho_a \sum_{\alpha,\alpha',\beta,\beta'} \langle\alpha|\,\mathbf{d}\,|\beta\rangle\langle\beta'|\,\mathbf{d}\,|\alpha'\rangle$$
$$\times \langle\alpha|\,\langle\beta|\exp[i(H_a - H_b)t/\hbar + \Phi_{ab}\, t]\,|\alpha'\rangle\,|\beta'\rangle \qquad (9\text{-}224)$$

Then performing the inverse Fourier transformation to obtain the intensity distribution, we find

$$I(\omega) = \frac{1}{\pi}\, Re \int_0^\infty \varphi(t) e^{-i\omega t}\, dt$$
$$= \frac{e^{-E_a/kT}}{\pi U(T)}\, Re \sum_{\alpha,\alpha',\beta,\beta'} \langle\alpha|\,\mathbf{d}\,|\beta\rangle\langle\beta'|\,\mathbf{d}\,|\alpha'\rangle\langle\alpha|\,\langle\beta| \int_0^\infty dt$$
$$\times \exp[i(H_a - H_b)t/\hbar - i\omega t + \Phi_{ab}\, t]\,|\alpha'\rangle\,|\beta'\rangle \qquad (9\text{-}225)$$

or introducing again the static ion field $F$,

$$I(\omega, F) = \frac{e^{-E_a/kT}}{\pi U(T)}\, Re \sum_{\alpha,\alpha',\beta,\beta'} \langle\alpha|\,\mathbf{d}\,|\beta\rangle\langle\beta'|\,\mathbf{d}\,|\alpha'\rangle\langle\alpha|\,\langle\beta|$$
$$\times \left[i\omega - \Phi_{ab}(F) - \frac{i}{\hbar}[H_a(F) - H_b(F)]\right]^{-1} |\alpha'\rangle\,|\beta'\rangle \qquad (9\text{-}226)$$

where we have made use of the fact that $\Phi_{ab}(F)$ is found to have a negative real part. Now if $W(F)$ is the probability of an ion field strength $F$, the final profile, averaged over all ion fields, is

$$I(\omega) = \frac{e^{-E_a/kT}}{\pi U(T)} \int_0^\infty W(F) Re \sum_{\alpha,\alpha',\beta,\beta'} \langle\alpha|\,\mathbf{d}\,|\beta\rangle\langle\beta'|\,\mathbf{d}\,|\alpha'\rangle$$
$$\times \langle\alpha|\,\langle\beta| \left[i\omega - \Phi_{ab}(F) - \frac{i}{\hbar}[H_a(F) - H_b(F)]\right]^{-1} |\alpha'\rangle\,|\beta'\rangle\, dF$$
$$(9\text{-}227)$$

Equation (9-227) is quite general and has been used by Griem, Cooper, and their coworkers in calculations of Stark-broadened line profiles. The result is valid so long as the $\Delta t$ used in the definition of $\Phi_{ab}$ can be chosen to include one complete collision, or, if the collisions overlap, they must be weak enough that their contributions to the iterative solution of the Schröd-inger equation are simply additive. Moreover, it is necessary that it be valid to treat the perturbers as classical particles. These validity criteria must be checked in each case.

## 9-8. Application to Hydrogen

As a specific example of the application of the theory developed above, let us now consider the Stark broadening of hydrogen lines which are of course of great importance in astrophysical problems.

### THE STARK PATTERN OF HYDROGEN LINES

To evaluate the line profile given by equation (9-227), we must know the shifts and intensities of the Stark components of hydrogen lines. In the absence of a perturbing field, each level of hydrogen is degenerate with $2n^2$ sublevels. Analyses by K. Schwarzschild (*Sitzber. Deutsch. Akad. Wiss. Berlin*, p. 584, 1916) and Epstein (*Ann. Phys.*, **50**, 489, 1916) showed that when a field is applied, these sublevels separate, and that since hydrogen has a permanent dipole moment, the splitting is linearly proportional to $F$ (linear Stark effect). We may write the shift for each component as in equation (9-153):

$$\Delta\lambda_k = C_k F$$

where

$$C_k = \frac{3h^7 c}{32\pi^6 m^3 e^9} \frac{n'^4 n^4}{(n^2 - n'^2)^2} \frac{X_k}{Z^5} \tag{9-228}$$

and $X_k$ is an integer

$$X_k = n(n_2 - n_1) - n'(n_2' - n_1') \tag{9-229}$$

Here $n'$ is the principal quantum number of the lower level, $n$ is the principal quantum number of the upper level, and $n_1 \geq 0$ while $n_2 \leq n - 1$, etc. By symmetry, $C_{-k} = -C_k$. An extensive tabulation of the quantities $X_k$ and $C_k$ has been given by Underhill and Waddell (Ref. 35) for lines in the Lyman, Balmer, Paschen, and Brackett series.

To each Stark component one can assign an oscillator strength $f_i$ and a weight $w_i$ depending upon the polarization of the component. These sum to yield the total $f$-value of the transition

$$f = \frac{1}{n'^2}\left[\sum_i w_i f_{+i} + \sum_{-i} w_{-i} f_{-i} + \sum_i w_{0,i} f_{0,i}\right] \tag{9-230}$$

The last term consists of all components with zero shift in the field, and it is usually denoted $f_0$. The first two terms are sums over all shifted components, and are usually written as $f_\pm$ (see Ref. 35, Table 2). The oscillator strengths of individual components follow from an expression derived by Gordon (see Ref. 35, Table 1). Since components with equal values of $C_k$ fall on top of one another, it is convenient to group them together into values $f_k$; these again are tabulated by Underhill and Waddell. In many applications, it is still more convenient to write

$$I_k \equiv \frac{f_k}{f} \tag{9-231}$$

so that the sum over all components is normalized to unity

$$\sum I_k = 1 \tag{9-232}$$

Typical Stark patterns for a few hydrogen lines are shown in Figure 9-3. Note that only every alternate line in a series has an unshifted component.

As one goes up the series, the lines on the average become much broader because strong components exist at larger values of $X_k$. This can be characterized by computing a mean shift

$$\overline{X} = \frac{\sum I_k X_k}{\sum I_k} \tag{9-233}$$

omitting the unshifted component from both sums. When this is done, we obtain the values listed in Table 9-4. We have listed also $\overline{C_2}$, a mean interaction constant, defined in terms of the relation $\overline{\Delta\omega} = \overline{C_2}/r^2$. But for linear Stark effect, $F = e/r^2$; thus substituting from equations (9-153) and (9-228), we find

$$\overline{C_2} = r^2\,\overline{\Delta\omega} = r^2\left(\frac{3h}{4\pi me}\,\overline{X}\right)\left(\frac{e^2}{r^2}\right) = \frac{3h\overline{X}}{4\pi m} = 1.738\overline{X} \tag{9-234}$$

where we have made use of equation (4-119). The two parameters $\overline{X}$ and $\overline{C}_2$ provide a caricature of the line profile which will be useful in later work.

Finally, one may compute the complete line profile, as given by equation (9-156); extensive tables are given by Underhill and Waddell (Ref. 35,

TABLE 9-4. Mean Interaction Constants and Line Shift Parameters for Hydrogen Lines

| *Line* | $L\alpha$ | $L\beta$ | $H\alpha$ | $H\beta$ | $H\gamma$ | $H\delta$ | $\lim n \gg n'$ |
|---|---|---|---|---|---|---|---|
| $\overline{X}$ | 2.0 | 4.0 | 2.24 | 5.96 | 11.8 | 15.9 | $\frac{1}{2}n(n-1)$ |
| $\overline{C}_2$ | 3.48 | 6.95 | 3.89 | 10.4 | 20.5 | 27.6 | $1.738\,\overline{X}$ |

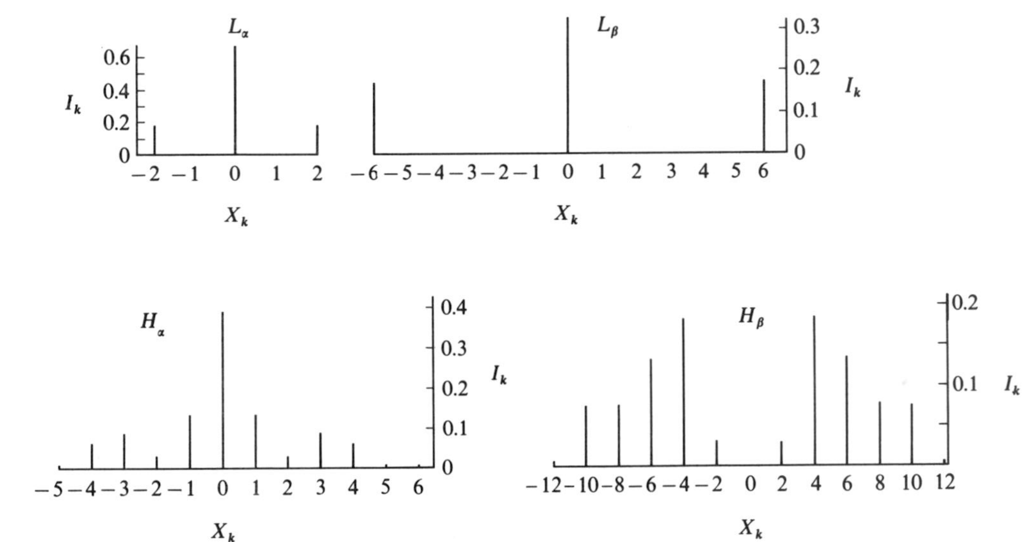


FIG. 9-3. Stark patterns for $L\alpha$, $L\beta$, $H\alpha$, and $H\beta$. Note that $H\beta$ lacks a central unshifted component.

Table 4) for a large number of hydrogen lines. These profiles allow for the broadening by ions only and thus have only limited usefulness since we now know that broadening by electrons is also important.

### VALIDITY CRITERIA FOR IONS AND ELECTRONS

Let us first inquire whether we expect the statistical or impact theories to be valid for broadening by ions and electrons. In the schematic picture discussed in Section 9-6, we showed that very roughly speaking for frequency separations $\Delta\omega$ less than some critical value $\Delta\omega_g$ we expect statistical broadening. Moreover, $\Delta\omega_g \sim \Delta\omega_w$, the shift caused by a perturber located at the Weisskopf radius. Now for $p = 2$, $\psi_p = \pi$ and

$$\Delta\omega_g \approx \Delta\omega_w = \left(\frac{v^p}{C_p \psi_p{}^p}\right)^{1/(p-1)} = \frac{v^2}{\pi^2 \bar{C}_2} \tag{9-235}$$

Converting to wavelength units,

$$\Delta\lambda_g \approx \frac{\lambda^2}{2\pi c} \frac{v^2}{\pi^2 \bar{C}_2} \tag{9-236}$$

Let us now evaluate $\Delta\lambda_g$ for ions and electrons under typical conditions. First, we calculate the average relative velocity.

$$v = \left[\frac{8kT}{\pi m_H}\left(\frac{1}{A_1} + \frac{1}{A_2}\right)\right]^{1/2} \tag{9-237}$$

with $A_1 = 1$, and $A_2 = 1$ for ions, and 1/1836 for electrons. We find the results listed in Table 9-5. Then using the values of $\bar{C}_2$ given in Table 9-4, we find for $\Delta\lambda_g$ (measured in Å) the results given in Table 9-6. Thus, we clearly see that to a good approximation we can expect that broadening by the ions will always be described by statistical broadening theory. This is even more the case when we realize that the very inner part of the line will usually be dominated by Doppler broadening. In the case of electrons, we see that

TABLE 9-5. Average Relative Velocities (cm/sec)

| *Particles* | $\frac{1}{A_1}+\frac{1}{A_2}$ | *T*: $2.5 \times 10^4$°K | *T*: $10^4$°K |
|---|---|---|---|
| H + electrons | 1837 | $9.8 \times 10^7$ | $6.2 \times 10^7$ |
| H + H$^+$ | 2 | $32.5 \times 10^5$ | $20.6 \times 10^5$ |
| H + heavy particle | 1 | $23.0 \times 10^5$ | $14.5 \times 10^5$ |

TABLE 9-6. Transition Wavelength $\Delta\lambda_g$ (Å) Between Statistical and Impact Broadening for Hydrogen Lines

| Line | Perturber | $T$ | |
|---|---|---|---|
| | | $2.5 \times 10^4$°K | $10^4$°K |
| Hα | Electrons | 580.0 | 230.0 |
| | Protons | 0.63 | 0.25 |
| Hβ | Electrons | 120.0 | 48.0 |
| | Protons | 0.13 | 0.05 |
| Hγ | Electrons | 48.0 | 19.0 |
| | Protons | 0.05 | 0.02 |
| Hδ | Electrons | 32.0 | 13.0 |
| | Protons | 0.03 | 0.01 |

over a large part of the line profile, we may expect the impact theory to be valid although at great distances from the line center, we must account for the transition to statistical broadening.

### THE ELECTRON BROADENING OF HYDROGEN LINES

Let us now apply the general results that we obtained in equation (9-227) for electron impact broadening to the hydrogen lines. The main effort is the calculation of an explicit form for the operator $\Phi_{ab}$.

If we assume the electron can be treated as a classical particle, the dominant term in the interaction between a perturber at $\mathbf{r}$ and radiating electron at $\mathbf{r}_a$ is

$$V_{cl}(t) = \frac{e^2\mathbf{r}\cdot\mathbf{r}_a}{|\mathbf{r}|^3} = \frac{e^2\mathbf{r}_a\cdot[\boldsymbol{\rho} + \mathbf{v}(t - t_0)]}{[\rho^2 + v^2(t - t_0)^2]^{3/2}} \tag{9-238}$$

where we have assumed explicitly that the perturber path is a straight line with impact parameter $\boldsymbol{\rho}$ and time of closest approach $t_0$. This expression should be inserted into equations (9-216) and (9-221) to calculate $\Phi_{ab}$. The thermal average is obtained averaging over the number of collisions at impact parameter and velocity $v$, namely, $2\pi N\rho\, d\rho v f(v)\, dv$. To simplify the calculation we introduce, following Griem (Ref. 14, p. 75), the following approximations and assumptions:

(a) We ignore the exponentials in equation (9-216) since they are very nearly unity because of degeneracy (see Griem, Kolb, and Shen, *Phys. Rev.*, **116**, 4, 1959, for discussion of this point).

(*b*) We assume that collisions are either completed in $\Delta t$ so that the integrations may formally be extended from $-\infty$ to $\infty$ or that they do not contribute at all if $t_0$ lies outside of $\Delta t$.

Then writing $\mathbf{r} = a_0 \mathbf{R}$, where $a_0$ is the Bohr radius, and ignoring those terms that integrate to zero, Griem obtains

$$\Phi_{ab} = -N\left(\frac{\hbar}{m}\right)^2 2\pi \int \rho \, d\rho \int dv f(v) v$$

$$\times \left\{\int_{-\infty}^{\infty} ds_1 \frac{\mathbf{R}_b \cdot (\boldsymbol{\rho} + \mathbf{v}s_1)}{(\rho^2 + v^2 {s_1}^2)^{3/2}} \int_{-\infty}^{\infty} ds_2 \frac{\mathbf{R}_b \cdot (\boldsymbol{\rho} + \mathbf{v}s_2)}{(\rho^2 + v^2 {s_2}^2)^{3/2}}\right.$$

$$+ \int_{-\infty}^{\infty} ds_1 \frac{\mathbf{R}_a \cdot (\boldsymbol{\rho} + \mathbf{v}s_1)}{(\rho^2 + v^2 {s_1}^2)^{3/2}} \int_{-\infty}^{\infty} ds_2 \frac{\mathbf{R}_a \cdot (\boldsymbol{\rho} + \mathbf{v}s_2)}{(\rho^2 + v^2 {s_2}^2)^{3/2}}$$

$$\left. - \int_{-\infty}^{\infty} ds_1 \frac{\mathbf{R}_b \cdot (\boldsymbol{\rho} + \mathbf{v}s_1)}{(\rho^2 + v^2 {s_1}^2)^{3/2}} \int_{-\infty}^{\infty} ds_1 \frac{\mathbf{R}_a \cdot (\boldsymbol{\rho} + \mathbf{v}s_1)}{(\rho^2 + v^2 {s_1}^2)^{3/2}} + \cdots \right\} \tag{9-239}$$

This form simplifies when the average over angles is performed. Making use of symmetry properties in this average, Griem obtains finally

$$\Phi_{ab} = -\frac{4\pi N}{3}\left(\frac{\hbar}{m}\right)^2 (\mathbf{R}_b \cdot \mathbf{R}_b + \mathbf{R}_a \cdot \mathbf{R}_a - 2\mathbf{R}_b \cdot \mathbf{R}_a) \int \frac{d\rho}{\rho} \int \frac{f(v)}{v} \, dv \tag{9-240}$$

The integral over $\rho$ diverges logarithmically both at large and small impact parameters. To avoid these divergences, one may introduce cutoffs in the integration. At small $\rho$, strictly speaking, one should retain higher order terms in the calculation of $\Phi_{ab}$. More usually, however, these strong, infrequent collisions are treated in a Lorentz-Weisskopf approximation to determine a cutoff $\rho_{\min}$. This cutoff is chosen such that the matrix element $\{U_b(\infty, -\infty)U_a^*(\infty, -\infty) - 1\}$ is of order unity. Now from equation (9-221) and assumption (a) above we can write

$$\Phi_{ab} = 2\pi N \iint \rho v f(v) \{U_b U_a^* - 1\} \, d\rho \, dv \tag{9-241}$$

By comparison with equation (9-240) we see that we should choose $\rho_{\min}$ such that

$$\rho_{\min}^2 \approx \frac{2}{3}\left(\frac{\hbar}{mv}\right)^2 (\mathbf{R}_b \cdot \mathbf{R}_b + \mathbf{R}_a \cdot \mathbf{R}_a - 2\mathbf{R}_b \cdot \mathbf{R}_a) \tag{9-242}$$

Thus, introducing the strong collisions as a contribution $Nv\pi\rho_{\min}^2$ to $\Phi_{ab}$ and terminating the integration at $\rho_{\max}$, we have

$$\Phi_{ab} = -N\pi \int v f(v) \, dv$$

$$\times \left[\rho_{\min}^2 + \frac{4}{3}\left(\frac{\hbar}{mv}\right)^2 \ln\left(\frac{\rho_{\max}}{\rho_{\min}}\right)(\mathbf{R}_b \cdot \mathbf{R}_b + \mathbf{R}_a \cdot \mathbf{R}_a - 2\mathbf{R}_b \cdot \mathbf{R}_a)\right] \tag{9-243}$$

This idea of using a $\rho_{\min}$ cutoff has been examined very carefully by Shen and Cooper (*Ap. J.*, **155**, 37, 1969). By working in the natural parabolic coordinates that can be used to describe hydrogen wave functions, they are able to sum the perturbation expansion to all orders and thus avoid the divergence at small impact parameters. They find that the usual strong collision cutoff is a reasonable approximation for the first few lines of a series but that significant modification is required for higher series members. They suggest an alternative cutoff procedure.

In equation (9-243) we have explicitly introduced the cutoff at large impact parameters $\rho_{\max}$. There are two considerations that determine $\rho_{\max}$. First, at impact parameters much larger than $\rho_D$, the Debye radius, the field produced by an electron is shielded by other charges in the plasma. Thus, one may set $\rho_{\max} = \rho_D$; from time to time, slightly different cutoffs have been suggested, but these differ from $\rho_D$ only by factors very little different from unity. Second, if the impact theory is to be valid, then at frequency separation $\Delta\omega$ from line center all collisions should be completed in times of the order $\Delta t \sim 1/\Delta\omega$. But $\Delta t \sim \rho/v$ so that this suggests a cutoff at $\rho_L = v/\Delta\omega \approx (kT/m)^{1/2}/\Delta\omega$ (Lewis, *Phys. Rev.*, **121**, 501, 1961). Physically, this accounts for the fact that in the very far line wings, even the electrons obey the quasi-static theory. In practice we must employ the *smaller* of the Debye or Lewis cutoffs, i.e.,

$$\rho_{\max} = \min(\rho_D, \rho_L) \tag{9-244}$$

Once $\Phi_{ab}$ is evaluated, we can carry out the (rather complicated) calculation of $I(\omega)$ via equation (9-227). Several such calculations are now available for hydrogen; we will try to summarize a few of the principal ones here.

Griem, Kolb, and Shen (*Phys. Rev.*, **116**, 4, 1959) carried out calculations for $L\alpha$, $L\beta$, $H\alpha$, $H\beta$, $H\gamma$, and $H\delta$. They allowed for electron broadening of the upper states only (except for $H\alpha$ where the broadening of the unshifted component of the lower state was included). The ion field distribution was approximated with the Ecker distribution functions. The cutoff $\rho_{\max}$ was taken to be $\rho_D$, and the Lewis cutoff was not accounted for. The strong collision term $N\pi v\rho_{\min}^2$ was ignored. The profiles are shown graphically in the paper cited and are given in tabular form in Reference 14. The profiles are given in the form of $S(\alpha)$ versus $\alpha$, in analogy with the pure quasi-static broadening functions, but now include the effects of electron impacts. In addition, tables are given for $R(N_e, T)$, defined such that if $a(\Delta\lambda)$ is the actual line profile at $\Delta\lambda$ from line center and if $a_{\text{ion}}(\Delta\lambda)$ is the quasi-static ion profile, then

$$a(\Delta\lambda) = a_{\text{ion}}(\Delta\lambda)[1 + R(N_e, T)\,\Delta\lambda^{1/2}] \tag{9-245}$$

This formula recognizes that while the ion broadening falls off as $\Delta\lambda^{-5/2}$, the electron broadening falls off as $\Delta\lambda^{-2}$ in the impact regime. Since the Lewis cutoff is not accounted for, however, this formula will yield incorrect results at large displacements from the line core and will result in too large an opacity. For higher series members, the detailed calculation of Stark profiles becomes very time-consuming even on a high-speed computer. Griem (*Ap. J.*, **132**, 883, 1960) suggested therefore an approximate method for estimating the profiles of high series members. Unfortunately, some of the approximations employed were not sufficiently accurate, leading to disagreement with experiments (see, e.g., Ferguson and Schlüter, *Ann. Phys.*, **22**, 351, 1963). This theory has now been replaced by a more satisfactory one by Griem (*Ap. J.*, **147**, 1092, 1967).

In a later paper, Griem, Kolb, and Shen (*Ap. J.*, **135**, 272, 1962) reconsidered the broadening of $H\beta$. Here they allowed for broadening of both the upper and lower levels and used the improved ion field distributions of Baranger and Mozer (*Phys. Rev.*, **118**, 626, 1960). The upper cutoff $\rho_{max}$ was taken to be 1.1 $\rho_D$, and again the Lewis cutoff was not included so that the results were again too large in the far wings.

Subsequently, Griem (*Ap. J.*, **136**, 422, 1962) suggested modifications to the earlier results in the form of interpolating formulae that attempt to allow approximately for the transition from the Debye cutoff to the Lewis cutoff and to incorporate the transition of the electron broadening from the impact regime to the quasi-static regime in the far line wings. These are, however, only interpolation formulae and are not completely reliable (particularly for higher series members), so more accurate calculations were necessary. Much improved calculations were given by Griem (*Ap. J.*, **147**, 1092, 1967), who derived expressions for the electron contribution to the line broadening, allowing for the Lewis cutoff, perturbations of the lower level, improved matrix elements, and higher-order interactions. This theory appears to be in much better agreement with experiment for the far line wings and for higher series members than the earlier results which it replaces.

Recently, quite extensive calculations for hydrogen lines have been carried out by Kepple (Ref. 21), incorporating most of the refinements neglected in the earlier work. These results are very useful although the densities considered are somewhat high for application in stellar atmospheres. Insofar as they can be applied, these results are the most definitive theoretical calculations now available.

The above results are all based upon strictly theoretical calculations. An alternative, semi-empirical approach has been suggested by Edmonds, Schlüter, and Wells (*Mem. R. A. S.*, **71**, 271, 1967). They developed a simple interpolation procedure that allows easy calculation of line profiles; their results show good agreement with experiments and have been successfully applied in stellar atmospheres calculations (see, e.g., Strom and Peterson, *Ap. J.*, **152**, 859, 1968).

OTHER BROADENING OF THE HYDROGEN LINES

In astrophysical applications the hydrogen lines are significantly affected by mechanisms other than the Stark effect. The effects of radiation damping and resonance damping (see Section 9-10 below) can be important at low electron densities. Each of these mechanisms result in a Lorentzian line profile. The core of the line is usually dominated by Doppler broadening. Assuming these mechanisms are all uncorrelated, we may account for them by a simple folding. Folding the Doppler profile with the Lorentz profiles due to radiation and resonance damping yields a Voigt profile, $H(a, v)$, where $a = (\Gamma_{res} + \Gamma_{rad})/4\pi\, \Delta\nu_D$ and $v = \Delta\nu/\Delta\nu_D$, as usual. This Voigt profile is then folded with the Stark profile $S(\alpha)$—allowing for both ion and electron broadening. Usually $S(\alpha)$ is normalized such that

$$\int_{-\infty}^{\infty} S(\alpha)\, d\alpha = 1$$

Thus to work in ordinary frequency units we define equation (9-57)

$$S^*(\Delta\nu) = \left(\frac{10^8 \lambda^2}{cF_0}\right) S(\alpha)$$

and the line profile is then given by

$$a_\nu(\Delta\nu) = \frac{\sqrt{\pi}\, e^2}{mc} f \int_{-\infty}^{\infty} S^*(\Delta\nu + v\, \Delta\nu_D) H(a, v)\, dv \tag{9-246}$$

This folding procedure is often tedious, requires considerable numerical work, but is nonetheless essential to astrophysical calculations.

## 9-9. Stark Broadening of Lines of Other Elements

HYDROGENIC IONS

Hydrogenic ions have Stark patterns essentially identical to hydrogen, though the energies involved are of course different. The Underhill and Waddell tables (Ref. 35) may be used to obtain the profile $S(\alpha)$ due to broadening by ions alone. If $Z$ is the charge of the ion and $S_{UW}$ denotes the value of $S$ given by Underhill and Waddell, then one finds (Ref. 35, p. vi)

$$S(\alpha) = Z^5 S_{UW}(Z^5 \alpha) \tag{9-247}$$

Thus, we see clearly that on a wavelength scale, the lines are narrower by a factor $Z^5$.

The effects of electron broadening for hydrogenic ions are similar to those for hydrogen although the expression for $\Phi_{ab}$ changes since now the perturber moves on a hyperbola around the positively charged ion instead of on a straight-line path. This calculation was carried out by Griem and Shen (*Phys. Rev.*, **122**, 1490, 1961) for $\lambda 3203$ and $\lambda 4686$ of He II. Unfortunately, calculations for other astrophysically important lines of He II (e.g., $\lambda\lambda 4200$, 4542, and 5412) are not yet available.

NEUTRAL HELIUM

The calculation of the broadening of neutral helium lines has been carried out by Griem, Baranger, Kolb, and Oertel (*Phys. Rev.*, **125**, 177, 1962). In broad outline, the calculation proceeds similarly to hydrogen. For isolated lines the profile can be written in the form

$$I(\omega) = \frac{w}{\pi} [(\omega - \omega_0 - d)^2 + w^2]^{-1} \tag{9-248}$$

where $w$ is a width and $d$ is a shift found from integrations over the interactions between the atom and electrons. This distribution is folded with the ion distribution, leading to a profile of the form

$$I(\omega) = \frac{w}{\pi} \int_0^\infty \frac{W(F)\, dF}{w^2 + (\omega - \omega_0 - d - 2\pi C_4 F^2/e^2)^2} \tag{9-249}$$

where $F$ is the field strength due to ions. Extensive tables for He I profiles are given in Reference 14 (Tables 4-4 and 4-5). Improved calculations for several multiplets have been given by Cooper and Oertel (*Phys. Rev.*, **180**, 286, 1969).

An essential complication for many lines of He I is the presence of overlapping forbidden components. For example, near the permitted transition $2^3P - 4^3D$, $\lambda 4471$, there is a forbidden component $2^3P - 4^3F$ at $\lambda 4470$. These components are actually observed in stellar spectra and should provide a good diagnostic tool for density determinations. Allowing for the forbidden components, Barnard, Cooper, and Shamey have recently obtained calculations for $\lambda 4471$ and $\lambda 4922$ (*Astron. Ap.*, **1**, 28, 1969).

OTHER LIGHT ATOMS

Proceeding in a way quite analogous to the He I calculations, Griem has given extensive tables describing the broadening of lines of several other ions, many of astrophysical interest (*Phys. Rev.*, **128**, 515, 1962, and Ref. 14, Tables 4-5 and 4-6).

## 9-10. Other Broadening Mechanisms

We have mentioned in our earlier discussions of impact broadening that collisions between identical (neutral) atoms gives rise to resonance broadening, which has an interaction proportional to $r^{-3}$, while collisions between an atom and nonidentical neutral atoms gives rise to *van der Waals broadening* proportional to $r^{-6}$. Both of these cases are of astrophysical interest for stars that are so cool that hydrogen is not strongly ionized (e.g., the sun), and therefore electrons and ions are much less numerous than hydrogen atoms.

## RESONANCE BROADENING

If we consider the resonance broadening of hydrogen, we have essentially two interacting dipoles, and if we write

$$\Delta\omega = C_3/r^3 \tag{9-250}$$

we have (Ref. 8, p. 231)

$$C_3 = \frac{e^2 f_{ij}}{4\pi m \nu_0} \tag{9-251}$$

where $f_{ij}$ is the absorption oscillator strength of the line. In this case the Weisskopf theory gives

$$\Gamma = 4\pi N C_3 = \frac{Ne^2}{m\nu_0} f_{ij} \tag{9-252}$$

while the Lindholm theory yields

$$\Gamma = 2\pi^2 N C_3 = \frac{\pi N e^2}{2m\nu_0} f_{ij}$$

A quantum mechanical calculation by Fursow and Wlassow (quoted in Ref. 8) yields

$$\Gamma = \frac{16\pi}{3} C_3 N = \frac{4}{3} \frac{Ne^2}{m\nu_0} f_{ij} \tag{9-253}$$

which differs only slightly from the above results.

Resonance broadening is usually important only for the first few members of a series because here the oscillator strengths $f_{ij}$ are largest and the Stark broadening the smallest. A discussion of the effects of resonance broadening of the Balmer lines in the solar spectrum has been given by Cayrel and Traving (*Z. Ap.*, **50**, 239, 1960). They show that the effects are important for $H\alpha$ but negligible for higher series members. They make use of an approximate formula for the absorption coefficient, obtained by adding the damping constants for electron and resonance broadening together. If more accurate results are required or complete profiles desired, the resonance damping should be used to define $a$ in a Voigt profile, which should then be folded with the complete Stark profile, as described previously.

VAN DER WAALS DAMPING

The usual situation of astrophysical interest is broadening resulting from collisions with hydrogen. Margenau (*Rev. Mod. Phys.*, **11**, 1, 1939) shows by a quantum mechanical calculation that in general the interaction of atom 1 in level $k$ with (an unlike) atom 2 in level $l$ is given by

$$V = \frac{2e^4}{3r^6} \sum_{k'} \sum_{l'} \frac{|r_{kk'}|_1{}^2 \, |r_{ll'}|_2{}^2}{(E_k - E_{k'})_1 + (E_l - E_{l'})_2} \tag{9-254}$$

Now for hydrogen, essentially all the atoms will be in the ground state; thus we may take $l = 0$. Also since all of the excited levels of hydrogen are well above the ground state ($n = 2$ lies 10.5 eV above $n = 1$), the term $(E_0 - E_{l'})_H$ will normally far outweigh the term $(E_k - E_{k'})$ for most other atoms, in which the energy levels are much more closely spaced. Thus we may neglect the term $(E_k - E_{k'})$ and split the sum up into two separate parts as follows:

$$V = -\frac{e^2}{r^6}\left[-\frac{2e^2}{3}\sum_{l'} \frac{|r_{0l'}|_H{}^2}{(E_0 - E_{l'})_H}\right] \sum_{k'} |r_{kk'}|^2 \tag{9-255}$$

The term in brackets can be identified with the polarizability of hydrogen, which numerically is $\alpha = 6.70 \times 10^{-25}$ cm$^3$. Further, in the second term, neglecting all matrix elements except those for the transition electron and using the Thomas-Kuhn sum rule, one may write

$$\sum_{k'} |r_{kk'}|^2 = \overline{r_k{}^2} = a_0{}^2 \overline{R_k{}^2} \tag{9-256}$$

We then have

$$V = -\frac{e\alpha a_0{}^2 \overline{R_k{}^2}}{r^6} \tag{9-257}$$

Both upper and lower levels will shift, and thus we write finally

$$\Delta\omega = \frac{e^2 \alpha a_0{}^2}{\hbar r^6}\,[\overline{R_U{}^2} - \overline{R_L{}^2}] = \frac{C_6}{r^6} \tag{9-258}$$

Numerically,

$$C_6 = 4.05 \times 10^{-33}[\overline{R_U{}^2} - \overline{R_L{}^2}] \tag{9-259}$$

In many cases, quantum mechanical results are not available for $\overline{R^2}$. As an approximation, one might assume that levels are hydrogenic and use the value

$$R^2_{n^*,l} = \frac{n^{*2}}{2Z^2}\,[5n^{*2} + 1 - 3l(l + 1)] \tag{9-260}$$

where $n^*$ is the effective principal quantum number of the level

$$n_i^* = Z\left(\frac{\chi_H}{\lambda_I - \lambda_i}\right)^{1/2} \tag{9-261}$$

Similar results hold for van der Waals broadening by helium but with $\alpha = 2.07 \times 10^{-25}$ cm$^3$.

To obtain a qualitative estimate of the effects of van der Waals broadening, we might consider the Na I transition $3^2S - 3^2P$ at $\lambda 5890$ and $\lambda 5896$, the D-lines of the solar spectrum. Applying the above hydrogenic estimates we find $\overline{R^2}(4s) = 19$ and $\overline{R^2}(4p) = 37$ so that $C_6 = 7.3 \times 10^{-32}$. Now $\Gamma_6 = 8.08 C^{2/5} v^{3/5} N$. To obtain an estimate, we assume $N_H = 10^{17}$ and $T = 5 \times 10^3$, which implies $v = 10.3 \times 10^5$, and thus $\Gamma_6 = 1.15 \times 10^9$, in fair agreement with direct estimates based on curve-of-growth studies (see Chapter 11). This may be compared with $\Gamma_R = 6.2 \times 10^7$. In addition, we may estimate $\Delta\lambda_g$, the validity boundary between statistical and impact broadening. From equation (9-181) we have

$$\Delta\lambda_g \approx \Delta\lambda_w = \frac{\lambda^2}{2\pi c}\left(\frac{8v}{3\pi}\right)^{6/5} C_6^{-1/5} \tag{9-262}$$

and for the conditions quoted above we find $\Delta\lambda_g \approx 40$ Å. Since the sodium *D*-lines are only a few angstroms wide, even including the wings, it is clear that they are dominated entirely by the impact broadening regime. The formulae used above to derive $\Gamma_6$ are obviously quite approximate and are subject to considerable error. Indeed, comparison with experiments (Kusch, *Z. Ap.*, **45**, 1, 1958) suggests that they may be too small by a factor of about 6. The quantum mechanical calculation of van der Waals broadening has been discussed by Griem (Ref. 14, p. 98), but unfortunately no detailed numerical results are yet available.

## 9-11. The Inglis-Teller Formula

As one approaches the series limit for any series of hydrogen lines, the lines become broader and also more closely spaced; eventually they merge together and appear to form a continuum. Since the Stark widths of the lines depend upon the electron density in the atmosphere, one might attempt to use this merging as a diagnostic tool to obtain estimates of $N_e$. The shift in energy of any particular Stark component can be written as

$$\Delta E = \tfrac{3}{2} a_0 e n (n_2 - n_1) F \tag{9-263}$$

where we have assumed that $n \gg n'$. The maximum shift occurs for $n_2 = n$,

$n_1 = 1$; thus

$$\Delta E_{\max} \approx \tfrac{3}{2} a_0 \, e n^2 \, F \tag{9-264}$$

On the other hand, the energy of the $n$th level is

$$E_n = \frac{-e^2}{2a_0 \, n^2} \tag{9-265}$$

and the spacing between levels is

$$\Delta E_n = E_{n+1} - E_n = \frac{e^2}{a_0 \, n^3} \tag{9-266}$$

Now if $\Delta E_{\max} = \Delta E_n/2$, the most strongly shifted components of levels $n$ and $n + 1$ will overlap, and the lines will merge. Combining equations (9-264) and (9-266), we find the field strength for merging is

$$F = \frac{e}{3{a_0}^2 n^5} \tag{9-267}$$

But in equation (9-146)

$$F = \beta F_0 = 2.60 e N_e^{2/3} \beta$$

so substitution into equation (9-267) yields

$$N_e^{2/3} = (7.8{a_0}^2 \beta n^5)^{-1} \tag{9-268}$$

We must now choose a value for $\beta$. We might use $\beta - 1.6$, which is the most probable field strength, or $\beta = 3.38$, which is the mean field strength. Then we find, for $\beta = 1.6$,

$$\log N_e = 23.19 - 7.5 \log n_{\max} \tag{9-269}$$

or, for $\beta = 3.38$,

$$\log N_e = 22.69 - 7.5 \log n_{\max} \tag{9-270}$$

Clearly the resulting value of $\log N_e$ is uncertain by at least $\pm 0.3$; thus we obtain only a rough estimate of the density. Further, realizing the fact that there exists a strong density gradient in an atmosphere and that the whole line transfer problem has been ignored, we see that there are surely even larger uncertainties involved. Nevertheless, even this rough formula serves vividly to demonstrate the wide range of conditions occurring in different kinds of stellar atmospheres. We list in Table 9-7 a few typical values of $N_e$

TABLE 9-7. Densities Estimated from the Inglis-Teller Formula

| *Star* | *Spectrum* | $n_{max}$ (*Balmer series*) | $\log N_e$ |
|---|---|---|---|
| Sirius | A2 V | 18 | 13.8 |
| $\alpha$ Cyg | A2 I | 29 | 12.2 |
| $\tau$ Sco | B0 V | 14 | 14.6 |
| White dwarf | DA | 8 | 16.4 |
| Pleione | Shell | 40 | 11.2 |

obtained from observations of the last visible Balmer line. Here we see that a range of 4 to 5 orders of magnitude is encountered! An alternative procedure for the estimation of electron densities from line merging has been suggested by Vidal (*J, Q. S. R. T.*, **6**, 575, 1966).

# 10 | The Redistribution Function

Let us now consider the question of redistribution, in angle and frequency, of photons scattered between bound atomic states. There are two primary mechanisms of astrophysical importance, and it is convenient to consider them in turn. First, if we consider a single atom in its own frame of reference, the form of redistribution that occurs will depend upon the detailed structure of the levels involved and upon the presence of perturbers. On the other hand, what is actually observed in a stellar atmosphere is an entire ensemble of atoms, moving at random with a thermal velocity distribution. When observed in the laboratory frame, the Doppler shift of a photon caused by the atoms' motion leads to redistribution in frequency. The final redistribution function is a combination of these two effects. A comprehensive discussion of this problem has been given recently by Hummer (*M. N.*, **125**, 21, 1962), whose results we will summarize here; the reader is referred to his paper for further details. An interesting analysis of the problem from a rather different conceptual point of view has been given by Henyey (*Ap. J.*, **103**, 332, 1946) and by Henyey and Grassberger (*Ap. J.*, **122**, 498, 1955).

## 10-1. Redistribution in the Atom's Frame

### GENERAL FORMULAE

Let us first consider the nature of redistribution in the rest-frame of the atom. We will write the redistribution function as the product of a *frequency redistribution function* $p(\xi', \xi)$, which gives the probability that a photon absorbed in frequency range $(\xi', \xi' + d\xi')$ is scattered into frequency range $(\xi, \xi + d\xi)$; and an angular *phase function* $g(\mathbf{n}', \mathbf{n})$, describing the probability of scattering from solid angle $d\omega'$ in direction $\mathbf{n}'$ into solid angle $d\omega$ in direction $\mathbf{n}$.

We assume that these functions are normalized such that

$$\int_{-\infty}^{\infty} p(\xi', \xi)\, d\xi' = 1 \tag{10-1}$$

and

$$\oint g(\mathbf{n}', \mathbf{n}) \frac{d\omega'}{4\pi} = 1 \tag{10-2}$$

Two useful forms of the angular phase function are those for isotropic scattering,

$$g_A(\mathbf{n}', \mathbf{n}) \equiv 1 \tag{10-3}$$

and for dipole scattering

$$g_B(\mathbf{n}', \mathbf{n}) = \tfrac{3}{4}(1 + \cos^2 \Theta) \tag{10-4}$$

where $\cos \Theta = \mathbf{n}' \cdot \mathbf{n}$. We write the absorption profile in the atom's rest-frame as $f(\xi')$, which is independent of direction, and normalized such that

$$\int_{-\infty}^{\infty} f(\xi')\, d\xi' = 1 \tag{10-5}$$

With these definitions, it is clear that the probability that a photon $(\xi', \mathbf{n}')$ will be absorbed is

$$f(\xi')\, d\xi' \frac{d\omega'}{4\pi}$$

while the probability that *if* a photon $(\xi', \mathbf{n}')$ is absorbed, *then* a photon $(\xi, \mathbf{n})$ will be emitted is

$$p(\xi', \xi)\, d\xi\, g(\mathbf{n}', \mathbf{n}) \frac{d\omega}{4\pi}$$

Thus the *joint probability* that a photon $(\xi', \mathbf{n}')$ is absorbed *and* a photon $(\xi, \mathbf{n})$ is emitted is

$$f(\xi')p(\xi', \xi)g(\mathbf{n}', \mathbf{n})\, d\xi'\, d\xi \frac{d\omega'}{4\pi} \frac{d\omega}{4\pi} \tag{10-6}$$

### RESULTS FOR SPECIFIC CASES

Let us now inquire further into the form of the function $p(\xi', \xi)$. Following Hummer, we will consider the following four categories: (a) Case I, zero natural line width; (b) Case II, radiation damping with coherence in the atom's frame; (c) Case III, complete redistribution in the atom's frame; and (d) Case IV, resonance scattering.

(a) Case I. In this case we assume that the idealized atom has two infinitely sharp states between which the transitions occur. Then we have simply

$$p(\xi', \xi) = \delta(\xi - \xi') \tag{10-7}$$

where $\delta$ is the Dirac function. Moreover, in this case

$$f(\xi')d\xi' = \delta(\xi' - \nu_0)d\xi' \tag{10-8}$$

where $\nu_0$ is the central frequency of the line. Clearly, in this case no redistribution occurs in the atom's frame. It is evident that this picture will not apply to any real atom since normally one or both of the levels will be broadened. Nevertheless, it is important to study the consequences of this limiting case in order to gain insight into the nature of the redistribution process.

(b) Case II. Here we envision an atom with an infinitely sharp lower (ground) state and an upper state whose finite lifetime leads to a broadening described by the usual Lorentz profile

$$f(\xi')\, d\xi' = \frac{\delta/\pi}{(\xi' - \nu_0) + \delta^2}\, d\xi' \tag{10-9}$$

If we assume that there are no additional perturbations of the atom while the electron is in the upper state, then there will be no reshuffling among substates of the upper state, and a return to the lower state will result in the emission of a photon of just the same frequency as was absorbed. Thus we again have

$$p(\xi', \xi) = \delta(\xi - \xi') \tag{10-10}$$

The physical situation in which this picture applies is for resonance lines in media of such low densities that collision broadening is negligible, for example, for the Lyman $\alpha$ line of hydrogen in the interstellar medium.

(c) Case III. The basic physical picture we have here is of an atom with a sharp lower state and with an upper state broadened by collision and radiation damping. If we assume that both the collisions and radiative decays result in Lorentz profiles with damping widths $\delta_C$ and $\delta_R$, respectively, we may then write the complete profile as

$$f(\xi')\, d\xi' = \frac{\delta/\pi}{(\xi' - \nu_0)^2 + \delta^2}\, d\xi' \tag{10-11}$$

where

$$\delta = \delta_R + \delta_C \tag{10-12}$$

The situation we are now considering differs markedly from Case II, despite the similarity of equations (10-9) and (10-11), because of the effects of collisions. While the radiative decays will, as before, be coherent, the collisions introduce a reshuffling of the electrons in the upper state before they return to the lower and thereby tend to destroy the coherence between absorbed and emitted photons. Indeed, if *only* collisions occurred, then the electrons would be completely randomly redistributed over the substates in the line, and the frequencies of the emitted photons would show no correlation with the frequency of those absorbed. The probability of emission at a particular frequency would be proportional to the number of substates available at that frequency and hence to the absorption profile itself; this situation is referred to as *complete redistribution.* When complete redistribution occurs we may then write

$$p(\xi', \xi)\, d\xi = f(\xi)\, d\xi = \frac{\delta/\pi}{(\xi - \nu_0)^2 + \delta^2}\, d\xi \tag{10-13}$$

which shows clearly that $p(\xi', \xi)$ is independent of $\xi'$. If complete redistribution occurs, then we see from equation (10-6) that in the atom's frame the joint probability of absorption at $\xi'$ and emission at $\xi$ is proportional to $f(\xi')f(\xi)$.

The first analysis of the situation in which both mechanisms occur was carried out by Zanstra (*M. N.*, **101**, 273, 1941, and *M. N.*, **106**, 225, 1946), who treated the radiating atom as a classical oscillator. He found that in this approximation a fraction $\delta_R/(\delta_R + \delta_C)$ of the photons was emitted coherently while a fraction $\delta_C/(\delta_R + \delta_C)$ was completely redistributed over the profile. Thus one would write

$$p(\xi', \xi)\, d\xi = \frac{\delta_R}{(\delta_R + \delta_C)}\delta(\xi' - \xi) + \frac{\delta_C}{(\delta_R + \delta_C)}\frac{\delta/\pi}{(\xi - \nu_0)^2 + \delta^2} \tag{10-14}$$

Now in the stellar atmospheres case, we usually have $\delta_C \gg \delta_R$; thus the approximation of complete redistribution [equation (10-13)] should be good. Moreover, Edmonds (*Ap. J.*, **121**, 418, 1955) has concluded that Zanstra's analysis actually leads to too large an estimate of the coherent component. He argues that the relevant time to consider is the period of fluctuation in the statistical perturbation field, instead of the time between collisions. In this case, the fraction of radiation that is coherent is reduced from Zanstra's estimate by a factor of 10 to 20. In addition, the radiation is not strictly coherent but is redistributed over a Lorentz profile of half-width $\delta_C$ centered on the incident frequency. Thus Edmonds' analysis indicates that complete redistribution for collision-broadened lines should be a good approximation indeed. Similarly, a calculation by Holstein (quoted by Hummer in Ref. 16, p. 20) leads to the conclusion that collisions effect complete redistribution in most cases of interest. Finally, there is a fair amount of experimental evidence that indicates that complete redistribution occurs when the line is collision broadened.

Therefore we will simply adopt complete redistribution, in the atom's frame, for collision-broadened lines, with the strong expectation that it is a valid picture. This is the case that is expected to be valid for most lines in the stellar atmospheres since pressure broadening is usually the dominant effect at atmospheric densities.

(d) Case IV. The term *resonance scattering* describes transitions between two radiation-broadened states and therefore applies to a large class of lines. This process results in the transition of an electron from the lower level $i$ to the upper level $j$ with a return to the lower level. The entire circuit must be treated as a single quantum mechanical process (see, e.g., Ref. 17, p. 198) and, as a result, the absorption and redistribution probabilities cannot be written separately but only in the combined form

$$f(\xi')p(\xi', \xi) = \frac{\delta_i \, \delta_j/\pi^2}{[(\xi - \xi')^2 + \delta_i^2][(\xi - \nu_0)^2 + \delta_j^2]} \tag{10-15}$$

Note that his product has two relative maxima, one when $\xi = \xi'$, and another when $\xi = \nu_0$. These may be understood in a rough way as follows: For atoms in the lower state, there will be a high probability of their being at the center of the frequency distribution of the substates. If a transition is made from this particular substate to an arbitrary substate of the upper level, the most probable return will be back to the center of the lower level. When this occurs, the emitted frequency $\xi$ is equal to the absorbed frequency $\xi'$. On the other hand, if the absorption occurs from some arbitrary substate of the lower level, it will most probably leave the electron at the center of the upper state, and again the most probable return will be to the center of the lower level. Hence, there is a conversion of photons from the absorbed frequency $\xi'$ to $\nu_0$, the central frequency of the line. We see therefore that in the atom's frame, scattering between two broadened levels may be expected to be partially coherent and partially noncoherent.

## 10-2. Redistribution by Doppler Shifts

### GENERAL FORMULAE

Let us now consider the effects upon photon redistribution of the Doppler shifts introduced by the motion of the absorbing atoms relative to the laboratory frame. As we shall see, these effects are quite important. Following Hummer's treatment, we will first derive very general formulae for the redistribution function and subsequently calculate explicit results for the four cases described above.

As discussed in Chapter 1, we describe redistribution by the function

$$R(\nu', \mathbf{n}'; \nu, \mathbf{n})\, d\nu'\, d\nu \frac{d\omega'}{4\pi} \frac{d\omega}{4\pi}$$

which gives the probabilities of scattering from $(\nu', \nu' + d\nu'; d\omega')$ to $(\nu, \nu + d\nu; d\omega)$. This function is normalized so that [equation (1-22)]

$$\oint\oint \int_0^\infty \int_0^\infty R(\nu', \mathbf{n}'; \nu, \mathbf{n})\, d\nu'\, d\nu \frac{d\omega'}{4\pi} \frac{d\omega}{4\pi} = 1$$

Now consider an atom moving with velocity $\mathbf{v}$. We wish to describe the absorption of a photon $(\nu', \mathbf{n}')$, as measured in the laboratory frame, with the subsequent emission of a photon $(\nu, \mathbf{n})$, also measured in the laboratory frame. Neglecting the aberration of directions in changing from the atom's frame to the laboratory frame, we have

$$\nu' = \xi' + \frac{\nu_0}{c} \mathbf{v} \cdot \mathbf{n}' \tag{10-16}$$

and

$$\nu = \xi + \frac{\nu_0}{c} \mathbf{v} \cdot \mathbf{n} \tag{10-17}$$

(see Figure 10-1). Now we showed above, equation (10-6), that the joint probability of absorption of a photon $(\xi', \mathbf{n}')$ with subsequent emission of a photon $(\xi, \mathbf{n})$ is

$$f(\xi')p(\xi', \xi)g(\mathbf{n}', \mathbf{n})\, d\xi'\, d\xi \frac{d\omega'}{4\pi} \frac{d\omega}{4\pi}$$

Transforming to the laboratory frame, we thus have

$$R_v(\nu', \mathbf{n}'; \nu, \mathbf{n}) = f\left(\nu' - \frac{\nu_0}{c} \mathbf{v} \cdot \mathbf{n}'\right) p\left(\nu' - \frac{\nu_0}{c} \mathbf{v} \cdot \mathbf{n}', \nu - \frac{\nu_0}{c} \mathbf{v} \cdot \mathbf{n}\right) g(\mathbf{n}', \mathbf{n}) \tag{10-18}$$

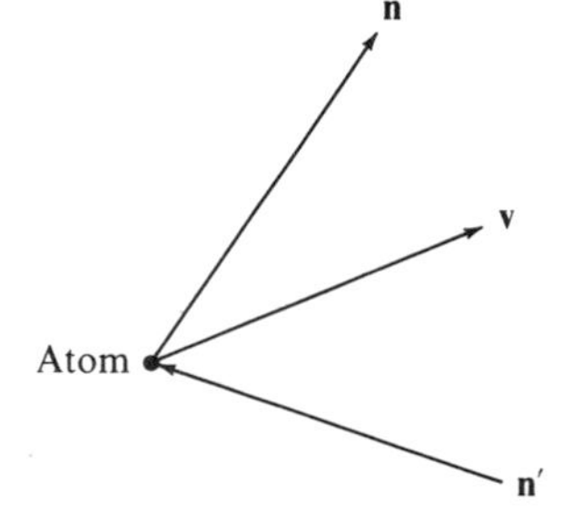


FIG. 10-1 Geometry of scattering process.

where the subscript $v$ implies that the redistribution is being effected by an atom of velocity $v$. To find the net effect of the complete ensemble of atoms, we must average over the velocity distribution, which we shall assume to be Maxwellian. To perform this average easily, we introduce an orthogonal triad of reference axes characterized by $(\mathbf{n}_1, \mathbf{n}_2, \mathbf{n}_3)$, where $\mathbf{n}_1$ and $\mathbf{n}_2$ are chosen to be coplanar with $\mathbf{n}'$ and $\mathbf{n}$ and with $\mathbf{n}_1$ bisecting the angle $\Theta$ between them (see Figure 10-2). Then we may write

$$\mathbf{n}' = \cos\frac{\Theta}{2}\,\mathbf{n}_1 + \sin\frac{\Theta}{2}\,\mathbf{n}_2 = \alpha\mathbf{n}_1 + \beta\mathbf{n}_2 \tag{10-19}$$

and

$$\mathbf{n} = \cos\frac{\Theta}{2}\,\mathbf{n}_1 - \sin\frac{\Theta}{2}\,\mathbf{n}_2 = \alpha\mathbf{n}_1 - \beta\mathbf{n}_2 \tag{10-20}$$

If we resolve $\mathbf{v}$ along these axes and write $v_i = \mathbf{v}\cdot\mathbf{n}_i$, the Maxwellian velocity distribution is simply

$$P(v_1, v_2, v_3)\,dv_1\,dv_2\,dv_3 = \left(\frac{m}{2\pi kT}\right)^{3/2} \exp[-m(v_1^2 + v_2^2 + v_3^2)/2kT]\,dv_1\,dv_2\,dv_3 \tag{10-21}$$

For convenience, we express velocities in dimensionless units:

$$\mathbf{u} = \left(\frac{m}{2kT}\right)^{1/2}\mathbf{v} \tag{10-22}$$

and introduce the Doppler width

$$w = \frac{\nu_0}{c}\left(\frac{2kT}{m}\right)^{1/2} \tag{10-23}$$

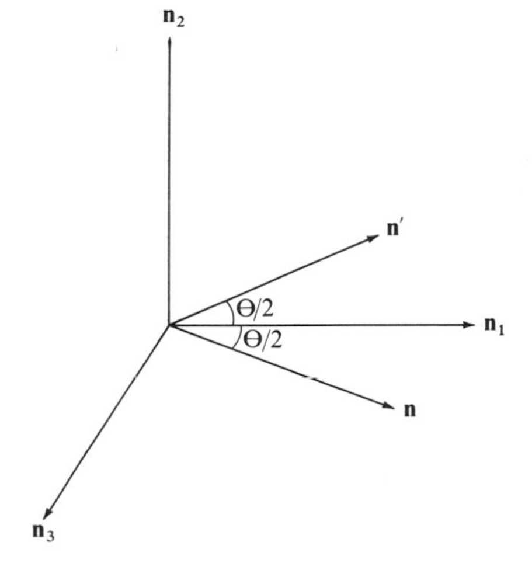


FIG. 10-2. Coordinate axes used in calculation of redistribution function. The vectors $\mathbf{n}_1$, $\mathbf{n}_2$, $\mathbf{n}$, and $\mathbf{n}'$ are coplanar. The vector $\mathbf{n}_1$ bisects the angle $\Theta$ $(0 \le \Theta \le 180°)$ between $\mathbf{n}$ and $\mathbf{n}'$.

We now average equation (10-18) over the velocity distribution of equation (10-21) to obtain, quite generally,

$$R(\nu', \mathbf{n}'; \nu, \mathbf{n}) = \frac{g(\mathbf{n}', \mathbf{n})}{\pi} \iint_{-\infty}^{\infty} \exp(-u_1^2 - u_2^2) f[\nu' - w(\alpha u_1 + \beta u_2)]$$

$$p[\nu' - w(\alpha u_1 + \beta u_2), \nu - w(\alpha u_1 - \beta u_2)]\, du_1\, du_2 \tag{10-24}$$

where we have carried out the integration over $u_3$ explicitly. An alternative form for $R$, which will prove useful, can be derived by choosing $\mathbf{n}_1$ to lie along $\mathbf{n}'$. Then

$$\mathbf{v} \cdot \mathbf{n}' = v_1 \tag{10-25}$$

and

$$\mathbf{v} \cdot \mathbf{n} = v_1 \cos\Theta + v_2 \sin\Theta = \alpha' v_1 + \beta' v_2 \tag{10-26}$$

We then find

$$R(\nu', \mathbf{n}'; \nu, \mathbf{n}) = \frac{g(\mathbf{n}', \mathbf{n})}{\pi} \int_{-\infty}^{\infty} du_1 e^{-u_1^2} f(\nu' - w u_1)$$

$$\times \int_{-\infty}^{\infty} du_2\, e^{-u_2^2} p[\nu' - w u_1, \nu - w(\alpha' u_1 + \beta' u_2)] \tag{10-27}$$

With the aid of these general formulae, we can now compute the redistribution functions for the four cases considered above. Note first, however, that in both Cases I and II the scattering in the atom's frame is coherent so that

$$p(\xi', \xi) = \delta(\xi - \xi') = \delta(\nu - \nu' + 2w\beta u_2) \tag{10-28}$$

In this case we see that $u_1$ no longer appears explicitly in the expression for $p$, a fact that simplifies the integrations indicated above. Inserting equation (10-28) into equation (10-24), we can now explicitly carry out the integration over $u_2$. In performing this integration, we transform to the variable $z = 2w\beta u_2$ in order to preserve normalization of the Dirac delta function. Then, for $\beta \neq 0$ we obtain

$$R(\nu', \mathbf{n}'; \nu, \mathbf{n})$$

$$= \frac{g(\mathbf{n}', \mathbf{n})}{\pi} \int_{-\infty}^{\infty} du_1 e^{-u_1^2} \int_{-\infty}^{\infty} \frac{dz}{2\beta w}$$

$$\times \exp(-z^2/4\beta^2 w^2) f\left(\nu' - w\alpha u_1 - \frac{3}{2}\right) \delta(\nu - \nu' + z)$$

$$= \frac{g(\mathbf{n}', \mathbf{n})}{2\pi\beta w} \exp\left[\frac{-(\nu - \nu')^2}{4\beta^2 w^2}\right] \int_{-\infty}^{\infty} e^{-u^2} f\left(\frac{\nu + \nu'}{2} - w\alpha u\right) du \tag{10-29}$$

If $\beta = 0$, then we have simply

$$R(\nu', \mathbf{n}'; \nu, \mathbf{n}) = \frac{g(\mathbf{n}', \mathbf{n})}{\pi} \int_{-\infty}^{\infty} du_1 e^{-u_1^2} f(\nu' - wu_1) \int_{-\infty}^{\infty} du_2\, e^{-u_2^2}\, \delta(\nu' - \nu)$$

$$= \frac{g(\mathbf{n}', \mathbf{n})}{\pi^{1/2}} \delta(\nu' - \nu) \int_{-\infty}^{\infty} e^{-u^2} f(\nu' - wu)\, du \tag{10-30}$$

RESULTS FOR SPECIFIC CASES

(a) Case I. Let us now consider the case of zero line width in the atom's frame so that [equation (10-8)]

$$f(\xi')d\xi' = \delta(\xi' - \nu_0)d\xi'$$

Then from equation (10-29) we have

$$R_{\mathrm{I}}(\nu', \mathbf{n}'; \nu, \mathbf{n})$$

$$= \frac{g(\mathbf{n}', \mathbf{n})}{2\pi\beta w} \exp\left[\frac{-(\nu - \nu')^2}{4\beta^2 w^2}\right] \int_{-\infty}^{\infty} e^{-u^2}\, \delta\left(\frac{\nu' + \nu - 2\nu_0}{2} - w\alpha u\right) du$$

$$= \frac{g(\mathbf{n}', \mathbf{n})}{2\pi\alpha\beta w^2} \exp\left[\frac{-(\nu - \nu')^2}{4\beta^2 w^2}\right] \exp\left[\frac{-(\nu + \nu' - 2\nu_0)^2}{4\alpha^2 w^2}\right] \tag{10-31}$$

Equation (10-31) may be written in a much more concise form by expressing frequency shifts in Doppler units and by defining

$$x = \frac{\nu - \nu_0}{w} \tag{10-32}$$

$$x' = \frac{\nu' - \nu_0}{w} \tag{10-33}$$

so that

$$R_{\mathrm{I}}(x', \mathbf{n}'; x, \mathbf{n}) = R_{\mathrm{I}}(\nu', \mathbf{n}'; \nu, \mathbf{n}) \frac{d\nu'}{dx'} \frac{d\nu}{dx} = w^2 R_{\mathrm{I}}(\nu', \mathbf{n}'; \nu, \mathbf{n}) \tag{10-34}$$

Also, we note that

$$2\alpha\beta = 2 \sin\frac{\Theta}{2} \cos\frac{\Theta}{2} = \sin\Theta \tag{10-35}$$

and

$$\alpha^2 + \beta^2 = 1 \tag{10-36}$$

We then find from equation (10-31) that

$$R_{\text{I}}(x', \mathbf{n}'; x, \mathbf{n}) = \frac{g(\mathbf{n}', \mathbf{n})}{\pi \sin \Theta} \exp[-x^2 - (x' - x \cos \Theta)^2 \csc^2 \Theta] \tag{10-37}$$

Thus, even though the scattering in the atom's frame was strictly coherent, in the laboratory frame there is redistribution. This is an extremely important point as Thomas (*Ap. J.*, **125**, 260, 1957) has emphasized. For example, suppose the incident radiation is isotropic and independent of frequency, $I(\nu', \mathbf{n}') \equiv I_0$. Then the radiation scattered is

$$j_x = \oint \frac{d\omega'}{4\pi} \int_{-\infty}^{\infty} I_{x'}(\mathbf{n}') R(x', \mathbf{n}'; x, \mathbf{n})\, dx'$$

$$= \frac{I_0 e^{-x^2}}{\pi} \oint \frac{g(\mathbf{n}', \mathbf{n})\, d\omega'}{4\pi} \int_{-\infty}^{\infty} \exp\left[\frac{-(x' - x \cos \Theta)^2}{\sin^2 \Theta}\right] \frac{dx'}{\sin \Theta} \tag{10-38}$$

Let $z = (x' - x \cos \Theta)/\sin \Theta$ so that $dz = dx'/\sin \Theta$. Then recalling that $g(\mathbf{n}', \mathbf{n})$ is normalized, we find

$$j_x = \frac{I_0 e^{-x^2}}{\pi} \int_{-\infty}^{\infty} e^{-z^2}\, dz = \frac{I_0 e^{-x^2}}{\pi^{1/2}} = I_0 \varphi_x \tag{10-39}$$

where $\varphi_x$ is the absorption profile at frequency $x$. This result shows that the scattered radiation is *completely redistributed* over the profile so that the emission and absorption profiles are identical. For emphasis we stress again that we find complete redistribution in the laboratory frame despite the assumption of strict coherence in the atom's frame. Thomas examines several other representative forms for the incident intensity distribution and finds that in the Doppler core ($x \lesssim 3$), over which the absorption profile varies by a factor of $10^4$, the emitted radiation departs from the absorption profile by only a factor of 4 or less! Thus within the Doppler core, complete redistribution appears to be an excellent approximation. In the wings of the line, one must account correctly for partial redistribution. For redistribution by pure Doppler broadening of two perfectly sharp levels, a simple argument by Spitzer (*Ap. J.*, **99**, 1, 1944) shows that in the far wings, the scattered radiation is about two-thirds noncoherent and one-third coherent. For very precise work, one certainly should use the correct redistribution function, but in practice, complete redistribution is a very useful first approximation. We shall return to this point again in Section 10-3.

(b) Case II. Here we have coherence in the atom's frame, and the absorption profile is [equation (10-9)]

$$f(\xi')\, d\xi' = \frac{\delta/\pi}{(\xi' - \nu_0)^2 + \delta^2}\, d\xi'$$

Substitution into equation (10-24) yields

$$R_{\text{II}}(\nu', \mathbf{n}'; \nu, \mathbf{n}) = \frac{g(\mathbf{n}', \mathbf{n})}{2\pi\alpha\beta w^2} \exp\left[\frac{-(\nu - \nu')^2}{4\beta^2 w^2}\right] \frac{\delta}{\pi\alpha w}$$

$$\times \int_{-\infty}^{\infty} e^{-u^2} \left[\left(\frac{\nu' + \nu - 2\nu_0}{2\alpha w} - u\right)^2 + \left(\frac{\delta}{\alpha w}\right)^2\right]^{-1} du \tag{10-40}$$

Transforming to dimensionless frequencies and recalling the definition of the Voigt function, equation (9-44),

$$H(a, v) \equiv \frac{a}{\pi} \int_{-\infty}^{\infty} \frac{e^{-u^2}\, du}{(v - u)^2 + a^2}$$

where $a \equiv \delta/w$, we have, finally,

$$R_{\text{II}}(x', \mathbf{n}'; x, \mathbf{n}) = \frac{g(\mathbf{n}', \mathbf{n})}{\pi \sin \Theta} \exp\left[\frac{-(x - x')^2 \csc^2 \dfrac{\Theta}{2}}{4}\right] a \sec \frac{\Theta}{2}$$

$$\times \int_{-\infty}^{\infty} e^{-u^2} \left[\left(\frac{x' + x}{2 \cos \dfrac{\Theta}{2}} - u\right)^2 + \left(\frac{a}{\cos \dfrac{\Theta}{2}}\right)^2\right]^{-1} du$$

$$= \frac{g(\mathbf{n}', \mathbf{n})}{\pi \sin \Theta} \exp\left[-\left(\frac{x - x'}{2}\right)^2 \csc^2 \frac{\Theta}{2}\right]$$

$$\times H\left(a \sec \frac{\Theta}{2}, \frac{x + x'}{2} \sec \frac{\Theta}{2}\right) \tag{10-41}$$

Thus the redistribution function can be expressed in terms of known functions. It is obvious that the above result is relatively complicated; in fact, this form of the redistribution function, accounting for both angle and frequency changes, requires more powerful methods of solution of the transfer equation than those we have described in this book. Recently, Monte Carlo methods have been applied successfully to this problem by Auer (*Ap. J.*, **153**, 783, 1968) and by Avery and House (*Ap. J.*, **152**, 493, 1968).

(c) Case III. For complete redistribution in the atom's frame, $p(\xi', \xi)$ is independent of $\xi'$. In this case equation (10-27) is a useful form, and we have

$$R_{\text{III}}(\nu', \mathbf{n}'; \nu, \mathbf{n}) = \frac{g(\mathbf{n}', \mathbf{n})}{\pi} \int_{-\infty}^{\infty} \frac{(\delta/\pi) e^{-u_1^2}\, du_1}{(\nu' - \nu_0 - w u_1)^2 + \delta^2}$$

$$\times \int_{-\infty}^{\infty} \frac{(\delta/\pi) e^{-u_2^2}\, du_2}{[\nu - w(u_1 \cos \Theta + u_2 \sin \Theta) - \nu_0]^2 + \delta^2}. \tag{10-42}$$

Converting to dimensionless units, we obtain

$$R_{\text{III}}(x', \mathbf{n}'; x, \mathbf{n}) = \frac{g(\mathbf{n}', \mathbf{n})}{\pi} \int_{-\infty}^{\infty} \frac{(a/\pi)e^{-u_1^2}\, du_1}{(x' - u_1)^2 + a^2}$$

$$\times \int_{-\infty}^{\infty} \frac{(a \csc^2 \Theta/\pi)e^{-u_2^2}\, du_2}{(x \csc \Theta - u_1 \operatorname{ctn} \Theta - u_2)^2 + a^2 \csc^2 \Theta}$$

$$= \frac{g(\mathbf{n}', \mathbf{n})}{\pi} \left(\frac{a \csc \Theta}{\pi}\right)$$

$$\times \int_{-\infty}^{\infty} \frac{e^{-u^2} H(a \csc \Theta, x \csc \Theta - u \operatorname{ctn} \Theta)\, du}{(x' - u)^2 + a^2} \tag{10-43}$$

Here the redistribution function cannot be expressed directly in simple functional form and must in general be obtained from numerical integrations.

(d) Case IV. Making use of equations (10-15) and (10-24), we have

$$R_{\text{IV}}(\nu', \mathbf{n}'; \nu, \mathbf{n})$$

$$= \frac{g(\mathbf{n}', \mathbf{n})}{\pi} \iint_{-\infty}^{\infty} e^{-(u_1^2 + u_2^2)} \frac{\delta_i\, \delta_j}{\pi^2} [(\nu' - \nu - 2\beta w u_2)^2 + \delta_i^2]^{-1}$$

$$\times [(\nu - \nu_0 + w\beta u_2 - w\alpha u_1)^2 + \delta_j^2]^{-1}\, du_1\, du_2 \tag{10-44}$$

and using equations (10-32) through (10-35), we obtain

$$R_{\text{IV}}(x', \mathbf{n}'; x, \mathbf{n}) = \frac{2g(\mathbf{n}', \mathbf{n})}{\pi \sin \Theta} \left(\frac{a_i \csc \dfrac{\Theta}{2}}{\pi}\right)$$

$$\times \int_{-\infty}^{\infty} \frac{e^{-u^2} H\left(a_j \sec \dfrac{\Theta}{2}, x \sec \dfrac{\Theta}{2} + u \tan \dfrac{\Theta}{2}\right)}{\left[(x' - x)\csc \dfrac{\Theta}{2} - 2u\right]^2 + \left(a_i \csc \dfrac{\Theta}{2}\right)^2}\, du \tag{10-45}$$

## 10-3. Angle-Averaged Redistribution Functions

### GENERAL FORMULAE

The equations derived in Section 10-2 describe both the angle and frequency redistribution of scattered radiation. In general, the amount of radiation scattered at frequency $\nu$ into solid angle $d\omega$ in direction $\mathbf{n}$ is an integral over angle and frequency of the specific intensity times the redistribution function [see, e.g., equation (1-23)]. To treat the radiation field in this much detail is usually too difficult, and some simplifications are desirable. In the theory of line formation the most important aspect of redistribution are changes in the photon's frequency. The angular distribution is less critical because radiation is efficiently trapped in the line and is nearly isotropic through most of the region of line formation. It is therefore very useful to introduce a frequency redistribution function averaged over angle [equation (1-24)]

$$R(\nu', \nu) = \oint\oint R(\nu', \mathbf{n}'; \nu, \mathbf{n}) \frac{d\omega'}{4\pi} \frac{d\omega}{4\pi}$$

Clearly this function is normalized such that

$$\iint_{-\infty}^{\infty} R(\nu', \nu)\, d\nu'\, d\nu = 1 \tag{10-46}$$

We may also write

$$R(x', x) = R(\nu', \nu) \frac{d\nu'}{dx'} \frac{d\nu}{dx} = w^2 R(\nu', \nu) \tag{10-47}$$

If we integrate over all emitted photons, we obtain the absorption profile

$$\varphi(\nu')\, d\nu' = d\nu' \int_{-\infty}^{\infty} R(\nu', \nu)\, d\nu$$

Similarly, if we integrate overall absorbed photons, we must obtain the emission profile

$$\psi(\nu)\, d\nu = d\nu \int_{-\infty}^{\infty} R(\nu', \nu)\, d\nu'$$

If the scattering process were assumed to be strictly coherent in the laboratory frame (a completely arbitrary assumption), then we would have

$$R(\nu', \nu) = \varphi(\nu')\delta(\nu' - \nu) \tag{10-48}$$

On the other hand, if we assume that complete redistribution occurs in the laboratory frame, then the absorption and emission probabilities are completely uncorrelated, and we would obtain (also recall our discussion in Section 1-2)

$$R(\nu', \nu) = \varphi(\nu')\varphi(\nu) \tag{10-49}$$

From the above formulae we can immediately verify that, as emphasized before, equation (10-49) implies that $\varphi_\nu \equiv \psi_\nu$. The assumption of complete redistribution is a great simplification and is sometimes physically quite realistic, for example, in the line core with pure Doppler broadening, or when the line is strongly pressure broadened. When a more detailed description is necessary, then the actual redistribution functions listed below should be employed.

The integration indicated in equation (1-24) could, in principle, be carried out directly by using the redistribution functions derived in the previous section. Such an approach, however, turns out to be rather complicated. It is simpler to follow the calculation of Hummer to obtain a general form for the angle-integrated redistribution function and then to derive specific forms for each of the cases of interests. We begin by recalling equation (10-18),

$$R_v(\nu', \mathbf{n}'; \nu, \mathbf{n}) = f(\nu' - w\mathbf{u} \cdot \mathbf{n}')p(\nu' - w\mathbf{u} \cdot \mathbf{n}', \nu - w\mathbf{u} \cdot \mathbf{n})g(\mathbf{n}', \mathbf{n})$$

where we have converted to Doppler units by using equations (10-32) through (10-35). We wish to fix $\nu'$ and $\nu$ and to integrate over all angles. Choose a triad $(\mathbf{n}_1, \mathbf{n}_2, \mathbf{n}_3)$ such that $\mathbf{u} = u\mathbf{n}_3$. Then we may write $\mathbf{u} \cdot \mathbf{n} = \mu u$, and $\mathbf{u} \cdot \mathbf{n}' = \mu' u$, where $\mu = \mathbf{n} \cdot \mathbf{n}_3$ and $\mu' = \mathbf{n}' \cdot \mathbf{n}_3$. An element of solid angle may then be written $d\omega = d\mu \, d\varphi$, where $\varphi$ is the azimuthal angle around $\mathbf{n}_3$. The phase function $g(\mathbf{n}', \mathbf{n})$ can be written quite generally as $g(\mu', \mu, \varphi)$. Then

$$R_u(\nu', \nu) = \frac{1}{16\pi^2} \int_0^{2\pi} d\varphi \int_{-1}^{1} d\mu' f(\nu' - w\mu' u)$$
$$\times \int_{-1}^{1} d\mu \, p(\nu' - w\mu' u, \nu - w\mu u) \int_0^{2\pi} d\varphi' g(\mu', \mu, \varphi') \qquad (10\text{-}50)$$

Now define

$$g(\mu', \mu) \equiv \frac{1}{4\pi} \int_0^{2} g(\mu', \mu, \varphi') \, d\varphi' \qquad (10\text{-}51)$$

Then

$$R_u(\nu', \nu) = \frac{1}{2} \int_{-1}^{1} d\mu' f(\nu' - w\mu' u) \int_{-1}^{1} d\mu \, g(\mu', \mu) p(\nu' - w\mu' u, \nu - w\mu u)$$
$$(10\text{-}52)$$

In the case of isotropic scattering, equations (10-3) and (10-51) imply

$$g_A(\mu', \mu) = \tfrac{1}{2} \qquad (10\text{-}53)$$

For dipole scattering one finds, using equations (10-4) and (10-51),

$$g_B(\mu', \mu) = \tfrac{3}{16}(3 - \mu^2 - \mu'^2 + 3\mu^2\mu'^2) \qquad (10\text{-}54)$$

For simplicity in what follows we will consider *isotropic scattering only*; formulae for dipole scattering are given by Hummer. In this more restrictive case we have

$$R_{A,u}(\nu', \nu) = \frac{1}{4}\int_{-1}^{1}\int_{-1}^{1} f(\nu' - w\mu' u)p(\nu' - w\mu' u, \nu - w\mu u)\, d\mu'\, d\mu \tag{10-55}$$

Now in Cases I and II the scattering in the atom's frame is coherent so that

$$p(\nu' - w\mu' u, \nu - w\mu u) = \delta[\nu' - \nu - wu(\mu' - \mu)] \tag{10-56}$$

Since the range of integration of $\mu'$ or $\mu$ is only $(-1, 1)$ it is clear that for certain values of $(\nu' - \nu)$, the singularity of the delta function will lie outside the range of integration, and $R_{A,u}(\nu', \nu)$ will, accordingly, be zero. Consider first the integration over $\mu$. Let $y = wu\mu$, and write

$$I = \frac{1}{wu}\int_{-wu}^{wu} \delta[y - (\nu - \nu' + wu\mu')]\, dy \tag{10-57}$$

The integral I will equal $1/wu$ if $-wu \leq \nu - \nu' + wu\mu' \leq wu$, and will be zero otherwise. Define $\Lambda(x)$ such that $\Lambda = 1$ if $-1 \leq x \leq 1$, and zero otherwise. Then equation (10-55) may be rewritten as

$$R_{A,u}(\nu', \nu) = \frac{1}{4wu}\int_{-1}^{1} f(\nu' - wu\mu')\Lambda\left[\mu' + \frac{(\nu - \nu')}{wu}\right] d\mu' \tag{10-58}$$

If $u$ is sufficiently small, then $|(\nu - \nu')/wu| > 1$ and $\Lambda$ will vanish for all values of $\mu'$. Physically, this states that there is a certain maximum frequency shift that can be caused by atoms of a given velocity, and the probability of redistribution beyond this maximum shift is zero. Thus there is a minimum value of $u$ for which scattering can occur from $\nu'$ to $\nu$. In what follows we will write

$$\bar{\nu} = \max(\nu, \nu') \tag{10-59}$$

and

$$\underline{\nu} = \min(\nu, \nu') \tag{10-60}$$

Now if $\nu > \nu'$, the requirement that the argument of the $\Lambda$-function fall in the range $(-1, 1)$ demands that

$$-1 + \frac{\nu - \nu'}{wu} = -1 + \frac{\bar{\nu} - \nu}{wu} \leq 1 \tag{10-61}$$

This inequality implies that

$$u_{\min} = \frac{\bar{\nu} - \underline{\nu}}{2w} = \frac{|\nu' - \nu|}{2w} \tag{10-62}$$

The same result is obtained if we assume $\nu' > \nu$; therefore equation (10-62) is general. For $u \leq u_{\min}$, $R_u$ will be zero. For $u \geq u_{\min}$, a contribution to $R_u$ will come from part of the range of integration of $\mu'$. To evaluate this range, suppose for definiteness that $\nu > \nu'$. Then we will obtain a contribution on the range

$$-1 \leq \mu' \leq 1 - \frac{\nu - \nu'}{wu} = 1 - \frac{\bar{\nu} - \underline{\nu}}{wu} \tag{10-63}$$

which implies

$$\bar{\nu} - wu \leq \underline{\nu} - wu\mu' \leq \underline{\nu} + wu \tag{10-64}$$

Or, since $\underline{\nu} = \nu'$ in this case, we can write

$$\bar{\nu} - wu \leq \nu' - wu\mu' \leq \underline{\nu} + wu \tag{10-65}$$

The choice $\nu' > \nu$ also leads to the same result; thus again equation (10-65) is general. Now introduce the *Heaviside function* $\Phi$, defined such that when $x > x_0$,

$$\Phi(x, x_0) = 1 \tag{10-66}$$

and when $x < x_0$,

$$\Phi(x, x_0) = 0 \tag{10-67}$$

Making the substitution $y = \nu' - wu\mu'$ and using equations (10-62), (10-65), (10-66), and (10-67), equation (10-58) may finally be written

$$R_{A,u}(\nu', \nu) = \frac{1}{4w^2u^2} \Phi\left(u - \frac{|\nu - \nu'|}{2w}, 0\right) \int_{\bar{\nu} - wu}^{\underline{\nu} + wu} f(y)\, dy \tag{10-68}$$

Finally, we average over a Maxwellian velocity distribution

$$P(u)\, du = \frac{1}{\pi^{3/2}} e^{-u^2} 4\pi u^2\, du \tag{10-69}$$

to obtain, for coherence in the atom's frame only,

$$R_A(\nu', \nu) = \frac{1}{w^2\pi^{1/2}} \int_{u_{\min}}^{\infty} du\, e^{-u^2} \int_{\bar{\nu} - wu}^{\underline{\nu} + wu} f(y)\, dy \tag{10-70}$$

### RESULTS FOR SPECIFIC CASES

(a) Case I. Here $f(y) = \delta(y - \nu_0)$ so that the integral over $y$ in equation (10-70) is nonzero only if $\underline{\nu} + wu \geq \nu_0 \geq \bar{\nu} - wu$. This implies that $u_{\min}$ now becomes effectively

$$u'_{\min} = \max(|x'|, |x|) \tag{10-71}$$

Clearly $u'_{\min} \geq u_{\min}$, as previously defined in equation (10-62); then we have simply

$$R_{\text{I},A}(x', x) = \frac{1}{\pi^{1/2}} \int_{u'_{\text{min}}}^{\infty} e^{-u^2}\, du = \frac{1}{2}\, \text{erfc}(u'_{\text{min}}) \tag{10-72}$$

where the complimentary error function is defined as

$$\text{erfc}(x) \equiv \frac{2}{\pi^{1/2}} \int_{x}^{\infty} e^{-z^2}\, dz \tag{10-73}$$

Thus, combining equations (10-71) and (10-72), we have

$$R_{\text{I},A}(x', x) = \tfrac{1}{2}\text{erfc}[\max(|x|, |x'|)] \tag{10-74}$$

The corresponding result for dipole scattering is given by Hummer. A plot of $R_{\text{I},A}(x', x)$ is shown in Figure 10-3. For the far wing of the line Hummer has

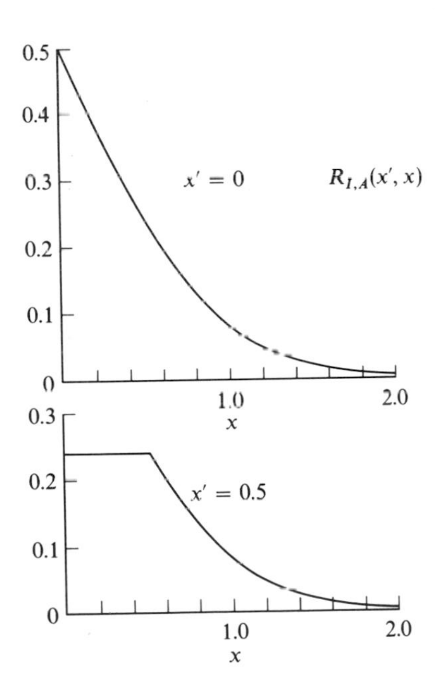


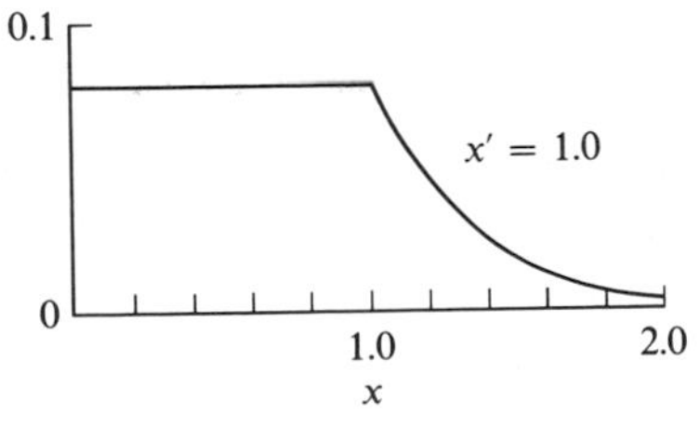


FIG. 10-3. Angle-averaged redistribution function for Doppler broadening with coherence in atom's frame. The angular redistribution is assumed to be isotropic.

shown that the asymptotic form of $R_{I,A}$ is

$$R_{\mathrm{I},A}(x', x) \propto \frac{e^{-\bar{x}^2}}{x} \tag{10-75}$$

where

$$\bar{x} \equiv \max(|x|, |x'|) \tag{10-76}$$

Equation (10-74) is quite obviously a great deal simpler than equation (10-37) and can be incorporated into transfer equations that can be solved with the methods described in this book. In particular, Avrett and Hummer (*M. N.*, **130**, 295, 1965) have used equation (10-74) to check the self-consistency of the assumption of complete redistribution for Case I in the solution of typical transfer problems. They find that while the complete redistribution approximation is by no means perfect, it is accurate to a few percentage points in the prediction of emergent intensities and in the evaluation of the line source function, particularly in the line core. These results indicate again that in the pure Doppler-broadening case, the assumption of complete redistribution is a useful starting approximation. In more precise work, one should, of course, employ the correct redistribution function.

(b) Case II. If we insert equation (10-9) into equation (10-70), we have

$$R_{\mathrm{II},A}(\nu', \nu) = \frac{1}{w^2\pi^{3/2}} \int_{u_{\min}}^{\infty} du\, e^{-u^2} \int_{\bar{\nu}-wu}^{\underline{\nu}+wu} \frac{\delta\, dy}{(y-\nu_0)^2+\delta^2} \tag{10-77}$$

Converting to Doppler units, writing $z = (y - \nu_0)/\delta = x/a$, and recalling from equation (10-62) that $u_{\min} = |x - x'|/2$, we have

$$\begin{aligned} R_{\mathrm{II},A}(x', x) &= \frac{1}{\pi^{3/2}} \int_{|x-x'|/2}^{\infty} du\, e^{-u^2} \int_{(\bar{x}-u)/a}^{(\underline{x}+u)/a} \frac{dz}{1+z^2} \\ &= \frac{1}{\pi^{3/2}} \int_{|x-x'|/2}^{\infty} e^{-u^2} \left[\tan^{-1}\left(\frac{\underline{x}+u}{a}\right) - \tan^{-1}\left(\frac{\bar{x}-u}{a}\right)\right] du \end{aligned} \tag{10-78}$$

The redistribution function for dipole scattering in this case is given by Hummer, who also shows that the asymptotic form in the wing is

$$R_{\mathrm{II},A}(x', x) \sim \frac{\exp[-(\bar{x}-\underline{x})^2/4]}{(\bar{x}-\underline{x})} \tag{10-79}$$

This redistribution function has been studied extensively by Jefferies and White (*Ap. J.*, **132**, 767, 1960) who give graphs of $R_{\mathrm{II},A}(x', x)/\varphi(x)$ for $a = 10^{-3}$. Their work shows clearly that in the core of the line, the scattering is strongly noncoherent so that complete redistribution is a good approximation while in the wings, the scattering is nearly coherent. Indeed, they show that to a good approximation one may write

$$R_{\text{II},A}(x', x) \approx a(x)\varphi(x)\,\delta(x - x') + [1 - a(x)]\varphi(x)\varphi(x') \tag{10-80}$$

where $a(x)$ is nearly zero for $x \lesssim 3$ and approximately unity for $x \gtrsim 3$. The fact that the scattering is nearly coherent in the wings can be understood by a simple physical argument. In the far wings (several Doppler widths from line center), the opacity is due primarily to the absorption by atoms nearly at rest because there are very few atoms with velocities high enough to have their central absorption peak shifted to these frequencies. Since the scattering is coherent in the atom's frame and the appropriate atoms are nearly at rest in the laboratory frame, the scattering in the laboratory frame is very nearly coherent as well.

(c) Case III. Since we do not now have coherence in the rest-frame of the atom, we may no longer employ equation (10-70). Rather, by substituting the profile given by equation (10-11) and the atomic redistribution function given in equation (10-13) into equation (10-55), we have directly

$$\begin{aligned} R_{A,u}(\nu', \nu) &= \frac{1}{4}\int_{-1}^{1} \frac{(\delta/\pi)\,d\mu'}{(\nu' - w\mu' u - \nu_0)^2 + \delta^2}\int_{-1}^{1} \frac{(\delta/\pi)\,d\mu}{(\nu - w\mu u - \nu_0)^2 + \delta^2} \\ &= \frac{1}{4\pi^2 w^2 u^2}\left[\tan^{-1}\left(\frac{x' + u}{a}\right) - \tan^{-1}\left(\frac{x' - u}{a}\right)\right] \\ &\quad \times \left[\tan^{-1}\left(\frac{x + u}{a}\right) - \tan^{-1}\left(\frac{x - u}{a}\right)\right] \end{aligned} \tag{10-81}$$

Now averaging over a Maxwellian velocity distribution and converting to Doppler units, we find

$$\begin{aligned} R_{\text{III},A}(x', x) &= \frac{1}{\pi^{5/2}}\int_0^\infty e^{-u^2}\left[\tan^{-1}\left(\frac{x' + u}{a}\right) - \tan^{-1}\left(\frac{x' - u}{a}\right)\right] \\ &\quad \times \left[\tan^{-1}\left(\frac{x + u}{a}\right) - \tan^{-1}\left(\frac{x - u}{a}\right)\right] du \end{aligned} \tag{10-82}$$

Hummer has discussed methods for calculating numerical values of $R_{\text{III},A}$ and has shown that the asymptotic form in the wing is

$$R_{\text{III},A}(x', x) \sim \frac{a^2}{x^2 x'^2} \tag{10-83}$$

This redistribution function falls off in the wings much less rapidly than in Cases I and II and changes the nature of the solution of the transfer equation.

In the past, it has been argued on intuitive grounds that since the redistribution is complete in the atom's frame and the Doppler motions are random, the redistribution must be complete in the observer's frame as well, and no correlations should exist between the absorbed and emitted photon frequencies. In contrast, equation (10-82) does *not* reduce to complete redistribution, and it is of interest to ask how this affects the solution of the transfer equation. A study of this problem has been carried out by Finn (*Ap. J.*, **147**, 1085, 1967), using the actual redistribution function given above. If complete redistribution were correct, then the ratio $R_{\text{III},A}(x', x)/\varphi(x')\varphi(x)$ would be identically unity. In fact, for $a = 10^{-3}$, Finn's calculations show this ratio may approach $10^3$ under some circumstances ($x = x' = 3$). As might be expected, this difference leads to a rather different source function compared with that obtained when one assumes complete redistribution. Nevertheless, Finn's work also shows that the emergent line profiles obtained on the assumption of complete redistribution differ only very slightly from those obtained using the correct redistribution function. Thus, while one probably should use equation (10-82) for the most precise work, it appears that the assumption of complete redistribution will still yield physically useful results in the first approximation.

(d) Case IV. If we substitute equation (10-15) into equation (10-55), we find

$$\begin{aligned} R_{A,u}(\nu', \nu) &= \frac{\delta_i \delta_j}{4\pi^2} \int_{-1}^{1} \frac{d\mu}{(\nu - w\mu u - \nu_0)^2 + \delta_j^2} \\ &\quad \times \int_{-1}^{1} \frac{d\mu'}{(\nu - \nu' - w\mu u - w\mu' u)^2 + \delta_i^2} \\ &= \frac{a_j}{4\pi^2 w^2 u^2} \int_{-1}^{1} d\mu \, [(x - u\mu)^2 + a_j^2]^{-1} \\ &\quad \times \left\{ \tan^{-1}\left[\frac{x - x' + u(1-\mu)}{a_i}\right] - \tan^{-1}\left[\frac{x - x' - u(1+\mu)}{a_i}\right] \right\} \end{aligned} \tag{10-84}$$

Thus, integrating over a Maxwellian velocity distribution and converting to Doppler units, we obtain

$$\begin{aligned} R_{\text{IV},A}(x', x) &= \frac{a_j}{\pi^{5/2}} \int_0^\infty du \, e^{-u^2} \int_{-1}^{1} d\mu \, [(x - u\mu)^2 + a_j^2]^{-1} \\ &\quad \times \left\{ \tan^{-1}\left[\frac{x - x' + u(1-\mu)}{a_i}\right] - \tan^{-1}\left[\frac{x - x' - u(1+\mu)}{a_i}\right] \right\} \end{aligned} \tag{10-85}$$

(e) Thomson Scattering by Electrons. As we have mentioned previously, although electron scattering can be treated as coherent so far as the continuum is concerned, the frequency redistribution due to electron Doppler shifts may need to be accounted for in precise treatments of line formation. For Thomson scattering, the cross-section is frequency-independent; thus we have simply

$$f(\xi')p(\xi', \xi) = \delta(\xi - \xi') \tag{10-86}$$

In this case equation (10-70) can be integrated directly with $f(y) \equiv 1$ to yield

$$\begin{aligned} R_{e,A}(\nu', \nu) &= \frac{1}{w^2 \pi^{1/2}} \int_{|\nu - \nu'|/2w}^{\infty} (2wu - |\nu - \nu'|) e^{-u^2}\, du \\ &= \frac{1}{w} \operatorname{ierfc} \left| \frac{\nu - \nu'}{2w} \right| \end{aligned} \tag{10-87}$$

The function ierfc is defined as

$$\operatorname{ierfc}(x) \equiv \int_x^{\infty} \operatorname{erfc}(z)\, dz = \frac{e^{-x^2}}{\pi^{1/2}} - x \operatorname{erfc}(x) \tag{10-88}$$

In equation (10-87), $w$ is an *electron* Doppler width. An expression for the more realistic assumption of dipole scattering has been derived by Hummer (*Ap. J. Letters*, **150**, L157, 1967). Note that $R_{e,A}$ is normalized such that

$$\int_{-\infty}^{\infty} R_{e,A}(\nu', \nu)\, d\nu' = \int_{-\infty}^{\infty} R_{e,A}(\nu', \nu)\, d\nu = 1 \tag{10-89}$$

A graph of $R_{e,A}(x', x) = wR_{e,A}(\nu', \nu)$ is shown in Figure 10-4. (An alternative derivation for $R_{e,A}$ by Auer may be found in *Ap. J.*, **153**, 245, 1968.) Studies

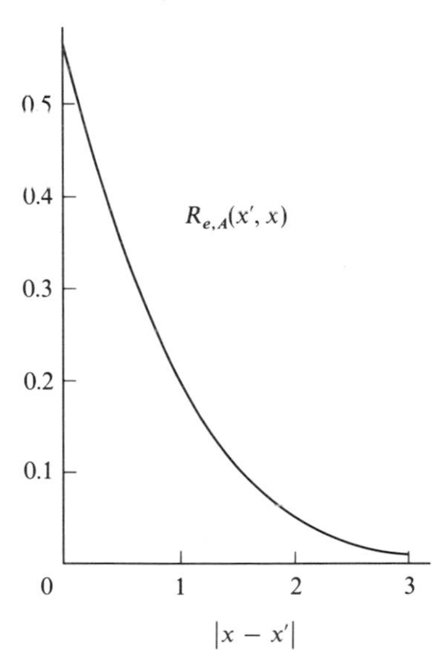


FIG. 10-4. Redistribution function for Thomson scattering by free electrons.

of line formation employing the function $R_{e,A}$ for the scattering by electrons have been carried out by Auer and Mihalas (*Ap. J.*, **153**, 245, 1968, and *Ap. J.*, **153**, 923, 1968). In a general way, the effects of the noncoherence of the electron scattering are found to be small under conditions resembling those appropriate to stellar atmospheres.

## 10-4. Summary

Before leaving the subject of redistribution, we should mention that the results obtained above are by no means rigorous. In particular, the effects of collisions have not been included properly, and it is assumed that the atom's velocity remains unaltered during the entire scattering process. Redistribution functions making the extreme opposite assumption that the atoms are completely redistributed in velocity over a Maxwellian distribution during the scattering process have been derived by Rees and Reichel (*J. Q. R. S. T.*, **8**, 1795, 1968). Their assumption, however, seems less realistic in the normal stellar atmospheres case than the ones employed in this chapter. The effects of recoil of the atom in the scattering process (of importance for lines like Lyman $\alpha$ in the interstellar medium) have been treated carefully by Field (*Ap. J.*, **129**, 551, 1959).

Thus, although the physical foundations of the theory should ultimately be improved upon, we may use the present results with fair confidence until that time. Moreover, we have seen that for the two cases of principal importance in the stellar atmospheres context, namely, pure Doppler broadening or Doppler plus pressure broadening, we may employ the much simpler approximation of complete redistribution as a good starting point for the solution of line transfer problems, even though it is not an exact description of the true situation. In fact, for simplicity of exposition in all further work in this book, we shall employ one of the two extreme assumptions of strict coherence (used in the now-obsolete classical theory) or of complete redistribution. In this way, we may gain important insight into the line formation problem with relative ease; more refined calculations will no doubt follow in the future work on line formation.

# 11 | Classical Treatments of Line Transfer

## 11-1. Characterization of the Problem

In this chapter we shall try to describe a few of the more important early approaches to the line formation problem. These will provide background for the more modern treatments to be discussed in subsequent chapters. It is important to be familiar with these older methods because of the great body of literature based upon them. Also, one must understand the analytical basis of these classical treatments in order to evaluate the reliability of spectroscopic diagnostics derived from them and to see more fully the differences inherent in recent work.

The usual classical procedure is to recognize at the outset the existence of two different line formation processes: scattering and absorption. We have discussed these two categories in Chapter 1 and have noted the physical differences between the two. It is usually supposed that a fraction $(1 - \varepsilon)$ of the photons absorbed are simply scattered so that the excited electron returns directly to its original lower level. If we assume that this scattering is isotropic and coherent (as was ordinarily done, even though we now know that complete redistribution is a better approximation), then we obtain an emission coefficient of the form

$$j_\nu^s = (1 - \varepsilon) l_\nu J_\nu \tag{11-1}$$

where

$$l_\nu = l\varphi_\nu = \frac{\pi e^2}{mc} f_{LU} n_L \left(1 - \frac{g_L n_U}{g_U n_L}\right)\varphi_\nu \tag{11-2}$$

is the line absorption coefficient. The remaining fraction $\varepsilon$ of the photons is assumed to be destroyed by various processes and converted to thermal energy. One then argues that this loss to thermal energy must be balanced exactly by thermal emissions, leading to an emission coefficient

$$j_\nu^{\,t} = \varepsilon l_\nu B_\nu \tag{11-3}$$

In addition, we may assume that there are contributions from continuum thermal emission and electron scattering. Thus we write

$$\frac{\mu}{\rho}\frac{dI_\nu}{dz} = -(\kappa + \sigma + l_\nu)I_\nu + \kappa B_\nu + \sigma J_\nu + \varepsilon l_\nu B_\nu + (1 - \varepsilon) l_\nu J_\nu \tag{11-4}$$

We have ignored the subscript $\nu$ on $\kappa$ and $\sigma$ because most lines are so narrow that these coefficients vary only negligibly over the line in comparison with the swift variation of $\varphi_\nu$. If we write

$$d\tau_\nu = -(\kappa + \sigma + l_\nu)\rho\, dz \tag{11-5}$$

and let

$$\eta_\nu = \frac{l_\nu}{(\kappa + \sigma)} \tag{11-6}$$

and

$$\rho = \frac{\sigma}{\kappa + \sigma} \tag{11-7}$$

we have

$$\mu \frac{dI_\nu}{d\tau_\nu} = I_\nu - \frac{(1 - \rho) + \varepsilon\eta_\nu}{1 + \eta_\nu} B_\nu - \frac{\rho + (1 - \varepsilon)\eta_\nu}{1 + \eta_\nu} J_\nu \tag{11-8}$$

Or if we write

$$\lambda_\nu = \frac{(1 - \rho) + \varepsilon\eta_\nu}{1 + \eta_\nu} \tag{11-9}$$

we have

$$\mu \frac{dI_\nu}{d\tau_\nu} = I_\nu - \lambda_\nu B_\nu - (1 - \lambda_\nu) J_\nu \tag{11-10}$$

The above equation as an approximation to the line transfer problem is due primarily to E. A. Milne and A. S. Eddington and bears their names. Excellent discussions of the derivation of this equation from the classical point of view may be found in the articles by Milne (Ref. 26, p. 169 ff.) and by Strömgren (*Z. Ap.*, **10**, 237, 1935).

Some improvement in the treatment can be obtained if we do not assume that the line scattering is coherent but allow for redistribution. Then we could write

$$\mu \frac{dI_\nu}{d\tau_\nu} = I_\nu - \frac{(1-\rho) + \varepsilon\eta_\nu}{1+\eta_\nu} B_\nu - \frac{\rho J_\nu}{1+\eta_\nu} - \frac{(1-\varepsilon)\eta_\nu}{(1+\eta_\nu)\varphi_\nu} \int R(\nu', \nu) J_{\nu'}\, d\nu' \tag{11-11}$$

Although the classical analyses mentioned above correctly specified $\varepsilon$ for a two-level atomic model, they failed to introduce noncoherent scattering, and practically all early work made use of equation (11-10) instead of (11-11). Moreover, a more complete derivation, using the equations of statistical equilibrium for more general atomic models (as will be described in later chapters) shows that other *kinds* of terms appear when multilevel problems are considered. Therefore, even equation (11-11) is seriously incomplete from a physical point of view.

## 11-2. Simplified Analytical Models

Let us consider the solution of the Milne-Eddington equation [equation (11-10)] under the simplifying assumptions that: (a) $\lambda_\nu$ is constant with depth; (b) $\varepsilon$ is constant with depth; (c) $\rho$ is constant with depth; and (d) the Planck function $B_\nu$ can be expanded linearly on the continuum optical depth scale $\tau$, i.e.,

$$B_\nu = a + b\tau = a + \frac{b}{1+\eta_\nu}\tau_\nu = a + p_\nu \tau_\nu \tag{11-12}$$

Under these assumptions, it is possible to obtain an exact solution (Chandrasekhar, *Ap. J.*, **106**, 145, 1947), but this differs only slightly from the approximate solution obtained below.

Taking the zero-order moment of equation (11-10), we obtain

$$\frac{dH_\nu}{d\tau_\nu} = J_\nu - (1-\lambda_\nu)J_\nu - \lambda_\nu B_\nu = \lambda_\nu(J_\nu - B_\nu)$$

Aside from the difference in the definition of $\lambda_\nu$, this equation is identical to that [equation (6-23)] discussed in Chapter 6. We showed there that the solution in the Eddington approximation may be written [equation (6-32)]

$$J_\nu = a + p_\nu \tau_\nu + \frac{p_\nu - \sqrt{3}\,a}{\sqrt{3}(1+\sqrt{\lambda_\nu})} \exp(-\sqrt{3\lambda_\nu}\,\tau_\nu)$$

and that the emergent flux is given by

$$H_\nu(0) = \frac{J_\nu(0)}{\sqrt{3}} = \frac{1}{3}\frac{p_\nu + \sqrt{3\lambda_\nu}\,a}{1 + \sqrt{\lambda_\nu}} \tag{11-13}$$

Again we see that thermalization occurs only at depths of the order $\lambda_\nu^{-1/2}$. Recalling the definition of $\lambda_\nu$, we see that this thermalization depth is of order $(1 - \rho)^{-1/2}$ in the continuum ($\eta_\nu = 0$) and $\varepsilon^{-1/2}$ in a strong line ($\eta_\nu \to \infty$). In both cases then the result predicts thermalization over a distance $p^{-1/2}$, where $p$ is the probability that in a particular absorption and emission process the photon is converted to thermal energy; thus the result is compatible with the random-walk arguments given before. Note that this result applies only for *coherent* scattering.

Equation (11-13) may be used to compute the profile of a line in a stellar atmosphere. In the continuum $\eta_\nu = 0$; then $\lambda_\nu = (1 - \rho)$, and equation (11-13) becomes

$$H_c(0) = \frac{1}{3}\frac{b + [3(1-\rho)]^{1/2}a}{1 + (1-\rho)^{1/2}} \tag{11-14}$$

Thus the depth in the line is, by equation (8-11),

$$A_\nu = 1 - \left(\frac{p_\nu + \sqrt{3\lambda_\nu}\,a}{1 + \sqrt{\lambda_\nu}}\right)\left(\frac{1 + \sqrt{1-\rho}}{b + a[3(1-\rho)]^{1/2}}\right) \tag{11-15}$$

In the particular case of no scattering in the continuum, we have $\rho = 0$, and equation (11-15) simplifies to

$$A_\nu = 1 - \frac{2(p_\nu + \sqrt{3\lambda_\nu}\,a)}{(1 + \sqrt{\lambda_\nu})(b + \sqrt{3}\,a)} \tag{11-16}$$

There are four important results of this classical theory, which provided the conceptual orientation of much of the early work on line formation. Let us discuss these briefly.

### PURE SCATTERING LINES

Consider the case where $\rho = 0$ so that there is no scattering in the continuum, while $\varepsilon = 0$ so that there is pure scattering in the lines. Then, $\lambda_\nu = 1/1(1 + \eta_\nu)$, and the residual flux in the line is given by

$$R_\nu = 2\left(\frac{b}{1+\eta_\nu} + a\sqrt{\frac{3}{1+\eta_\nu}}\right)\left[\left(1 + \sqrt{\frac{1}{1+\eta_\nu}}\right)(\sqrt{3}\,a + b)\right]^{-1} \tag{11-17}$$

If we consider the case of a very strong line and take the limit as $\eta_\nu \to \infty$, we obtain $R_\nu = H_\nu(0)/H_c(0) = 0$. This result implies that a very strong line that is formed by scattering can be completely dark in the core, in contrast to a line in pure absorption, as we will show now.

#### PURE ABSORPTION LINES

If again we assume $\rho = 0$, but now assume $\varepsilon = 1$ (LTE in the line), we have $\lambda_\nu \equiv 1$, and we find

$$R_\nu = \left(\frac{b}{1+\eta_\nu} + \sqrt{3}\,a\right)(b + \sqrt{3}\,a)^{-1} \tag{11-18}$$

In this case, as $\eta_\nu \to \infty$, the flux in the line does *not* go to zero but approaches a finite value so that the central depth of the line is

$$A_0 = A_\nu(\varepsilon = 1, \eta_\nu \to \infty) = \frac{b}{b + \sqrt{3}\,a} \tag{11-19}$$

This is what one might expect since as $\eta_\nu \to \infty$, we see only the surface layers of the star and hence any thermal sources present there. The emergent flux in the line core is then determined by the surface value of the Planck function, which will be nonzero. In contrast, in the scattering case, the photons are constantly being scattered out of the pencil of radiation and destroyed by thermalization so that we expect little if any emergent flux.

It is of interest to estimate the residual intensity for lines in the solar spectrum on the basis of this model. We write, on a *mean* optical depth scale,

$$B_\nu(\bar\tau) = B_\nu(T_0) + \left.\frac{dB_\nu}{d\bar\tau}\right|_0 \bar\tau = B_0 + B_1\bar\tau \tag{11-20}$$

We may now write

$$\left.\frac{dB_\nu}{d\bar\tau}\right|_0 = \left.\frac{dB_\nu}{dT}\right|_0 \left.\frac{dT}{d\bar\tau}\right|_0 \tag{11-21}$$

and from the gray relation

$$T^4 = \frac{3}{2}\,T_0^{\,4}\left(\bar\tau + \frac{2}{3}\right) \tag{11-22}$$

we have

$$\left.\frac{dT}{d\bar\tau}\right|_0 = \frac{3}{8}\,T_0 \tag{11-23}$$

We can also easily find

$$\left.\frac{dB_\nu}{dT}\right|_0 = B_\nu(T_0)\,\frac{u_0}{T_0(1 - e^{-u_0})} \equiv B_\nu(T_0)\,\frac{X_0}{T_0} \tag{11-24}$$

where $u_0 \equiv h\nu/kT_0$. Thus, combining equations (11-20), (11-21), (11-23), and (11-24), we have

$$B_1 = \tfrac{3}{8} X_0 B_\nu(T_0) = \tfrac{3}{8} X_0 B_0 \tag{11-25}$$

Now at the particular wavelength under consideration, we have written equation (11-12):

$$B_\nu(\tau) = a + b\tau = a + b\left(\frac{\kappa}{\bar{\kappa}}\right)\bar{\tau}$$

Thus by comparison with equations (11-20) and (11-25), we have

$$a = B_0 \tag{11-26}$$

while

$$b = \left(\frac{\bar{\kappa}}{\kappa}\right)B_1 = \frac{3}{8}X_0\left(\frac{\bar{\kappa}}{\kappa}\right)B_0 \tag{11-27}$$

and

$$p_\nu = \left(\frac{\bar{\kappa}}{\kappa}\right)\frac{3}{8}\frac{X_0}{1+\eta_\nu}B_0 \tag{11-28}$$

Thus equation (11-19) may be rewritten as

$$A_0 = \left[1 + \sqrt{3}\left(\frac{\kappa}{\bar{\kappa}}\right)\frac{8}{3X_0}\right]^{-1} \tag{11-29}$$

For the sun, $T_0 \sim 4800°\text{K}$, and if we take $\lambda \sim 5000$ Å, we have $u_0 \sim 6$, $X_0 \sim 6$, and $\kappa \approx \bar{\kappa}$, so that

$$A_0 = \left(1 + \frac{4}{3\sqrt{3}}\right)^{-1} = 0.56 \tag{11-30}$$

This value is in rough agreement with the depths of some of the stronger lines observed in this region of the solar spectrum. Some lines, however, are much deeper, particularly resonance lines such as the sodium D-lines, which led to the conceptual identification of resonance lines as "scattering" lines and subordinate lines from higher levels as "absorption" lines. Such an identification is not intuitively unreasonable since we expect that in a resonance line, the most probable route of exit from the upper state is, in fact, a simple scattering to the lower state. In contrast, for subordinate lines, a large number of other possibilities will, in general exist, and the photons may be effectively removed from the line. It must be emphasized again, however, that this kind of treatment is only very schematic. In particular, no line is a pure scattering line or a pure absorption line. The radiation field in a line is the result of an integration over all sources and sinks through a large depth range in the atmosphere.

### THE CENTER-TO-LIMB VARIATION OF LINES

From the solution of the Milne-Eddington equation given above, we may compute the variation of line intensity from center to limb. In general

$$I_\nu(0, \mu) = \int_0^\infty S_\nu(\tau_\nu)\exp(-\tau_\nu/\mu)\frac{d\tau_\nu}{\mu}$$

$$= \int_0^\infty B_\nu \exp(-\tau_\nu/\mu)\frac{d\tau_\nu}{\mu} + \int_0^\infty (J_\nu - B_\nu)(1-\lambda_\nu)\exp(-\tau_\nu/\mu)\frac{d\tau_\nu}{\mu} \tag{11-31}$$

Now under the assumptions made above, we find from equations (6-32) and (11-12)

$$I_\nu(0, \mu) = a + p_\nu \mu + \frac{p_\nu - \sqrt{3}\,a}{\sqrt{3}(1+\sqrt{\lambda_\nu})}\frac{(1-\lambda_\nu)}{(1+\sqrt{3\lambda_\nu}\,\mu)} \tag{11-32}$$

or from equations (11-9) and (11-28) we have

$$I_\nu(0, \mu) = a\left\{1 + \frac{3}{8}\left(\frac{\bar{\kappa}}{\kappa}\right)\frac{X_0}{1+\eta_\nu}\mu + \frac{\eta_\nu(1-\varepsilon)}{(1+\eta_\nu)(1+\sqrt{3\lambda_\nu}\,\mu)}\left[\frac{\frac{3}{8}\left(\frac{\bar{\kappa}}{\kappa}\right)\frac{X_0}{1+\eta_\nu} - \sqrt{3}}{\sqrt{3}(1+\sqrt{\lambda_\nu})}\right]\right\} \tag{11-33}$$

In the continuum $\eta_\nu = 0$; so

$$I_c(0, \mu) = a\left[1 + \frac{3}{8}\left(\frac{\bar{\kappa}}{\kappa}\right)X_0\,\mu\right]$$

and thus, from equation (8-4),

$$r_\nu(\mu) = \left\{1 + \frac{3}{8}\left(\frac{\bar{\kappa}}{\kappa}\right)\frac{X_0\,\mu}{1+\eta_\nu} + \frac{\eta_\nu(1-\varepsilon)}{(1+\sqrt{3\lambda_\nu}\,\mu)(1+\eta_\nu)} \times \left[\frac{\frac{3}{8}\left(\frac{\bar{\kappa}}{\kappa}\right)\frac{X_0}{1+\eta_\nu} - \sqrt{3}}{\sqrt{3}(1+\sqrt{\lambda_\nu})}\right]\right\}\left[1 + \frac{3}{8}\left(\frac{\bar{\kappa}}{\kappa}\right)X_0\,\mu\right]^{-1} \tag{11-34}$$

Consider now the pure absorption case for which $\varepsilon = 1$, $\lambda_\nu = 1$. Then

$$r_\nu(\mu) = \frac{1 + \dfrac{3}{8}\left(\dfrac{\bar{\kappa}}{\kappa}\right)\dfrac{X_0}{1 + \eta_\nu}\mu}{1 + \dfrac{3}{8}\left(\dfrac{\bar{\kappa}}{\kappa}\right)X_0 \mu} \tag{11-35}$$

Here we see that as we approach the limb $\mu \to 0$ and $I_\nu \to I_c$ so that the line vanishes. This is compatible with the physical picture we sketched before in that as we approach the limb, we see only the surface value of the Planck function at all frequencies; thus contrast between line and continuum disappears. On the other hand, if we set $\varepsilon = 0$ and take the limit as $\eta_\nu \to \infty$, $\lambda_\nu \to 0$, we find $r_\nu(\mu) \equiv 0$ so that the cores of pure scattering lines always remain dark, even at the limb. In the actual solar spectrum some lines do weaken toward the limb while others do not, or they weaken only slightly. This kind of observed behavior again led to the classification of lines into the "absorption" or "scattering" categories, though in some cases the results were in conflict with those based on central intensities. A few representative values of the variation of the depth across the limb are listed in Table 11-1. These show very clearly the behavior described above. In this table $X_0(\bar{\kappa}/\kappa) = 8$. Negative values imply emission (see below). Some studies (e.g., that by Houtgast, reviewed by Spitzer in *Ap. J.*, **98**, 107, 1943) have shown that neither model described here is adequate because the effects of noncoherent scattering are dominant. In any case, this whole approach is quite schematic and at best provides only a very rough pictorial guide.

### THE SCHUSTER MECHANISM

In the above discussion we have assumed that the continuum is purely thermal. When scattering is accounted for, several interesting effects are found. These were first discussed by Schuster (*Ap. J.*, **21**, 1, 1905) in one of the fundamental papers of radiative transfer theory. To emphasize the role of continuum scattering, suppose we take $\rho = 1$; then $\lambda_\nu = \varepsilon\eta_\nu/(1 + \eta_\nu)$, and equation (11-18) becomes

$$R_\nu = \left[\frac{1}{1 + \eta_\nu} + \left(\frac{a}{b}\right)\left(\frac{3\varepsilon\eta_\nu}{1 + \eta_\nu}\right)^{1/2}\right]\left[1 + \left(\frac{\varepsilon\eta_\nu}{1 + \eta_\nu}\right)^{1/2}\right]^{-1} \tag{11-36}$$

First, consider the case with $\varepsilon = 0$. Then

$$R_\nu = \frac{1}{1 + \eta_\nu} \tag{11-37}$$

and we obtain a dark line. This occurs since both line and continuum are formed by scattering, and the scattering length in the line is longer. On the other hand, suppose we take $\varepsilon = 1$. Then we find

$$R_\nu = \left[\frac{1}{1+\eta_\nu} + \left(\frac{a}{b}\right)\left(\frac{3\eta_\nu}{1+\eta_\nu}\right)^{1/2}\right]\left[1 + \left(\frac{\eta_\nu}{1+\eta_\nu}\right)^{1/2}\right]^{-1} \tag{11-38}$$

and in the line core, as $\eta_\nu \to \infty$,

$$R_\nu = \frac{\sqrt{3}}{2}\left(\frac{a}{b}\right) \tag{11-39}$$

TABLE 11-1. Center-to-Limb Variation in the Milne-Eddington Model

| | | $a_\nu(\mu)$ | | | | |
|---|---|---|---|---|---|---|
| $\varepsilon$ | $\eta_\nu$ | $\mu = 1$ | 0.5 | 0.3 | 0 | $A_\nu$ (*Flux*) |
| 1.0 | 0.01 | 0.007 | 0.006 | 0.005 | 0.000 | 0.006 |
| | 0.1 | 0.068 | 0.055 | 0.043 | 0.000 | 0.058 |
| | 1.0 | 0.375 | 0.300 | 0.237 | 0.000 | 0.317 |
| | 10.0 | 0.682 | 0.545 | 0.431 | 0.000 | 0.576 |
| | 100.0 | 0.743 | 0.594 | 0.469 | 0.000 | 0.628 |
| | $\infty$ | 0.750 | 0.600 | 0.474 | 0.000 | 0.634 |
| | 0.01 | 0.007 | 0.005 | 0.004 | −0.002 | 0.006 |
| | 0.1 | 0.066 | 0.051 | 0.037 | −0.019 | 0.054 |
| 0.3 | 1.0 | 0.378 | 0.306 | 0.246 | 0.026 | 0.322 |
| | 10.0 | 0.723 | 0.633 | 0.565 | 0.334 | 0.653 |
| | 100.0 | 0.798 | 0.713 | 0.648 | 0.438 | 0.731 |
| | $\infty$ | 0.808 | 0.723 | 0.659 | 0.452 | 0.741 |
| | 0.01 | 0.007 | 0.005 | 0.004 | −0.003 | 0.006 |
| | 0.1 | 0.066 | 0.049 | 0.035 | −0.024 | 0.053 |
| 0.1 | 1.0 | 0.379 | 0.308 | 0.250 | 0.035 | 0.324 |
| | 10.0 | 0.751 | 0.687 | 0.639 | 0.483 | 0.700 |
| | 100.0 | 0.847 | 0.799 | 0.765 | 0.658 | 0.809 |
| | $\infty$ | 0.860 | 0.815 | 0.783 | 0.684 | 0.824 |
| | 0.01 | 0.007 | 0.005 | 0.003 | −0.004 | 0.006 |
| | 0.1 | 0.066 | 0.049 | 0.034 | −0.027 | 0.053 |
| 0.0 | 1.0 | 0.379 | 0.310 | 0.252 | 0.039 | 0.325 |
| | 10.0 | 0.778 | 0.732 | 0.698 | 0.589 | 0.742 |
| | 100.0 | 0.931 | 0.920 | 0.912 | 0.885 | 0.922 |
| | $\infty$ | 1.000 | 1.000 | 1.000 | 1.000 | 1.000 |

Thus the line will appear in absorption *or emission*, depending upon the ratio $(a/b)$. The more shallow the temperature gradient, the brighter the line compared with the continua. The reason is that since $\varepsilon = 1$, the source function in the line is everywhere equal to the thermal value while in the continuum, the presence of scattering terms causes a drop below the local thermal source term. If the ratio $(a/b)$ is sufficiently large, the line appears in emission for all values of $\eta_\nu$ (i.e., at all frequencies in the profile). If the value of $(a/b)$ is not quite so large, then the line will appear in emission, but with a central reversal. For example, if we take just the critical value $(a/b) = 2/\sqrt{3}$, we find the values of $R_\nu$ given in Table 11-2. A slightly smaller value of $(a/b)$ produces a line with

TABLE 11-2. Schuster Mechanism Line Profile with $\left(\frac{a}{b}\right) = \frac{2}{\sqrt{3}}$

| $\eta_\nu$ | $R_\nu$ |
|---|---|
| 0 | 1.00 |
| 1 | 1.12 |
| 3 | 1.06 |
| 10 | 1.022 |
| $\infty$ | 1.00 |

a central residual intensity of less than 1.0, with faint emission in the wings. This interplay of scattering and absorption which gives rise to either absorption or emission lines depending upon the thermal gradient is known as the *Schuster mechanism.* A thorough discussion of the various possible cases has been given by Schuster and more recently by Gebbie and Thomas (*Ap. J.*, **154**, 285, 1968).

## 11-3. The Curve of Growth

Using equation (11-15), one could, in principle, compute the profile of a spectrum line and by integration over frequency determine its equivalent width. Such a procedure is quite laborious, however, and requires the use of a computer. It is therefore very instructive to consider a simple model which allows the equivalent width to be computed analytically. In this way we can construct what is known as a *curve of growth*, which gives the equivalent width directly in terms of the number of absorbing atoms that produce the line.

To begin with, we assume that the line formation occurs in a layer to which we can assign a unique temperature and electron pressure. This kind of assumption is often valid in laboratory work, but in the stellar atmospheres situation there are usually strong gradients in both temperature and pressure. Thus the choice of an appropriate temperature and pressure is difficult and, at best, can refer only to some ill-defined mean value of the atmosphere. With this choice, we can compute (assuming LTE) the populations of the atomic levels and, hence, the continuous opacity $\kappa_c$ and the line opacity [equation (4-19)]

$$l_{ij} = \frac{\pi e^2}{mc} f_{ij} N_i(1 - e^{-h\nu/kT})$$

We assume that the line profile is given by a Voigt profile [equation (9-44)]

$$H(a, v) = \frac{a}{\pi} \int_{-\infty}^{\infty} \frac{e^{-y^2}\, dy}{(v - y)^2 + a^2}$$

where $v \equiv \Delta\nu/\Delta\nu_0$, $a \equiv \Gamma/4\pi\, \Delta\nu_D$, and $\Delta\nu_D = \nu\xi_0/c$, and we write

$$l_\nu = l_0\, H(a, v) \tag{11-40}$$

where

$$l_0 = \frac{l_{ij}}{\sqrt{\pi}\, \Delta\nu_D} \tag{11-41}$$

We assume that $\Delta\nu_D$ and the parameter $a$ are fixed in the region of line formation and that the ratio $\eta_\nu = l_\nu/\kappa_c$ is independent of depth. The assumption of depth-independent $\eta_\nu$ is actually a fairly good approximation for some spectrum lines, e.g., Mg II $\lambda$ 4481 and Si II $\lambda\lambda$4128, 4131. Indeed, for these lines both $l_\nu$ and $\kappa_c$ may vary over orders to magnitude in the atmosphere while their ratio remains almost constant. In general, for a line in LTE we have

$$F_\nu = 2 \int_0^\infty B_\nu[T(\tau)] E_2\left[\int_0^\tau (1 + \eta_\nu)\, dt\right](1 + \eta_\nu)\, d\tau \tag{11-42}$$

As before, we assume

$$B_\nu[T(\tau)] = B_0 + B_1\tau$$

where $\tau$ is the continuum optical depth. Then having assumed $\eta_\nu$ is constant with depth, we can write

$$F_\nu = 2 \int_0^\infty (B_0 + B_1\tau) E_2[(1 + \eta_\nu)\tau](1 + \eta_\nu)\, d\tau$$

$$= 2B_0 \int_0^\infty E_2(x)\, dx + \frac{2B_1}{1 + \eta_\nu} \int_0^\infty E_2(x) x\, dx \tag{11-43}$$

or

$$F_\nu = B_0 + \frac{2B_1}{3(1 + \eta_\nu)} \tag{11-44}$$

Clearly

$$F_c = B_0 + \tfrac{2}{3}B_1 \tag{11-45}$$

so that

$$A_\nu = \frac{\eta_\nu}{(1+\eta_\nu)\left(1+\dfrac{3}{2}\dfrac{B_0}{B_1}\right)} \tag{11-46}$$

It is very convenient to work in terms of $A_0$, the central depth of the line. Taking the limit as $\eta_\nu \to \infty$, we find

$$A_0 = \left(1+\frac{3}{2}\frac{B_0}{B_1}\right)^{-1} \tag{11-47}$$

so that we may rewrite equation (11-46) as

$$A_\nu = \frac{A_0\,\eta_\nu}{1+\eta_\nu} \tag{11-48}$$

We may now compute the equivalent width by integration over frequency to obtain

$$W_\nu = \int_0^\infty A_\nu\,d\nu = 2A_0\,\Delta\nu_D \int_0^\infty \frac{\eta(v)}{1+\eta(v)}\,dv \tag{11-49}$$

where we have assumed that the line is symmetrical about it central frequency. But

$$\eta(v) = \frac{l_0}{\kappa}H(a, v) = \eta_0 H(a, v) \tag{11-50}$$

Thus

$$\frac{W}{2A_0\,\Delta\nu_D} = \int_0^\infty \frac{\eta_0 H(a, v)}{1+\eta_0 H(a, v)}\,dv \tag{11-51}$$

Before we attempt to evaluate this integral, let us describe qualitatively how the line develops as more and more atoms absorb. At the start, with only a few absorbing atoms, we expect only the Doppler core (where the opacity is highest) to contribute significantly to the line strength, the wings being too weak to add much to the equivalent width. As more and more atoms absorb in the line, the core, at some point, becomes essentially opaque while the contribution from the wings remains negligible, and the addition of more absorbers does little to increase the equivalent width of the line. In this situation, the line is said to be *saturated*. Finally, when enough absorbers are present, the opacity in the wings becomes appreciable, and the equivalent width again increases as the contribution of the line wings grows. This is shown in Figure

11-1. On the basis of this discussion, we see that we should expect three basically distinct regimes in the curve of growth; we will treat these individually.

We can schematize the Voigt profile function as [equation (9-61)]

$$H(a, v) \sim e^{-v^2} + \frac{a}{\pi^{1/2} v^2}$$

where we assume that the first term applies only in the core for $v \leq v^*$, and the second only in the wings for $v \geq v^*$, where $v^*$ is defined as the transition point at which the two terms are equal.

Consider now the contribution from the core only, and write $\eta_\nu = \eta_0 e^{-v^2}$, where we assume $\eta_0 \ll 1$. Then equation (11-51) becomes

$$\frac{W}{2A_0\,\Delta\nu_D} = \eta_0 \int_0^\infty \frac{e^{-v^2}\,dv}{1 + \eta_0 e^{-v^2}} = \eta_0 \int_0^\infty e^{-v^2}(1 - \eta_0 e^{-v^2} + \cdots)\,dv \quad (11\text{-}52)$$

or

$$\frac{W}{2A_0\,\Delta\nu_D} = \frac{\eta_0\sqrt{\pi}}{2}\left(1 - \frac{\eta_0}{\sqrt{2}} + \frac{\eta_0^2}{\sqrt{3}} - \cdots\right) \quad (11\text{-}53)$$

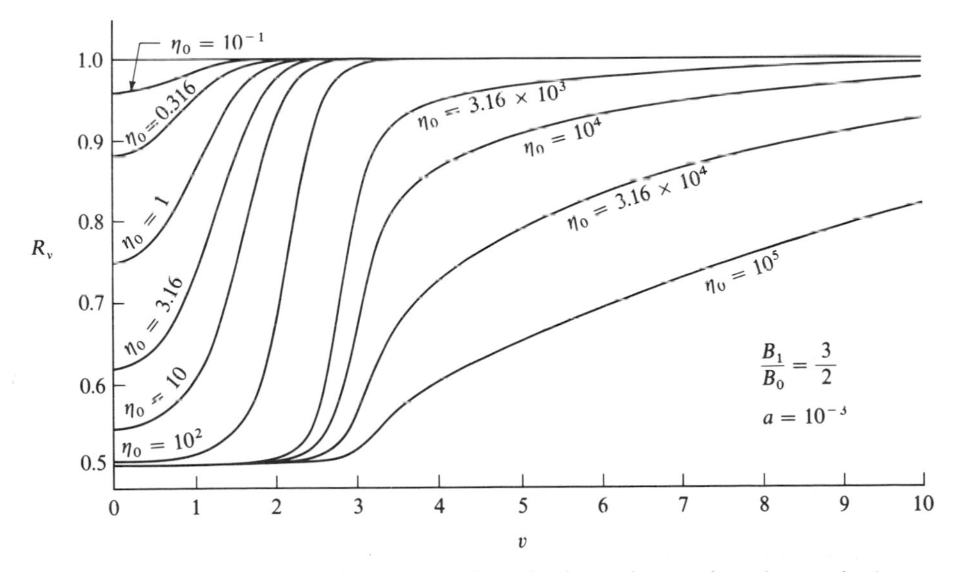


FIG. 11-1. Development of a spectrum line with increasing number of atoms in the line of sight. The line is assumed to be formed in pure absorption. For $\eta_0 \lesssim 1$, the line strength is proportional to the number of absorbers. For $30 \lesssim \eta_0 \lesssim 10^3$ the line is saturated, but the wings have not yet begun to develop. For $\eta_0 \gtrsim 10^4$ the line wings are strong and contribute the dominant part of the equivalent width.

For small values of $\eta_0$ (weak lines), the linear term is clearly dominant so that the equivalent width of the line is directly proportional to the number of absorbers present. This is known at the *linear part* of the curve of growth.

In the saturation part of the curve of growth, $\eta_0$ is large but not yet large enough that the damping wings contribute to the equivalent width of the line. Then again we have $\eta(v) = \eta_0 e^{-v^2}$, and if we let $u = v^2$, $dv = du/2\sqrt{u}$, then we may write

$$\frac{W}{2A_0\,\Delta\nu_D} = \frac{1}{2}\int_0^\infty \frac{\eta_0 e^{-u}\,du}{(1+\eta_0 e^{-u})\sqrt{u}} \tag{11-54}$$

Let $\eta_0 = e^\alpha$ so that

$$\frac{W}{2A_0\,\Delta\nu_D} = \frac{1}{2}\int_0^\infty \frac{du}{(1+e^{u-\alpha})\sqrt{u}}$$

$$= \frac{1}{2}\int_0^\infty \frac{du}{\sqrt{u}}[1 - e^{u-\alpha} - e^{2(u-\alpha)} - \cdots] \tag{11-55}$$

The series as written is not strictly convergent, but if $\alpha$ is large, we can replace the upper limit of the integral by $\alpha$, and we obtain

$$\frac{W}{2A_0\,\Delta\nu_D} \approx \sqrt{\ln \eta_0}\left[1 - \frac{\pi^2}{24(\ln \eta_0)^2} - \frac{7\pi^4}{384(\ln \eta_0)^4} - \cdots\right] \tag{11-56}$$

This expansion is only semiconvergent and must always be truncated after a finite number of terms. In practice, the series is useful for $\eta_0 \gtrsim 55$. From equation (11-56) we see very clearly that on the *saturation* or *flat part* of the curve of growth, the equivalent width grows extremely slowly with increasing numbers of absorbers, namely, $W \propto \sqrt{\ln \eta_0}$.

Finally, for very large numbers of absorbers, the line wings become strong enough to provide the dominant contribution to the equivalent width. Here we use the approximation $H(a, v) \sim a/\sqrt{\pi}v^2$ so that writing $C = \eta_0 a/\sqrt{\pi}$, we find

$$\frac{W}{2A_0\,\Delta\nu_D} = \int_0^\infty \frac{dv}{1 + v^2/C} = \frac{\pi}{2}\sqrt{C} = \frac{\sqrt{\pi a \eta_0}}{2} \tag{11-57}$$

Thus $W \propto \eta_0^{1/2}$, giving rise to what is called the *damping* or *square root part* of the curve of growth. The entire curve of growth is shown in Figure 11-2. It is important to note that the larger the value of the damping parameter $a$, the sooner the wings contribute strongly to the equivalent width, and hence the sooner the damping part of the curve rises away from the flat part. This effect is shown clearly in the Figure 11-2. Numerous curves of growth have been computed by various authors. A particularly useful set has been published by

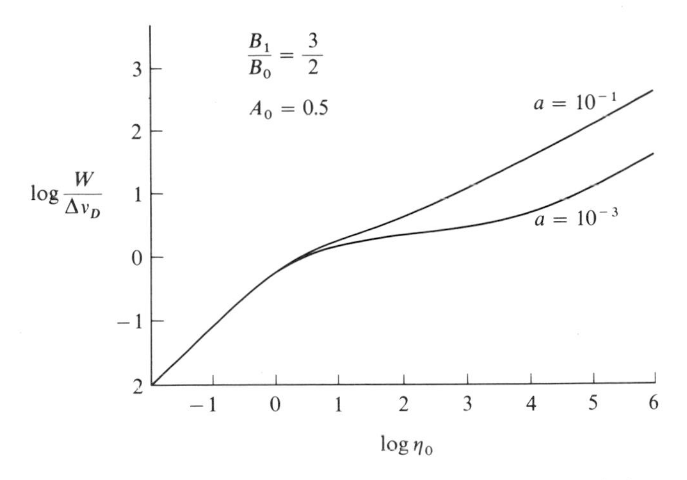


FIG. 11-2. Curves of growth for pure absorption lines. Note that the larger the value of $a$, the sooner the square root part of the curve rises away from the flat part.

Wrubel for a wide range of the temperature parameters $B_0$ and $B_1$ and under different assumptions concerning the transfer problem (see *Ap. J.*, **109**, 66, 1949; *Ap. J.*, **111**, 157, 1950; and *Ap. J.*, **119**, 51, 1954).

## 11-4. The Empirical Curve of Growth

The curve of growth has long been one of the astronomer's favorite tools for performing an analysis of a stellar atmosphere. The literature of the various applications of this approach is enormous, and we will refer to only a few typical studies. The reasons for the popularity of the curve of growth are that it provides estimates of several key parameters with great speed and ease and makes use of equivalent widths alone, which are relatively well determined observationally even in cases where profiles would be impossible to measure.

As described above, the theoretically computed curve gives $\log (W_\nu/\Delta\nu_D) = \log(W_\lambda/\Delta\lambda_D) = \log(W_\lambda c/\lambda\xi_0)$ as a function of $\log \eta_0$. Here

$$\eta_0 = \frac{\sqrt{\pi}\, e^2}{mc} f \frac{N_{i,r}(1 - e^{-h\nu/kT})}{\Delta\nu_D\, k_c(1 - e^{-h\nu/kT})} = \frac{\sqrt{\pi}\, e^2}{mc} f \frac{\lambda N_{i,r}}{\xi_0\, k_c} \tag{11-58}$$

and $\xi_0$ is the total random velocity of the atoms forming the line. The population $N_{i,r}$ is assumed to be that given by the Boltzmann equation, i.e.,

$$N_{i,r} = N_r \frac{g_i e^{-\chi_i/kT}}{U_r(T)} \tag{11-59}$$

where $N_r$ is the abundance of the $r$th ion of the element under consideration, $g_i$ is the statistical weight of the lower level, $\chi_i$ is the excitation potential, and $U_r(T)$ is the partition function.

Now normally in a stellar spectrum, we will observe lines from several multiplets of a given ion and one or more ionization stages of a given element for each of several elements. Let us first consider all of the lines of a given ion of a single element. Because of the Boltzmann factor in the excitation equilibrium, each multiplet has its own individual curve of growth. Thus we may write

$$\log \eta_0 = \log(gf\lambda) - \theta\chi - \log C \tag{11-60}$$

where $\theta \equiv 5040/T$, and

$$C \equiv \frac{N_r}{U_r(T)} \frac{\sqrt{\pi}\, e^2}{mc} \frac{1}{\xi_0 k_c} \tag{11-61}$$

Clearly, the value of $\eta_0$ will be affected by the choice of the $\theta$ used to describe the temperature in the atmosphere and will be different for each line. To construct an empirical curve of growth, we plot for each line the value of $\log(W_\lambda/\lambda)$ versus $\log(gf\lambda)$. Now if we assume that there is a *unique* relation between $\eta_0$ and $W_\lambda$, we should attempt as well as possible to force all points for different lines to define a single curve. To do this, we adjust the value of $\theta$ and choose that value of $\theta$ which yields the smallest scatter around a unique curve. This value is called the *excitation temperature*, $\theta_{\text{exc}}$, and is assumed to be the characteristic temperature of the line forming region. It should be realized that it may not always be possible to obtain a unique excitation temperature since, as one might well expect, different lines will actually be formed in different regions of the atmosphere. For example, one would expect that lines from levels with high excitation potentials will be formed deeper in the atmosphere and hence at higher temperatures. Similarly, we would expect the average excitation temperature to be higher for higher stages of ionization. Such refinements can be accounted for by more elaborate calculations that use model atmospheres but are normally ignored in curve of growth work. In addition, there are complications that arise in practice: There are errors both in the values of $W_\lambda$ and $f$, and these introduce scatter in the curve; also, there may be observable lines arising from only a limited range of excitation potentials so that $\theta_{\text{exc}}$ may not be very well determined over this short baseline.

Once the empirical curve, corrected for excitation effects, has been established, it may be compared with theoretical curves. To superimpose the two curves normally requires a shift both in abscissa and ordinate of the empirical curve relative to the theoretical. From this fitting procedure we can deduce three essential bits of information.

(a) The ordinate of the empirical curve is $\log(W_\lambda/\lambda)$ while that of the theoretical curve is $\log(W_\lambda/\Delta\lambda_0) = \log(W_\lambda/\lambda) - \log(\xi_0/c)$. Thus the difference in the ordinates of the two curves, when superimposed, yields directly $\log(\xi_0/c)$ so that we determine $\xi_0$, the velocity parameter. The derived value can be compared with the most probable thermal velocity appropriate to the excitation temperature, $\xi_{therm} = (2kT_{exc}/m)^{1/2}$. It is usually found that $\xi_0$ as implied by the curve of growth exceeds $\xi_{therm}$, sometimes by a large factor. To interpret this situation, it has been customary to postulate the existence of some additional nonthermal motion, which is usually referred to as *turbulence*. If these small-scale mass motions have a Gaussian distribution around some most probable speed $\xi_{turb}$, then we should have

$$\xi_0 = \left(\frac{2kT_{exc}}{M} + \xi_{turb}^2\right)^{1/2} \tag{11-62}$$

Turbulent velocities have been derived for many stars by this mode of analysis; the most dramatic examples are the supergiants, where velocities as high as 64 km/sec have been obtained. It should be realized, however, that such diagnoses are not on entirely firm ground since the introduction of a velocity field drastically affects the details of line formation. Indeed, with the very high velocities sometimes derived, one must inquire whether the excitation state of the gas can be affected by the interchange of energy between mass motions and atomic excitation. Very little work has been done on this difficult problem, and our knowledge of stellar velocity fields remains rudimentary to the point that we merely recognize their existence.

(b) The difference in abscissae between the empirical and theoretical curve yields

$$\log \eta_0 - [\log(gf\lambda) - \chi\theta_{exc}] = \log C \tag{11-63}$$

To proceed further, we need an estimate of the electron pressure. This may be derived either from a theoretical model or from the profiles of the Balmer lines. If we know $p_e$, we may compute $k_c$, and since we know $\xi_0$, we find $N_r$, the number of atoms in ionization state $r$ directly from $C$. By use of Saha's equation, we can correct this number to $N$, the number of atoms in all ionization states. More precisely, we obtain $N/N_H$ since $k_c$ is usually dominated by hydrogen and thus is proportional to $N_H$. In short, the horizontal shift between the two curves of growth yields the abundance of the element. This method of analysis has been applied to a wide variety of stars and has shown that certain stars have abundances that differ markedly from solar values, which in turn are fairly typical of many stars. The uncertainties in this type of analysis should be clear. We have explicitly assumed that the Saha and Boltzmann relations are valid, have ignored depth variations of the relevant parameters (usually called *stratification effects*), and have used a very schematic picture of the transfer problem.

(c) The horizontal and vertical shifts described above make use of the linear and flat parts of the curve, respectively. For a given set of theoretical curves, a comparison between the observed and computed damping parts determines the value of $a = \Gamma/4\pi\,\Delta\nu_D$ and hence $\Gamma$. This may in turn be compared with the value predicted by theory.

Let us now turn to a brief discussion of a few typical results. A very interesting study using curves of growth was carried out by Wright (*Pub. Dom. Ap. Obs.*, **8**, 1, 1948), who analyzed the sun and three other solar-type stars. This study is by no means the most recent available, but since it is a classic example of the procedure, we will consider it here. From an extensive set of equivalent width measurements and laboratory f-values, Wright constructed an empirical curve of growth, using lines of Fe I and Ti I. In all, some 75 lines of Fe I with $0 \leq \chi \leq 1.6$ eV and 137 lines of Ti I with $0 \leq \chi \leq 2.5$ eV were used to obtain the curve shown in Figure 11-3. Slightly different excitation temperatures were found for these two ions, namely, $T_{\text{exc}} = 4850°\text{K} \pm 150°\text{K}$ for Fe I and $T_{\text{exc}} = 4550°\text{K} \pm 150°\text{K}$ for Ti I. As can be seen, the curve is quite well defined, though it is true that the linear part depends mainly upon lines of Ti I while the damping part depends mainly upon Fe I lines. It would be more satisfactory if lines of each ion were found both in the linear and damping regions of the curve. The vertical shift in the curve relative to a theoretical curve by Menzel yields a velocity parameter $\xi_0 = 1.6$ km/sec. Since the mean thermal velocity in the solar atmosphere is of the order 1.2 km/sec, this implies a turbulent velocity of the order 1 km/sec. By a fit to the damping part, Wright obtains $\log a = -1.4$. Adopting a mean wavelength of 4500 Å, this yields $\Gamma = 1.7 \times 10^9\ \text{sec}^{-1}$, which is very nearly a factor of 10 larger than the classical damping constant $\Gamma_C$. It is clear that the main source of broadening must be collisions, and since hydrogen is largely unionized in the solar photosphere, the most likely process is the van der Waals interaction between hydrogen and the radiating atom. We found previously that, approximately, [equation (9-77)]

$$\Gamma_6 = 8.1 C_6^{2/5} v^{3/5} N_{\text{H}}$$

where [equation (9-259)]

$$C_6 \approx \left(\frac{13.6}{\chi_{\text{I}} - \chi_i}\right)^2 \times 10^{-32}$$

For a typical value $\chi_{\text{ion}} - \chi_i = 6$ eV, $C_6 \approx 52. \times 10^{-32}$. Typical values of $v$ and $N_{\text{H}}$ are $10^6$ cm/sec and $10^{17}\ \text{cm}^{-3}$, so we compute $\Gamma_6 \approx 10^9$, which is in basic agreement with the empirical value. Interestingly, Wright notes that the slope of the damping portion of his curve of growth is actually closer to 0.6 than 0.5. One possible explanation is that lines of widely differing strengths are formed in different layers so that the parameters describing them (e.g., damping widths) are not all identical, as is assumed by the curve of growth method.

One very important application that can be made using the solar curve of growth is to perform abundance analyses *differentially* with respect to the

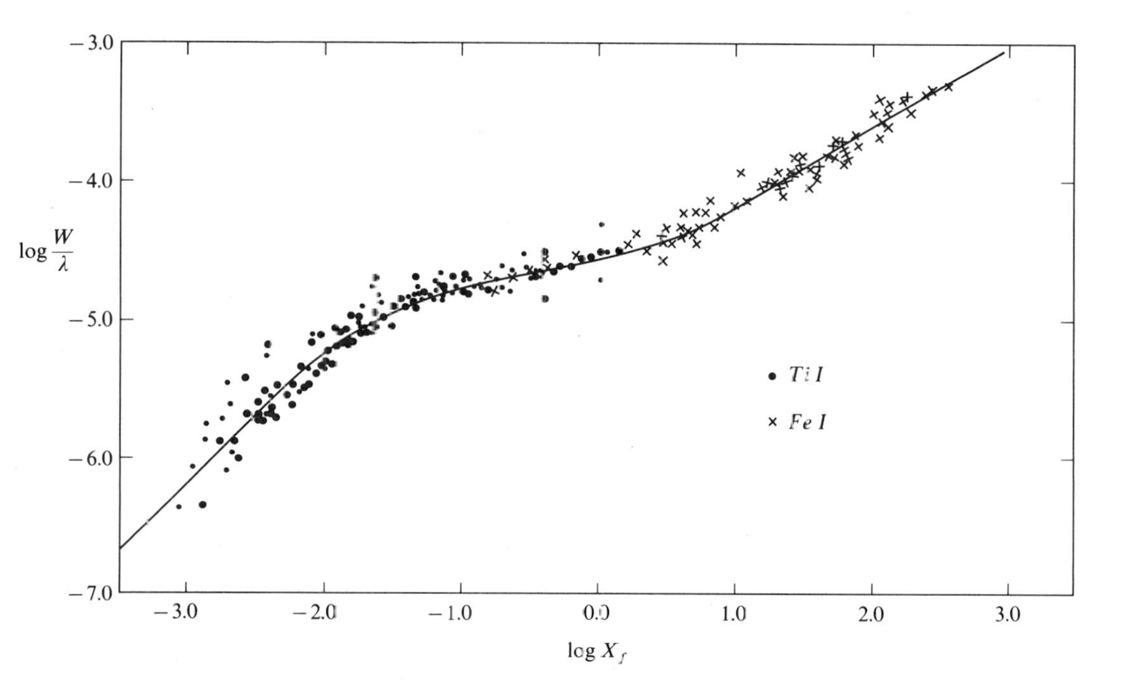


FIG. 11-3. Empirical curve of growth for solar Fe I and Ti I lines. Abscissa is based upon laboratory *f*-values. (From K. O. Wright, *Pub. Dom. Ap. Obs.*, **8**, 1, 1948.)

sun. A good example of an abundance analysis of G-type subdwarfs (extremely metal deficient stars) was done by Aller and Greenstein (*Ap. J. Supp. No. 46*, **5**, 139, 1960); and one of Population II K-giants was done by Helfer, Wallerstein, and Greenstein (*Ap. J.*, **129**, 700, 1959). An advantage of this approach is that the oscillator strengths, which in many cases are poorly known, cancel out. Also, since the temperatures of these stars are close to the solar value, one might expect their atmospheric structure to be at least roughly similar. If this is the case, then other effects such as blending of lines, stratification, departures from LTE, etc., might also cancel to first order.

The fundamental assumption in this approach is that the stellar curve of growth $\log(W_c/\lambda\xi_0)_*$ versus $\log \eta_0{}^*$ is identical to the solar curve of growth $\log(W_c/\lambda\xi_0)_\odot$ versus $\log \eta_0{}^\odot$. Now in practice, we do not know $\eta_0{}^*$ or $\xi_0{}^*$, and thus we cannot construct the stellar curve directly. What is known for each line is $\log(W/\lambda)_*$ and the value of $\log \eta_0{}^\odot$ for that line from the observed solar equivalent width. Now from the definition of $\eta_0$ we clearly have

$$\log \frac{\eta_0{}^\odot}{\eta_0{}^*} = \log \frac{N_r{}^\odot}{N_r{}^*} - \chi_{r,s}(\theta_{\text{exc}}^\odot - \theta_{\text{exc}}^*) + \log\left(\frac{\xi_0{}^* k_c{}^* U^*}{\xi_0{}^\odot k_c{}^\odot U^\odot}\right) \tag{11-64}$$

If we define

$$[X] \equiv \log \frac{X^\odot}{X^*} \tag{11-65}$$

for any quantity $X$, then we may rewrite equation (11-64) as

$$[\eta_0] = [N_r] + \chi_{r,s}\,\Delta\theta - [\xi_0] - [k_c] - [U] \tag{11-66}$$

Thus, if we were to plot $\log(W/\lambda)_*$ for lines of a given ion, say, Fe I versus $\log \eta_0{}^\odot$, instead of $\log \eta_0{}^*$, lines of different excitation potentials will scatter around a mean curve if there is a difference in excitation temperature between star and sun. Therefore, we plot $\log(W/\lambda)_*$ against $(\log \eta_0{}^\odot - \chi_{r,s}\,\Delta\theta)$ and choose $\Delta\theta$ to minimize the scatter. In the G-subdwarf analysis cited above, $\Delta\theta$ was found to be essentially zero (or at least $\Delta\theta < 0.05$) while for the K-giants, $\Delta\theta$ was of the order 0.25 to 0.35. After the effect of $\Delta\theta$ is eliminated, we superimpose the empirical curve for the star upon the solar curve, in which we have plotted $\log(Wc/\lambda\xi_0)_\odot$ versus $\log \eta_0{}^\odot$. The vertical shift gives directly $[\xi_0]$ while the horizontal shift yields the average value of $\delta = [N_r] - [\xi_0 k_c U]$ for the ion under consideration. In the case of the G-subdwarfs the velocity parameter is found to correspond closely to a pure thermal value with no turbulence; in the K-giants there is appreciable turbulence. Since the temperatures of the stars are close to the solar value, the partition function ratio is normally set to unity, and the value of $\delta$ will depend primarily upon abundances, differences in ionization, and differences in the continuous opacity $k_c$. Since in the temperature range under consideration $k_c$ is due mainly to $H^-$, the value of $[k_c]$ will essentially equal $[p_e]$. We determine $[p_e]$ from an analysis of the ionization equilibrium using information from two stages of ionization of the same element *E*. If we have observations for two stages of ionization, we may write

$$\delta_0{}^E = [N_0]^E - [\xi_0\, k_c] \tag{11-67}$$

and

$$\delta_1{}^E = [N_1]^E - [\xi_0\, k_c] \tag{11-68}$$

so that

$$\Delta \equiv \delta_1{}^E - \delta_0{}^E = [N_1]^E - [N_0]^E = \log\left(\frac{N_1{}^{\odot}}{N_0{}^{\odot}}\right) - \log\left(\frac{N_1{}^*}{N_0{}^*}\right) \tag{11-69}$$

Since $\log(N_1{}^{\odot}/N_0{}^{\odot})$ is assumed known, the observed value of $\Delta$ yields $\log(N_1{}^*/N_0{}^*)$. On the other hand, Saha's equation gives $\log(N_1{}^*p_e/N_0{}^*)$ since $\theta^*_{\text{exc}}$ is known. We may thus determine $\log p_e{}^*$ and hence $[p_e]$. This determination may be carried out for several different elements and a mean value taken. Knowledge of $[p_e]$ allows us to evaluate $[k_c]$, and hence finally

$$\log\frac{N^{\odot}}{N^*} = \log\frac{N^{\odot}}{N_r{}^{\odot}} + \log\frac{N^*}{N_r{}^*} + \log\frac{N_r{}^{\odot}}{N_r{}^*}$$

$$= \log\frac{N^{\odot}}{N_r{}^{\odot}} + \log\frac{N^*}{N_r{}^*} + \delta + [\xi_0\, k_c\, U] \tag{11-70}$$

which gives the ratio of abundance of the element in the star to the solar abundance. The results of the two analyses mentioned above lead to the very striking conclusion that the abundances of the heavy elements in the stars studied are lower than the solar abundance by a factor of the order of $10^2$! This is a fact of tremendous significance in the construction of a picture of the evolution of the Galaxy. While we have emphasized the uncertainties of the curve of growth approach, it must also be recognized that such large factors as these can hardly be in serious enough error to change the basic qualitative conclusion that some stars in the Galaxy have much smaller heavy element abundances than the sun.

## 11-5. The Method of Weighting Functions

The curve of growth approach explicitly assumes that line formation can be described by using the physical conditions prevalent at a single representative point in the atmosphere. This is of course a very crude approximation, and it is very desirable to try to incorporate the variation of the relevant parameters with depth. A simple method of doing this, which requires only a minimum of computation, is the *method of weighting functions*. This method was quite popular before the availability of large-scale high-speed computers and is now primarily of historical interest.

The first approach we describe is applicable only to weak lines. For simplicity, assume the line is formed in pure absorption (LTE). Let $\kappa$ denote the absorption coefficient in the continuum and $l_\nu$ denote the line absorption coefficient. Define

$$d\tau = -\kappa\rho\, dz \tag{11-71}$$

$$d\tau_l = -l_\nu \rho\, dz \tag{11-72}$$

and

$$d\tau_\nu = -(\kappa + l_\nu)\rho\, dz \tag{11-73}$$

Then

$$H_\nu = \frac{1}{2}\int_0^\infty S_\nu(\tau_\nu)E_2(\tau_\nu)\, d\tau_\nu = \frac{1}{2}\int_0^\infty B_\nu(\tau_\nu)E_2(\tau_\nu)(d\tau + d\tau_l) \tag{11-74}$$

Now, if $l_\nu \ll \kappa$, we can perform the expansion

$$E_2(\tau_\nu) \approx E_2(\tau) + \left.\frac{dE_2}{d\tau}\right|_\tau \tau_l = E_2(\tau) - E_1(\tau)\tau_l \tag{11-75}$$

Then, neglecting terms of higher than first order in $l_\nu$, we obtain

$$H_\nu \approx \frac{1}{2}\int_0^\infty B_\nu(\tau)E_2(\tau)\, d\tau + \frac{1}{2}\int_0^\infty B_\nu(\tau)E_2(\tau)\, d\tau_l - \frac{1}{2}\int_0^\infty B_\nu(\tau)E_1(\tau)\tau_l\, d\tau \tag{11-76}$$

Now if we define

$$\varphi(\tau) \equiv \int_\tau^\infty B_\nu(t)E_1(t)\, dt \tag{11-77}$$

it is clear that

$$\int_0^\infty B_\nu(\tau)E_1(\tau)\tau_l\, d\tau = -\Big|_0^\infty \tau_l\varphi(\tau) + \int_0^\infty \varphi(\tau)\, d\tau_l = \int_0^\infty \varphi(\tau)\, d\tau_l \tag{11-78}$$

Thus

$$H_\nu \approx \frac{1}{2}\int_0^\infty B_\nu(\tau)E_2(\tau)\, d\tau + \frac{1}{2}\int_0^\infty [B_\nu(\tau)E_2(\tau) - \varphi(\tau)]\, d\tau_l \tag{11-79}$$

so that

$$A_\nu = \int_0^\infty G_1(\tau)\, d\tau_l = \int_0^\infty G_1(\tau)\frac{l_\nu}{\kappa}\, d\tau \tag{11-80}$$

where

$$G_1(\tau) \equiv [\varphi(\tau) - B_\nu(\tau)E_2(\tau)] \Big/ \int_0^\infty B_\nu(\tau)E_2(\tau)\,d\tau \tag{11-81}$$

The function $G_1$ is called the *weighting function*; it gives a measure of how much each layer contributes to the observed flux. In the case of the sun, where one can observe the variation of the profile over the disk, a similar discussion shows that we can write

$$a_\nu(\mu) = \int_0^\infty g_1(\tau, \mu)\frac{l_\nu}{\kappa}e^{-\tau_l/\mu}\frac{d\tau}{\mu} \tag{11-82}$$

where

$$g_1(\tau, \mu) = \left[\int_\tau^\infty B_\nu(\tau)e^{-\tau/\mu}\frac{d\tau}{\mu} - B_\nu(\tau)e^{-\tau/\mu}\right] \Big/ \int_0^\infty E_\nu(\tau)e^{-\tau/\mu}\frac{d\tau}{\mu} \tag{11-83}$$

The main advantage of using weighting functions is that they vary only relatively slowly with frequency, and thus a single function can be used for a large number of lines. A tabulation of the weighting functions has been given by Andrews and Mugglestone (*Ap. J. Supp. No. 63*, **6**, 481, for 1962) several representative wavelengths in a selection of typical solar models.

The above expressions may be integrated over frequency to yield the equivalent width of the line. A convenient form of performing this integration was suggested by Pecker (*Ann. d'Ap.*, **14**, 383, 1951), who introduced a *saturation function* to account for the fact that the cores and wings of the line are formed in different regions of the atmosphere. Thus, we assume that $l_\nu = l_0\,H(a, v)$ where $a$ is *independent of depth* in the atmosphere. Then we may write from equation (11-82)

$$a_\nu(\mu) = \int_0^\infty g_1(\tau, \mu)e^{-\tau_c H(a, v)/\mu}\frac{l_0}{\kappa}H(a, v)\frac{d\tau}{\mu} \tag{11-84}$$

where $\tau_c$ is the optical depth at line center. Integrating over frequency, we obtain

$$W_\lambda(\mu) = \int_0^\infty a_\nu(\mu)\,d\lambda = \int_0^\infty g_1(\tau, \mu)\frac{l_0}{\kappa}\Delta\lambda_D\left[\int_\infty^\infty H(a, v)e^{-\tau_c H(a, v)/\mu}\,dv\right] \tag{11-85}$$

Thus, if we define the saturation function to be

$$\Psi\left(\frac{\tau_c}{\mu}, a\right) \equiv 2\int_0^\infty H(a, v)e^{-\tau_c H(a, v)/\mu}\,dv \tag{11-86}$$

we have

$$\frac{W_\lambda(\mu)}{\Delta\lambda_D} = \int_0^\infty g_1(\tau, \mu)\Psi\left(\frac{\tau_c}{\mu}, a\right)\frac{l_0}{\kappa}\frac{d\tau}{\mu} \tag{11-87}$$

As in the case of the weighting function, the saturation function can be computed once and for all. An extensive table has been given by Oliinyk, Babii, and Sakhman (*Sov. Astron.—AJ*, **11**, 648, 1967). The saturation function approach is valid for weak and medium-strong lines. For very strong lines, the exact choice of damping parameter becomes relatively critical, and the assumption that $a$ is depth-independent is no longer a good approximation. The Pecker method of computing equivalent widths has been applied extensively. One of the more prominent applications is the analysis of abundance of elements in the solar atmosphere by Goldberg, Müller, and Aller (*Ap. J. Supp. No. 45*, **5**, 1, 1960). A very thorough discussion of this method has been given by Aller (Ref. 13, Chap. 4).

## 11-6. The Model Atmosphere Method

The approaches described above are all limited by various kinds of simplifying assumptions. If we retain the basic physical assumptions, with modern computers it is then possible to eliminate most of the mathematical approximations and to obtain numerical solutions that allow fully for the depth variation of the relevant physical parameters. We will mention here a few typical computations made assuming LTE.

The basic requirement is to choose a model atmosphere that most closely resembles the stellar atmosphere to be considered. Once this choice is made, the computation of LTE line opacities, profiles, and the solution of the transfer equation is completely straightforward. To carry out an abundance analysis, for example, one replaces the curve of growth with a detailed computation of the equivalent width of each individual line as a function of the abundance, relative to hydrogen of the atom under consideration. Knowledge of the observed equivalent width of each line thus leads to an estimate of the abundance which, when suitably weighted, can be averaged to yield a fairly reliable abundance. Although the fundamental criteria in the choice of the model may involve fitting data such as the observed continuum energy distribution and the hydrogen line profiles, there is usually some ambiguity in the choice. The results of the abundance analysis can often be used to narrow the range of uncertainty, much in the same way as atmospheric properties may be inferred from a curve of growth analysis. For example, after the analysis is completed, we may examine the derived abundances for many lines to see if they show a correlation with excitation potential; such a correlation would be expected if the temperatures in the atmosphere are incorrectly chosen. The elimination of any such correlations may, therefore, lead to an improved choice of $T_{\text{eff}}$ for the model. Similarly, identical abundances should be derived, for a given element, from all ionization states. Differences that are found between ionization states contain information about errors in the choice of temperatures and pressures, and thus about $T_{eff}$ and $\log g$ for the model. Finally, one can examine the derived abundances to see if they show a correlation against equivalent width. If, for example, stronger lines systematically lead to larger abundances, we may have underestimated the velocity parameter; thus we may derive information about turbulent velocities to be included in the model. Naturally, errors in observed equivalent widths, *f*-values, and so on will introduce scatter; hence the above procedure is not always completely unambiguous. Nevertheless, an examination for such correlations may lead to a significantly better choice of the model. A good example of this procedure is the analysis of the two bright A-stars, Vega and Sirius, by Strom, Gingerich, and Strom (*Ap. J.*, **146**, 880, 1966). These interesting results show that Vega has very nearly solar abundances of the heavy elements while Sirius has abundances of a factor of 4 to 10 higher. In many respects Sirius mildly resembles the group of A stars called metallic-line A stars because of the strengths of the metal lines in their spectra.

Another useful function of direct numerical computation is to carry out comparisons of calculated and observed line profiles, for example, of the hydrogen lines (remembering, however, the probable inadequacy of the basic physical theory). An example of this kind of comparison is shown in Figure 11-4 for H$\gamma$ in Vega (Mihalas, *Ap. J. Supp. No. 114*, **13**, 1, 1966). The solid curve is the profile observed by Oke. The dots are the computed profile from a model with $T_{eff} = 10{,}000°K$ and $\log g = 4$, while the circles are from a model with $T_{eff} = 9150°K$ and $\log g = 4$. The agreement is satisfactory. A comparison of computed and observed energy distributions suggest $T_{eff} = 9600°K$. Direct computation can also show the merging of the hydrogen lines in stellar spectra. In Figure 11-5 we show the energy distribution for a model with $T_{eff} = 10{,}000°K$ and $\log g = 4$, displaying H$\alpha$, H$\beta$, H$\gamma$, H$\delta$, and H$\varepsilon$. Beyond H$\varepsilon$ (the range $2.5 \leq 1/\lambda \leq 2.75$) the spectrum is quite chopped up, and

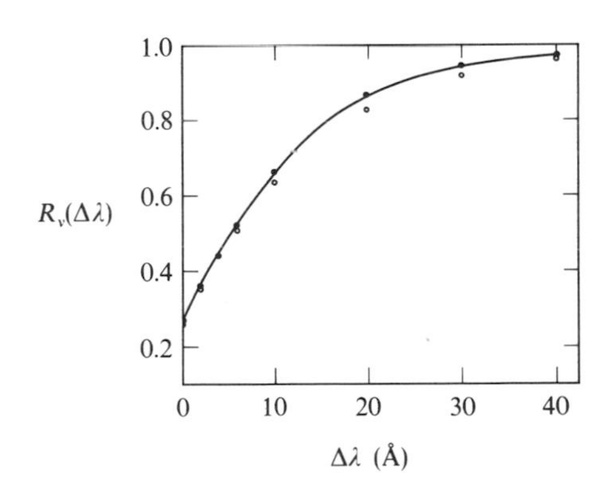


FIG. 11-4. Comparison of observed profile of $H\gamma$ in Vega with results of LTE model computations for models with $T_{eff} = 10{,}000°K$ and $\log g = 4$ (filled dots) and $T_{eff} = 9150°K$ and $\log g = 4$ (open dots).

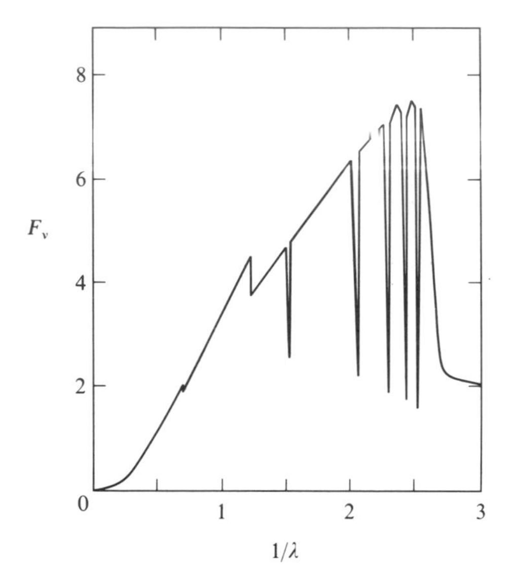


FIG. 11-5. Emergent energy distribution for a hydrogen line-blanketed model with $T_{\text{eff}} = 10{,}000°\text{K}$ and log $g = 4$. Ordinate gives $F_\nu$ in arbitrary units; abscissa gives $1/\lambda$, where $\lambda$ is in microns. The lines $H\alpha$ through $H\varepsilon$ are shown explicitly; for the sake of clarity, higher series lines are omitted in this plot.

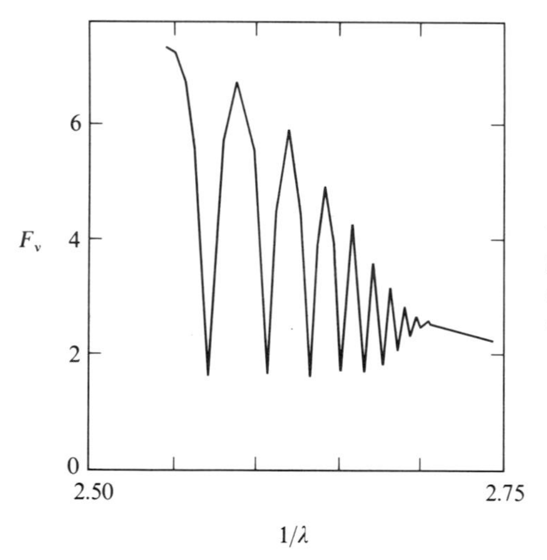


FIG. 11-6. Detailed energy distribution in region of higher Balmer-series lines for a model with $T_{\text{eff}} = 10{,}000°\text{K}$ and log $g = 4$. Note that the lines merge at $H$ 16.

only the upper envelope is shown. The details of this frequency range are shown on a larger scale in Figure 11-6. Here we see that the lines merge at H 16 and form a pseudocontinuum. Figure 11-7 shows the same region of the spectrum but for log $g = 3$. Because the gravity is lower, the electron densities are also lower, and the Stark-broadened lines are narrower; in fact, here the

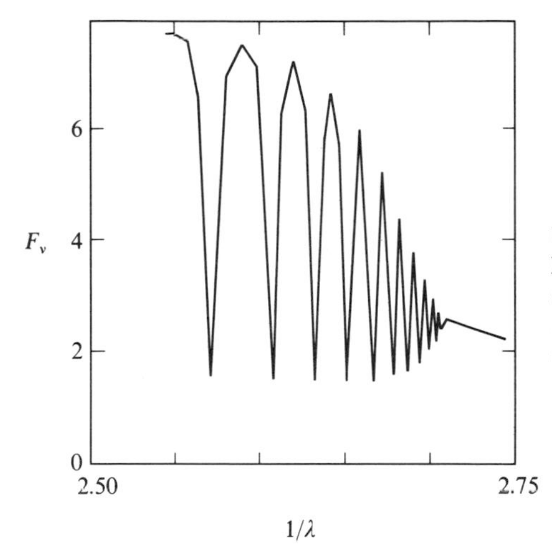


FIG. 11-7. Detailed energy distribution in region of higher Balmer-series lines for a model with $T_{eff} = 10{,}000°K$ and $\log g = 3$. Note that the lines merge at $H$ 18 and are much narrower than those shown in Figure 11-6.

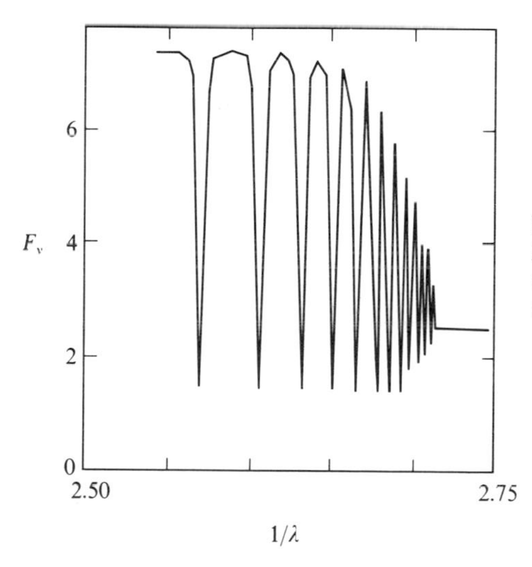


FIG. 11-8. Detailed energy distribution in region of higher Balmer-series lines for a model with $T_{eff} = 10{,}000°K$ and $\log g = 2$. Note that the lines do not merge until at least $H$ 20.

series does not merge until H 18. At $\log g = 2$, Figure 11-8, the lines are narrower still, and we see lines up to or beyond H 20 ($n = 20$ was the highest upper level included in the computation).

The numerical computation of line strengths has also been applied in the interpretation of stellar spectra of special interest. For example, Baschek has carried out a model atmosphere analysis of the metal-deficient stars HD

140283 (*Z. Ap.*, **48**, 95, 1959) and Wilson 10367 (*Z. Ap.*, **61**, 27, 1965). Hardorp has carried out an analysis of the peculiar (high $He^3/He^4$ ratio, high phosphorous abundance) B star 3 Cen A (*Z. Ap.*, **63**, 137, 1966). The peculiar A star (high Mn abundance) 53 Tauri has been analyzed by Auer, Mihalas, Aller, and Ross (*Ap. J.*, **145**, 153, 1966). A very large number of similar analyses for other stars may be found in the literature. The results we have discussed here at least outline the broad areas that have been studied, and it is time to turn to the problem of constructing a more general theory, using a detailed analysis of the various atomic processes.

# 12 | Non-LTE Line Transfer: The Two-Level Atom

Much of the progress in our understanding of the physics of line formation has come from detailed study of the combined solutions of the transfer equation and statistical equilibrium equations: the so-called non-LTE approach. In this chapter we shall consider some very schematic line formation problems, which are simple enough to be solved readily but which nevertheless contain a good enough description of the physically important processes to yield insight into the situation. It will be obvious that some of the assumptions that we will make are violated in actual stellar atmospheres and that, in general purely numerical solutions of the very complicated equations will be required. On the other hand, our real goal is to *understand* the answers, not merely obtain them, and this can be done only with a clear grasp of the prototype cases we will discuss here.

## 12-1. Diffusion Lengths

One of the most important differences between the LTE and non-LTE approaches is the way in which we account for the coupling between the gas and radiation field. In the LTE approach, it is *assumed* that the two thermodynamic variables $T$ and $N_e$ are sufficient to describe completely the excitation and ionization state of the gas. In this sense, it is a purely local theory, inasmuch as the state of the gas at a point is assumed to be independent of the state of the gas at other points. (This remark is strictly true only when the temperature structure is given. If the temperature structure is determined by a condition of radiative equilibrium, then there is a kind of coupling among all points in the atmosphere, but not in the sense we are concerned with here.) We have emphasized previously that in fact the excitation and ionization state of the gas is strongly influenced by the radiation field, which in turn is determined by the state of the gas through the entire atmosphere, so that the two problems of radiative transfer and statistical equilibrium are inextricably coupled and must be considered simultaneously.

It is easy to see some of the broad qualitative outlines of the problem by considering typical photon diffusion lengths. A clear description of the physical situation has been given by Avrett and Loeser (*Smithsonian Ap. Obs. Special Report No. 201*, 1966) whose discussion we will summarize here. For simplicity, let us consider an atom with a ground state and a single excited state and inquire about the relative populations of these two states. Suppose that the mean geometrical distance a photon travels in the medium before it is absorbed by an atom is $l$. After absorption, the photon will most often simply be re-emitted (scattered), but occasionally it will be destroyed by a collisional de-excitation of the excited atom. Let the characteristic length for photon destruction be $L$. If $A_{21}$ is the spontaneous radiative de-excitation rate and $C_{21}$ is the collisional de-excitation rate, then we expect that $L \gg l$ if $A_{21} \gg C_{21}$ while $L \approx l$ if $C_{21} \gg A_{21}$. Since photons are scattered independently of any reference to temperature, the coupling of the radiation field to a kinetic temperature will occur only through the collisions. This is a very important point because the fact that photons travel a distance $L$ before destruction implies that the radiation field at a given point is determined, at least in part, by photons having their origins within a sphere of radius $L$ centered at the point. If there are strong variations of kinetic temperatue within this *interaction sphere*, then the radiation field cannot, in general, be expected to be consistent with the local kinetic temperature. Thus, for example, if the intensity of the field were greater than that which would occur in a *hohlraum* at the local kinetic temperature, then the upper level of the atoms would be overpopulated compared with thermodynamic equilibrium at the local temperature.

Now at great depths in an atmosphere the collision rates become large because of increasing densities, and the destruction length grows ever smaller. If it becomes sufficiently small ($\sim l$), then the material contained in the interaction sphere will be very homogeneous, and we expect local thermodynamic equilibrium to occur. However, as we approach the surface of the atmosphere, $L$ will increase because of decreasing densities, and ultimately we expect $L \gg l$. If we consider a point such that the boundary of the atmosphere is contained within the interaction sphere of radius $L$, then we clearly may expect the intensity of radiation to be lower than in the case where the boundary were not there because no contribution is made to the radiation field from points outside the boundary. Of course the radiation field scattered by points between the boundary and the point under consideration is also lower, thus amplifying the effect still further. As we shall see below, the source function in the line is dominated by the scattering term $\int \varphi_\nu J_\nu \, d\nu$; thus we expect that at depths $\lesssim L$ from the boundary, the source function $S$ will be less than $B_\nu(T_e)$ [assuming, as is usually the case, that $B_\nu(T_e)$ is roughly constant or increases inward in the region of line formation].

If, in addition, there are other atoms that can absorb in a continuum at the line frequency and emit generally elsewhere, then the above picture may be modified. Let $P$ denote the distance a photon travels before it is destroyed by this photoionization process. If $P \gg L$, then the collisional destruction and production of photons dominates, and our remarks above are essentially unchanged. If, however, $P \ll L$, then we expect that the range over which $S$ is less than $B_\nu$ will extend to depths $\lesssim P$ from the boundary of the atmosphere. If $P$ decreases to the order of $l$, we expect that local thermodynamic equilibrium will occur again.

Another point of extreme importance in the process of line formation is the fact that a line is more transparent at some frequencies than at others. Since the *optical* depth interval corresponding to $l$ will always be $\approx 1$, it is clear that the characteristic lengths $l$ and $L$ can vary widely between wings and core. Thus, as we go from core to wing, the geometric depth in the atmosphere at which a photon can still "see" the boundary (i.e., contains the boundary within its interaction sphere) steadily increases, although an upper limit is ultimately set by the photoionization length $P$ in the continuum. If the scattering process in the line is assumed to be coherent, then this change in thermalization depth applies frequency by frequency. On the other hand, we have seen that a more realistic approximation for the Doppler core or for pressure-broadened lines is that of complete redistribution. In the case of complete redistribution, a photon absorbed at a given point in the atmosphere will be redistributed statistically over the entire line profile upon emission. Those photons redistributed from core to wing may then escape through the boundary at points where the core itself is so opaque that it does not yet contain the boundary within its interaction sphere. In this case, the radiation field in the line as a whole responds to the fact that the boundary can be seen at some frequencies, even if not at others. This coupling in frequency-space strongly changes the nature of the photon diffusion in a line. Since the effects of the boundary now extend deeper in the atmosphere, we may expect the source function in the noncoherent case to lie below $B_\nu$ to greater depths in the atmosphere. Also, in a general way, we expect the effects of the boundary upon the source function to extend to greater depth for lines with highly developed wings.

We have been able to outline here many of the qualitative features of the line formation problem with the aid of simple diffusion-length arguments. We will now proceed to make these statements more quantitative with detailed calculations.

## 12-2. The Two-Level Atom Without Continuum

### THE SOURCE FUNCTION

Consider a schematic atomic model consisting of two levels between which radiative and collisional transitions can occur. This model is obviously very incomplete, but it nevertheless provides a fairly good description of the real situation for some lines. In particular, lines excited from the ground state for which coupling to continuum is weak are well described by this model.

If we assume that there are no other opacity or emission sources at the line frequency, then the equation of transfer becomes

$$\mu \frac{dI_\nu}{dz} = -N_1 B_{12} \frac{h\nu}{4\pi} \varphi_\nu I_\nu + N_2 \frac{h\nu}{4\pi} \varphi_\nu (A_{21} + B_{21} I_\nu) \tag{12-1}$$

where we have assumed that the emission and absorption profiles are identical (complete redistribution). Then, if we define

$$d\tau = -(N_1 B_{12} - N_2 B_{21}) \frac{h\nu}{4\pi} dz \tag{12-2}$$

we have

$$\mu \frac{dI_\nu}{d\tau} = \varphi_\nu (I_\nu - S_l) \tag{12-3}$$

where

$$S_l = \frac{N_2 A_{21}}{N_1 B_{12} - N_2 B_{21}} \tag{12-4}$$

Or, making use of the Einstein relations, we have

$$S_l = \frac{2h\nu^3}{c^2} \left( \frac{N_1 g_2}{N_2 g_1} - 1 \right)^{-1} \tag{12-5}$$

Since the factor $\nu^3$ varies only slowly compared with the (usually) sharply peaked profile $\varphi_\nu$, $S_l$ is often referred to as a *frequency-independent* source function. Equations (12-4) and (12-5) are the *implicit* forms for the source function in the sense defined in Section 4-1. Now the statistical equilibrium equation governing the populations of the two levels is

$$N_1 \left( B_{12} \int \varphi_\nu J_\nu \, d\nu + N_e \Omega_{12} \right) = N_2 \left( A_{21} + B_{21} \int \varphi_\nu J_\nu \, d\nu + N_e \Omega_{21} \right) \tag{12-6}$$

Hence

$$\frac{N_1 g_2}{N_2 g_1} = \frac{A_{21} + B_{21} \int \varphi_\nu J_\nu \, d\nu + N_e \Omega_{21}}{B_{21} \int \varphi_\nu J_\nu \, d\nu + N_e \Omega_{21} e^{-h\nu_{12}/kT}} \tag{12-7}$$

where we have made use of the detailed balancing result that

$$\Omega_{12} = \frac{N_2^* \Omega_{21}}{N_1^*} = \frac{g_2}{g_1} \Omega_{21} e^{-h\nu_{12}/kT} \tag{12-8}$$

Thus, substituting equation (12-7) into equation (12-5), we obtain, with the help of the Einstein relations,

$$S_l = \frac{\int \varphi_\nu J_\nu \, d\nu + \dfrac{N_e \Omega_{21}}{A_{21}} \left(\dfrac{2h\nu_{12}^3}{c^2}\right) e^{-h\nu_{12}/kT}}{1 + \dfrac{N_e \Omega_{21}}{A_{21}} (1 - e^{-h\nu_{12}/kT})} \tag{12-9}$$

Defining

$$\varepsilon' \equiv \frac{N_e \Omega_{21}}{A_{21}} (1 - e^{-h\nu_{12}/kT}) \tag{12-10}$$

equation (12-9) can be written

$$S_l = \frac{\int \varphi_\nu J_\nu \, d\nu + \varepsilon' B_\nu}{1 + \varepsilon'} \tag{12-11}$$

Since it incorporates the statistical equilibrium equation directly, we refer to equation (12-11) as the *explicit* form of the source function, in the sense used in Chapter 4. In this form, we see that the source function contains a *non-coherent scattering term* $\int \varphi_\nu J_\nu \, d\nu$ and a *thermal source* term $\varepsilon' B_\nu$. Physically, this thermal source term represents photons that are created by collisional excitations, followed by a radiative de-excitation. The term $\varepsilon'$ in the denominator is a *sink term*, which, physically speaking, represents those photons that are destroyed by collisional de-excitation. Thus these two terms describe completely the coupling of the radiation field to the local thermal state of the gas. The scattering term may be viewed as a *reservoir term*, which represents the end result of the cumulative contributions of the source and sink terms over the entire interaction sphere.

Clearly, if densities are made sufficiently large, then we may eventually find $N_e \Omega_{21} \gg A_{21}$ so that $\varepsilon' \gg 1$ and then $S_l \to B_\nu(T_e)$. However, in virtually all situations of astrophysical interest, $\varepsilon' \ll 1$ (even as small as $10^{-7}$ in some cases). Thus it is clear that in general the source function cannot be expected to have a value close to that of the Planck function. Actually, this state of affairs was partially recognized in the classical theory by the division of lines into "scattering" lines and "absorption" lines. But it is important to realize that even if the term $\varepsilon' B_\nu$ is small compared with $\int \varphi_\nu J_\nu \, d\nu$, it *cannot be discarded* since ultimately—at depths greater than $L$ as defined above—the intensity must couple to the local thermal radiation field, i.e., even though the thermal source and sink terms are small locally compared with the scattering term, when integrated through the entire interaction sphere they accumulate to a point of importance. If one discarded the thermal terms (as was done in some early treatments of "scattering" lines), then the transfer equation would be homogeneous and the scale of the solution would be unknown; this scale is in fact fixed by the (small) thermal terms.

## THE TRANSFER EQUATION IN INTEGRAL FORM

Having obtained an expression for the source function, let us now consider the solution of the transfer equation. It is very convenient to work in terms of Doppler widths (or collision widths for a Lorentz profile) in the line profile. Let $x$ be a dimensionless frequency variable in these units, measured with respect to line center. Then the profile functions may be written as

$$\varphi(x) = \frac{e^{-x^2}}{\sqrt{\pi}} \tag{12-12}$$

for a Doppler profile,

$$\varphi(x) = \frac{a}{\pi^{3/2}} \int_0^\infty \frac{e^{-y^2} \, dy}{(x-y)^2 + a^2} \tag{12-13}$$

for a Voigt profile, and

$$\varphi(x) = \frac{1}{\pi} \frac{1}{1+x^2} \tag{12-14}$$

for a Lorentz profile. All of the above profile functions are normalized such that

$$\int_{-\infty}^{\infty} \varphi(x) \, dx = 1 \tag{12-15}$$

In addition, it is convenient to define

$$\varepsilon \equiv \frac{\varepsilon'}{1+\varepsilon'} \tag{12-16}$$

so that we may rewrite equation (12-11) as

$$S_l = (1-\varepsilon) \int_{-\infty}^{\infty} \varphi_x J_x \, dx + \varepsilon B \tag{12-17}$$

Now the transfer equation

$$\mu \frac{dI_x}{d\tau} = \varphi_x(I_x - S_l) \tag{12-18}$$

has the formal solution

$$J_x(\tau) = \frac{1}{2}\int_0^\infty S_l(t)E_1\left|\int_\tau^t \varphi_x(t')\,dt'\right|\varphi_x(t)\,dt \tag{12-19}$$

In the important case of a depth-independent profile function (which we shall assume for the present discussion), equation (12-19) reduces to

$$J_x(\tau) = \frac{1}{2}\varphi_x\int_0^\infty S_l(t)E_1\,|(t-\tau)\varphi_x|\,dt \tag{12-20}$$

If we now substitute this expression for $J_x$ into equation (12-17), we obtain an integral equation for $S_l$, namely,

$$S_l(\tau) = \frac{(1-\varepsilon)}{2}\int_{-\infty}^\infty dx\,\varphi_x{}^2\int_0^\infty S_l(t)E_1\,|(t-\tau)\varphi_x|\,dt + \varepsilon B$$

$$= (1-\varepsilon)\int_0^\infty dt\,S_l(t)\left[\frac{1}{2}\int_{-\infty}^\infty \varphi_x{}^2E_1\,|(t-\tau)\varphi_x|\,dx\right] + \varepsilon B \tag{12-21}$$

or

$$S_l(\tau) = (1-\varepsilon)\int_0^\infty S_l(t)K_1\,|t-\tau|\,dt + \varepsilon B \tag{12-22}$$

The function $K_1$ in equation (12-22) is called the *kernel function* and clearly is defined as

$$K_1(s) = \frac{1}{2}\int_{-\infty}^\infty \varphi_x{}^2E_1(s\varphi_x)\,dx = \int_0^\infty \varphi_x{}^2E_1(s\varphi_x)\,dx \tag{12-23}$$

The form of the solution for the source function depends intimately upon the detailed behavior of the kernel function, and we must therefore turn to a consideration of its properties.

### THE NATURE OF THE KERNEL FUNCTION

The mathematical properties of the kernel function have been discussed by Avrett and Hummer (*M. N.*, **130**, 295, 1965) whose results we will summarize briefly here. The reader is referred to their paper for details.

As we will see later, it is useful also to define an auxiliary kernel function $K_2(s)$ such that

$$K_2(s) = \int_{-\infty}^\infty E_2(s\varphi_x)\varphi_x\,dx \tag{12-24}$$

Clearly

$$\frac{dK_2(s)}{ds} = -\int_{-\infty}^{\infty} E_1(s\varphi_x)\varphi_x^2\, dx = -2K_1(s) \tag{12-25}$$

or, alternatively,

$$\int_0^s K_1(s')\, ds' = \frac{1}{2}[1 - K_2(s)] \tag{12-26}$$

where we have made use of the fact that $K_2(0) = 1$.

The most outstanding property of the kernel function in line formation problems with noncoherent scattering is its long-range nature. Recall that for coherent scattering the appropriate kernel function is $E_1|t-\tau|$, which falls off strongly, proportional to $\exp(-|t-\tau|)/|t-\tau|$, thereby severely limiting the range of depth points that effectively couple together in the scattering process. For noncoherent scattering this is no longer the case, and much larger depth intervals in the atmosphere can influence one another. To see this clearly, let us derive the asymptotic behavior of $K_1(s)$ for large $s$.

Consider first the case of Doppler profiles. Then combining equations (12-12) and (12-23), we have

$$K_1(s) = \frac{1}{\pi}\int_0^{\infty} e^{-2x^2} E_1\left(\frac{se^{-x^2}}{\sqrt{\pi}}\right) dx \tag{12-27}$$

Let

$$u = \frac{se^{-x^2}}{\sqrt{\pi}} = \lambda e^{-x^2}$$

Then

$$x = (\ln \lambda - \ln u)^{1/2}$$

and

$$dx = -\frac{du}{2ux}$$

so that

$$K_1(s) = \frac{1}{2s^2\sqrt{\ln \lambda}}\int_0^{\lambda} \frac{E_1(u)u\, du}{(1 - \ln u/\ln \lambda)^{1/2}} \tag{12-28}$$

Since we desire the asymptotic behavior for $s \gg 1$, we estimate the value of the integral by assuming $\lambda \to \infty$. Thus we replace the upper limit by $\infty$ and set the term in the denominator to unity. Then

$$K_1(s) \sim \frac{1}{2s^2\sqrt{\ln \lambda}}\int_0^{\infty} E_1(u)u\, du = \frac{E_3(0)}{2s^2\sqrt{\ln \lambda}}$$

or

$$K_1(s) \sim \frac{1}{4s^2[\ln(s/\pi^{1/2})]^{1/2}} \tag{12-29}$$

which shows clearly the slow falloff of the kernel. Consider now the case of a Lorentz profile. Here

$$K_1(s) = \frac{1}{\pi^2}\int_0^\infty E_1\left(\frac{s/\pi}{1+x^2}\right)\frac{dx}{(1+x^2)^2} \tag{12-30}$$

Now for large $s$, the $E_1$ term will be small unless $x \gg 1$. Therefore, to estimate $k_1$, we write

$$K_1(s) \sim \int_0^\infty E_1\left(\frac{s}{\pi x^2}\right)\frac{dx}{\pi^2 x^4} \tag{12-31}$$

Now let

$$u = \frac{s}{\pi x^2}$$

Then

$$dx = \frac{-x\,du}{2u}$$

so that

$$K_1(s) \sim \frac{1}{2\pi^{1/2}s^{3/2}}\int_0^\infty E_1(u)\sqrt{u}\,du$$

But by integrating by parts, we find

$$\int_0^\infty E_1(u)\sqrt{u}\,du = \frac{2}{3}\int_0^\infty e^{-u}u^{1/2}\,du = \frac{2}{3}\Gamma\left(\frac{3}{2}\right) = \frac{\sqrt{\pi}}{3} \tag{12-32}$$

Thus

$$K_1(s) \sim \frac{1}{6s^{3/2}} \tag{12-33}$$

which is of even a longer range than the Doppler kernel, as expected from our discussion of diffusion lengths. By an almost identical argument we may show that for a Voigt profile the asymptotic behavior of the kernel function is given by

$$K_1(s) \sim \frac{\sqrt{a}}{6s^{3/2}} \tag{12-34}$$

The corresponding asymptotic forms for $K_2(s)$ may be estimated directly using the relation

$$K_2(s) = 2\int_s^\infty K_1(s')\,ds' \tag{12-35}$$

Thus, for a Doppler profile

$$K_2(s) \sim \frac{1}{2s[\ln(s/\pi^{1/2})]^{1/2}} \tag{12-36}$$

while for a Lorentz profile

$$K_2(s) \sim \frac{2}{3s^{1/2}} \tag{12-37}$$

and for a Voigt profile

$$K_2(s) \sim \frac{2}{3}\left(\frac{a}{s}\right)^{1/2} \tag{12-38}$$

As mentioned above, the long range of the noncoherent kernel functions compared with the coherent case comes about because photons can be redistributed into the transparent line wings, where they are transported freely and thereby couple together the radiation fields of distant points.

Actually, the above forms remain valid only if the opacity drops to zero somewhere in the line. If we suppose that there is a nonzero minimum opacity $\varphi_0$, then from equation (12-23) we see that

$$K_1(s) < \int_0^\infty \varphi_x{}^2 E_1(s\varphi_0)\,dx \sim \frac{e^{-s\varphi_0}}{s\varphi_0} \tag{12-39}$$

for large $s$. Thus the kernel function will behave like equation (12-28), equation (12-33), or equation (12-34)—whichever is appropriate—for $s \lesssim 1/\varphi_0$ and will fall off exponentially beyond that point. This interesting behavior has two important implications. First, if we suppose that the minimum value of the opacity is set by some source other than the line (say, an overlapping continuum), then we expect the line scattering terms to dominate the solution to about optical depth unity in the continuum but not much beyond. This behavior is again in agreement with what we would expect from the diffusion-length arguments given above. Second, if the minimum value is set by a truncation of the line using a finite frequency band (as is necessary in practice), then we expect the scattering terms in the solution to have their true behavior only over depths less than $1/\varphi_0$. In practical applications one must be certain that this cutoff lies below the depth at which the solution properly thermalizes to the local Planck function, or a spurious thermalization at too shallow a depth will be found.

DEPTH VARIATION OF THE SOURCE FUNCTION

Let us now consider the behavior of the source function for the case of an atmosphere in which $\varepsilon$ and $B$ are constant with depth. Naturally, this is a very schematized case, but it provides very useful insight. We consider first the variation at great depth. Since we expect that $S_l$ ultimately will approach $B$, we might suppose that at great depth it will vary only slowly and that to the first order we may take it outside the integral and rewrite equation (12-22) as

$$
\begin{aligned}
S_l(\tau) &\approx (1-\varepsilon)S_l(\tau)\int_0^\infty K_1\,|t-\tau|\,dt + \varepsilon B \\
&= (1-\varepsilon)S_l(\tau)[1-\tfrac{1}{2}K_2(\tau)] + \varepsilon B
\end{aligned}
\tag{12-40}
$$

or

$$
S_l(\tau) \approx B\left[1 + \frac{(1-\varepsilon)}{2\varepsilon}K_2(\tau)\right]^{-1} \tag{12-41}
$$

From this result we see that $S_l$ will lie below $B$ and will approach $B$ closely only when $K_2(\tau) \ll \varepsilon$, which occurs at large $\tau$. Beyond this depth we say that the solution has *thermalized.* From a detailed study of the upper and lower bounds on the source function obtained by a Neumann series expansion, Avrett and Hummer suggest that the *thermalization length* $\Lambda$ be chosen such that

$$
K_2(\tfrac{1}{2}\Lambda) = \varepsilon \tag{12-42}
$$

For $\tau \gg \Lambda$, we expect the source function to agree closely with the thermal source $B$, and for $\tau < \Lambda$ we expect $S_l$ and $B$ to differ significantly. Since we are most interested in the case $\varepsilon \ll 1$, we may employ the asymptotic forms for $K_2$ that we obtained above. We thus find for a Doppler profile

$$
\Lambda \sim \frac{C}{\varepsilon} \tag{12-43}
$$

while for a Lorentz profile

$$
\Lambda \sim \frac{8}{9\varepsilon^2} \tag{12-44}
$$

and for a Voigt profile

$$
\Lambda \sim \frac{8a}{9\varepsilon^2} \tag{12-45}
$$

In equation (12-43), $C$ is a number of order unity which depends implicitly upon $\varepsilon$. We note again that these results stand in marked contrast to the case of coherent scattering where by random-walk arguments we conclude that $\Lambda \sim \varepsilon^{-1/2}$. Note also that the stronger the line wings, the greater the thermalization length, as we expected on the basis of diffusion-length arguments.

Let us now turn to the problem of determining the source function at shallow and intermediate depths. For the simplified constant-properties case that we are considering, it is easiest to derive the solution by using the method of discrete ordinates. Thus, we replace integrals by quadrature sums:

$$\int_{-\infty}^{\infty} \varphi_x f(x)\, dx \rightarrow \sum_{i=-n}^{n} a_i f(x_i) \tag{12-46}$$

and

$$\int_{-1}^{1} f(\mu)\, d\mu \rightarrow \sum_{j=-m}^{m} b_j f(\mu_j) \tag{12-47}$$

with the usual symmetry requirements:

$$x_i = -x_{-i} \qquad a_i = a_{-i}$$

$$\mu_j = -\mu_{-j} \qquad b_j = b_{-j}$$

Equation (12-18) then reduces to the coupled equations

$$\lambda_i \mu_j \frac{dI_{ij}}{d\tau} = I_{ij} - \left(\frac{1-\varepsilon}{2}\right) \sum_{i'=-n}^{n} a_{i'} \sum_{j'=-m}^{m} b_{j'} I_{i'j'} - \varepsilon B \tag{12-48}$$

where we have written

$$\lambda_i = 1/\varphi(x_i) \tag{12-49}$$

Considering first the homogeneous equation ($B \equiv 0$) and seeking a solution of the form

$$I_{ij} = g_{ij} e^{-k\tau} \tag{12-50}$$

we find that

$$I_{ij} = \frac{C}{1 + k\lambda_i \mu_j} e^{-k\tau} \tag{12-51}$$

where the constants $k$ are the roots of the characteristic equation

$$2(1-\varepsilon) \sum_{i=1}^{n} a_i \sum_{j=1}^{m} \frac{b_j}{1 - k^2 \lambda_i^2 \mu_j^2} = 1 \tag{12-52}$$

For $\varepsilon = 0$, the root $k^2 = 0$ is a solution. For $\varepsilon > 0$, $k^2$ must be greater than 0. As in our earlier treatment of the discrete-ordinate method (see Section 2-3), we may delimit the roots by the poles $k^2 = 1/\lambda_i^2 \mu_j^2$, and if we label these poles in the order of *decreasing* values of $(\lambda_i \mu_j)$, we have

$$0 < k_1^2 < (\lambda\mu)_1^{-2} < k_2^2 < \cdots < k_{mn}^2 < (\lambda\mu)_{mn}^{-2} \tag{12-53}$$

which shows that the roots may be determined rapidly by a systematic search on the appropriate finite intervals. As can be seen, it is convenient to label the roots with a subscript $\alpha$, with $1 \leq \alpha \leq mn$. By an analysis of the characteristic function quite similar to that used in deriving equation (2-91), one may readily show that

$$\prod_{\alpha=1}^{mn} (k_\alpha \lambda_i \mu_j)^2 = \varepsilon \tag{12-54}$$

We will make use of this result below. Now since $B$ is assumed constant, it is easy to see that a particular solution of equation (12-48) is $I_{ij} = B$. Thus, the general solution for the semi-infinite case must be of the form

$$I_{ij} = B\left(1 + \sum_{\alpha=1}^{mn} \frac{L_\alpha e^{-k_\alpha \tau}}{1 + k_\alpha \lambda_i \mu_j}\right) \tag{12-55}$$

where, as before, we have discarded the ascending exponentials to keep the solution bounded at infinity. The surface boundary condition is, as usual, $I_{ij}(-\mu_j) = 0$. If we define

$$\mathscr{S}(x) \equiv 1 + \sum_{\alpha=1}^{mn} \frac{L_\alpha}{1 - k_\alpha x} \tag{12-56}$$

then the surface boundary condition may be expressed as

$$\mathscr{S}(\lambda_i \mu_j) = 0 \qquad (i = 1, \ldots, n)$$

$$(j = 1, \ldots, m) \tag{12-57}$$

This set of linear equations may be solved numerically for the $L_\alpha$'s. Now the source function is given directly by

$$\begin{aligned} S_l(\tau) &= \frac{(1-\varepsilon)}{2} \sum_{i=-n}^{n} a_i \sum_{j=-m}^{m} b_j I_{ij}(\tau) + \varepsilon B \\ &= (1-\varepsilon)B + B \sum_{\alpha=1}^{mn} L_\alpha e^{-k_\alpha \tau} \left[\frac{(1-\varepsilon)}{2} \sum_{i=-n}^{n} \sum_{j=-m}^{m} \frac{a_i b_j}{1 + k_\alpha \lambda_i \mu_j}\right] + \varepsilon B \end{aligned} \tag{12-58}$$

But the term in brackets is simply unity, by use of the characteristic equation (12-52). Hence

$$S_l(\tau) = B\left(1 + \sum_{\alpha=1}^{mn} L_\alpha e^{-k_\alpha \tau}\right) \tag{12-59}$$

A result of particular interest is the value of the source function at the surface. We see that

$$S_l(0) = B\left(1 + \sum_{\alpha=1}^{mn} L_\alpha\right) = B\mathscr{S}(0) \tag{12-60}$$

Now by an analysis analogous to that used to obtain equation (2-97), one may show that

$$\mathscr{S}(x) = \frac{\prod_{\alpha=1}^{mn} k_\alpha \prod_{i=1}^{n} \prod_{j=1}^{m} (\lambda_i \mu_j - x)}{\prod_{\alpha=1}^{mn} (1 - k_\alpha x)} \tag{12-61}$$

so that

$$\mathscr{S}(0) = \prod_{\alpha=1}^{mn} (k_\alpha \lambda_i \mu_j) = \sqrt{\varepsilon} \tag{12-62}$$

by virtue of equation (12-54). Substitution into equation (12-60) then yields the extremely important result that

$$S_l(0) = \sqrt{\varepsilon}\, B \tag{12-63}$$

which is independent of the order of approximation in the quadrature sums and of the form of the profile and, hence, is completely general. We see clearly that for $\varepsilon \ll 1$, there will be very significant differences between the surface value of the source function in the line and its limiting thermal value. An important physical point to realize is that the rate at which photons are being created locally in the line is $\varepsilon B$ while the source function near the surface exceeds this value by a factor of $\varepsilon^{-1/2}$. From this fact we infer that $S_l$ near the surface is controlled primarily by photons which are being fed in from the line wings. These photons originate at depths where $S_l(\tau) \gg S_l(0)$. We see therefore that the surface value of the source function has very little to do with the local value of the thermal source term.

Let us mention in passing that the quadrature formulae must be well chosen if the results are to be reliable. The quadrature over angle may be performed with the usual double-Gauss points and weights. The quadrature over frequency requires a bit more care. In principle the integration extends over an infinite interval. In the case of a Doppler profile, special quadrature formulae (Gauss-Hermite) exist but do not yield reliable results since they tacitly assume a polynomial representation, and such representations must always diverge at infinity. Therefore it is better to choose a quadrature on a finite frequency band. However, as we showed before, if $x_0$ is the maximum value of the frequency in this band, then the scattering solution will be reliable only to depths $\tau_{max} \sim 1/\varphi(x_0)$. For a semi-infinite atmosphere, $x_0$ should be chosen large enough that $\tau_{max}$ is several times the thermalization length. If we choose such a finite band, then care must be taken to preserve normalization of the profile so that spurious sources or sinks of photons are not introduced. In pactice, this is done by scaling the quadrature weights and choosing

$$a_i = \frac{A_i}{\sum_{j=-n}^{n} A_j \varphi(x_j)} \tag{12-64}$$

where the $A_i$ are the original weights.

Several solutions of the two-level problem were obtained by Avrett and Hummer. In Figure 12-1 we show $S_l(\tau)$ for a semi-infinite atmosphere and a Doppler profile with various values of $\varepsilon$. We see that in each case $S_l(0) = \varepsilon^{1/2}$ ($B$ taken to be unity) and that $S_l$ approaches $B$ when $\tau \sim 1/\varepsilon$. In Figure 12-2

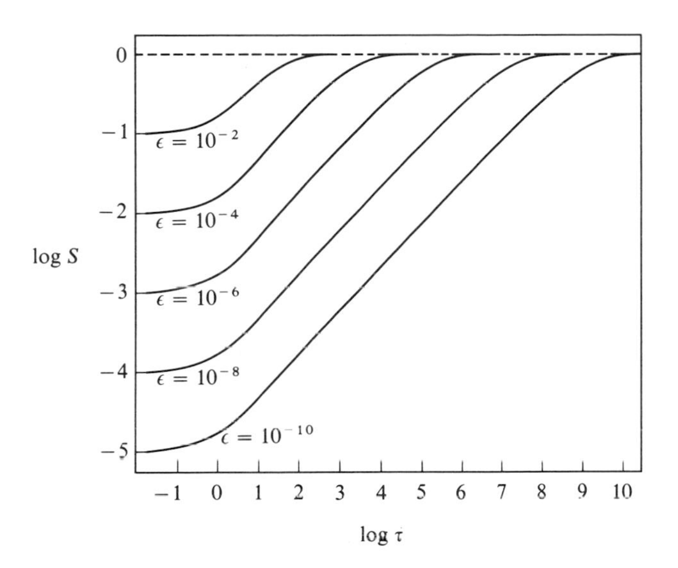


FIG. 12-1. Line source functions for Doppler broadening and a semi-infinite atmosphere. (From E. H. Avrett and D. G. Hummer, *M. N.*, **130**, 295, 1965.)

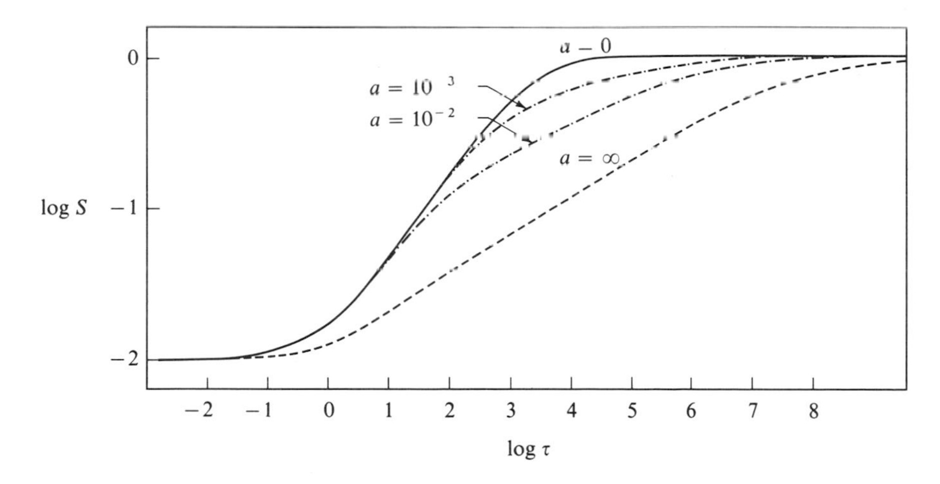


FIG. 12-2. Line source functions in a semi-infinite atmosphere for $\varepsilon = 10^{-4}$ and Voigt profiles, ranging from a pure Doppler profile to a pure Lorentz profile. (From E. H. Avrett and D. G. Hummer, *M. N.*, **130**, 295, 1965.)

we show $S_l(\tau)$ when $\varepsilon = 10^{-4}$ and with Voigt profiles ranging from a pure Doppler profile ($a = 0$) to a pure Lorentz profile ($a = \infty$). We see clearly the increase in thermalization length from $1/\varepsilon$ to $1/\varepsilon^2$, in agreement with our analysis above.

The important case of finite atmosphere, of total thickness $T$, has also been studied in some detail by Avrett and Hummer. They find that two physically distinct situations occur, in which the atmosphere can be characterized as *effectively thick* or *effectively thin*. If $T$ exceeds the thermalization length, then the solution at the upper boundary ($\tau = 0$) will be "unaware" of the lower boundary surface, and the value of the source function will be close to the semi-infinite atmosphere value for the corresponding value of $\varepsilon$. If, however, $T$ is much less than the thermalization length, then the solution never thermalizes and is nearly proportional to the local creation term $\varepsilon B$. Thus $S_l/\varepsilon B$ is nearly proportional to the local creation term $\varepsilon B$. Thus $S_l/\varepsilon B$ is nearly independent of $\varepsilon$ and depends primarily upon $T$. This situation is shown very clearly in Figure 12-3, which gives $S_l(\tau)$ for various values of $\varepsilon$ in an atmosphere with $T = 10^4$. The dotted curves give the semi-infinite atmosphere solution for $\varepsilon = 10^{-4}$. We see that for $\varepsilon \geq 10^{-4}$, the solutions do indeed resemble the semi-infinite case while for $\varepsilon < 10^{-4}$ they fall far below the appropriate semi-infinite surface values. In Figure 12-4 we show the emer-

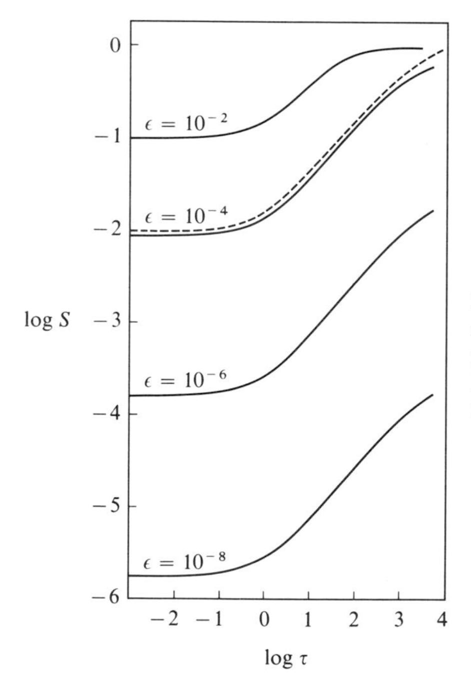


FIG. 12-3. Line source functions in infinite atmospheres with $T = 10^4$. Dashed curve corresponds to semi-infinite atmosphere solution with $\varepsilon = 10^{-4}$. (From E. H. Avrett and D. G. Hummer, *M. N.*, **130**, 295, 1965.)

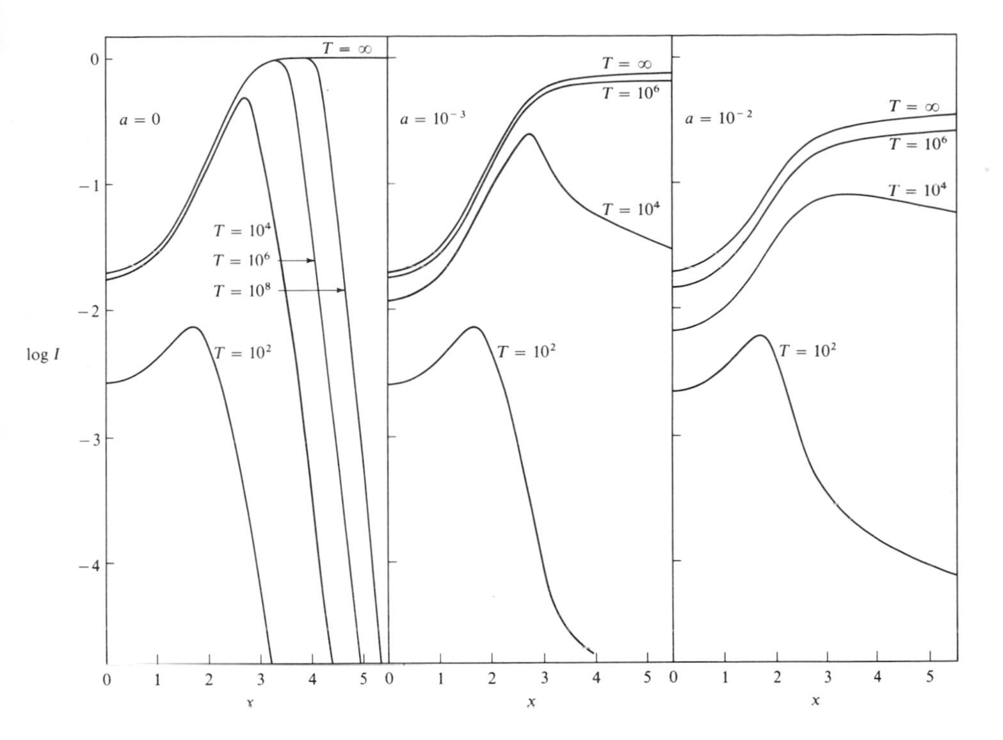


FIG. 12-4. Emergent intensity for finite and semi-infinite atmospheres with Doppler and Voigt profiles and with $\varepsilon = 10^{-4}$. (From E. H. Avrett and D. H. Hummer, *M. N.*, **130**, 295, 1965.)

gent intensities obtained by Avrett and Hummer for $\varepsilon = 10^{-4}$ and various values of $a$ and $T$. In the semi-infinite case we obtain, in each case, an absorption line, with central intensity independent of $a$. For infinite atmospheres, the intensity goes to zero for large $x$, as the wing becomes completely transparent. For smaller $x$, the intensity increases and for sufficiently thick atmospheres saturates at the semi-infinite value. Finally, for very small $x$ the scattering in the core leads to a self-reversal.

### SCALING RULES

An interesting scaling rule for the solution of the two-level problem has been given by Hummer and Stewart (*Ap. J.*, **146**, 290, 1966). They observe that the solutions for various values of $\varepsilon$ can all be written as

$$S_\varepsilon(\tau) \approx \varepsilon^{1/2} f(\tau) \tag{12-65}$$

for $0 \le \tau \le \tau_\varepsilon$, where $f(\tau)$ depends upon the profile function $\varphi_x$ but is independent of $\varepsilon$, and $\tau_\varepsilon$ is of the order of 0.01 $\Lambda$. For $\tau \gg \Lambda$, the solution $S_\varepsilon(\tau) \to B$ (which is taken to be unity for the present discussion). As $\varepsilon \to 0$, the scaling law written above holds for ever-increasing depths in the atmosphere, and this observation suggests that it is useful to write

$$S_H(\tau) = \lim_{\varepsilon \to 0} \frac{S_\varepsilon(\tau)}{\varepsilon B} \tag{12-66}$$

Here $S_H(\tau)$ is the solution of the homogeneous equation

$$S_H(\tau) = \int_0^\infty S_H(t) K_1 |t - \tau| \, dt \tag{12-67}$$

normalized such that $S_H(0) = 1$. The function $S_H(\tau)$ is unbounded at infinity since regardless of how small $\varepsilon$ becomes, $S_\varepsilon$ must approach $B$ at sufficiently great depths. For $\varepsilon = 0$ one might thus imagine that the solution corresponds to an infinitely strong source placed at infinitely great depth. For a Doppler profile the asymptotic form of $S_H$ was found by Ivanov (*Sov. Astron.—AJ*, **6**, 793, 1963)

$$S_H(\tau) \sim \frac{4\tau^{1/2}}{\pi} \left[ \ln\left( \frac{\tau}{\pi^{1/2}} \right) \right]^{1/4} \tag{12-68}$$

while for a profile with asymptotic form $\varphi(x) \sim Cx^{-\alpha}$ with $\alpha > 1$, Hummer and Stewart find

$$S_H(\tau) \sim \frac{\tau^\gamma}{\Gamma(\gamma + 1)} \left[ \left( \frac{1 + 2\gamma}{1 - 2\gamma} \right) \left( \frac{\sin \pi\gamma}{\pi} \right) C^{2\gamma - 1} \right]^{1/2} \tag{12-69}$$

where $\gamma \equiv (\alpha - 1)/2\alpha$. For a Lorentz or Voigt profile, $C$ is $1/\pi$ or $a/\pi$, respectively, and $\gamma = 1/4$. For a profile of the above form the thermalization length found from the asymptotic expression for $K_2$ is

$$\Lambda \sim 2C^{\beta/\alpha} \left[ \frac{2\Gamma\left(1 - \dfrac{1}{\alpha}\right)}{2\alpha - 1} \right]^\beta \varepsilon^{-\beta} \tag{12-70}$$

where $\beta \equiv \alpha/(\alpha - 1)$; this expression agrees with our previous result for $\alpha = 2$. While $S_\varepsilon(\tau)$ will depart from $\varepsilon^{1/2} S_H(\tau)$ for sufficiently large depths, Hummer and Stewart show that for the inverse power-law profile function mentioned above,

$$\varepsilon^{1/2} S_H(\Lambda) = \left[ \frac{2^{\gamma + 1}\Gamma(2\gamma)}{\Gamma(\gamma)\Gamma(\gamma - 1)\Gamma^2(\gamma + 1)} \right]^{1/2} \tag{12-71}$$

which is independent of $C$ and close to unity for the values of $\alpha$ that are of physical interest. Therefore $\varepsilon^{1/2}S_H(\tau)$ actually provides a very good estimate of $S_\varepsilon(\tau)$ for $\tau \lesssim \Lambda$ and is of the right order even at $\tau = \Lambda$. An improved estimate of $S_\varepsilon(\tau)$ is given by the empirical formula

$$\frac{1}{S_\varepsilon(\tau)} \approx \frac{1}{B}\left(\frac{1}{\varepsilon S_H^{\,2}(\tau)} + 1\right)^{1/2} \tag{12-72}$$

which is accurate to a few percentage points.

### THE EFFECTS OF AN OVERLAPPING CONTINUUM

We have assumed thus far that the only opacity source is the line itself. In almost all situations of astrophysical interest, however, the line will be superimposed upon other opacity sources, such as continua from other levels of the atom or from other atomic species. As we have indicated above, the nature of the transfer problem is altered significantly from the situation of a line alone. A discussion of this problem has been given by Hummer (*M. N.*, **138**, 73, 1968), whose results we will now summarize.

For the sake of simplicity, we will assume that the line and continuum opacities $l$ and $\kappa_c$, the line profile function $\varphi_x$, the line thermalization parameter $\varepsilon$, the ratio $r \equiv \kappa_c/l$, and the thermal source $B$ are all independent of depth in the atmosphere.

The transfer equation then reads

$$\frac{\mu}{\rho}\frac{dI_x}{dz} = -(\kappa_c + l\varphi_x)I_x + l\varphi_x S_l + \kappa_c B \tag{12-73}$$

where, as before [equation (12-17)], we write

$$S_l = (1 - \varepsilon)\int_{-\infty}^{\infty} \varphi_x J_x \, dx + \varepsilon B$$

We measure optical depths on the line opacity scale (in a general way we will regard the continuum merely as a perturbation) and define

$$d\tau = -\rho l dz \tag{12-74}$$

so that

$$\mu \frac{dI_x}{d\tau} = (\varphi_x + r)I_x - \varphi_x S_l - rB \tag{12-75}$$

or

$$\mu \frac{dI_x}{d\tau} = (\varphi_x + r)(I_x - S_x) \tag{12-76}$$

where

$$S_x \equiv \frac{(1-\varepsilon)\varphi_x}{\varphi_x + r} \int_{-\infty}^{\infty} J_{x'} \varphi_{x'}\, dx' + \frac{\varepsilon\varphi_x + r}{\varphi_x + r} B \tag{12-77}$$

If we now define

$$\xi_x = \frac{r + \varepsilon\varphi_x}{r + \varphi_x} = \varepsilon + \frac{(1-\varepsilon)r}{r + \varphi_x} \tag{12-78}$$

then

$$S_x = (1 - \xi_x) \int_{-\infty}^{\infty} J_{x'} \varphi_{x'}\, dx' + \xi_x B \tag{12-79}$$

We see that $\xi_x$ is the parameter describing the coupling or the radiation field at frequency $x$ to the thermal sources in line and continuum. The formal solution of the transfer equation yields

$$\begin{aligned} J_x(\tau) &= \frac{1}{2}\int_0^{\infty} S_x(t) E_1 |(\varphi_x + r)(t-\tau)| (\varphi_x + r)\, dt \\ &= \frac{\varphi_x}{2}\int_0^{\infty} S_l(t) E_1 |(\varphi_x + r)(t-\tau)|\, dt + \frac{rB}{2}\int_0^{\infty} E_1 |(\varphi_x + r)(t-\tau)|\, dt \end{aligned} \tag{12-80}$$

It is clear that this expression could be substituted into equation (12-79) for $S_x$, but the resulting integral equation would be two-dimensional ($x$ and $\tau$) and thus difficult to solve. An alternative approach is to calculate the line source function $S_l$; $S_x$ could subsequently be obtained from $S_l$ via equation (12-79) if it were desired. Now from equation (12-80) we see that the frequency-averaged mean intensity is given by

$$\begin{aligned} \int_{-\infty}^{\infty} \varphi_x J_x\, dx &= \frac{1}{2}\int_0^{\infty} dt\, S_l(t) \int_{-\infty}^{\infty} E_1 |(\varphi_x + r)(t-\tau)| \varphi_x^2\, dx \\ &\quad + \frac{rB}{2}\int_0^{\infty} dt \int_{-\infty}^{\infty} E_1 |(\varphi_x + r)(t-\tau)| \varphi_x\, dx \\ &= \int_0^{\infty} S_l(t) K_{1,r}^{*} |t-\tau|\, dt + rB \int_0^{\infty} L_{1,r}^{*} |t-\tau|\, dt \end{aligned} \tag{12-81}$$

where we have defined

$$K_{1,r}^{*}(s) \equiv \frac{1}{2}\int_{-\infty}^{\infty} E_1[(\varphi_x + r)s]\varphi_x^2\, dx \tag{12-82}$$

and

$$L^*_{1,r}(s) \equiv \frac{1}{2}\int_{-\infty}^{\infty} E_1[(\varphi_x + r)s]\varphi_x \, dx \tag{12-83}$$

Substituting equation (12-81) into the definition of $S_l$, we have

$$S_l(\tau) = (1 - \varepsilon)\int_0^{\infty} S_l(t)K^*_{1,r}|t - \tau| \, dt + B\left[\varepsilon + r(1 - \varepsilon)\int_0^{\infty} L^*_{1,r}|t - \tau| \, dt\right] \tag{12-84}$$

A very instructive simplification in form can be obtained if we rewrite this equation in terms of the mean coupling constant

$$\bar{\xi} \equiv \int_{-\infty}^{\infty} \xi_x \, dx = \varepsilon + r(1 - \varepsilon)\int_{-\infty}^{\infty} \frac{\varphi_x \, dx}{\varphi_x + r} = \varepsilon + r(1 - \varepsilon)F(r) \tag{12-85}$$

Note in passing that in the limit, as $r \to 0$, $F(r)$ becomes infinite. Now

$$(1 - \bar{\xi}) = (1 - \varepsilon)[1 - rF(r)] \tag{12-86}$$

and

$$(\bar{\xi} - \varepsilon) = (1 - \varepsilon)rF(r) \tag{12-87}$$

Thus if we define

$$K_{1,r}(s) = \frac{K^*_{1,r}(s)}{1 - rF(r)} \tag{12-88}$$

and

$$L_{1,r}(s) \equiv \frac{L^*_{1,r}(s)}{F(r)} \tag{12-89}$$

we may rewrite equation (12-84) as

$$S_l(\tau) = (1 - \bar{\xi})\int_0^{\infty} S_l(t)K_{1,r}|t - \tau| \, dt + (\bar{\xi} - \varepsilon)B\int_0^{\infty} L_{1,r}|t - \tau| \, dt + \varepsilon B \tag{12-90}$$

In this expression the first term obviously represents scattering in the line, with the probability that the photon is lost by all processes given by $\bar{\xi}$. The second term represents photons fed into the line from the continuum while the third term represents photons created by collisional processes. Thus we see that $\bar{\xi}$ replaces $\varepsilon$ as the effective coupling constant between radiation and gas. The behavior of the source function will, as in the pure line case, be determined by the properties of the kernel functions. A few of their most important properties will be mentioned here, but a more complete discussion may be found in Hummer's paper.

First, it is easy to show that

$$\lim_{r \to 0} K_{1,r}(\tau) = K_1(\tau) \tag{12-91}$$

where $K_1$ is the kernel function previously introduced in the pure line case while

$$\lim_{r \to 0} L_{1,r}(\tau) = 0 \tag{12-92}$$

For $1 \ll \tau \ll r^{-1}$, $K_{1,r}(\tau)$ is nearly independent of $r$ and has the same asymptotic form as $K_1(\tau)$ while $L_{1,r}(\tau)$ has the asymptotic form

$$L_{1,r}(\tau) \sim \frac{1}{2F(r)\tau[\ln(\tau/\pi^{1/2})]^{1/2}} \tag{12-93}$$

For $\tau \gg r^{-1}$, both $L_{1,r}$ and $K_{1,r}$ fall off more rapidly than $\exp(-r\tau)/\tau$, as we would expect from our previous discussion of $K_1$ for a nonzero minimum $\varphi$. Physically, this result merely shows the domination by the continuum when its optical depth is greater than unity.

Again, in analogy to the pure line case, it is useful to define

$$K_{2,r}(\tau) = \frac{1}{1 - rF(r)} \int_{-\infty}^{\infty} \frac{{\varphi_x}^2}{\varphi_x + r} E_2[(\varphi_x + r)\tau]\, dx \tag{12-94}$$

so that

$$\lim_{r \to 0} K_{2,r}(\tau) = K_2(\tau) \tag{12-95}$$

where $K_2$ is defined in equation (12-24). For $\tau > r^{-1}$, this kernel again falls off as $\exp(-r\tau)/\tau$ while for $1 \ll \tau \ll r^{-1}$, the behavior is again independent of $r$, and $K_{2,r}(\tau) \approx K_2(\tau)$.

As in the case of a line without overlapping continuum, we expect the source function to fall below the thermal value near the boundary and to approach it at depth. Hummer shows that the appropriate definition of the thermalization length is now

$$K_{2,r}\left(\frac{\Lambda}{2}\right) = \xi \tag{12-96}$$

which obviously is completely compatible with the definition for $r = 0$. From our previous results for the asymptotic form of $K_2$ we have the estimates

$$\Lambda \sim \frac{1}{\xi} \tag{12-97}$$

for a Doppler profile, and

$$\Lambda \sim \frac{8}{9\xi^2} \tag{12-98}$$

for a Lorentz profile while

$$\Lambda \sim \frac{8a}{9\xi^2} \tag{12-99}$$

for a Voigt profile. Note however that an upper limit on $\Lambda$ is $r^{-1}$ since at greater depths thermalization in the continuum occurs. Thus $\Lambda$ is given by the smaller of these two estimates. A brief table of $\xi$ and $\Lambda$ for $\varepsilon = 10^{-6}$ and various values of $r$ and $a$ is given in Table 12-1 (the factors of order unity have been ignored). It is easy to see the dramatic effect that even a weak continuum has upon thermalization.

TABLE 12-1. Mean Coupling Parameter $\xi$ and Thermalization Length $\Lambda$ for a Line with Overlapping Continuum ($\varepsilon = 10^{-6}$)

| $r$ | $a$ = 0 | | $a = 10^{-3}$ | | $a = 10^{-2}$ | |
|---|---|---|---|---|---|---|
| | $\xi$ | $\Lambda$ | $\xi$ | $\Lambda$ | $\xi$ | $\Lambda$ |
| 0 | $10^{-6}$ | $1 \times 10^6$ | $10^{-6}$ | $10^9$ | $10^{-6}$ | $10^{10}$ |
| $10^{-7}$ | $1.79 \times 10^{-6}$ | $5.6 \times 10^5$ | $1.87 \times 10^{-5}$ | $2.9 \times 10^6$ | $5.71 \times 10^{-5}$ | $3.1 \times 10^7$ |
| $10^{-6}$ | $8.26 \times 10^{-6}$ | $1.2 \times 10^5$ | $5.72 \times 10^{-5}$ | $3.1 \times 10^5$ | $1.78 \times 10^{-4}$ | $3.2 \times 10^5$ |
| $10^{-5}$ | $6.69 \times 10^{-5}$ | $1.5 \times 10^4$ | $1.85 \times 10^{-4}$ | $2.9 \times 10^4$ | $5.63 \times 10^{-4}$ | $3.2 \times 10^4$ |
| $10^{-4}$ | $5.85 \times 10^{-4}$ | $1.7 \times 10^3$ | $7.84 \times 10^{-4}$ | $1.6 \times 10^3$ | $1.83 \times 10^{-3}$ | $3.0 \times 10^3$ |
| $10^{-3}$ | $4.98 \times 10^{-3}$ | $2.0 \times 10^2$ | $5.23 \times 10^{-3}$ | $3.7 \times 10^1$ | $7.25 \times 10^{-3}$ | $1.9 \times 10^2$ |

*Source:* D. G. Hummer, *M. N.*, **138**, 73, 1968.

To estimate the source function at the surface, we observe that the present transfer equation is the same as for a line without continuum, but with $\varepsilon$ replaced by $\xi$ and $B$ replaced by

$$B_{\text{eff}}(\tau) = \xi^{-1}B\left[\varepsilon + (\xi - \varepsilon)\int_0^\tau L_{1,r}|t - \tau|\, dt\right] \tag{12-100}$$

We therefore expect that there exists a relation of the form

$$S_l(0) = \xi^{1/2}\bar{B}_{\text{eff}} \tag{12-101}$$

where $\bar{B}_{\text{eff}}$ is an appropriate average value of $B_{\text{eff}}(\tau)$.

It is clear from the above that if $r \ll \varepsilon$, the line will be controlled by collisions, as it was before. On the other hand, if $(1 - \varepsilon)rF(r) > \varepsilon$, either because $r > \varepsilon$ or because extensive line wings insure that the inequality holds even though $r < \varepsilon$, then both $\xi$ and $B_{\text{eff}}$ are dominated by continuum terms, and the line is *continuum controlled.* This term implies that the continuum will set the value of $S_l(0)$, even though in the line core the line opacity and line source function are larger than the continuum terms. This is still another manifestation of the fact that the source function near the surface is controlled by the line wings, as we remarked upon earlier.

Numerical solutions for several illustrative cases have been obtained by Hummer using the method of discrete ordinates. His results for $S_l(\tau)$ for a semi-infinite atmosphere in which $\varepsilon = 10^{-6}$ and $B = 1$ are shown in Figure 12-5 for a Doppler profile and various values of $r$. Also shown by arrows are the thermalization depths and the value of $\bar{\xi}^{1/2}$, which would be $S_l(0)$ if $B_{\text{eff}}$ were everywhere unity. For this case, the agreement is good. Results for Voigt profiles are shown in Figure 12-6. Here the estimate $S_l(0) = \bar{\xi}^{1/2}$ is too large, in the worst case by about 40%. A small improvement results if one adopts

$$\bar{B}_{\text{eff}} = \Lambda^{-1} \int_0^{\Lambda} B_{\text{eff}}(\tau)\, d\tau \tag{12-102}$$

and much better agreement is obtained if one takes

$$\bar{B}_{\text{eff}} = \left(\frac{\Lambda}{3}\right)^{-1} \int_0^{\Lambda/3} B_{\text{eff}}(\tau)\, d\tau \tag{12-103}$$

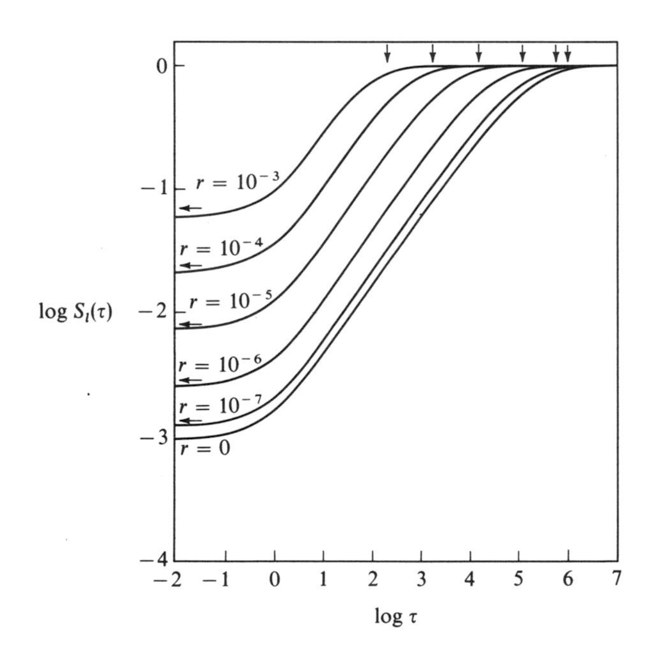


FIG. 12-5. Line source functions for two-level atom with $\varepsilon = 10^{-6}$ and overlapping continuum in semi-infinite atmosphere. (From D. G. Hummer, *M. N.*, **138**, 73, 1968.)

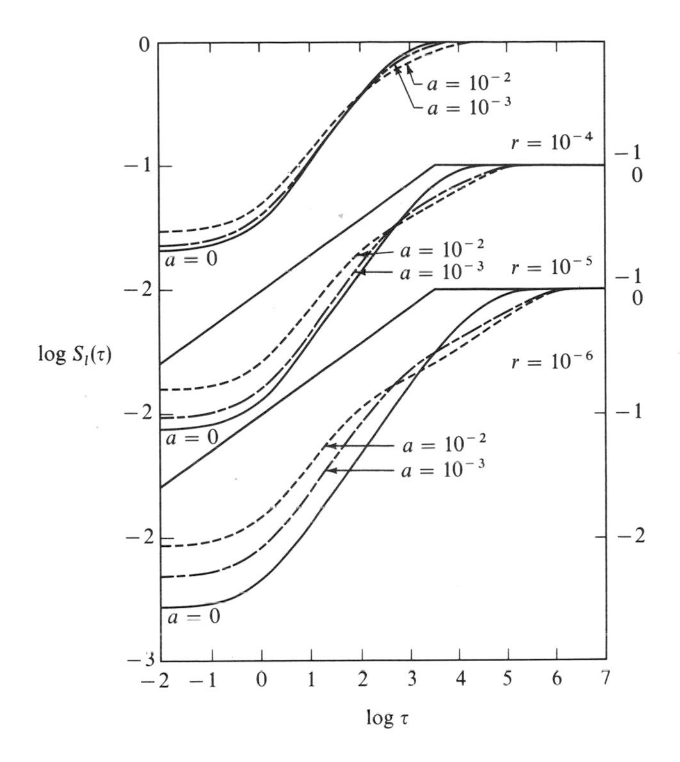


FIG. 12-6. Line source functions for two-level atom with $\varepsilon = 10^{-6}$ and overlapping continuum in semi-infinite atmosphere, showing effects of changing line profiles. (From D. G. Hummer, *M.N.*, **138**, 73, 1968.)

which accounts for the strong decrease in the weighting function at depth. A more accurate approximation, which reproduces the numerical results very closely, is given by Hummer.

## 12-3. The Two-Level Atom with Continuum

### THE SOURCE FUNCTION

The discussion given in the preceding sections is predicated entirely on the assumption of a very schematic and admittedly restrictive atomic model. Naturally, the true situation for any real atom is much more complicated. To introduce some of the physically important effects and at the same time retain the greatest possible analytical simplicity, we now consider an atomic model consisting of two bound states and a continuum. The addition of the continuum increases the number of physical processes that may take place. In the two-level case without continuum, the only processes that could occur were photoexcitations or collisional excitations from the lower to upper state and radiative or collisional de-excitations from the upper to lower state and radiative or collisional de-excitations from the upper to lower state. In the present case, we may, in addition, have photoionizations and collisional ionizations from each bound level to the continuum, and from the continuum to the bound levels we may expect radiative recombinations and three-body collisional recombinations. Thus, it is clear that we have a much more general and flexible model, which should embrace a considerable amount of new information. Basically, this model should be relevant to the resonance lines of an atom; it may also apply to subordinate lines if transitions to other bound levels can be shown to be essentially in detailed balance.

As we did before, we assume that complete redistribution is valid so that we may again write equation (12-5):

$$S_l = \left(\frac{2h\nu^3}{c^2}\right)\left[\frac{N_1 g_2}{N_2 g_1} - 1\right]^{-1}$$

The statistical equilibrium equations for this atomic model are now

$$\begin{aligned} N_1\left(B_{12}\int \varphi_\nu J_\nu\,d\nu + N_e\Omega_{12} + R_{1\kappa} + N_e\Omega_{1\kappa}\right) \\ - N_2\left(A_{21} + B_{21}\int \varphi_\nu J_\nu\,d\nu + N_e\Omega_{21}\right) \\ = N_1{}^*(R_{\kappa 1} + N_e\Omega_{1\kappa}) \end{aligned} \tag{12-104}$$

for the first level,

$$\begin{aligned} -N_1\left(B_{12}\int \varphi_\nu J_\nu\,d\nu + N_e\Omega_{12}\right) \\ + N_2\left(A_{21} + B_{21}\int \varphi_\nu J_\nu\,d\nu + N_e\Omega_{21} + R_{2\kappa} + N_e\Omega_{2\kappa}\right) \\ = N_2{}^*(R_{\kappa 2} + N_e\Omega_{2\kappa}) \end{aligned} \tag{12-105}$$

for the second level, and

$$\begin{aligned} N_1(R_{1\kappa} + N_e\Omega_{1\kappa}) + N_2(R_{2\kappa} + N_e\Omega_{2\kappa}) \\ = N_1{}^*(R_{\kappa 1} + N_e\Omega_{1\kappa}) + N_2{}^*(R_{\kappa 2} + N_e\Omega_{2\kappa}) \end{aligned} \tag{12-106}$$

for the continuum.

Note that the third equation is redundant because it is merely the sum of the first two. In the above equations we have used equation (5-35),

$$R_{i\kappa} = 4\pi \int_{\nu_i}^{\infty} \frac{\alpha_i(\nu)J_\nu}{h\nu}\,d\nu$$

and equation (5-38),

$$R_{\kappa i} = 4\pi \int_{\nu_i}^{\infty} \frac{\alpha_i(\nu)}{h\nu} [B_\nu(1 - e^{-h\nu/kT}) + J_\nu e^{-h\nu/kT}]\, d\nu$$

If we solve equations (12-104) and (12-105) directly for $N_1/N_2$ and make use of the Einstein relations as well as the relation $N_1{}^*\Omega_{12} = N_2{}^*\Omega_{21}$, we find

$$\left(\frac{g_2 N_1}{g_1 N_2} - 1\right) = \frac{\begin{gathered}[A_{21} + N_e\Omega_{21}(1 - e^{-h\nu/kT})][N_1{}^*(R_{\kappa 1} + N_e\Omega_{1\kappa}) + N_2{}^*(R_{\kappa 2} + N_e\Omega_{2\kappa})] - \\ \left[(R_{2\kappa} + N_e\Omega_{2\kappa})N_1{}^*(R_{\kappa 1} + N_e\Omega_{1\kappa}) - (R_{1\kappa} + N_e\Omega_{1\kappa})\frac{g_1 N_2{}^*}{g_2}(R_{\kappa 2} + N_e\Omega_{2\kappa})\right]\end{gathered}}{\begin{gathered}(B_{21}\int \varphi_\nu J_\nu\, d\nu + N_e\Omega_{21} e^{-h\nu/kT})[N_1{}^*(R_{\kappa 1} + N_e\Omega_{1\kappa}) + N_2{}^*(R_{\kappa 2} + N_e\Omega_{2\kappa})] + \\ \frac{g_1}{g_2}(R_{1\kappa} + N_e\Omega_{1\kappa})N_2^*(R_{\kappa 2} + N_e\Omega_{2\kappa})\end{gathered}} \tag{12-107}$$

from which we obtain

$$S_l = \frac{\begin{gathered}\int \varphi_\nu J_\nu\, d\nu + \frac{N_e\Omega_{21}}{A_{21}}(1 - e^{-h\nu/kT})B_\nu(T) + \left(\frac{2h\nu^3}{c^2}\right)\frac{g_1}{g_2 A_{21}} \times \\ \left[\frac{(R_{1\kappa} + N_e\Omega_{1\kappa})N_2{}^*(R_{\kappa 2} + N_e\Omega_{2\kappa})}{N_1{}^*(R_{\kappa 1} + N_e\Omega_{1\kappa}) + N_2{}^*(R_{\kappa 2} + N_e\Omega_{2\kappa})}\right]\end{gathered}}{\begin{gathered}1 + \frac{N_e\Omega_{21}}{A_{21}}(1 - e^{-h\nu/kT}) + \\ \frac{1}{A_{21}}\left[\frac{(R_{2\kappa} + N_e\Omega_{2\kappa})N_1{}^*(R_{\kappa 1} + N_e\Omega_{1\kappa}) - g_1(R_{1\kappa} + N_e\Omega_{1\kappa})N_2{}^*(R_{\kappa 2} + N_e\Omega_{2\kappa})/g_2}{N_1^*(R_{\kappa 1} + N_e\Omega_{1\kappa}) + N_2{}^*(R_{\kappa 2} + N_e\Omega_{2\kappa})}\right]\end{gathered}} \tag{12-108}$$

An expression of this form was first studied extensively by Thomas (*Ap. J.*, **125**, 260, 1957). The various terms entering the numerator and denominator can be given a simple physical explanation. First consider the numerator. The first term is simply the scattering term that describes the photon diffusion through the medium and acts as a reservoir term. The second term is a source term, giving the rate at which photons are created by collisional excitation. The third term is a source term, describing the rate at which electrons leave the ground state to the continuum, times the fraction that recombines to the upper state and thus becomes available for radiative decay to the ground state. In the denominator, the second term accounts for those photons destroyed by collisional de-excitations of atoms in the upper state. The third term accounts for line photon losses by photoionization from the upper state, followed by recombination to the lower state (corrected for the inverse process). We thus find all of the relevant physical processes displayed in this source function.

For ease of visualization we may rewrite equation (12-108) in the form

$$S_l = \frac{\int \varphi_\nu J_\nu \, d\nu + \varepsilon B_\nu(T) + \eta B^*}{1 + \varepsilon + \eta} \tag{12-109}$$

where the definition of $\varepsilon$ is the same as before, the definition of $\eta$ is obvious from the above, and

$$B^* \equiv \frac{2h\nu^3}{c^2} \left[ \frac{N_1{}^* g_2 (R_{2\kappa} + N_e \Omega_{2\kappa})(R_{\kappa 1} + N_e \Omega_{1\kappa})}{N_2{}^* g_1 (R_{1\kappa} + N_e \Omega_{1\kappa})(R_{\kappa 2} + N_e \Omega_{2\kappa})} - 1 \right]^{-1} \tag{12-110}$$

The term $B_\nu(T_e)$ is the value of the Planck function at the local electron temperature. In contrast, $B^*$ is a source whose value is set by the radiation field in the continua of the two levels. For example, if we imagine that both continua are so opaque that $J_\nu \rightarrow B_\nu(T_e)$, then $R_{2\kappa} \rightarrow R_{2\kappa}^* = R_{\kappa 2}^*$ while $R_{1\kappa} \rightarrow R_{1\kappa}^* = R_{\kappa 1}^*$ so that $B^* \rightarrow B_\nu(T_e)$. In this opaque situation, we thus find $S_l \rightarrow B_\nu(T_e)$, as expected. At the opposite extreme, we can imagine both continua to be completely transparent (even at large optical depths in the line), in which case $B^*$ is essentially fixed at some characteristic value $B_\nu(T_r)$, where $T_r$ is the "temperature" of the radiation field and is in general, not equal to $T_e$. The parameter $T_r$ will, in addition, generally be different in the two continua.

## CLASSIFICATION OF LINES

Close examination of equation (12-109) allows us to discriminate among different kinds of line transfer problems. If, for example, the terms with $\varepsilon$ are much larger than those with $\eta$, we have what Thomas has designated *collision-dominated* lines (*Ap. J.*, **125**, 260, 1957). On the other hand, if the terms in $\eta$ exceed those in $\varepsilon$, we have what Thomas has characterized as *photoionization-dominated* lines. Clearly we may also have cases where $\varepsilon$ dominates $\eta$ in the numerator while $\eta$ dominates $\varepsilon$ in the denominator, and vice versa. These lines are designated *mixed-domination* lines. As we shall see below, the behavior of these cases is very different in terms of their emergent profiles, particularly when a temperature rise occurs in the outermost layers of the atmosphere (a *chromosphere*). The recognition of these classes of lines represents a considerable advance over the earlier, ill-defined, classical division of lines into "absorption" and "scattering" lines and has led to important insights into the line formation problem.

The category to which any particular line belongs will depend upon the details of the structure of the ion from which it arises via the relevant cross-sections and, to a lesser extent, upon the structure of the atmosphere since certain rates will depend upon atmospheric parameters such as temperature and density. Indeed, different lines of even the same ion will fall into different classes. Thus, no comprehensive a priori classification is easy, and each case must be examined in turn. Nevertheless, it is possible to make certain very broad groupings, as has been done by Thomas. We list his broad categorizations in Table 12-2 and emphasize that they are intended only as guidelines.

TABLE 12-2. Broad Classes of Line Source Functions in a Solar-Type Atmosphere

| *Collision dominated* | *Photoionization dominated* |
|---|---|
| Resonance lines of singly ionized metals ($Ca^+$, $Sr^+$, etc.) | Resonance lines of neutral metals |
| Resonance lines of H and other nonmetals (C, N, O, etc.) | Hydrogen Balmer lines |

*Source:* Adapted from R. N. Thomas, *Ap. J.*, **125**, 260, 1957.

For example, even though neutral sodium would normally fall in the photoionization-dominated category, it turns out to have an unusually large collision cross-section, and as a result the sodium *D*-lines are actually collision dominated (see, e.g., the discussions by Johnson and by Mugglestone, Ref. 16, pp. 333 and 347). A somewhat more refined classification has recently been given by Thomas (Ref. 33, p. 174).

### BEHAVIOR OF THE SOURCE FUNCTION IN THE PRESENCE OF A CHROMOSPHERE

From the form of the source function given in equation (12-109), we see that the source and sink terms in a collision-dominated line couple directly to the local electron kinetic temperature. In contrast, photoionization-dominated lines couple to some characteristic radiation temperature, which is set at some point much deeper in the atmosphere where the continua begin to become opaque. In a qualitative way then, we expect collision-dominated source functions to be at least partly responsive to local temperature conditions and the profiles of such lines to contain some kind of information about these conditions. On the other hand, we expect photoionization-dominated lines to be very insensitive to changes in the local value of $T_e$. This strong dichotomy of behavior was very clearly demonstrated in the pioneering study of Jefferies and Thomas (*Ap. J.*, **129**, 401, 1959). They showed that these considerations lead naturally to emission cores in lines like the $H$ and $K$ lines of $Ca^+$—as are observed in the solar spectrum (and in other late-type stars)—and an absence of emission cores in lines like the Balmer lines of hydrogen—again in agreement with observation. They considered a schematic temperature distribution which had a uniform gradient inward in the photosphere, a minimum value at some characteristic depth in the outer layers ($T_{\min} \approx 4000°K$ at $\tau_c \sim 10^{-2}$), and then a steep outward gradient mimicking the sharp temperature rise in the chromosphere. The continuum source function was thus assumed to be of the form

$$S_c(\tau) = B_\nu(T_e) = S_1(1 + \alpha\tau + \beta e^{-\tau}) \tag{12-111}$$

where $\tau$ is the optical depth at line center. The transfer equation in the Eddington approximation,

$$\frac{1}{3}\frac{d^2 J_\nu}{d\tau^2} = (\varphi_\nu + r)^2 \left[ J_\nu - \frac{\varphi_\nu S_l + r S_c}{\varphi_\nu + r} \right] \tag{12-112}$$

was solved using the method of discrete ordinates for a Doppler profile, choosing typical values of $\varepsilon$, $\eta$, $r$, and $B^*$. We show in Figure 12-7 some of the results for a collision-dominated line ($\eta \equiv 0$) for $\varepsilon = 10^{-4}$, $r = 10^{-4}$, and various choices for the continuum source function. Note that at great depth, $S_l$ thermalizes to $S_c$ but that as the wings of the line begin to become transparent ($\tau_c \lesssim 1$), the line source function lies below $S_c$. Proceeding outward,

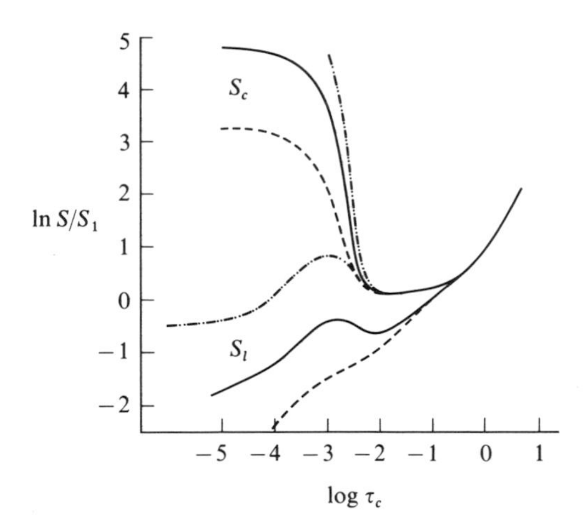


FIG. 12-7. Line source functions for a collision-dominated line in a semi-infinite atmosphere with a chromospheric temperature rise. Upper curves show continuum source function, and lower curves show corresponding line source functions. In all cases $\varepsilon = 10^{-4}$ and $r = 10^{-4}$. (From J. T. Jefferies and R. N. Thomas, *Ap. J.*, **129**, 401, 1959; by permission.)

the continuum source function rises very steeply, following the assumed rise of $T_e(\tau)$. The line source function tries to follow this rise through collisional coupling, but ultimately the effects of scattering dominate, and at the surface the source function decreases to a value some 5 to 6 orders of magnitude below the local thermal value. The hump on the $S_l$ curve leads to an emission bump in the line profile, as shown in Figure 12-8. This bump, as mentioned above, is in qualitative agreement with observations of the solar *H*- and *K*-lines. The *H*- and *K*-lines are also observed to be in emission in many late-type stars and provide valuable clues to the structure of their chromospheres. One of the most fascinating empirical results was the discovery by Wilson and Bappau (*Ap. J.*, **125**, 661, 1957) that the widths of the emission components are correlated closely with the luminosity of the stars, and several theoretical studies have been carried out to develop an understanding of this effect. As was early emphasized by Jefferies and Thomas (*Ap. J.*, **131**, 695, 1960), such an understanding must be based on an accurate physical picture of line formation. Their analysis shows that the results are influenced by many parameters; the gradient of $T_e$, the depth variations of $\varepsilon$, of $r$, of the Doppler width $\Delta\nu_D$, and the presence of any turbulent velocities. This problem has been attacked with increasingly general approaches by several workers (see, e.g., the work of Athay and Skumanich in *Solar Phys.*, **3**, 181, 1968, and *Solar Phys.*, **4**, 176, 1968; and of Linsky, Ref. 24).

In contrast to the behavior shown by the collision-dominated case, Jefferies and Thomas (*Ap. J.*, **129**, 401, 1959) find the behavior shown in Figure 12-9 for a photoionization-dominated line with $\varepsilon = 0$, $\eta = 10^{-2}$, and $r = 10^{-4}$. Here we see that the line source function shows a uniform decrease

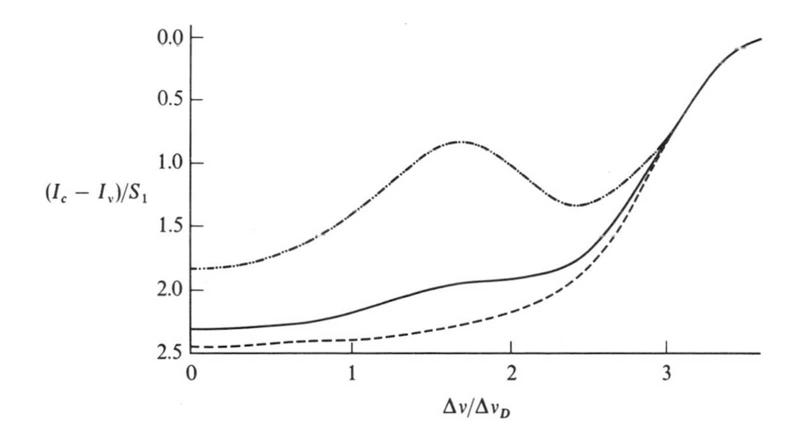


FIG. 12-8. Line profiles for a collision-dominated line in a semi-infinite atmosphere with a chromospheric temperature rise. Curves are coded to correspond with those of Figure 12-7. (From J. T. Jefferies and R. N. Thomas, *Ap. J.*, **129**, 401, 1959; by permission.)

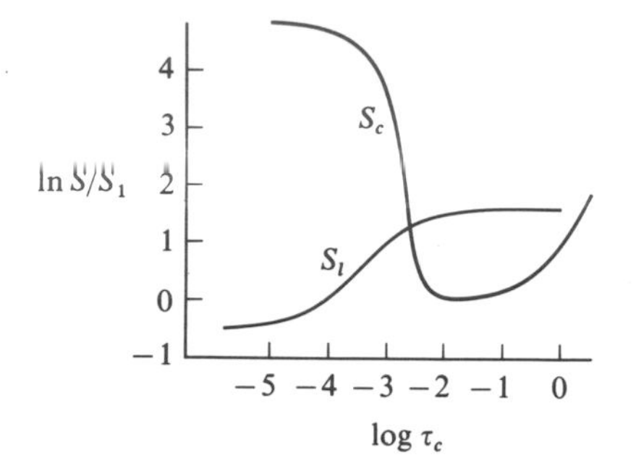


FIG. 12-9. Source function for a photo-ionization-dominated line in a semi-infinite atmosphere with a chromospheric temperature rise. (From J. T. Jefferies and R. N. Thomas, *Ap. J.* **129**, 401, 1959; by permission.)

outward with a strong drop at the surface—characteristic of scattering. The source function shows no sensitivity to the variation of $T_e$, the overall scale of the solution being set by $B^*$ only. At great depth the line source function $S_l$ will, of course, thermalize to $S_c$. Note that $S_l$ at first lies *above* $S_c$ (as we proceed outward) but then because it fails to respond to the sharp rise in $S_c$ at the boundary, it lies *below* $S_c$ at the surface. In this case, the line appears as a pure absorption line with no central emission. This type of behavior correlates quite well (at least qualitatively) with the empirical source function in the solar Balmer lines deduced by Athay and Thomas (*Ap. J.*, **127**, 96, 1958), as shown in Figure 12-10. The parameter $T_{ex}$ used in the figure is defined by

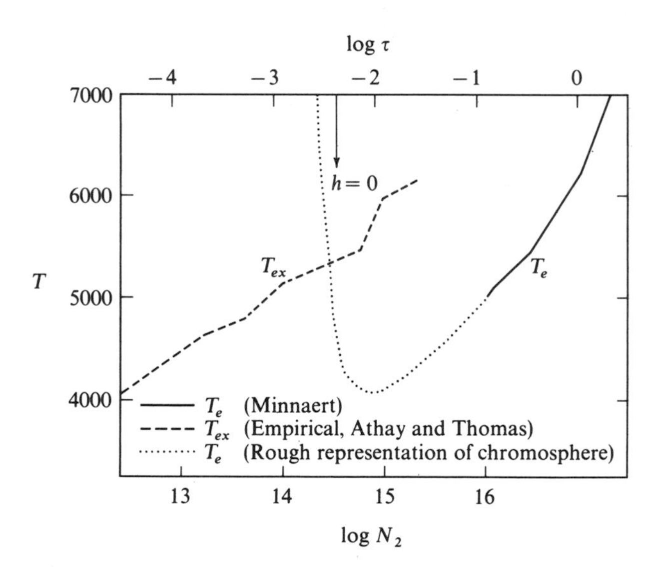


FIG. 12-10. Empirical excitation temperature deduced from Balmer lines compared with electron temperature distribution in solar atmosphere. Note qualitative resemblance to Figure 12-9. (From J. T. Jefferies and R. N. Thomas, *Ap. J.*, **129**, 401, 1959; by permission.)

the relation $S_l = B_\nu(T_{ex})$. We see that the empirical $T_{ex}$ (or, equivalently, $S_l$) lies below $T_e$ (or $S_c$) in the outermost regions of the atmosphere, then rises above in the region of minimum $T_e$, and finally appears to approach $T_e$ at depth.

## 12-4. Effects of Doppler-Width Variations

We have thus far assumed that most of the parameters describing the line formation are constant with depth, the exception being a variation of $B_\nu(T_e)$, which we discussed briefly. When other depth variations are allowed, one must generally solve the transfer problem by using one of the numerical techniques described previously. A particularly interesting case to study is that of variations of the Doppler width of the line profile; this problem has been treated by Hummer and Rybicki (*J, Q. S. R. T.*, **6**, 661, 1966).

Consider a Doppler-broadened line, and take as the frequency variable

$$x = \frac{\nu - \nu_0}{\Delta_0} \tag{12-113}$$

where $\Delta_0$ is the Doppler width at $\tau = 0$. Assume that the temperature varies with depth, and write

$$\delta = \frac{\Delta(T)}{\Delta_0} = \left[\frac{T(\tau)}{T_0}\right]^{1/2} \tag{12-114}$$

and

$$\varphi(x, \delta) = \frac{1}{\sqrt{\pi}\,\delta} e^{-x^2/\delta^2} \tag{12-115}$$

Then for a two-level atom we have

$$S(\tau) = (1 - \varepsilon) \int_{-\infty}^{\infty} \varphi(x, \delta) J_x \, dx + \varepsilon B_\nu(T) \tag{12-116}$$

where $B_\nu$ is the usual Planck function. Hummer and Rybicki solve the transfer equation in the form

$$\mu \frac{dI(\tau, x, \mu)}{d\tau} = \varphi(x, \delta)[I(\tau, x, \mu) - S(\tau)] \tag{12-117}$$

using the Riccati transformation method. To study the effects of temperature variations, they consider an isothermal atmosphere at 5000°K with an added rise or drop in temperature near the boundary. Accordingly, they write

$$T(\tau) = 5000 + \frac{a}{\exp[b(\tau - \tau_0)] + 1} \tag{12-118}$$

where $a$ is chosen to give the desired boundary temperature and $\tau_0$ is the effective thickness of the layer with the altered temperature. To start, they took $b = 2$ and $\tau_0 = 3$. The run of $T(\tau)$ for the two extreme cases that were considered is shown in Figure 12-11. They considered two lines: $L_\alpha$ and a line for which $h\nu = 1$ eV. Figure 12-11 also shows the run of $S(\tau)$ for $L\alpha$ in the cases where $\delta \equiv 1$ and $\delta$ is variable. Clearly, variations in the Doppler width lead to significant changes in the source function, which are much larger than the effects of the temperature change by itself with the Doppler width held constant. This result is shown in Table 12-3 which gives surface values of $S$ and $I$. These results can be understood physically when we recall that the surface value of the source function is fixed by the radiation intercepted in the wings of the line. When the temperature near the surface drops, the line profile becomes narrower and the line source function decreases. When the temperature rises, the opposite effect occurs.

In the above example, the region over which the temperature variation occurs is not thick. It is very instructive to consider the case of a 5000°K

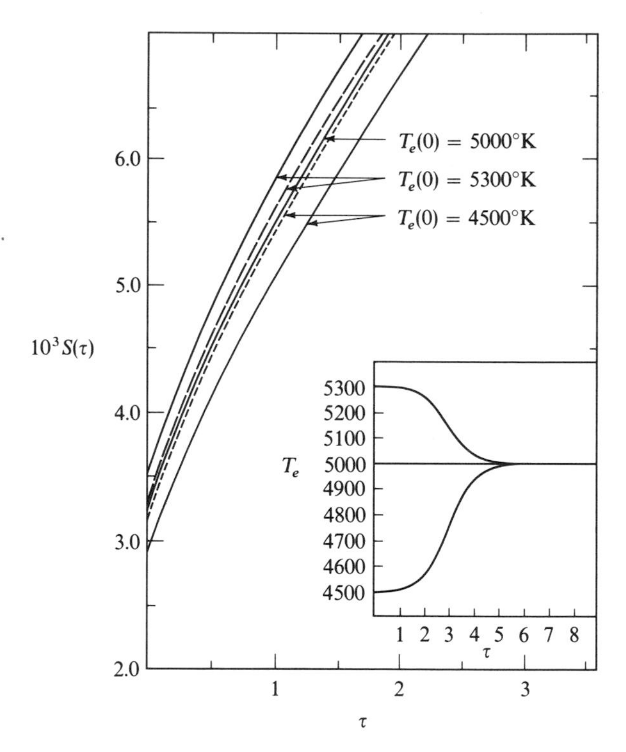


FIG. 12-11. Lyman $\alpha$ source functions for $N_e = 10^{10}$ and temperature distribution shown in inset. Dashed lines correspond to constant Doppler-width solutions. (From D. G. Hummer and G. Rybicki, *J.Q.S.R.T.*, **6**, 661, 1966; by permission.)

TABLE 12-3. Surface Values of Planck and Source Functions, and Line-Center Normally Emergent Intensities for Lyman α and for a 1 eV Line[a]

| | $h\nu_0 = 10.2$ eV | | | | $h\nu_0 = 1.0$ eV | | | |
|---|---|---|---|---|---|---|---|---|
| $T_e(0)$ | $B(0)$ | $10^3S(0)^b$ | $10^3S(0)$ | $10^3I_0(0)$ | $B(0)$ | $10^3S(0)^b$ | $10^3S(0)$ | $10^3I_0(0)$ |
| 4500 | 0.072 | 3.23 | 2.93 | 5.91 | 0.771 | 4.10 | 3.73 | 7.49 |
| 4600 | 0.128 | 3.23 | 2.99 | 6.00 | 0.816 | 4.10 | 3.80 | 7.61 |
| 4700 | 0.221 | 3.24 | 3.05 | 6.10 | 0.861 | 4.10 | 3.87 | 7.72 |
| 4800 | 0.373 | 3.24 | 3.12 | 6.20 | 0.907 | 4.10 | 3.95 | 7.84 |
| 4900 | 0.617 | 3.25 | 3.19 | 6.30 | 0.953 | 4.11 | 4.03 | 7.96 |
| 5000 | 1.00 | 3.26 | 3.26 | 6.41 | 1.00 | 4.11 | 4.11 | 8.08 |
| 5100 | 1.59 | 3.27 | 3.34 | 6.53 | 1.05 | 4.11 | 4.19 | 8.20 |
| 5200 | 2.49 | 3.30 | 3.43 | 6.65 | 1.09 | 4.11 | 4.27 | 8.33 |
| 5300 | 3.82 | 3.34 | 3.53 | 6.79 | 1.14 | 4.12 | 4.36 | 8.45 |

[a] *B, S,* and *I* are expressed in units of the appropriate Planck function at 5000°K.
[b] $\delta = 1$.
*Source:* D. G. Hummer and G. B. Rybicki, *J. Q. S. R. T.*, **6**, 661, 1966; by permission.

atmosphere with a surface zone which can be more or less thick, rising to 10,000°K. This model provides a caricature of a chromospheric temperature rise. The same analytical form is used for $T(\tau)$, but with $b = 1$. Several assumed variations in $B_\nu$ for $L\alpha$ are shown in Figure 12-12. The resulting source functions are shown in Figure 12-13. As the thickness of the high-temperature

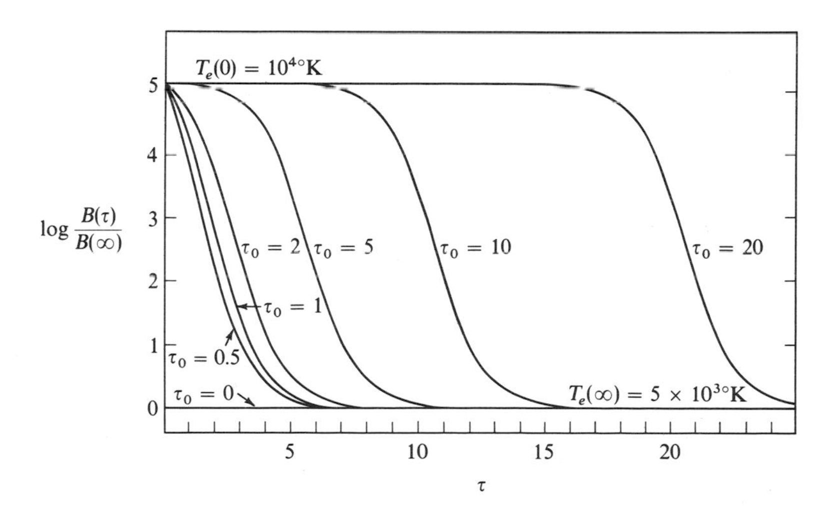


FIG. 12-12. Assumed Planck functions for Lyman α. (From D. G. Hummer and G. Rybicki, *J.Q.S.R.T.*, **6**, 661, 1966; by permission.)

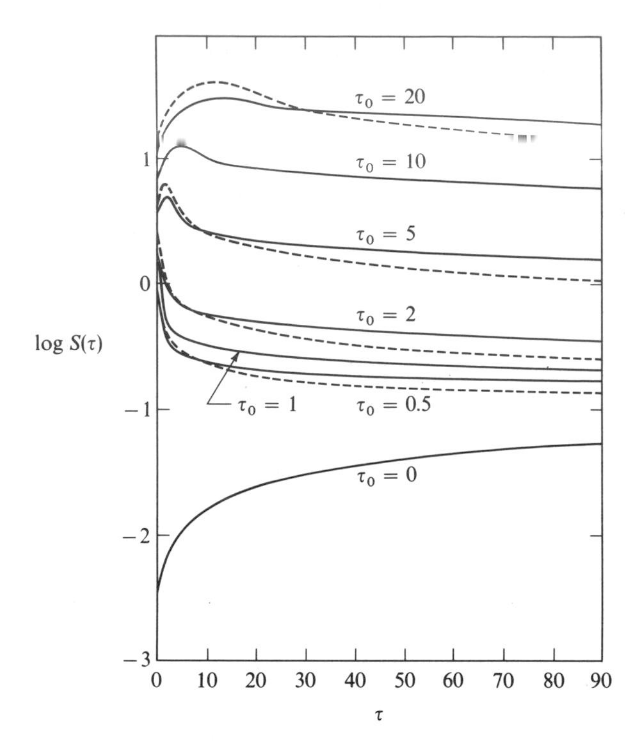


FIG. 12-13. Lyman $\alpha$ source functions for the Planck functions shown in Figure 12-12—$T_e(0) = 10^4$°K, $T_e(\infty) = 5 \times 10^3$°K, and $N_e = 10^{10}$. Dashed lines give results for constant Doppler width. (From D. G. Hummer and G. Rybicki, *J.Q.S.R.T.*, **6**, 661, 1966; by permission.)

layer increases, very large changes in the source function are found, which extend to depths much beyond $\tau_0$. For $\tau_0 = 0.5$, 1, and 2, $S(\tau)$ increases right to the surface without showing the characteristic drop of the other cases. This behavior is caused by the dominance of the creation term $\varepsilon B$ when the region is too thin for the scattering term to build to an appreciable value. Note particularly that near the surface, the variable width case lies below the $\delta = 1$ solution while at great depth it lies above it. This may be understood physically in terms of the variation of the wings in the emission profile. The wings are stronger at the surface and thus allow the radiation to escape more efficiently there while in the cooler lower regions, the radiation flowing out is impeded by the extra opacity in the wings in the surface layers, which now act as partial reflectors. Hummer and Rybicki have called this phenomenon the *reflector effect*. If we now consider the 1 eV line, the variations in $B_\nu$ caused by the temperature rise are small, and if $\delta = 1$, the two extreme cases $\tau_0 = 0$ and $\tau_0 = 20$ scarcely differ (see Figure 12-14). Here, virtually all of

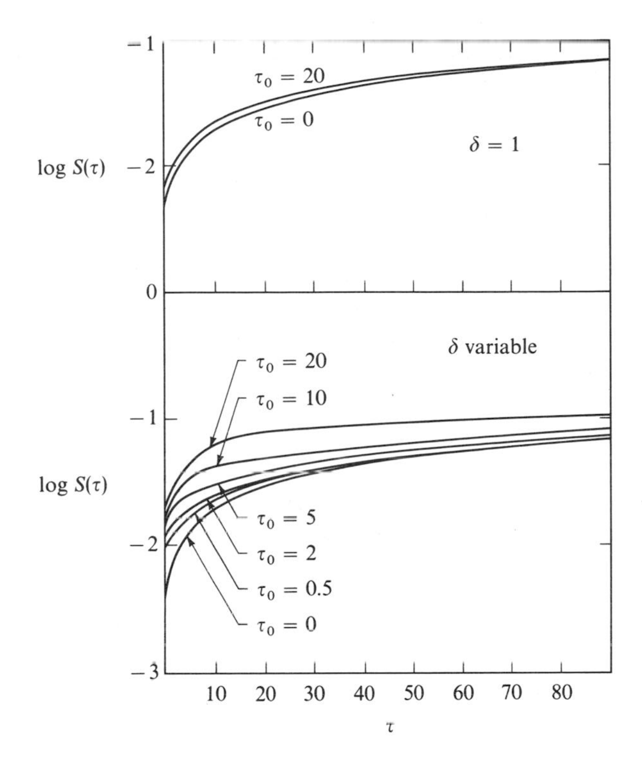


FIG. 12-14. Source functions for 1 eV line—$T_e(0) = 10°$K, $T_e(\infty) = 5 \times 10^3$°K, and $N_e = 10^{10}$. Upper panel gives results for constant Doppler width; lower panel gives results for variable Doppler width. (From D. G. Hummer and G. Rybicki, *J.Q.S.R.T.*, **6**, 661, 1966; by permission.)

the effect is due to variations in the opacity in the wings, as described earlier. These two different responses show a natural division of lines in a nonisothermal atmosphere into two categories: those with $h\nu/kT \gg 1$ and those with $h\nu/kT \lesssim 1$. In the former case, the dominant effect is a response to the large variations in the Planck function; in the latter case, variations in the Doppler width are more important. This study clearly demonstrates the dependence of the result upon all the details of the line formation process and shows the importance of properly accounting for depth variations.

## 12-5. Comments on LTE Diagnostics

Before turning to the more difficult multilevel problem, it seems worthwhile to summarize and stress a few of the important differences between the results of LTE and non-LTE analysis from the point of view of diagnosing the run of physical variables in a stellar atmosphere. We will mention here only a few of the more important points and refer the reader to Thomas' book (Ref. 33) for a much fuller discussion.

Consider first the central intensities of strong lines. If the line is assumed to be formed in LTE, then the central intensity $R_0$ is given, at least roughly, by $R_0 = B_\nu(T_0)/B_\nu(\tau_c \approx 1)$, i.e., the central intensity of the line is essentially fixed by the boundary temperature. If the observed central intensity of the line is deeper than calculated, the LTE analysis demands that the boundary temperature be lowered. As mentioned above, not all strong lines show the same central intensities in actual stellar spectra, and thus no one value of $T_0$ can generally be found that is compatible with all lines. In contrast, the simplified non-LTE analysis we have studied shows that $R_0$ is almost entirely uncorrelated with $T_0$ but instead is fixed by the details of the line transfer over a thermalization length in the line. To be sure, the division of lines into "scattering" lines and "absorption" lines in the classical theory alleviates this difficulty, but we must recall that actually the scattering is not coherent, that the classical division is essentially ad hoc, and that it provides no framework for specifying the coupling constants. Indeed, the classical theory does not really even recognize clearly the photoionization-dominated case. In short, interpretations of the temperature structure of the atmosphere on the basis of the LTE picture must be regarded with great caution.

A related point is that a line calculated in LTE is almost always weaker than when deviations from LTE are allowed. The most extreme example is that of an atmosphere with constant $T$ and $N_e$. Here LTE predicts no depression whatsoever in the continuum while the non-LTE theory leads to an absorption line with central intensity dependent upon the parameters $\varepsilon$, $\eta$, and $B^*$. If departures from LTE are really very important in stellar atmospheres, then we must of course be skeptical of abundances deduced by LTE considerations.

Another problem lies in the introduction of the "turbulence" parameter. This requires careful consideration since the position of the flat part of the curve of growth may be altered when non-LTE effects are allowed. While there is little doubt of some kind of mass motions in the most extreme cases, turbulent velocities of the order of thermal velocities, or smaller, can be affected strongly by the precise position of the flat part of the curve of growth and may be in error. A beginning has been made on this problem by Athay and Skumanich (*Ap. J.*, **152**, 211, 1968), but since they do not apply their results to stellar spectra, much work remains to be done on this important question.

Another contrast occurs when we consider the case of a chromospheric temperature rise. Here the LTE assumption would lead to emission cores in all strong lines. In contrast, the non-LTE theory distinguishes between collision-dominated cases, which may have emission cores (with self-reversals), and photoionization-dominated cases, which show no emission.

We emphasize that aside from the case of the sun, where several fairly complete analyses have been carried out, only the first steps have been taken in applying the non-LTE approach to the analysis of stellar atmospheres. In view of the differences found in the illustrative cases described above, we should not be too surprised if many of our present LTE interpretations are changed when a more complete theory is employed. Certainly it is possible that LTE is an adequate approximation in some cases; we only remark that this must be *shown* and not merely *assumed.*

# 13 | Non-LTE Line Transfer: The Multilevel Atom

The atomic models discussed in the previous chapter are obviously very restrictive, and it clearly is of prime importance to consider more realistic approximations. Of course, with a larger number of levels, the number of interactions that can occur greatly increases, and both the physical and mathematical nature of the problem becomes much more complex. In particular, the radiation field in one transition may now affect that in other transitions; this type of interaction is commonly referred to as *interlocking*. The effects of interlocking couple together transitions in such a way that a given photon no longer belongs strictly to a specific transition but to a certain extent belongs to the ensemble of radiation fields of the entire set of transitions of the atom. Photons, under certain conditions, may now freely transfer back and forth between transitions and their corresponding spectral regions. Thus, in a real physical sense the photons are members of a collective pool rather than individual, distinct species; this point has been emphasized clearly by Jefferies (Ref. 20, Chap. 8).

In the following section we shall consider the multilevel problem from essentially the same point of view as was used for the two-level problem. Generally speaking, one is then employing a formalism in which interlocking effects are assumed to be minor. Under certain circumstances this approach can be used with success, but as we shall see in Section 13-2, there exist situations where interlocking effects dominate with the result that the physical nature of the problem is profoundly different. A much more general approach, which has proven successful in very complicated problems, will then be described in Section 13-3.

## 13-1. The Equivalent Two-Level-Atom Approach

### FORMULATION

In our study of the two-level atom we made use of the statistical equilibrium equations to eliminate the population ratios appearing in the line source function, and we thus found an expression of the general form

$$S_l = \frac{\int \varphi_\nu J_\nu \, d\nu + \alpha}{1 + \beta} \tag{13-1}$$

where $\alpha$ and $\beta$ depend upon the possible ways of creating or destroying a photon. A similar approach could be adopted for an atom with an arbitrary number of levels. In principle, the statistical equilibrium equations

$$\sum_{j=1}^{n} A_{ij} N_j = B_i \qquad (i = 1, \ldots, n) \tag{13-2}$$

could be solved explicitly for each $N_j$ and ratios taken to form source functions for all transitions. The source function would again have the same form as equation (13-1), and the quantities $\alpha$ and $\beta$ could be expressed in terms of expansions of cofactors of the rate matrix **A** (see, e.g., the results of Kalkofen, Ref. 16, p. 187). In practice, this procedure is *very* tedious, and the resulting source functions are nonlinear in the radiation fields of all transitions other than the one under consideration. Moreover, the algebra very quickly becomes completely unmanageable.

A similar (though not exactly equivalent) procedure, which is a good deal simpler, is to use only the two statistical equilibrium equations involving the upper and lower levels of the transition that are being considered as reference levels. Rates to and from all other levels are expressed as *net* rates. The presumption here is that the dominant members among the source-sink terms for the line involve those two levels and the continuum, i.e., that interlocking effects are minor. Thus for the lower level we write

$$\begin{aligned}
N_L\Big(B_{LU} \int \varphi_\nu J_\nu \, d\nu + N_e \Omega_{LU} + \sum_{i<L} A_{Li} Z_{Li} \\
+ \sum_{L<j\neq U} N_e \Omega_{Lj} Y_{Lj} + R_{L\kappa} + N_e \Omega_{L\kappa}\Big) \\
- N_U\Big(A_{UL} + B_{UL} \int \varphi_\nu J_\nu \, d\nu + N_e \Omega_{UL}\Big) \\
= N_L^*(R_{\kappa L} + N_e \Omega_{\kappa L}) + \sum_{L<j\neq U} N_j A_{jL} Z_{jL} + \sum_{i<L} N_i N_e \Omega_{iL} Y_{iL}
\end{aligned} \tag{13-3}$$

where

$$R_{L\kappa} \equiv 4\pi \int_{\nu_L}^{\infty} \frac{\alpha_L(\nu) J_\nu}{h\nu} \, d\nu \tag{13-4}$$

$$R_{\kappa L} \equiv 4\pi \int_{\nu_L}^{\infty} \frac{\alpha_L(\nu)}{h\nu} [B_\nu(1 - e^{-h\nu/kT}) + J_\nu e^{-h\nu/kT}] \, d\nu \tag{13-5}$$

$$Z_{ij} \equiv 1 - \frac{\int \varphi_\nu J_\nu \, d\nu}{S_{ij}} \tag{13-6}$$

and

$$Y_{ij} \equiv \left(1 - \frac{b_j}{b_i}\right) \tag{13-7}$$

Similarly, for the upper level we may write

$$\begin{aligned} N_U\Big(A_{UL} &+ B_{UL} \int \varphi_\nu J_\nu \, d\nu + N_e \Omega_{UL} \\ &+ \sum_{U>i\neq L} A_{Ui} Z_{Ui} + \sum_{U<j} N_e \Omega_{Uj} Y_{Uj} + R_{U\kappa} + N_e \Omega_{U\kappa}\Big) \\ &- N_L\left(B_{LU} \int \varphi_\nu J_\nu \, d\nu + N_e \Omega_{LU}\right) \\ &= N_U^*(R_{\kappa U} + N_e \Omega_{U\kappa}) + \sum_{j>U} N_j A_{jU} Z_{jU} + \sum_{U>i\neq L} N_i N_e \Omega_{iU} Y_{iU} \end{aligned} \tag{13-8}$$

By direct solution of equations (13-3) and (13-8) we can calculate the ratio $N_L/N_U$, and hence

$$S_{LU} = \frac{2h\nu^3}{c^2} \left(\frac{g_U N_L}{g_L N_U} - 1\right)^{-1} \tag{13-9}$$

By use of the Einstein relations, the appropriate detailed balancing results, and by grouping terms, we find

$$S_{LU} = \frac{\int \varphi_\nu J_\nu \, d\nu + (\varepsilon + \theta) B_\nu(T_e)}{1 + \varepsilon + \eta} \tag{13-10}$$

where

$$\varepsilon = \frac{N_e \Omega_{UL}}{A_{UL}} (1 - e^{-h\nu_{UL}/kT}) \tag{13-11}$$

$$\eta = \frac{a_2 a_3 - g_L a_1 a_4 / g_U}{A_{UL}(a_2 + a_4)} \tag{13-12}$$

and

$$\theta = \frac{N_L^* a_1 a_4 (1 - e^{-h\nu_{UL}/kT})}{N_U^* A_{UL}(a_2 + a_4)} \tag{13-13}$$

and where, in turn,

$$a_1 = R_{L\kappa} + N_e \Omega_{L\kappa} + \sum_{i<L} A_{Li} Z_{Li} + \sum_{L<j \neq U} N_e \Omega_{Lj} Y_{Lj} \tag{13-14}$$

$$a_2 = N_L^*(R_{\kappa L} + N_e \Omega_{L\kappa}) + \sum_{L<j \neq U} b_j N_j^* A_{jL} Z_{jL} + \sum_{i<L} b_i N_i^* \Omega_{iL} Y_{iL} \tag{13-15}$$

$$a_3 = R_{U\kappa} + N_e \Omega_{U\kappa} + \sum_{U>i \neq L} A_{Ui} Z_{Ui} + \sum_{U<j} N_e \Omega_{Uj} Y_{Uj} \tag{13-16}$$

and

$$a_4 = N_U^*(R_{\kappa U} + N_e \Omega_{U\kappa}) + \sum_{U<j} b_j N_j^* A_{jU} Z_{jU} + \sum_{U>i \neq L} b_i N_i^* N_e \Omega_{iU} Y_{iU} \tag{13-17}$$

We see that equation (13-10) very strongly resembles the corresponding expression for a two-level atom [equation (12-109)], except for the rates coupling to other levels. For this reason we refer to this approach as the *equivalent two-level atom formulation*. Indeed, if the net rates to all other levels were zero, then equation (13-10) would reduce identically to equation (12-109).

In actual model computations, one must allow for the effects of overlapping continua. This may be done for example by using the factoring procedure described in Chapter 7. There we had written [combining equations (7-61) and (7-63)]

$$\gamma_\nu \int \Phi_\nu J_\nu \, d\nu + \varepsilon_\nu B_\nu = \left( \frac{N_U N_L^*/N_U^*}{N_L - g_L N_U/g_U} \right) B_\nu = S_{LU}(1 - e^{-h\nu/kT})^{-1} \tag{13-18}$$

Recalling the definition of $\Phi_\nu$, as given in equation (7-64), we have for a line

$$\Phi_\nu = B_{LU} \varphi_\nu \tag{13-19}$$

Thus combining equations (13-10), (13-18), and (13-19), we have

$$\gamma_\nu = [B_{LU}(1 - e^{-h\nu/kT})(1 + \varepsilon + \eta)]^{-1} \tag{13-20}$$

and

$$\varepsilon_\nu = \frac{\varepsilon + \theta}{(1 - e^{-h\nu/kT})(1 + \varepsilon + \eta)} \tag{13-21}$$

With this definition of $\gamma_\nu$ and $\varepsilon_\nu$, the remaining formalism developed in Chapter 7 remains unaltered.

To carry out the calculation, one must make initial guesses for the net radiative brackets appearing in equations (13-14) through (13-17). Typically, one sets all brackets to zero as a first estimate (lines in detailed balance). Then the transfer equation [in the form of equation (7-65)] may be solved, and having the radiation field, the net radiative brackets may be re-evaluated. This iterative procedure may then be repeated until convergence is obtained. If in fact the net rates to other bound levels are small compared with the direct rates $R_{L\kappa}$, $R_{U\kappa}$, $\Omega_{LU}$, $\Omega_{L\kappa}$, $\Omega_{U\kappa}$, etc., then the iterations should converge without difficulty since the basic physical assumptions implied in the equivalent two-level atom approach are met. On the other hand, if strong coupling exists between line transitions (see Section 13-2), then convergence may be slow. Moreover, as has been pointed out by Avrett (Ref. 29, p. 27), situations may arise where iteration procedures of this type may converge upon *inconsistent* solutions; a technique that avoids such difficulties is described in Section 13-3.

### APPLICATIONS

A number of very interesting cases of multilevel transfer problems have been studied using variants of the procedure described above. We cannot discuss all of these in depth here, but let us at least mention some of the results and recommend that the reader examine the original papers for details. For example, an instructive schematic non-LTE problem, in which the radiation field and level populations for a hydrogen atom consisting of 3 bound levels and a continuum in a constant-temperature, constant-density atmosphere, has been considered by Johnson and Klinglesmith (Ref. 16, p. 221), Kalkofen and Avrett (Ref. 16, p. 249), and Cuny (Ref. 16, p. 275). In this model one deals with the three continua $(1 \rightarrow \kappa)$, $(2 \rightarrow \kappa)$, and $(3 \rightarrow \kappa)$, and the three lines $L\alpha(1 \rightarrow 2)$, $L\beta(1 \rightarrow 3)$, and $\mathrm{H}\alpha(2 \rightarrow 3)$. The results by Johnson and Klinglesmith and by Cuny are in good agreement, even though quite different methods of solving the transfer equation were employed. By contrast, the results by Kalkofen and Avrett showed very poor convergence properties. This fact is of interest since the former authors expressed the line source functions in terms of net radiative brackets, as we have done above, while the latter authors employed the radiation fields directly. As emphasized by Cayrel (Ref. 16, p. 295), the former approach is often preferable since the net radiative brackets may be well determined even when the radiation field itself is far from its final value. Results for $S_{ij}(\tau)/S_{ij}(\infty)$ are shown in Figure 13-1; all source functions are plotted against $\tau_{L\alpha}$. The departure coefficients are shown in Figure 13-2. Note that at $\tau_{L\alpha} = 10^6$, where $S_{12} \rightarrow S_{12}(\infty)$, $L\alpha$ is in detailed balance; then $N_1$ couples to $N_2$ and $b_1 \rightarrow b_2$. At the same point we

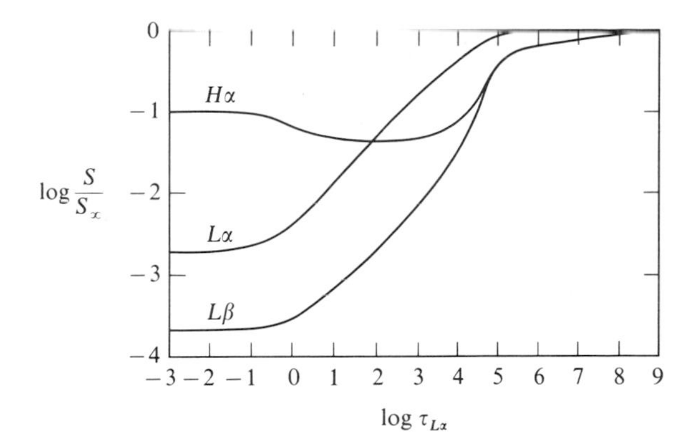


FIG. 13-1. Source functions for Lyman $\alpha$, Lyman $\beta$, and Balmer $\alpha$ in a constant-temperature, constant-density atmosphere. (From Y. Cuny, Harvard-Smithsonian Conference on Stellar Atmospheres, *Proceedings of the Second Conference*, S.A.O. Special Report No. 174. Cambridge, Smithsonian Astrophysical Observatory, 1965, p. 293.)

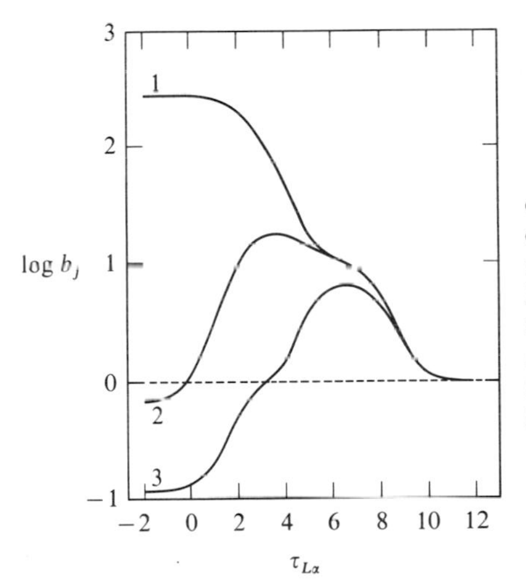


FIG. 13-2. Non-LTE departure coefficients for a three-level hydrogen atom in a constant-temperature, constant-density atmosphere. (From H. R. Johnson and D. A. Klinglesmith, Harvard-Smithsonian Conference on Stellar Atmospheres, *Proceedings of the Second Conference*, S.A.O. Special Report No. 174. Cambridge, Smithsonian Astrophysical Observatory, 1965, p. 246.)

find $S_{13}/S_{13}(\infty) \rightarrow S_{23}/S_{23}(\infty)$. This result is expected since, aside from stimulated emission factors, we may write $S_{13}/S_{13}(\infty) = b_3/b_1$ and similarly for the other transitions. Thus the fact that $b_1 = b_2$ guarantees the above-mentioned relation for the source functions. The drop in the surface values of the resonance-line source functions because of scattering is quite evident. Even more striking is the rise in the $H\alpha$ source function, which is due to photons that are being fed into $H\alpha$ from $L\beta$. Optical depth unity in the continuum occurs at $\tau_{L\alpha} \approx 10^9$, and we see that all level populations rapidly approach LTE beyond this point.

The calculation described above is quite restrictive because of the constant-properties atmosphere that is employed. Recently a more realistic computation of the formation of the solar hydrogen lines has been carried out by several authors (Ref. 29, p. 171), who used detailed temperature and density distributions for the solar chromosphere. Linsky has made similar calculations for the calcium H and K lines (Ref. 24), employing a variety of multilevel model atoms to study the effects of level coupling. Johnson and Poland (Ref. 29, p. 415) have computed departure coefficients and line profiles for He I lines, using model atmospheres for early-type stars.

Non-LTE hydrogen line profiles have been obtained by Peterson and Kalkofen (Ref. 29, p. 433). They construct model atmospheres in radiative and statistical equilibrium under the assumption that the lines are in detailed balance. The resulting temperature structure is then used in the solution of the line transfer problem. To compute Balmer line profiles, they set the Lyman lines in detailed balance and solved for the source functions of $H\alpha$, $H\beta$, and $P\alpha$. The resulting $H\alpha$ profile is shown in Figure 13-3. Here we see that the non-LTE profile is much shallower in the wing and deeper in the core than the LTE profile. The assumption of detailed balance in the lines leads to incorrect population ratios near the surface; indeed, the $H\alpha$ profile computed with these assumptions has an *emission* core. In the final non-LTE profile calculation, the populations are allowed to adjust themselves to be consistent with the radiation field (holding temperatures fixed), and the emission core disappears. The non-LTE profile for $\Delta\lambda \lesssim 1$ Å may still not be completely reliable since the effects of the lines were not included in the determination of the temperature and density structure of the model; in particular, the hump on the profile immediately outside the core is probably spurious (see

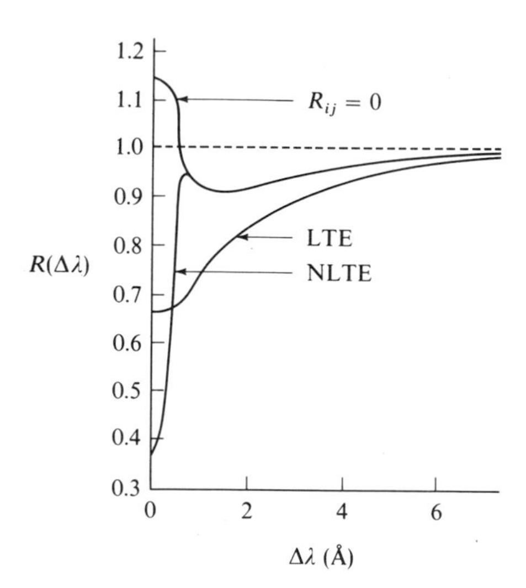


FIG. 13-3. Profile of $H\alpha$ in LTE and non-LTE model atmospheres with $T_{\text{eff}} =$ 15,000°K and log $g = 2$. The upper curve assumes that the lines are in detailed balance; this assumption leads to a spurious emission core. Also, the hump on the final NLTE solution is spurious (see Figure 13–12). (From D. M. Peterson and W. Kalkofen, National Center for Atmosphere Research, *Resonance Lines in Astrophysics*. Boulder, The Center, 1968, p. 433.)

also Figure 13-12) and is likely to be a residual effect of the detailed balance assumption. The behavior in the line wing follows from the fact that level 3, as described in Chapter 7, is overpopulated while level 2 is underpopulated. Thus, the non-LTE value of the source function

$$S_{23} \approx \left(\frac{2h\nu^3}{c^2}\right)\frac{N_3}{N_2} = \left(\frac{2h\nu^3}{c^2}\right)\left(\frac{N_3^*}{N_2^*}\right)\left(\frac{b_3}{b_2}\right) \tag{13-22}$$

(neglecting stimulated emissions) will exceed the LTE value. In contrast, their calculations show that the profile of $H\gamma$ is essentially unaffected by non-LTE effects (see also Figure 13-13). Thus a comparison of $H\alpha$ with $H\gamma$ profiles provides a relatively good indicator of non-LTE effects. A recent study by Peterson and Strom (*Ap. J.*, **157**, 1341, 1969) has shown that the non-LTE profiles are in substantially better agreement with observation.

The formation of $L\alpha$ and $H\alpha$ in early-type stars, without the assumption of detailed balance in the lines, has been considered by Auer and Mihalas (*Ap. J.*, **156**, 157, 1969, and *Ap. J.*, **156**, 681, 1969), using the formulation described in Chapter 7 and in equations (13-20) and (13-21). The primary goal of this work was actually the determination of the effects of lines upon the temperature structure of the atmosphere, but we defer discussion of this point until Chapter 14. It is appropriate to mention here, however, that enforcement of the radiative equilibrium constraint using the equivalent two-level atom form for the line source functions is difficult. In the case of $L\alpha$, the solution converged only slowly; in the case of $H\alpha$ the solution was unstable. In fact, such difficulties motivated the development of the complete linearization scheme described in Section 13-3. We mention them here simply to stress that multilevel coupling can prove complex not only in the solution of the transfer problem for a given temperature-density structure but can be even more troublesome in the general case where the temperature structure is regarded as unknown.

The $L\alpha$ source function for a model with $T_{\text{eff}} = 15{,}000°\text{K}$ and $\log g = 4$ is shown in Figure 13-4. Here $B_\nu$ in the non-LTE atmosphere is much larger than in the LTE atmosphere because of the temperature rise in the outer layers that is caused by heating in the Lyman continuum (see Chapter 7). Nevertheless, $S_{12}$ is strongly decoupled from the Planck function, and the non-LTE line profile (Figure 13-5) ends up much deeper than in LTE. Again this emphasizes the weak dependence of line source functions upon local thermal conditions. The corresponding results for $H\alpha$ are shown in Figures 13-6 and 13-7. Here we see that $S_{23}$ lies above $B_\nu$ in a certain depth range, consistent with the classification of $H\alpha$ as a photoionization-dominated line. The line profile in Figure 13-7 is quite schematic because only a simple Doppler profile was considered. Nevertheless, the result is in qualitative agreement with that obtained by Peterson and Kalkofen shown in Figure 13-3.

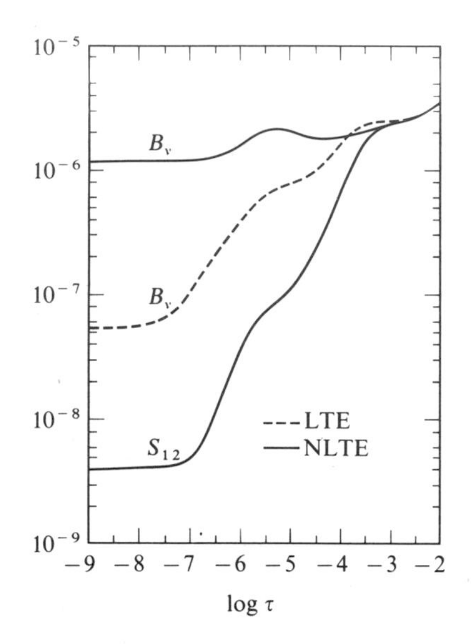


FIG. 13-4. Lyman $\alpha$ source function in a non-LTE model atmosphere with $T_{\text{eff}} =$ 15,000°K and log $g = 4$ in radiative equilibrium.

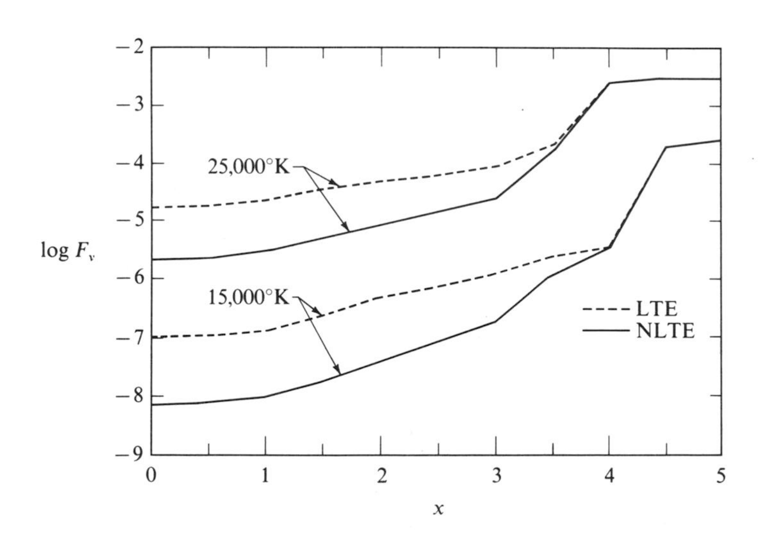


FIG. 13-5. Lyman $\alpha$ line profile for LTE and non-LTE models. The lines are assumed to have depth-independent Doppler profiles.

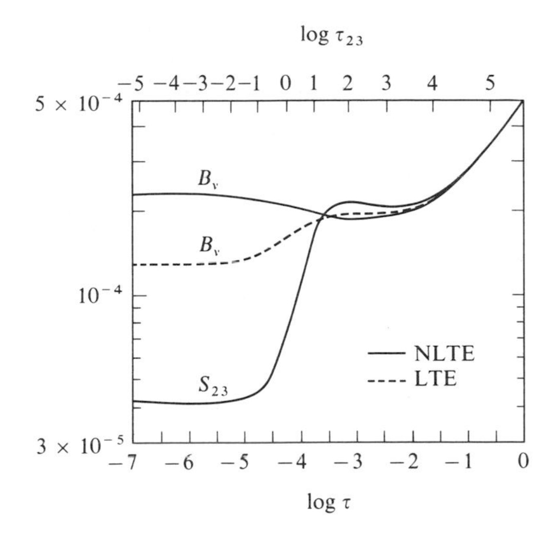


FIG. 13-6. Balmer α source function in a non-LTE model atmosphere in radiative equilibrium.

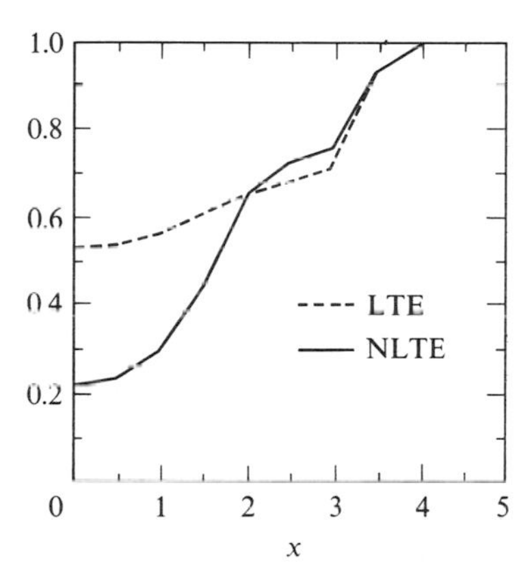


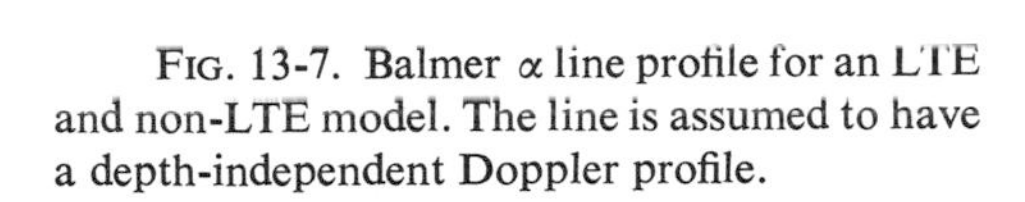

FIG. 13-7. Balmer α line profile for an LTE and non-LTE model. The line is assumed to have a depth-independent Doppler profile.

# 13-2. Effects of Level Couplings: Source Function Equality in Multiplets

## THERMALIZATION LENGTHS AND CONVERSION LENGTHS

As we emphasized above, one might expect the equivalent two-level approach to work best when coupling terms to other levels are small. But situations exist where this is not the case, and it is important to understand the effects of strong coupling among transitions. To be specific, we consider the case of a three-level atom consisting of a ground state and two close-lying upper states. We assume that radiative and collisional transitions can occur between states 1 and 2 and between 1 and 3 but that only collisional transitions occur between states 2 and 3. This model resembles the actual physical situation for a *resonance doublet* where the upper states are separated by fine-structure splitting. For example, the sodium *D*-lines arise from transitions faom a $^2S_{1/2}$ ground state to the upper states $^2P_{1/2}$ and $^2P_{3/2}$.

Consider first the case where the $2 \leftrightarrow 3$ collision rate is zero, i.e., the *uncoupled case*. Then clearly the transitions $1 \leftrightarrow 2$ and $1 \leftrightarrow 3$ can take place without reference to one another, and the two lines are formed independently. Our previous work has shown that each line will have a source function that falls below the local Planck function near the surface and that each line will thermalize to the Planck function at depths given by

$$\Lambda_{12} \sim A_{21}/C_{21} \tag{13-23}$$

and

$$\Lambda_{13} \sim A_{31}/C_{31} \tag{13-24}$$

As usual, we have made here the simplifying assumption of complete redistribution over a Doppler profile and of no strong gradients. Thus, in general, the run of the source functions for the lines will be different through the atmosphere, and at any given depth point they will not have the same value. On the other hand, suppose that collisions occur very rapidly between levels 2 and 3. In this case, photons may transfer back and forth between the two line transitions. Thus, from the point of view of photons in the lines, the two upper states become, for all practical purposes, identical to a single state with their combined statistical weights. In this case, the source functions for the two lines *should be equal* at a given physical point in the atmosphere. Usually, some intermediate situation will pertain, and we expect that the source functions will be equal only over a certain range in the atmosphere. Thus, in addition to the two thermalization lengths, there is a characteristic *conversion length* beyond which photons will be transferred back and forth between the lines, thus leading to source function equality. By analogy with the definition of thermalization lengths given above, it seems reasonable to suggest a conversion length of order

$$\Lambda_{32} \sim A_{31}/C_{32} \tag{13-25}$$

(see Jefferies, *Ap. J.*, **132**, 775, 1960; and Ref. 16, p. 177). From equations (13-23), (13-24), and (13-25), we see that we can expect the source functions to be independent over the ranges $\Lambda_{12}$ and $\Lambda_{13}$ only if $C_{31} > C_{32}$ *and* $C_{21} > C_{23}$. If either of these conditions is violated, it then becomes more likely that conversion will occur than that a collisional destruction will take place. By analogy with the two-level problem, we can estimate the depths at which photons can emerge without being either collisionally destroyed *or* converted as

$$\Lambda_{12}^* \sim (A_{21} + C_{21} + C_{23})/(C_{21} + C_{23}) \tag{13-26}$$

and

$$\Lambda_{13}^* \sim (A_{31} + C_{31} + C_{32})/(C_{31} + C_{32}) \tag{13-27}$$

These are the depths from which photons in the $1 \leftrightarrow 2$ and $1 \leftrightarrow 3$ transitions, respectively, can retain their identity and emerge. Let $z_{12}^*$ be the *physical depth* corresponding to $\Lambda_{12}^*$ and $z_{13}^*$ be that corresponding to $\Lambda_{13}^*$. Then the two groups of photons may propagate independently only over a depth $z^*$ equal to the minimum of $z_{12}^*$ and $z_{13}^*$.

From a mathematical point of view, the strong coupling we have described here makes the usefulness of iterations in an equivalent two-level approach less clear. It now appears more attractive to consider the two lines and their effects upon one another *simultaneously*.

### OBSERVATIONAL INDICATIONS OF SOURCE FUNCTION EQUALITY

Before we consider the details of the theory, let us discuss briefly some of the observational evidence that indicates that source function equality actually occurs. An extensive set of precise observations of the sodium *D*-line at various positions across the disk of the sun has been obtained by Waddell (*Ap. J.*, **136**, 223, 1962).

The emergent intensity at any particular frequency and position of the disk is given simply by

$$I_\nu(0, \mu) = \int_0^\infty \frac{S_c(\tau) + \eta_\nu S_l(\tau)}{1 + \eta_\nu} \exp\left[-\int_0^\tau (1 + \eta_\nu)\, dt/\mu\right](1 + \eta_\nu)\, dt/\mu \tag{13-28}$$

where $\tau$ is the continuum optical depth scale. Now in the very core of the line we expect $\eta_\nu \gg 1$ so that to good accuracy we may write

$$I_\nu(0, \mu) = \int_0^\infty S_l(\tau)\exp(-\tau\varphi_\nu/\mu)\varphi_\nu\, d\tau/\mu \tag{13-29}$$

where $\tau$ is now the line optical depth scale. For the sodium *D*-lines, the oscillator strength in the $1 \rightarrow 3$ transition is twice that of the $1 \rightarrow 2$ transition; thus we have

$$I_{12}(0, \mu, \nu) = \int_0^\infty S_{12}(\tau_{12}) \exp\left(\frac{-\varphi_\nu \tau_{12}}{\mu}\right) \frac{\varphi_\nu \, d\tau_{12}}{\mu} \tag{13-30}$$

and

$$I_{13}(0, \mu, \nu) = \int_0^\infty S_{13}(\tau_{12}) \exp\left(\frac{-2\varphi_\nu \tau_{12}}{\mu}\right) \frac{2\varphi_\nu \, d\tau_{12}}{\mu} \tag{13-31}$$

Thus, clearly $I_{13}(0, \mu, \nu) = I_{12}(0, \mu/2, \nu)$ *if $S_{12}$ and $S_{13}$ have a common depth dependence*. In essence, we are compensating for the higher opacity in one of the lines by increasing the path length in the other. When this kind of comparison was carried out, Waddell found very good agreement between the cores of the two $D$-lines, as shown in Figures 13-8a through 13-8c. By way of contrast, Figure 13-8d shows the inverse comparison $I_{13}(0, \mu/2, \nu)$ and $I_{12}(0, \mu, \nu)$, demonstrating the genuine significance of the agreement found in the other figures. The disagreement in the wings is due to the increasing importance of the continuum, which invalidates the assumptions used in equation (13-29). This striking observational result provided the motivation to examine in detail the conditions under which strict source function equality can occur (aside from the trivial case of LTE).

### ANALYSIS FOR THE SODIUM D-LINES

A first analysis by Waddell (*Ap. J.*, **136**, 231, 1962) showed that a sufficient condition to guarantee source function equality was that $C_{32} \gg A_{31}$ or $C_{23} \gg A_{21}$. In a later discussion he concluded that this condition was also necessary (*Ap. J.*, **138**, 1147, 1963) even though estimates of collision rates seemed to show that this could not actually be the case for the $D$-lines in the solar atmosphere. Waddell's conclusion was subsequently questioned by Athay, whose analysis (*Ap. J.*, **140**, 1579, 1964) led to the less stringent condition $C_{32} \gg A_{31}Z_{31}$. Athay's criterion is a bit more difficult to apply a priori because the variation of the net radiative bracket with depth must be computed from a solution of the transfer problem. On the other hand, it does account for cancellation of spontaneous emissions by absorptions and shows that source function equality will ultimately be attained at some depth. Waddell's criterion will be valid if one demands equality at *all* depths (right up to the surface); Athay's expression is more in the spirit of Jefferies' conversion length and when satisfied guarantees equality below that point, though deviations may be large closer to the surface.

A very clear discussion of the problem has been given by Avrett (*Ap. J.*, **144**, 59, 1966), whose conclusions we will summarize here. For the sake of simplicity, we will ignore the difference in line frequencies and neglect stimulated emission terms; if desired, these terms may be accounted for, but the algebra becomes much more complicated.

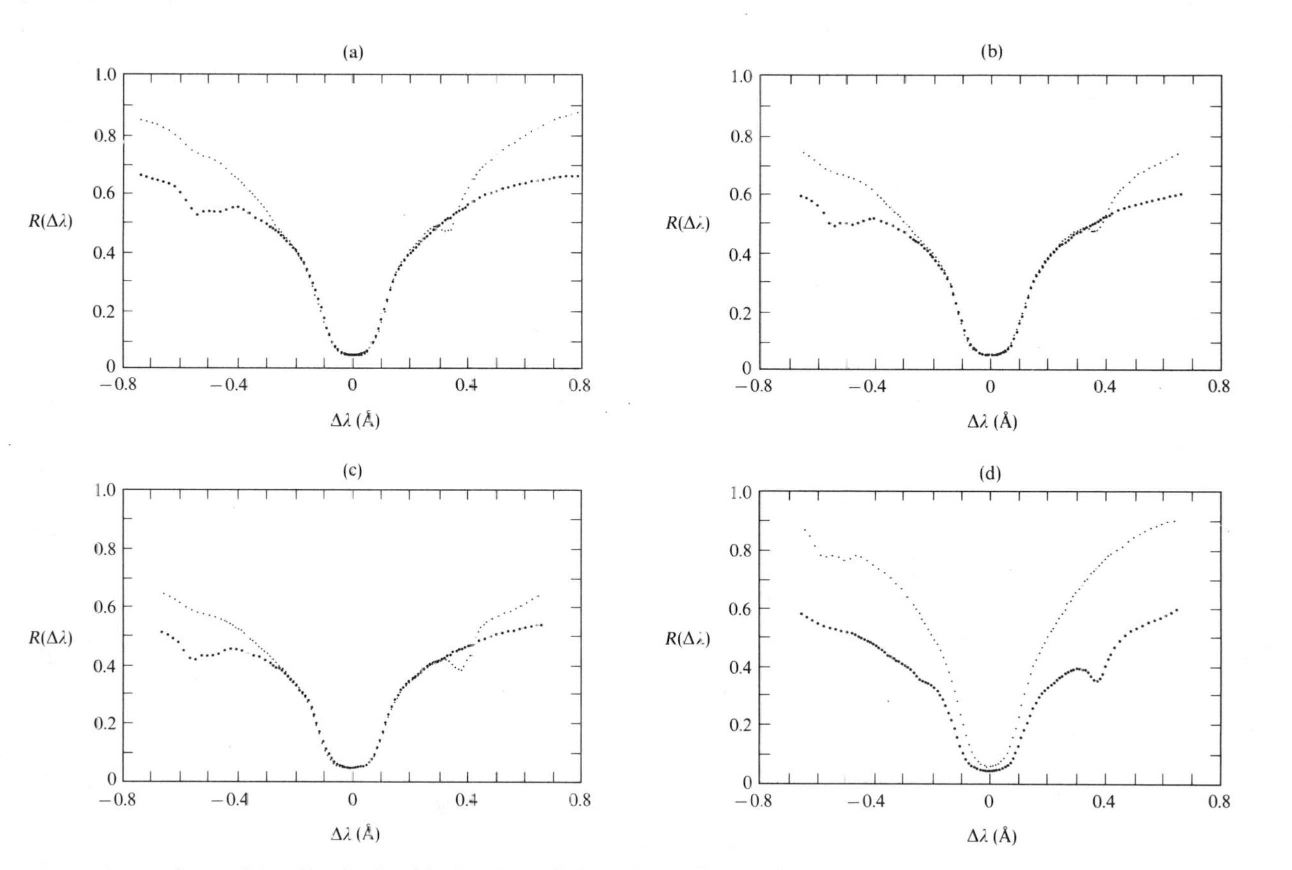


FIG. 13-8. Comparison of $I_{12}(0, \mu/2, \nu)$ with $I_{13}(0, \mu, \nu)$ for solar sodium *D*-lines: (a) comparison for $\mu = 1.0$; (b) comparison for $\mu = 0.8$; (c) comparison for $\mu = 0.6$; and (d) inverse comparison of $I_{12}(0, \mu, \nu)$ with $I_{13}(0, \mu/2, \nu)$ for $\mu = 1$. Note complete lack of agreement in this case. (From J. H. Waddell, *Ap. J.*, **136**, 223, 1962; by permission.)

The statistical equilibrium equations for the case under study are

$$N_2[A_{21} + N_e(\Omega_{21} + \Omega_{23})] = N_1(B_{12}\bar{J}_{12} + N_e\Omega_{12}) + N_3 N_e \Omega_{32} \tag{13-32}$$

and

$$N_3[A_{31} + N_e(\Omega_{31} + \Omega_{32})] = N_1(B_{13}\bar{J}_{13} + N_e\Omega_{13}) + N_2 N_e \Omega_{23} \tag{13-33}$$

where

$$\bar{J}_{ij} \equiv \int \varphi_\nu(ij) J_\nu \, d\nu \tag{13-34}$$

The source functions we desire are

$$S_{12} = \frac{2h\nu^3}{c^2} \frac{g_1 N_2}{g_2 N_1} \tag{13-35}$$

and

$$S_{13} = \frac{2h\nu^3}{c^2} \frac{g_1 N_3}{g_3 N_1} \tag{13-36}$$

Making use of the Einstein relations and detailed balancing results for collision rates, we find

$$S_{12}\left(1 + \frac{N_e\Omega_{21}}{A_{21}} + \frac{N_e\Omega_{23}}{A_{21}}\right) = \bar{J}_{12} + \left(\frac{N_e\Omega_{21}}{A_{21}}\right)B_\nu + \left(\frac{N_e\Omega_{23}}{A_{21}}\right)S_{13} \tag{13-37}$$

Defining

$$\varepsilon_{12} \equiv \frac{N_e\Omega_{21}}{A_{21}} \tag{13-38}$$

and

$$\eta_{12} \equiv \frac{N_e\Omega_{23}}{A_{21}} \tag{13-39}$$

we thus have

$$S_{12} = \frac{\bar{J}_{12} + \varepsilon_{12} B_\nu + \eta_{12} S_{13}}{1 + \varepsilon_{12} + \eta_{12}} \tag{13-40}$$

By a completely analogous analysis we find

$$S_{13} = \frac{\bar{J}_{13} + \varepsilon_{13} B_\nu + \eta_{13} S_{12}}{1 + \varepsilon_{13} + \eta_{13}} \tag{13-41}$$

where now

$$\varepsilon_{13} \equiv \frac{N_e \Omega_{31}}{A_{31}} \tag{13-42}$$

and

$$\eta_{13} \equiv \frac{N_e \Omega_{32}}{A_{31}} \tag{13-43}$$

These expressions for the source functions display, in a particularly transparent way, the linear dependence between $S_{12}$ and $S_{13}$ and show clearly that as $\eta_{12} \gg 1$ or $\eta_{13} \gg 1$, $S_{12} \to S_{13}$, as we have argued previously. It is evident that in this case the simultaneous solution for both source functions is mandatory. The question of interest now is whether the $\eta$'s must always be greater than unity to give source function equality. If we take $\tau_{13}$ as the reference optical depth, then the appropriate equations of transfer become

$$\mu \frac{dI_{13}(\tau, \mu, \nu)}{d\tau} = \varphi_\nu[I_{13}(\tau, \mu, \nu) - S_{13}(\tau)] \tag{13-44}$$

and

$$\mu \frac{dI_{12}(\tau, \mu, \nu)}{d\tau} = \gamma\varphi_\nu[I_{12}(\tau, \mu, \nu) - S_{12}(\tau)] \tag{13-45}$$

where

$$\gamma = \frac{B_{12}}{B_{13}} \tag{13-46}$$

As usual, we obtain solutions of the form

$$\overline{J_{13}(\tau)} = \int_0^\infty S_{13}(t) K_1 |t - \tau| \, dt \tag{13-47}$$

and

$$\overline{J_{12}(\tau)} = \int_0^\infty S_{12}(t) K_1 |t - \tau| \, dt \tag{13-48}$$

where $K_1$ is the standard kernel function, equation (12-23),

$$K_1(s) = \frac{1}{2} \int_0^\infty E_1(\varphi_x s) \varphi_x^2 \, dx$$

Substitution of equations (13-47) and (13-48) into equations (13-40) and (13-41) leads to a pair of linear integral equations for $S_{12}$ and $S_{13}$, which may be solved numerically. Now whenever $\eta_{12} \gg \varepsilon_{12}$ or $\eta_{13} \gg \varepsilon_{13}$, we have

$S_{12} \approx S_{13}$, and direct iteration between equations (13-40) and (13-41) is very inefficient. An alternative procedure should then be followed. Avrett multiplies equation (13-40) by $\eta_{13}$, and equation (13-42) by $\eta_{12}$ and subtracts. Rearranging terms, we then find

$$S_{13} = \frac{\bar{J}_{13} + \varepsilon_{13} B^*}{1 + \varepsilon_{13} + \eta_{13}} \tag{13-49}$$

where

$$B^* = B_\nu - \frac{\eta_{13}\varepsilon_{12}}{\eta_{12}\varepsilon_{13}}\left(S_{12} - B_\nu + \frac{S_{12} - \bar{J}_{12}}{\varepsilon_{12}}\right) \tag{13-50}$$

$B^*$, defined in this way, is insensitive to errors in $S_{12}$. We start therefore with $B^* = B_\nu$, calculate $S_{13}$ from equation (13-49), substitute this value into equation (13-40), and solve to obtain $S_{12}$. We may then insert this value into equation (13-50), recalculate $S_{13}$, and iterate until convergence is obtained. Avrett finds that this procedure converges quickly.

In the particular case of the sodium $D$-lines, certain relations exist among the various parameters. Specifically, $A_{21} = A_{31}$ and $\Omega_{21} = \Omega_{31}$; hence $\varepsilon_{12} = \varepsilon_{13}$. Also, $g_1 = 2$, $g_2 = 2$, and $g_3 = 4$; hence $\gamma = 1/2$ and $\eta_{13} = \frac{1}{2}\eta_{12}$. The energy difference between levels 2 and 3 has been ignored. Avrett discusses solutions for typical values of these parameters. For example, in Figure 13-9 we show his solutions for $B_\nu = 1$, $\varepsilon = 10^{-4}$, and $\eta = 0$, $10^{-4}$, $10^{-3}$, $10^{-2}$, $10^{-1}$, and 1. We see that while $\eta_{13}$ would have to be larger than 1 to guarantee source function equality to the very surface, much smaller values of $\eta_{13}$ yield agreement for $\tau \gtrsim 1$, which is the significant region in determining the emergent intensity. We show the emergent specific intensity for $\mu = 1.0$, 0.8, and 0.6 in Figure 13-10. Clearly, the profiles agree closely when $\eta = 10^{-3}$ and are nearly indistinguishable when $\eta = 10^{-2}$. This result is more or less consistent with Athay's analysis since $Z_{21}$ and $Z_{31}$ turn out to be of order $10^{-2}$. Avrett's calculation shows clearly that, practically speaking, source function equality can occur even for $\eta \ll 1$. Avrett also examines the effects of a schematized temperature gradient; he finds that the value of $\eta$ required for equality decreases (compared with the $B_\nu \equiv 1$ case) when $B_\nu(\tau)$ increases inward and increases when $B_\nu(\tau)$ decreases inward. However, these results may depend upon the specific chosen forms of $B_\nu(\tau)$ and should not be regarded as general.

A generalization of the simultaneous approach has been developed by Kalkofen (Ref. 29, p. 1), who shows that the source functions for an arbitrary number of lines arising from a common lower level may be determined simultaneously. If stimulated emissions can be neglected, the solution may be effected without iteration; when stimulated emissions are important, the coupling is nonlinear, and iterations are required.

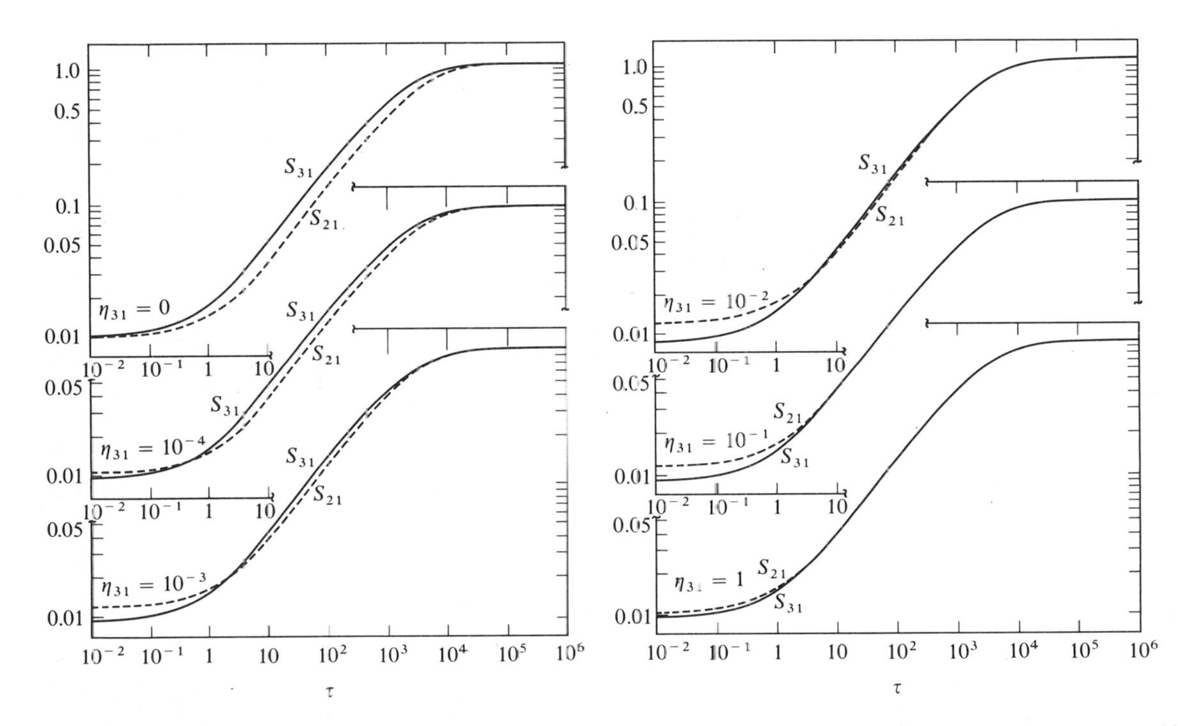


FIG. 13-9. Source functions for idealized multiplet in a semi-infinite atmosphere, assuming $B = 1$, $\varepsilon_{21} = \varepsilon_{31} = 10^{-4}$, and $\eta_{31} = \frac{1}{2}\eta_{21}$. (From E. H. Avrett, *Ap. J.*, **144**, 59, 1966; by permission.)

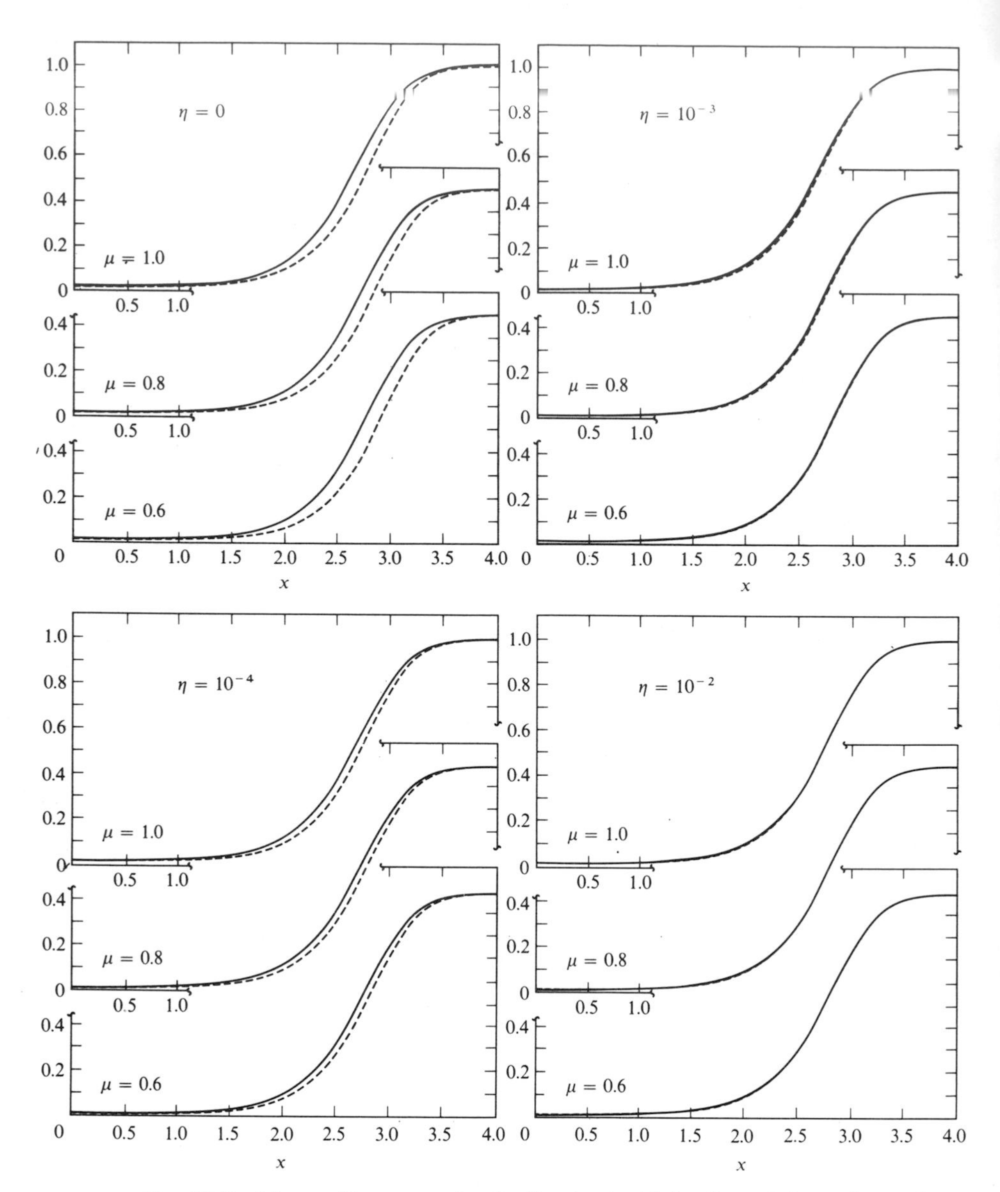


FIG. 13-10. Line profiles comparing $I_{12}(0, \mu/2, \nu)$ with $I_{13}(0, \mu, \nu)$ for various values of $\eta$ at $\mu = 1.0, 0.8$, and $0.6$. Solid curves give $I_{13}$; dashed curves give $I_{12}$. (From E. H. Avrett, *Ap. J.*, **144**, 59, 1966; by permission.)

As Jefferies has shown (Ref. 20, Chap. 9), the fact that source function equality may occur in multiplets is of profound importance in the development of a non-LTE method of analysis of line profile data. In keeping with the remainder of this book, in which we have emphasized theoretical predictions rather than analysis of observations, we will not discuss these analytical methods here but merely urge the reader to study Jefferies' work with care.

## 13-3. The Complete Linearization Approach

The approaches we have described above are not completely satisfactory. The equivalent two-level atom formulation is ill-chosen when strong coupling occurs between lines. The simultaneous approach can avoid this difficulty when the lines arise from a common lower level but is difficult to apply in a completely general case. In addition, these approaches may either fail to converge or converge to an inconsistent solution unless great care is taken (Avrett, Ref. 29, p. 27). Finally, the procedures we have described above are difficult to apply when the constraint of radiative equilibrium is to be enforced.

In this section we will describe a method developed by Auer and Mihalas (*Ap. J.*, **158**, 641, 1969) that allows the solution of multilevel transfer problems that are subject to the constraints of statistical, hydrostatic, and radiative equilibrium. We will defer a discussion of the results for the temperature distribution until Chapter 14 but will include, at the present time, the radiative equilibrium constraint for the sake of completeness and generality.

The basic feature of this approach is that it attempts to account at the outset for all possible couplings among the physical variables resulting from the constraints and to allow for the interaction of each variable at a given point in the atmosphere with every other variable at all other points. This is done by a linearization procedure in such a way that all variables at all depths remain strictly self-consistent to first order at each stage of the calculation.

At any given point in the atmosphere, the complete solution consists of knowledge of the radiation field, the bound level populations, the total particle density, the electron density, and the temperature. For simplicity we will assume that the atmosphere is composed of pure hydrogen so that the ion density equals the electron density. As before, we discretize the depth, angle, and frequency variables. Suppose we choose $NJ$ frequency angle points and consider a model atom with $NL$ bound levels. The solution at depth point $d$ is then a vector

$$\psi_d = (P_1, \ldots, P_{NJ}, N_1, \ldots, N_{NL}, N_{\mathrm{H}}, N_e, N, T), (d = 1, \ldots, ND) \tag{13-51}$$

Here we have again written

$$P_i = \tfrac{1}{2}[I(\nu_i, \mu_i) + I(\nu_i, -\mu_i)] \tag{13-52}$$

and have for convenience introduced $N_H$ as the sum of all hydrogen atoms, whether in bound states or as ions. To determine the solution $\psi_d$, we require a total of $NJ + NL + 4$ equations. Of these, $NJ$ are transfer equations, which we write as difference equations, and $NL$ are statistical equilibrium equations. The remaining four may be regarded as specifying $T$ (radiative equilibrium), $N$ (hydrostatic equilibrium), $N_H$ (a sum over the statistical equilibrium equations), and $N_e$ (particle conservation). The basic difficulty is that these equations are *nonlinear*. Therefore, we must generally solve them by iteration. One of the most efficient iteration schemes is solution by *linearization* (i.e., replace all variables $x$ by $x_0 + \delta x$ and retain only terms of zero and first order in the $\delta$'s). If we linearize the equations, then given a current estimate of the solution $\psi_0$, we may solve for a correction $\delta\psi$ determined in such a way that the vector $\psi = \psi_0 + \delta\psi$ satisfies the constraint equations more closely. Since the transfer equations in difference equation form and all of the constraint equations (using quadrature sums) are *algebraic* equations, we may employ the same elimination scheme as described in Chapter 6. In this elimination scheme, the entire set of equations is treated as a whole, subject to boundary conditions. Thus, interaction from depth to depth is automatically accounted for, and the convergence properties are global. Let us now develop the required formulae in detail.

Consider first the equations of transfer. Since, in general, there are large variations in the opacity from frequency to frequency, there is really no uniquely preferred optical depth scale. Any depth parameter that allows sufficient control over step-sizes is acceptable. In the present formulation we will adopt the mass in a square-centimeter column as the depth variable; as we shall see below, this choice affords certain advantages. We may now write the difference equation form of the transfer equation as

$$\frac{\mu^2 P_{d-1}}{\Delta\tau_{d-1/2}\,\Delta\tau_d} - \frac{\mu^2 P_d}{\Delta\tau_d}\left(\frac{1}{\Delta\tau_{d+1/2}} + \frac{1}{\Delta\tau_{d-1/2}}\right) + \frac{\mu^2 P_{d+1}}{\Delta\tau_{d+1/2}\,\Delta\tau_d}$$

$$= P_d - \frac{j_d}{\chi_d} + \frac{N_e \sigma_e J_d}{\chi_d} \tag{13-53}$$

where the opacity is defined by

$$\chi_d = \sum_{i}^{k-1} \sum_{j>i}^{k-1} \alpha_{ij}(\nu)(N_i - g_i N_j/g_j) + \sum_{i=1}^{k-1} \alpha_{ik}(\nu)(N_i - N_i^{*} e^{-h\nu/kT})$$

$$+ N_e^2 \alpha_{kk}(\nu)(1 - e^{-h\nu/kT}) + N_e \sigma_e \tag{13-54}$$

the emissivity is given by

$$j_\nu = \frac{2h\nu^3}{c^2}\left\{\sum_{i=1}^{k-1}\sum_{j>i}^{k-1}\alpha_{ij}(\nu)\frac{g_i N_j}{g_j} + e^{-h\nu/kT}\left[\sum_{i=1}^{k-1}\alpha_{ik}(\nu)N_i^* + N_e^2\alpha_{kk}(\nu)\right]\right\} \tag{13-55}$$

and we have written

$$\Delta\tau_{d-1/2} \equiv \tfrac{1}{2}(\omega_{d-1} + \omega_d)(m_d - m_{d-1}) \tag{13-56}$$

$$\Delta\tau_{d+1/2} \equiv \tfrac{1}{2}(\omega_d + \omega_{d+1})(m_{d+1} - m_d) \tag{13-57}$$

and

$$\Delta\tau_d = \tfrac{1}{2}(\Delta\tau_{d-1/2} + \Delta\tau_{d+1/2}) \tag{13-58}$$

Here

$$\omega_d \equiv \frac{\chi_d}{\rho_d} \tag{13-59}$$

where

$$\rho_d = m_H N_H \tag{13-60}$$

is the mass density. An equation of the form of (13-53) applies at each angle-frequency point; for the sake of brevity we have suppressed all subscripts other than those for depths.

At the surface, the boundary condition for the radiation field is

$$\frac{2\mu(P_2 - P_1)}{(\omega_1 + \omega_2)(m_2 - m_1)} = P_1 \tag{13-61}$$

while at depth ($d = ND$) we have

$$\frac{2\mu(P_d - P_{d-1})}{(\omega_d + \omega_{d-1})(m_d - m_{d-1})} = I_+ - P_d \tag{13-62}$$

By an analysis similar to that used to derive equation (6-209), we may write

$$I_+ = B + \frac{\pi\mu}{4\sigma T^3\chi}\frac{\partial B}{\partial T}Z \tag{13-63}$$

and requiring that

$$\sum_i w_i\mu_i(I_+ - P_i) = \mathscr{H} \tag{13-64}$$

we find

$$I_+ = B + \frac{3\mu}{\chi}\frac{\partial B}{\partial T}\left[\frac{\mathscr{H} + \sum_i w_i\mu_i(P_i - B_i)}{\sum_i \frac{w_i}{\chi_i}\frac{\partial B_i}{\partial T}}\right] \tag{13-65}$$

Linearization of the transfer equation yields

$$\frac{\mu^2\,\delta P_{d-1}}{\Delta\tau_d\,\Delta\tau_{d-1/2}} - \frac{\mu^2\,\delta P_d}{\Delta\tau_d}\left(\frac{1}{\Delta\tau_{d+1/2}} + \frac{1}{\Delta\tau_{d-1/2}}\right) - \delta P_d + \frac{N_e\sigma_e}{\chi_d}\,\delta J_d$$

$$+ \frac{\mu^2\,\delta P_{d+1}}{\Delta\tau_d\,\Delta\tau_{d+1/2}} + a_d\,\delta\omega_{d-1} + b_d\,\delta\omega_d + c_d\,\delta\omega_{d+1}$$

$$+ \frac{\delta j_d}{\chi_d} + \frac{\sigma_e J_d}{\chi_d}\,\delta N_e - \left(\frac{j_d}{\chi_d^{\,2}} + \frac{N_e\sigma_e J_d}{\chi_d}\right)\delta\chi_d$$

$$= \mu^2\beta_d + P_d - \frac{j_d}{\chi_d} - \frac{N_e\sigma_e J_d}{\chi_d} \tag{13-66}$$

where

$$a_d \equiv \frac{\mu^2}{\omega_d + \omega_{d-1}}\left(\gamma_d + \frac{1}{2}\,\beta_d\,\frac{\Delta\tau_{d-1/2}}{\Delta\tau_d}\right) \tag{13-67}$$

$$c_d \equiv \frac{\mu^2}{\omega_{d+1} + \omega_d}\left(\alpha_d + \frac{1}{2}\,\beta_d\,\frac{\Delta\tau_{d+1/2}}{\Delta\tau_d}\right) \tag{13-68}$$

and

$$b_d \equiv a_d + c_d \tag{13-69}$$

and where in turn

$$\alpha_d \equiv \frac{P_d - P_{d+1}}{\Delta\tau_{d+1/2}\,\Delta\tau_d} \tag{13-70}$$

$$\gamma_d \equiv \frac{P_d - P_{d-1}}{\Delta\tau_d\,\Delta\tau_{d-1/2}} \tag{13-71}$$

and

$$\beta_d \equiv \alpha_d + \gamma_d \tag{13-72}$$

The quantities $\delta P_i$, etc., are elements of the correction vector $\boldsymbol{\delta\psi}_d$ while quantities without a $\delta$-prefix are current estimates of the solution vector $\boldsymbol{\psi}_0$. In addition, it is to be understood that changes in composite quantities such as $\delta\chi$ are expanded in terms of changes in level populations and temperature. Thus, we write

$$\delta\chi = \frac{\partial\chi}{\partial T}\,\delta T + \frac{\partial\chi}{\partial N_e}\,\delta N_e + \sum_{i=1}^{NL}\frac{\partial\chi}{\partial N_i}\,\delta N_i \tag{13-73}$$

and similarly for $\delta j$ and $\delta\omega$. Expressions for the derivatives may easily be found from equations (13-54), (13-55), and (13-59). We thus see that each of the $NJ$ equations of the form of equation (13-66) involve the $(NJ + NL + 4)$ components of $\delta\psi_d$. The linearized form for the boundary condition equations is similar.

Let us now consider the hydrostatic equilibrium equation. If we assume the perfect gas law and write the flux appearing in the radiation pressure gradient in terms of the gradient of the mean intensity [equation (6-95)], then the difference-equation form of the hydrostatic equilibrium equation is

$$\frac{k(N_d T_d - N_{d-1} T_{d-1})}{(m_d - m_{d-1})} + \frac{4\pi}{c} \frac{\sum_i w_i \mu_i^2 (P_{d,i} - P_{d-1,i})}{(m_d - m_{d-1})} = g \qquad (13\text{-}74)$$

Here we see that the choice of mass as the independent variable is advantageous, for if the radiation pressure term is small, then the gas pressure is simply a linear function of the depth variable. The upper boundary condition is obtained by assuming the pressure is zero when $m$ is zero so that

$$\frac{kN_1 T_1}{m_1} + \frac{4\pi}{c} \sum w_i \mu_i P_{1,i} = g \qquad (13\text{-}75)$$

where we have made use of equation (6-99). In linearized form equation (13-74) becomes

$$\begin{aligned} T_d\, \delta N_d + N_d\, \delta T_d - T_{d-1}\, \delta N_{d-1} - N_{d-1}\, \delta T_{d-1} \\ + \frac{4\pi}{ck} \sum_i w_i \mu_i^2 (\delta P_{d,i} - \delta P_{d-1,i}) \\ = \frac{g}{k}(m_d - m_{d-1}) + N_{d-1} T_{d-1} - N_d T_d \\ + \frac{4\pi}{ck} \sum_i w_i \mu_i^2 (P_{d-1,i} - P_{d,i}) \end{aligned} \qquad (13\text{-}76)$$

with a similar result for the boundary condition.

The condition of radiative equilibrium requires that

$$\sum_i w_i [(\chi_{d,i} - N_e \sigma_e) P_{d,i} - j_{d,i}] = 0 \qquad (13\text{-}77)$$

The sum extends over all angles and frequencies. In linearized form we have

$$\begin{aligned} \sum_i w_i [(\chi_{d,i} - N_e \sigma_e)\, \delta P_{d,i} + (\delta \chi_{d,i} + \sigma_e\, \delta N_e) P_{d,i} - \delta j_{d,i}] \\ = \sum_i w_i [j_{d,i} - (\chi_{d,i} - N_e \sigma_e) P_{d,i}] \end{aligned} \qquad (13\text{-}78)$$

Since we have defined $N_H$ to be the density of hydrogen in all forms, the particle conservation equation may be written simply

$$N = N_H + N_e \qquad (13\text{-}79)$$

so that

$$\delta N - \delta N_{\mathrm{H}} - \delta N_e = N_{\mathrm{H}} + N_e - N \tag{13-80}$$

Finally, we have $NL + 1$ equations of statistical equilibrium (the last being a defining equation for $N_{\mathrm{H}}$). Thus, if we write

$$\mathbf{N} = (N_1, \ldots, N_{NL}, N_{\mathrm{H}}) \tag{13-81}$$

then the statistical equilibrium equations are simply

$$\mathbf{AN} = \mathbf{B} \tag{13-82}$$

where for $i \leq NL$ we have

$$A_{ij} = -(R_{ji} + N_e \Omega_{ji}), \qquad (j < i) \tag{13-83}$$

$$A_{ii} = \sum_{j<i} \frac{N_j^*}{N_i^*}(R_{ij} + N_e \Omega_{ij}) + \sum_{j>i}^{k}(R_{ij} + N_e \Omega_{ij}) \tag{13-84}$$

$$A_{ij} = -\left(\frac{N_i^*}{N_j^*}\right)(R_{ji} + N_e \Omega_{ij}), \qquad (j > i) \tag{13-85}$$

and

$$B_i = N_e^2 \varphi_i(T)(R_{\kappa i} + N_e\, \Omega_{i\kappa}) \tag{13-86}$$

while for $i = NL = 1$ we have

$$A_{ij} = 1, \qquad (j < i) \tag{13-87}$$

$$A_{ii} = -1 \tag{13-88}$$

and

$$B_i = -N_e - \sum_{j=NL+1}^{n_{\max}} N_e^2 \varphi_j(T) \tag{13-89}$$

In the above equations the terms $R_{ij}$, etc., have the meaning given in Chapter 5. As in equation (3-56) we have written

$$N_i^* = N_e^2 \varphi_i(T) = N_e^2 \left(\frac{2\pi mkT}{h^2}\right)^{-3/2} \frac{g_i}{2} \exp\left[\frac{(\chi_I - \chi_i)}{kT}\right]$$

We may now write $(NL + 1)$ linearized equations in the form

$$\frac{\partial \mathbf{N}}{\partial T}\delta T + \frac{\partial \mathbf{N}}{\partial N_e}\delta N_e + \sum_{i=1}^{NJ} \frac{\partial \mathbf{N}}{\partial P_i}\delta P_i - \boldsymbol{\delta}\mathbf{N} = 0 \tag{13-90}$$

where for any variable $x$ we calculate the derivative from equation (13-82) in the form

$$\frac{\partial \mathbf{N}}{\partial x} = \mathbf{A}^{-1}\left(\frac{\partial \mathbf{B}}{\partial x} - \frac{\partial \mathbf{A}}{\partial x}\mathbf{N}\right) \tag{13-91}$$

Expressions for $(\partial \mathbf{A}/\partial x)$ and $(\partial \mathbf{B}/\partial x)$ are straightforward (if laborious) to derive and have been given by Auer and Mihalas (*Ap. J.*, **158**, 641, 1969). Equation (13-90) has a rather attractive form since it contains only one $\delta N_i$ in each row and thus is partially diagonalized.

Although the coefficients are now more complicated, the above linearized equations are still all of the standard form

$$-\mathbf{A}_d\,\delta\boldsymbol{\psi}_{d-1} + \mathbf{B}_d\,\delta\boldsymbol{\psi}_d - \mathbf{C}_d\,\delta\boldsymbol{\psi}_{d+1} = \mathbf{Q}_d \tag{13-92}$$

Thus the solution proceeds as described in Chapter 6. The matrix size is now $(NN \times NN)$ where $NN = (NJ + NL + 4)$; thus, for a multilevel model atom including several transitions, the matrix size becomes rather large.

Let us emphasize again that all possible interactions among all variables at a given depth point are included, as is evident from equations (13-66), (13-76), (13-78), (13-80), and (13-90). Moreover, the coupling from one depth point to the next and, ultimately, to the boundary conditions is accounted for completely in equation (13-92) and the elimination procedure. We thus see that a change in any variable at any point in the atmosphere will have its proper effect (to the first order) upon all variables at all points.

To construct a model, one must have a first estimate of the solution vectors $\boldsymbol{\psi}_d$ at all depths in the atmosphere. Experience has shown that the following procedure is efficient. As a first guess, one may construct the density and temperature distribution as a function of mass using a gray temperature distribution on a Rosscland-mean optical depth scale. If we then assume LTE, the values of $\mathbf{N}$, $\chi$, and $j$ may be determined as a function of depth, and holding the temperature and level populations fixed, one may solve for the radiation field. In subsequent iterations the correction vectors $\delta\boldsymbol{\psi}$ are computed and applied for all variables. Since LTE obtains at depth, a good procedure is first to compute an LTE model allowing for bound-free transitions only and then a non-LTE model. The bound-bound transitions may then be added. To start with, we set the radiation field in the lines equal to the Planck function, and holding the temperature and density structure fixed, we perform an iteration to determine the radiation field. Subsequently, full corrections are applied to all variables. It is worth mentioning that although we have included the constraints of hydrostatic and radiative equilibrium, these could be dropped if the temperature and density structure are regarded as given. Thus, one would drop the perturbations $\delta T$ and $\delta N$ from all equations and delete equations (13-76) and (13-78). Then the method could be used to solve problems of the type considered in the preceding sections of this chapter.

A number of relatively complicated models have been successfully constructed with this method. For example, non-LTE models have been obtained for models in which departures from LTE are allowed in the first five levels of a sixteen-level hydrogen atom, allowing for bound-free transitions, and all lines that couple levels 2 through 5 (i.e., $H\alpha$, $H\beta$, $H\gamma$, $P\alpha$, $P\beta$, and $B\alpha$). Departures are allowed in the Lyman continuum since it, of course, may influence the line cores, but the Lyman lines are so opaque that they may be set into detailed balance. Experience has shown that convergence in all transitions is obtained (including *all* constraints) in six or less iterations and that the method is remarkably stable.

TABLE 13-1. Properties of Non-LTE Models Including $H\alpha$, $H\beta$, $H\gamma$, $P\alpha$, $P\beta$, and $B\alpha$

| $T_{eff}$ | log $g$ | *Theory* | $T_0$ | $T_{min}$ | *Lyman jump* | *Balmer jump* | *Paschen jump* |
|---|---|---|---|---|---|---|---|
| 12,500 | 4 | LTE, no lines | 8279 | | 10.3 | 1.07 | 0.181 |
| 12,500 | 4 | NLTE, no lines | 9563 | 9012 | 11.8 | 1.07 | 0.184 |
| 12,500 | 4 | NLTE, 6 lines | 11,230 | 9020 | 10.5 | 1.07 | 0.184 |
| 12,500 | 3 | LTE, no lines | 8331 | | 9.94 | 0.999 | 0.164 |
| 12,500 | 3 | NLTE, no lines | 9613 | 9059 | 11.8 | 0.981 | 0.178 |
| 12,500 | 3 | NLTE, 6 lines | 11,339 | 9077 | 10.6 | 0.982 | 0.178 |
| 12,500 | 2.5 | LTE, no lines | 8391 | | 9.59 | 0.909 | 0.147 |
| 12,500 | 2.5 | NLTE, no lines | 9609 | 9081 | 11.8 | 0.872 | 0.168 |
| 12,500 | 2.5 | NLTE, 6 lines | 11,260 | 9107 | 10.6 | 0.873 | 0.168 |
| 15,000 | 4 | LTE, no lines | 9495 | | 8.90 | 0.812 | 0.136 |
| 15,000 | 4 | NLTE, no lines | 11,010 | 10,298 | 10.4 | 0.807 | 0.141 |
| 15,000 | 4 | NLTE, 6 lines | 12,845 | 10,410 | 9.94 | 0.808 | 0.141 |
| 15,000 | 3 | LTE, no lines | 9577 | | 8.51 | 0.735 | 0.121 |
| 15,000 | 3 | NLTE, no lines | 11,081 | 10,417 | 10.5 | 0.718 | 0.136 |
| 15,000 | 3 | NLTE, 6 lines | 12,936 | 10,647 | 10.0 | 0.719 | 0.136 |
| 15,000 | 2.5 | LTE, no lines | 9662 | | 8.17 | 0.633 | 0.101 |
| 15,000 | 2.5 | NLTE, no lines | 11,075 | 10,442 | 10.5 | 0.599 | 0.128 |
| 15,000 | 2.5 | NLTE, 6 lines | 12,852 | 10,496 | 10.1 | 0.600 | 0.128 |

Let us briefly consider a few of the results of these calculations. In Table 13-1 we list some of the interesting properties of the models. Again we defer comment upon the temperature structure until Chapter 14. We note that, as was found in Chapter 7, the continuum radiation field in the visible region of the spectrum is little affected by departures from LTE. On the other hand, substantial changes occur in the Lyman continuum radiation field. If such changes persist to higher effective temperatures (for which computations are only now in progress), they may be relevant to our understanding of the excitation of gaseous nebulae by early-type stars. In Figure 13-11 we show the run of level populations with depth for a model with $T_{eff} = 15{,}000°K$ and $\log g = 3$; actual number populations are shown since the

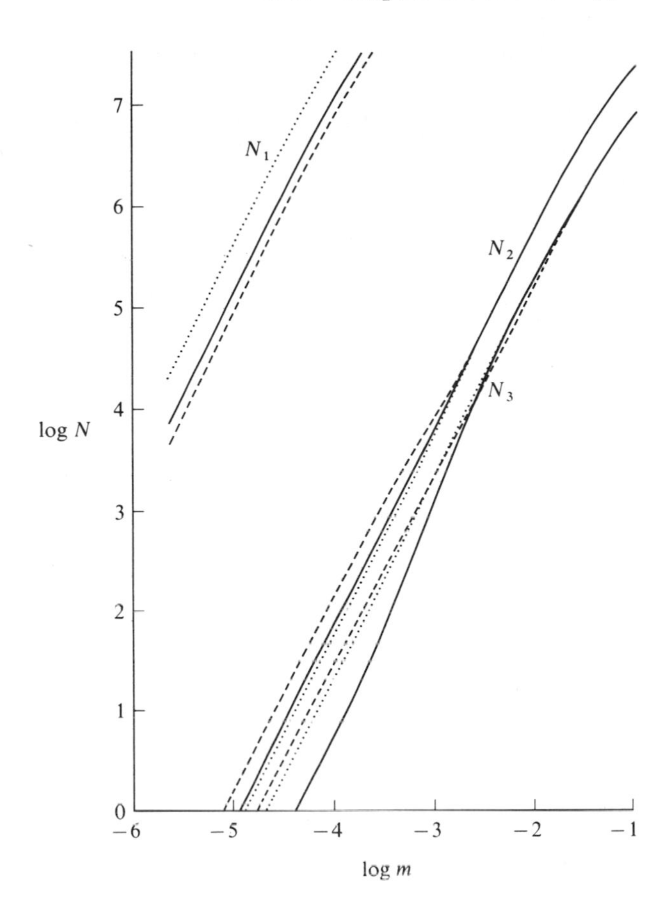


FIG. 13-11. Populations of first three levels of hydrogen in model atmospheres with $T_{\text{eff}} = 15{,}000°\text{K}$ and $\log g = 3$. *Dashed lines*: LTE model; *dotted lines*: non-LTE model with lines in detailed balance; and *solid lines*: non-LTE model including $H\alpha$, $H\beta$, $H\gamma$, $P\alpha$, $P\beta$, and $B\alpha$.

*b*-factors are strongly affected by temperature changes and thus are harder to interpret. Even with number densities, we must recall that the upper boundary condition upon the total pressure implies that the product $N \times T$ must remain constant in the outermost layers.

Consider first the comparison between the LTE solution without lines (dashed line) and the non-LTE solution without lines (dotted line). Accounting for the temperature rise alone (see Table 13-1), one would expect all densities to be smaller by a factor of about 0.87. Note, however, that the ground state population *increases* by a factor of 5, reflecting a real over-population effect. Level 2 is everywhere underpopulated relative to LTE by a factor of 0.45 near the boundary. This is too large an effect to be explained by the temperature rise alone and therefore represents a real depopulation effect. Moreover, the underpopulation persists to depths where no temperature changes occur; these results are in agreement with those of Chapter 7. At masses greater than $10^{-3}$, level 3 is overpopulated, again in agreement with the results of Chapter 7. Near the surface it is underpopulated; the effect there is mainly (though not completely) due to the temperature rise. Now consider the non-LTE results including the effects of lines (solid lines) compared with the non-LTE results without lines. Again the additional temperature rise would account for population decreases of a factor of 0.86. This explains about half of the drop found in $N_1$, the rest being due to increased ionization. The population of level 2 *rises*, reflecting the effect of cascades from higher states, which now occur freely in the line transitions. Previously, such cascades were suppressed by setting the lines into detailed balance. The population of level 3 decreases by a factor of 0.25; this is an effect of efficient decay into level 2. Similar results are found for levels 4 and 5, which have not been plotted in order to avoid confusion in the figure.

In Figure 13-12 we show the $H\alpha$ profile; in a general way the agreement with the results of Peterson and Kalkofen (Figure 13-3) is excellent. We see, however, that there is no indication of a hump on the non-LTE profile. In Figure 13-13 we show the $H\gamma$ profile; here we see that the results of the LTE and non-LTE calculations are indistinguishable. Note the spurious rise in the line core for the detailed balance model. In Figure 13-14 we have plotted the profile of $P\alpha$; here the non-LTE model gives a deeper line than LTE. This occurs because $n = 3$ is overpopulated in those layers where the line wing is formed. In Figure 13-15 we show the profile of $B\alpha(n = 4 \rightarrow n = 5)$. Here the complete non-LTE model predicts an emission core. This emission arises from recombinations to $n = 5$ with cascade to $n = 4$. This feature is

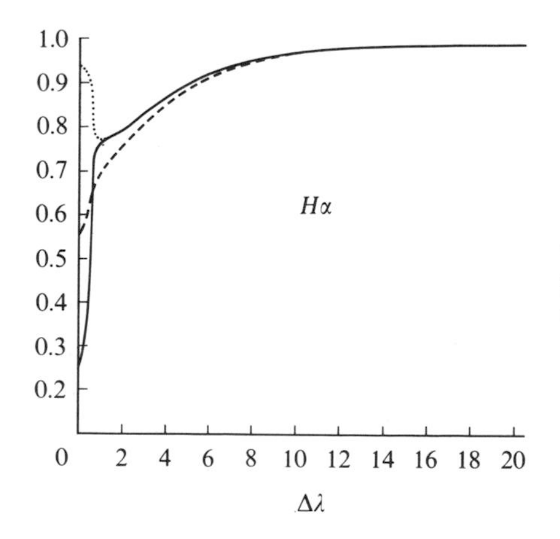


FIG. 13-12. $H\alpha$ profile in models with $T_{\text{eff}} = 15{,}000°\text{K}$ and $\log g = 3$. Coding of lines same as in Figure 13-11. Compare with Figure 13-3.

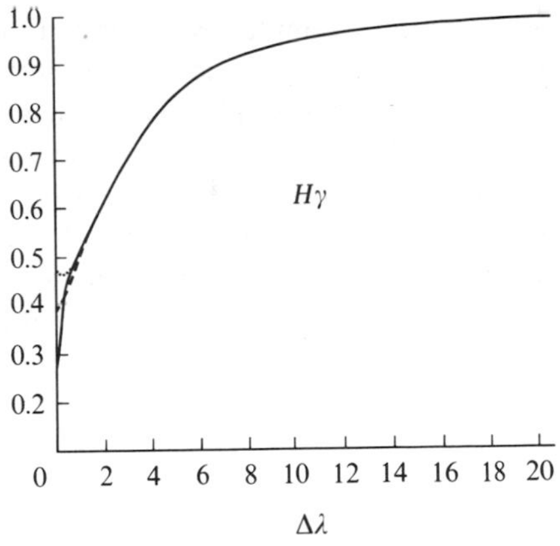

1.0
0.9
0.8
0.7
0.6
0.5
0.4
0.3
0.2
$H\gamma$
0 2 4 6 8 10 12 14 16 18 20
$\Delta\lambda$

FIG. 13-13. $H\gamma$ profile in models with $T_{eff} = 15{,}000°K$ and $\log g = 3$. Coding of lines same as in Figure 13-11.

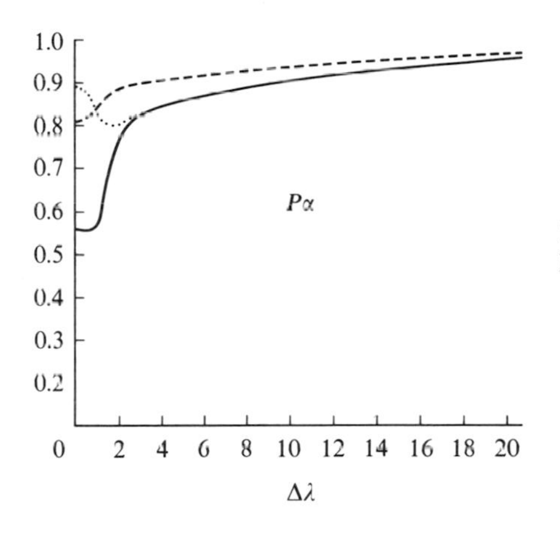


FIG. 13-14. $P\alpha$ profile in models with $T_{eff} = 15{,}000°K$ and $\log g = 3$. Coding of lines same as in Figure 13-11.

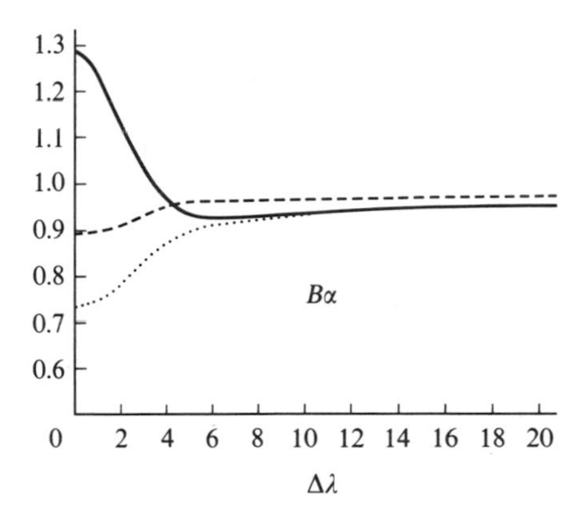


FIG. 13-15. $B\alpha$ profile in models with $T_{eff} = 15{,}000°K$ and $\log g = 3$. Coding of lines same as in Figure 13-11.

fairly strongly coupled with the temperature rise in the outermost layers and may provide a diagnostic tool for determining the temperature distribution in stellar atmospheres. The $B\alpha$ line lies at 4.05 $\mu$, which is in a window of the terrestrial absorption spectrum and should be accessible to observation.

While considerable progress has been made toward the solution of the multilevel problem and many physically important aspects have emerged, much work remains to be done before our understanding will be fully developed. It may well be expected that this area of stellar atmospheres research will remain especially active for some time to come.

# 14 | Effects of Line Blanketing on Stellar Atmospheres

Up to this point our discussion has treated continuum and line formation as more or less unrelated. We must now consider their interaction and study the effects of spectrum lines upon the physical conditions in a stellar atmosphere and upon the emergent flux distribution.

## 14-1. Qualitative Expectations

It is very useful to separate the total effects of line blanketing into three basic categories. In the first place, it is clear that the lines are darker than the nearby continuum, and therefore the flux that is emergent in any given band will be decreased by the presence of lines. This effect is usually called the *blocking effect* since it results from the simple blocking out of radiation. The blocking effect has very important consequences upon observationally measured colors of stars and must be accounted for carefully when comparing stars in which line strengths are known to differ. Several empirical studies of this effect have been made, and methods have been developed to allow for it (see, e.g., Wildey, Burbidge, Sandage, and Burbidge, *Ap. J.*, **135**, 94, 1962).

A second effect of line blanketing is to cause a *backwarming* in the deeper layers of the atmosphere. Again, we compare an atmosphere in which there are strong lines with one without strong lines. Very little flux is transported in the opaque lines, and virtually all of the transport occurs in the continuum. Suppose we require that the same total flux be transported in both atmospheres. The frequency band available to the continuum is decreased by the presence of the lines; therefore the flux per unit frequency interval must increase. This fact implies a rise in the local temperature in the blanketed case relative to the unblanketed case. Moreover, since we require that the same total flux must escape from the star, it is evident that the flux in the continuum between the lines must be higher than it is when the lines are absent.

A third effect of line blanketing is to alter the boundary temperature of the atmosphere. The amount by which the boundary temperature changes is a sensitive function of the mode of line formation and is a question of considerable importance. For example, consider the simple classical picture in which the lines have a source function of the form

$$S = (1 - \varepsilon)l_\nu J_\nu + \varepsilon l_\nu B_\nu \tag{14-1}$$

where $\varepsilon$ is the parameter deciding the partitioning between photons "absorbed" and "scattered" (coherently). Then the radiative equilibrium condition is

$$\int_0^\infty (\kappa_\nu + \sigma_\nu + l_\nu)J_\nu \, d\nu = \int_0^\infty \kappa_\nu B_\nu \, d\nu + \int_0^\infty \sigma_\nu J_\nu \, d\nu + \varepsilon \int_0^\infty l_\nu B_\nu \, d\nu + (1 - \varepsilon) \int_0^\infty l_\nu J_\nu \, d\nu \tag{14-2}$$

or

$$\int_0^\infty \kappa_\nu B_\nu \, d\nu = \int_0^\infty \kappa_\nu J_\nu \, d\nu + \varepsilon \int_0^\infty l_\nu(J_\nu - B_\nu) \, d\nu \tag{14-3}$$

Now if we assume the lines are very opaque, we would expect that, at least approximately, $J_\nu = \frac{1}{2}B_\nu(T_0)$ at the boundary in the lines. Hence

$$\int_0^\infty \kappa_\nu B_\nu(T_0) \, d\nu = \int_0^\infty \kappa_\nu J_\nu \, d\nu - \frac{\varepsilon}{2} \int_0^\infty l_\nu B_\nu(T_0) \, d\nu \tag{14-4}$$

Thus we see that the presence of lines will tend to *lower* the boundary temperature compared with the case of no lines. The effect can be large if $\varepsilon = 1$ (LTE) but can be made vanishingly small if $\varepsilon \to 0$ (lines formed by scattering). Note however that this argument is not really complete because we have ignored the change in $J_\nu$ in the continuum bands. What is required is a solution of the transfer problem that includes both lines and continuum, subject to the overall requirement of radiative equilibrium. Let us now consider a few examples of this kind of calculation.

## 14-2. The Picket Fence Model

A very simple and illuminating treatment of line blanketing is the *picket fence* model proposed by Chandrasekhar (*M. N.*, **96**, 21, 1936) and developed further by Münch (*Ap. J.*, **104**, 87, 1946). We have used this model in Chapter 6 in the discussion of the Lyman continuum with a somewhat different interpretation of some of the quantities involved. Let us now examine this approach in the context of line formation.

The transfer equation we consider is

$$-\frac{\mu}{\rho}\frac{dI_\nu}{dz} = (\kappa_\nu + l_\nu)I_\nu - (\kappa_\nu + \varepsilon l_\nu)B_\nu - (1-\varepsilon)l_\nu J_\nu \tag{14-5}$$

For simplicity we shall assume: (a) the continuum opacity is frequency-independent, i.e., $\kappa_\nu \equiv \kappa$; (b) the lines have a constant opacity ratio $\eta = l/\kappa$ to this continuum and square profiles of finite width; (c) the lines are distributed at random and uniformly throughout the spectrum so that within a given frequency band, a fraction $w_1$ contains continuum only and a fraction $w_2$ contains continuum plus lines. Pictorially, we are considering the problem sketched in Figure 14-1, which shows why the name "picket fence" is appropriate. For the frequencies in the continuum only we may write

$$\mu\frac{dI_\nu^{(1)}}{d\tau} = I_\nu^{(1)} - B_\nu \tag{14-6}$$

while in the lines we have

$$\mu\frac{dI_\nu^{(2)}}{d\tau} = (1+\eta)I_\nu^{(2)} - (1-\varepsilon)\eta J_\nu^{(2)} - (1+\varepsilon\eta)B_\nu \tag{14-7}$$

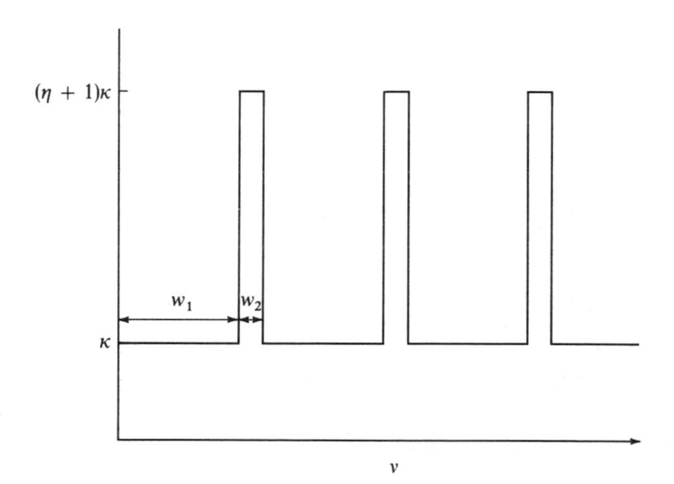


FIG. 14-1. Picket fence model. Lines are assumed to be a factor of $\eta$ more opaque than continuum and occur with a probability $w_2 = 1 - w_1$ in any frequency band.

Integrating over all frequencies and accounting for the relative probabilities of line and continuum, we find

$$\mu \frac{dI^{(1)}}{d\tau} = I^{(1)} - w_1 B \tag{14-8}$$

and

$$\mu \frac{dI^{(2)}}{d\tau} = (1 + \eta)I^{(2)} - (1 - \varepsilon)\eta J^{(2)} - w_2(1 + \varepsilon\eta)B \tag{14-9}$$

This pair of equations must be solved simultaneously with the requirement of radiative equilibrium

$$J^{(1)} + (1 + \varepsilon\eta)J^{(2)} = w_1 B + w_2(1 + \varepsilon\eta)B \tag{14-10}$$

### LTE LINE FORMATION

If we consider first the case of $\varepsilon = 1$, we find

$$\mu \frac{dI^{(1)}}{d\tau} = I^{(1)} - w_1 B \tag{14-11}$$

$$\mu \frac{dI^{(2)}}{d\tau} = (1 + \eta)I^{(2)} - w_2(1 + \eta)B$$

$$= \gamma I^{(2)} - w_2 \gamma B \tag{14-12}$$

and

$$B = \frac{J^{(1)} + (1 + \eta)J^{(2)}}{w_1 + (1 + \eta)w_2} = \frac{J^{(1)} + \gamma J^{(2)}}{w_1 + \gamma w_2} \tag{14-13}$$

This set of equations is precisely the same as we considered in the case of the Lyman continuum. We showed there that the solution has the form [equation (6-241)]

$$B(\tau) = \frac{\frac{3}{4}F}{(w_1 + w_2/\gamma)}\left(\tau + Q + \sum_{\alpha=1}^{2n-1} L_\alpha e^{-k_\alpha \tau}\right)$$

and the boundary value of $B$ is fixed at [equation (6-252)]

$$B(0) = \frac{\sqrt{3}}{4} F[(w_1 + \gamma w_2)(w_1 + w_2/\gamma)]^{-1/2} = \frac{\sqrt{3}}{4} F\left(\frac{\kappa_R}{\kappa_P}\right)^{1/2}$$

which implies [equation (6-253)]

$$\frac{T(0)}{T_{\text{eff}}} = \left(\frac{\sqrt{3}}{4}\right)^{1/4}\left(\frac{\kappa_R}{\kappa_P}\right)^{1/8}$$

Clearly, large values of $\eta$ lead to low values of $T(0)$. This is shown in Figure 14-2, in which we plot $B(T)/F$ for the standard gray case and for one of Münch's solutions with $\varepsilon = 1$, $w_1 = 0.8$, $w_2 = 0.2$, and $\gamma = (1 + \eta) = 10$. In this case, $B(0)/F$ drops from 0.433 to 0.286.

## SCATTERING LINES

In the more general case, we may write

$$\mu \frac{dI^{(1)}}{d\tau} = I^{(1)} - w_1 B \tag{14-14}$$

$$\mu \frac{dI^{(2)}}{d\tau} = \gamma I^{(2)} - (\gamma - \lambda)J^{(2)} - \lambda w_2 B \tag{14-15}$$

where

$$\lambda \equiv 1 + \varepsilon\eta \tag{14-16}$$

The condition of radiative equilibrium requires that

$$B = \frac{J^{(1)} + \lambda J^{(2)}}{w_1 + \lambda w_2} - \sigma(J^{(1)} + \lambda J^{(2)}) \tag{14-17}$$

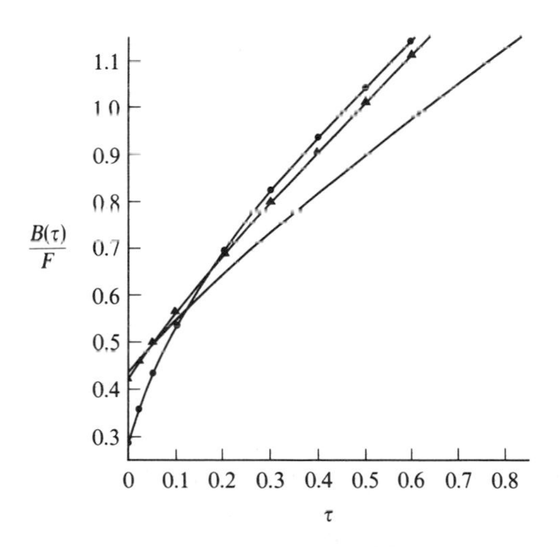


FIG. 14-2. Run of integrated Planck function with depth in picket fence models. *Straight line*: unblanketed solution, $\eta \equiv 1$; *dotted line*: $\gamma = \eta + 1 = 10$, $w_1 = 0.8$, $w_2 = 0.2$, and $\varepsilon = 1$ (LTE); and *triangled line*: $\gamma = \eta + 1 = 10$, $w_1 = 0.8$, $w_2 = 0.2$, and $\varepsilon = 0$ (pure scattering). (From G. Münch, *Ap. J.*, **104**, 87, 1946; by permission.)

Thus the system of equations to be solved becomes

$$\mu \frac{dI^{(1)}}{d\tau} = I^{(1)} - w_1 \sigma J^{(1)} - w_1 \lambda \sigma J^{(2)} \tag{14-18}$$

and

$$\mu \frac{dI^{(2)}}{d\tau} = \gamma I^{(2)} - (\gamma - w_1 \lambda \sigma) J^{(2)} - w_2 \lambda \sigma J^{(1)} \tag{14-19}$$

This system may be solved by the usual discrete-ordinate method. Münch shows that the characteristic equation becomes

$$1 - w_1 \sigma G - \left(1 - \frac{w_1 \lambda \sigma}{\gamma}\right) H - w_1 \sigma \left(1 - \frac{\lambda}{\gamma}\right) GH = 0 \tag{14-20}$$

where

$$G \equiv \frac{1}{2} \sum_{i=-n}^{n} \frac{a_i}{1 + k\mu_i} \tag{14-21}$$

and

$$H \equiv \frac{1}{2} \sum_{i=-n}^{n} \frac{a_i}{1 + k\mu_i/\gamma} \tag{14-22}$$

Equation (14-20) has $2n - 1$ positive roots $k_\alpha$. The solution for $B(\tau)$ is given by

$$B(\tau) = \frac{\frac{3}{4}F}{(w_1 + w_2/\gamma)} \left[ \tau + Q + \sum_{\alpha=1}^{2n-1} L_\alpha e^{-k_\alpha \tau} \left( \frac{G_\alpha}{1 - G_\alpha} + \left(\frac{\lambda}{\gamma}\right) \frac{H_\alpha}{H_\alpha - 1} \right) \sigma \right] \tag{14-23}$$

where the $2n - 1$ constants $L_\alpha$ and the constant $Q$ are determined by the boundary conditions

$$\frac{1}{w_1} \sum_{\alpha=1}^{2n-1} \frac{L_\alpha}{1 - G_\alpha} \frac{1}{1 - k_\alpha \mu_i} + Q = \mu_i \tag{14-24}$$

and

$$\frac{1}{\gamma w_2} \sum_{\alpha=1}^{2n-1} \frac{L_\alpha}{H_\alpha - 1} \frac{1}{1 - k_\alpha \mu_i/\gamma} + Q = \frac{\mu_i}{\gamma} \tag{14-25}$$

A solution for $w_1 = 0.8$, $w_2 = 0.2$, $\gamma = 10$, and $\varepsilon = 0$ obtained by Münch is shown in Figure 14-2. Here one finds that the boundary temperature lies only slightly below the gray value, with $B(0)/F = 0.4308$ instead of the gray result 0.4330. Thus we see that *the lines, when formed by scattering, have practically no influence at all upon the boundary temperatures.* Of course, the backwarming effect is still present, as shown, since the frequency band in which the radiation flows freely is still restricted.

In his paper Münch considers, in addition, the case of scattering lines formed in a finite layer on top of a semi-infinite absorbing atmosphere. He also treats the case of pure absorbing lines in an atmosphere in which the probability $w_2$ is a function of frequency. This kind of model is found to reproduce the observed solar flux quite satisfactorily, in view of the approximations made.

## 14-3. Numerical Approaches in LTE

The model described above is purposely made quite simple so that the analysis will be easy. For other cases where more realistic frequency and depth variations are to be included, it is convenient to resort to direct numerical computation. We will describe here some typical results obtained under the assumption that the lines are formed in LTE.

There have been basically two approaches to the numerical computations thus far. In the so-called *direct approach* one simply includes enough frequency points to describe the profiles of the lines under consideration. In this way, the full frequency and depth variation of the absorption coefficient can be included. The disadvantage, of course, is that the time required to compute a model goes up with the number of frequency points, and the procedure quite soon becomes prohibitively time-consuming. Nevertheless, the spectra of certain stars are dominated by just a few lines, and in this case the procedure remains practicable.

An example of this kind of calculation is the model for a B 1.5 V star computed by Mihalas and Morton (*Ap. J.*, **142**, 253, 1965), which includes the effects of the stronger lines in the ultraviolet on the range $912\ \text{Å} \leq \lambda \leq 1600\ \text{Å}$. The computed emergent flux is shown in Figure 14-3. Figure 14-4 shows details in the strongly blanketed region $912\ \text{Å} \leq \lambda \leq 1130\ \text{Å}$, wherein we find strong lines from hydrogen (the Lyman series) and ions of C, N, Cl, Si, and Fe. The integrated emergent flux of the blanketed model corresponds to $T_{\text{eff}} = 21{,}900°\text{K}$. The flux distribution of an unblanketed model with the same effective temperature is also shown, and we see that the effects of line blanketing have been dramatic. The flux in the continuum between the lines lies far above the unblanketed continuum, and the energy deficit in the ultraviolet caused by line blocking is compensated by a large rise in the continuum in the visible. In fact, the energy distribution in the visible region closely approximates that of an unblanketed model with $T_{\text{eff}} = 24{,}000°\text{K}$. This shows that if we had fitted the observed flux of a star in the visible region to an unblanketed model, we would have made an error of some 2100°K (or more) in our estimate of the effective temperature. The effects of blanketing upon

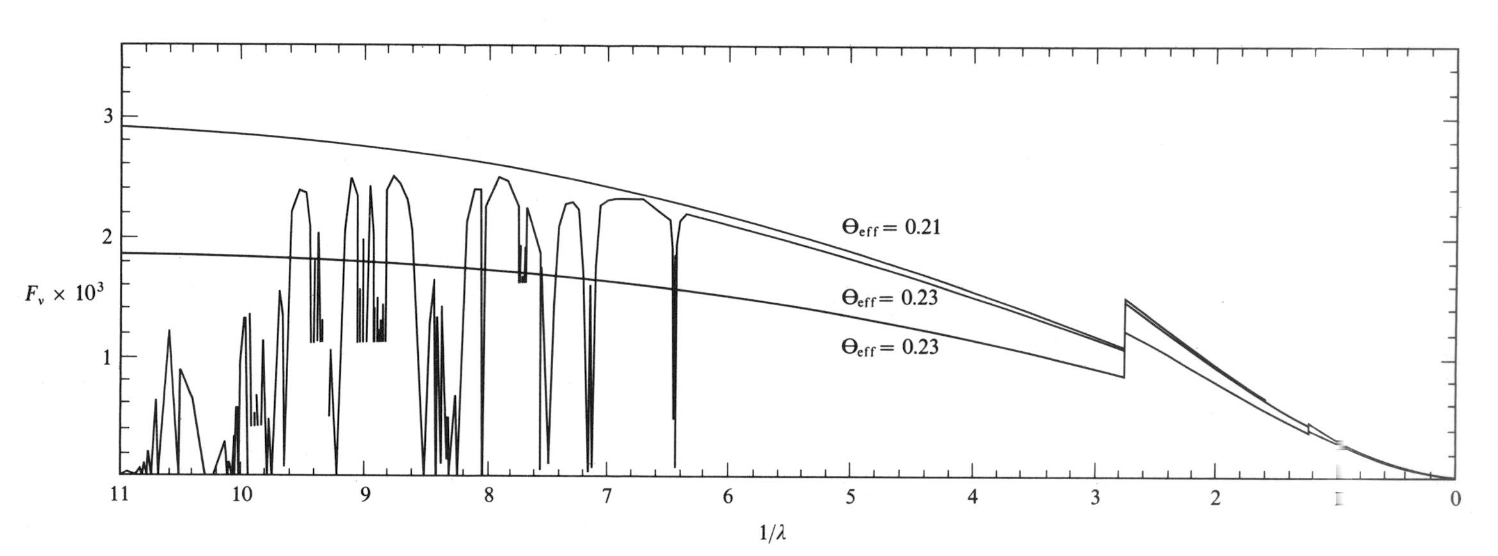


FIG. 14-3. Flux from blanketed and unblanketed LTE models. Blanketed model corresponds to $\theta_{eff} = 0.23$. Abscissa: $1/\lambda$, where $\lambda$ is in microns; ordinate: $F_\nu \times 10^3$.

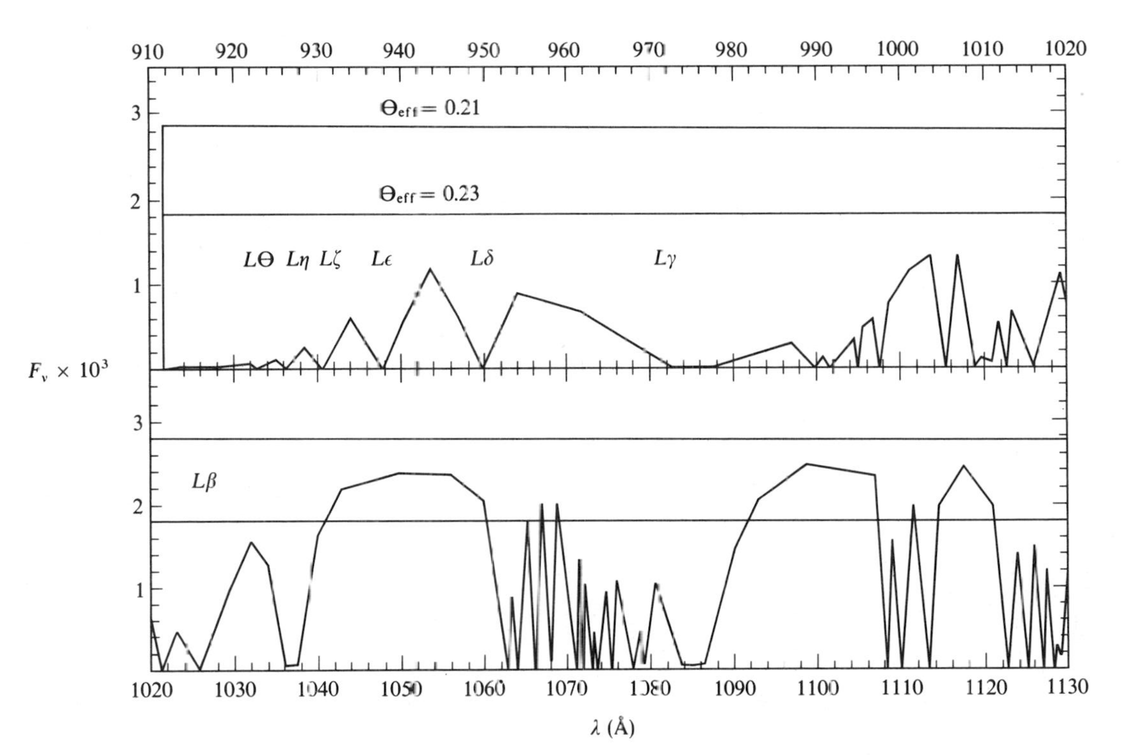


FIG. 14-4. Details of flux distribution from blanketed model on range 910 Å $\leq \lambda \leq$ 1130 Å. Only Lyman lines are labeled, though these are by no means the strongest lines in the spectrum.

both the effective temperature scale and the bolometric correction scale have important implications in the comparison of stellar interiors models with observations of the mass luminosity relation and in the Hertzsprung-Russell diagram, as has been discussed by Morton and Adams (*Ap. J.*, **151**, 611, 1968). Similar models have been constructed at O5 and B0 V by Hickock and Morton (*Ap. J.*, **152**, 195, 1968). At cooler temperatures the blanketing by Balmer lines is important; we have already described some models that allow for this (*Ap. J. Supp. No. 114*, **13**, 1, 1966).

The direct approach is quite straightforward but becomes too laborious when the number of lines becomes very large. We are faced with this fact when we come to consider solar-type stars, where literally thousands of weak to strong lines of neutral and singly ionized metals blanket the spectrum. An impression of the complexity of the situation can be gained from even a cursory inspection of spectrograms of the sun. The situation becomes even more extreme as we pass to the late-type stars (K and M), in which the spectrum is strongly blanketed by the *millions* of lines in the absorption bands of molecules, such as $H_2O$ (Auman, *Ap. J. Supp. No. 127*, **14**, 171, 1967). We must therefore, resort to some alternative procedure. An example of one possible approach is a simple numerical analogue of the picket fence model. We shall refer to this approach as the *statistical method.* Here one attempts to replace the complicated frequency variation of the lines in a band (see Figure 14-5) by a smooth opacity distribution. To do this, we must consider a narrow enough band to assure that the exact position of the line is not important. At several values of $l_i$ we might then count the fraction of the band covered by lines with opacity $l_\nu \geq l_i$ and plot a graph of this fraction versus $l_i$. The result is a smooth curve that can be approximated by just a few subintervals containing appropriate opacities (see Figure 14-6). This procedure may be carried

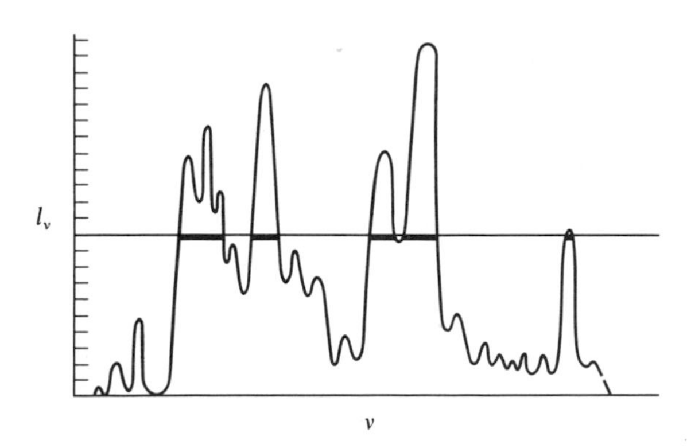


FIG. 14-5. Schematic opacity distribution of overlapping spectrum lines. A large number of frequencies would be required to specify the run of the opacity in detail.

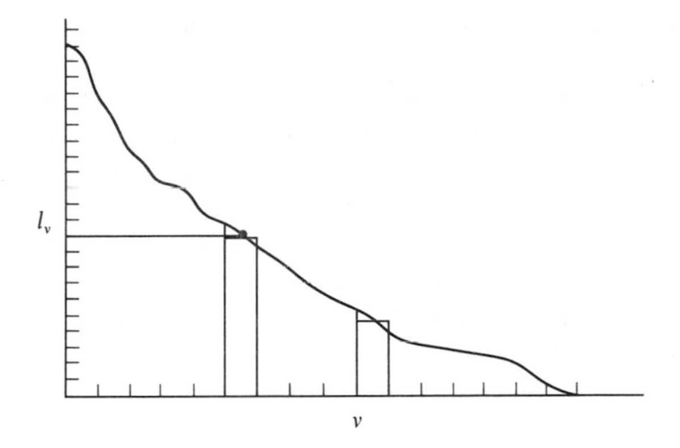


FIG. 14-6. A schematic smoothed transformation of the opacity run shown in Figure 14-5.

out as a function of temperature and pressure, and thus it allows, at least roughly, for the variation of line strengths through the atmosphere. An excellent example of this type of calculation was done by Strom and Kurucz (*J. Q. S. R. T.*, **6**, 591, 1966), who computed a blanketed model of the F5 IV star Procyon. These authors form a statistical blanketing coefficient allowing for the effects of some 30,000 lines grouped into 200 Å to 300 Å bands. We show in Figure 14-7 the comparison of an unblanketed model, their blanketed

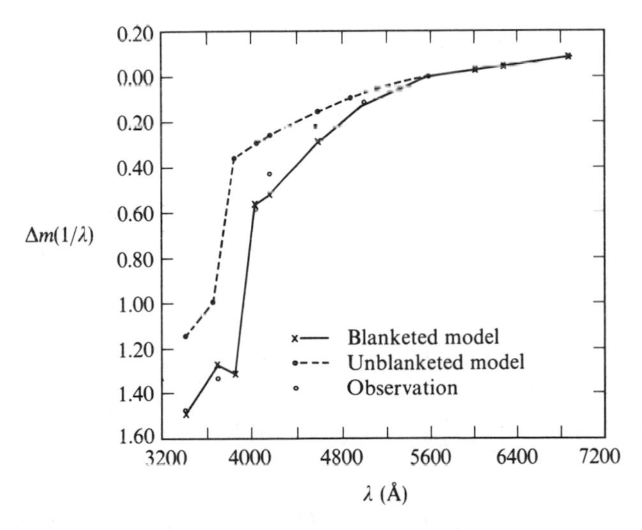


FIG. 14-7. Unblanketed and blanketed energy distribution with $T_{eff} = 6500°K$ and $\log g = 4$, compared with observation of Procyon. (From. S. E. Strom and R. L. Kurucz, *J.Q.S.R.T.*, **6**, 591, 1966; by permission.)

model, and observations. It is clear that the authors obtain satisfactory agreement with the observations. Since the lines are assumed to be formed in LTE, they find a drop in the boundary temperature of the model, which, in addition, leads to a drop in the central intensity for strong lines from about 27% of the continuum to about 11% of the continuum, a result of importance in abundance analyses.

One difficulty with the statistical approach described above is that it tacitly assumes that the relative strengths of lines in the band are independent of depth and that no lines appear or disappear. Actually, this is not quite realistic. On the other hand, so long as the relative strengths change significantly only on a scale large compared with a unit optical depth interval in the continuum, this is not too bad an approximation. If, however, a marked change in the spectrum occurs abruptly at some depth as, for example, might occur across a strong shock front in the outer layers of an atmosphere, the approximation breaks down. Under such circumstances it is critical to the transfer problem whether a line in the upper region lies (in frequency space) atop a line or a continuum band in the lower region, and similarly for the continuum. In this case, one must employ either the direct approach or a statistical approach that allows in some way for the changes in position of lines in frequency space.

## 14-4. Non-LTE Line-Blanketing Effects

As we have emphasized above, the effects of line blanketing upon the boundary temperature depend strongly upon the form of the line source function, and we have discussed the differences in the effects of "absorption" and "scattering" lines. This kind of analysis has been extended by Thomas (*Ap. J.*, **141**, 333, 1965) to include lines with two-level-atom type source functions, assuming complete redistribution. Thus for the $j$th line we write $l_\nu(j) = l_j \varphi_\nu$ and use the frequency-independent source function

$$S_{l,j} = \frac{\int \phi_\nu J_\nu \, d\nu + \varepsilon_j' B_j}{1 + \varepsilon_j'} = (1 - \varepsilon_j) \int \phi_\nu J_\nu \, d\nu + \varepsilon_j B_j \tag{14-26}$$

Radiative equilibrium then requires

$$\begin{aligned}\int \kappa_\nu B_\nu(T_0) \, d\nu &= \int \kappa_\nu J_\nu \, d_\nu - \int l_\nu [S_l - J_\nu] \, d\nu \\ &= \int \kappa_\nu J_\nu \, d\nu - \sum l_j \left[(1 - \varepsilon_j) \int \varphi_\nu J_\nu \, d\nu + \varepsilon_j B_j\right] \\ &\quad + \sum l_j \int \varphi_\nu J_\nu \, d\nu \end{aligned} \tag{14-27}$$

or

$$\int \kappa_\nu B_\nu(T_0)\, d\nu = \int \kappa_\nu J_\nu\, d\nu - \sum l_j \varepsilon_j \left[ B_j(T_0) - \int \varphi_\nu J_\nu\, d\nu \right] \tag{14-28}$$

As before, for LTE, $\varepsilon_j = 1$, and at $\tau = 0$, for an opaque line, $J_\nu \approx \frac{1}{2} B_\nu(T_0)$ so that

$$\int \kappa_\nu B_\nu(T_0)\, d\nu = \int \kappa_\nu J_\nu\, d\nu - \tfrac{1}{2} \sum l_j B_j(T_0) \tag{14-29}$$

which implies that a large drop in the boundary temperature can occur. On the other hand, for $\varepsilon_j \ll 1$, as $\tau \to 0$,

$$B_j(T_0) - \int \varphi_\nu J_\nu\, d\nu \to \frac{1 + \sqrt{\varepsilon_j}}{1 - \varepsilon_j} B_j(T_0) \approx B_j(T_0) \tag{14-30}$$

and thus

$$\int \kappa_\nu B_\nu(T_0)\, d\nu = \int \kappa_\nu J_\nu\, d\nu - \sum l_j \varepsilon_j B_j(T_0) \tag{14-31}$$

which implies a much smaller drop. This result is not surprising in view of our previous discussion, inasmuch as the two-level atom picture leads to a scattering line near the surface.

As before, we can object to this analysis on the grounds that changes in the continuum terms are neglected; we shall see that such changes occur and have important implications. Moreover, the analysis assumes that the line strength (i.e., $l_j$) is unchanged; in general, in the non-LTE case there are significant alterations in level populations and accordingly in the opacity of lines. To meet these objections, one must again carry out a self-consistent calculation for both continuum and lines. Such a calculation is, however, extraordinarily difficult because of the weak coupling between the radiation field in the lines and the local thermal field (small $\varepsilon$), the high opacities of the lines, the long thermalization lengths, and the complexity of the interlocking effects. One must therefore use great care in the computation. Problems of this kind have been treated successfully by Auer and Mihalas (*Ap. J.*, **156**, 157, 1969, and *Ap. J.*, **156**, 681, 1969), who used the linearization approaches described in Chapters 7 and 13. In the first calculation cited, a two-level hydrogen atom was considered, allowing for L$\alpha$. In LTE the boundary temperature in a model with $T_{\text{eff}} = 15{,}000$°K without $L\alpha$ is about 9400°K; inclusion of $L\alpha$ lowers the temperature to about 7800°K, a drop of about 1600°K. On the other hand, in the non-LTE model the boundary temperature without $L\alpha$ is about 10,350°K because of heating in the Lyman continuum. When $L\alpha$ is included, the boundary temperature drops to about 9800°K, a drop only one-third as large as that in LTE. The temperature structure in the non-LTE $L\alpha$ case is relatively complex, as seen in Figure 14-8. The second calculation emphasizes the importance of coupling between lines and continuum. Here a three-level hydrogen atom was studied, allowing for the

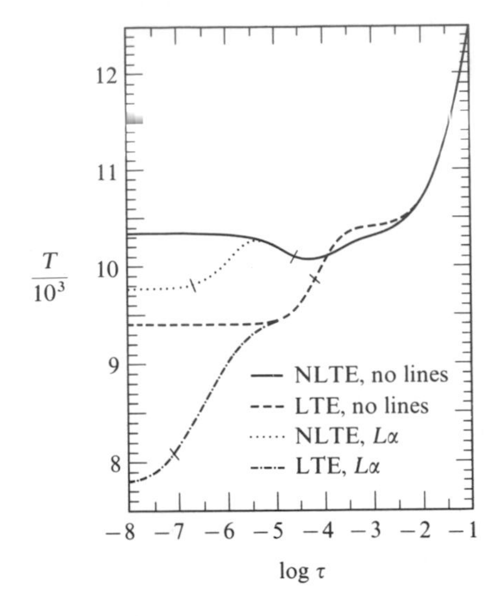


FIG. 14-8. Temperature distribution for LTE and non-LTE models with $T_{eff} = 15{,}000°K$ and $\log g = 4$, including the effects of Lyman $\alpha$.

effects of $H\alpha$. The Lyman lines were ignored since they are formed in the very outermost layers of the atmosphere; $H\alpha$, on the other hand, is formed at about the same depth as the Lyman continuum and can have an important interaction with it. The calculations show that the effect of $H\alpha$ in LTE is to drop the boundary temperature from about 9400°K to about 8900°K (see Figure 14-9). In the non-LTE case, however, inclusion of $H\alpha$ *raises* the boundary temperature from 10,500°K to 11,200°K. The line itself provides a negative (cooling) contribution to the energy balance equation. At the same time it provides an efficient channel for atoms to fall into the $n = 2$ state, where the radiation field in the continuum provides strong heating. In this case the direct cooling effect of the line is outweighed by the indirect effects of the line upon the continuum energy balance and emphasizes the importance of carrying out a physically consistent analysis.

The curves shown in Figure 14-9 are based on somewhat simplified models. More accurate calculations have been carried out by Auer and Mihalas, allowing for six lines, as described in Chapter 13. Results from these calculations are summarized in Table 13-1. The interpretation is the same as given above, though the actual boundary temperatures are slightly different. One point of interest is that higher lines of the Balmer series enhance the heating effect initiated by $H\alpha$ by allowing still higher levels to cascade into the $n = 2$ state. This is illustrated in Figure 14-10, which shows the change in the temperature structure caused by adding successive line transitions. We see that the effect of $H\alpha$ alone is comparable with that of the additional transitions $H\beta$ and $P\alpha$. Including the still higher transitions $H\gamma$, $P\beta$, and $B\alpha$ leads to only a small additional rise.

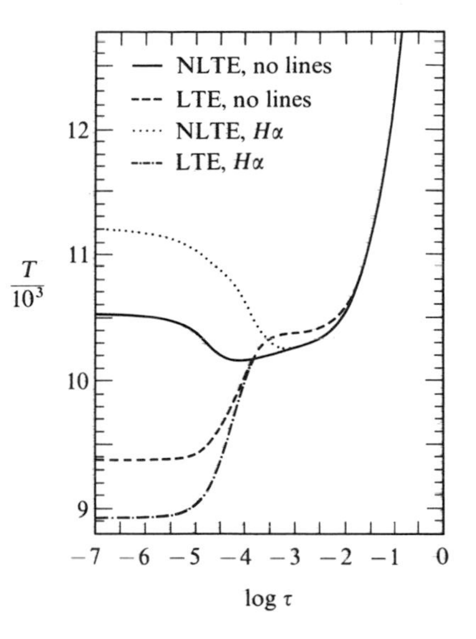


FIG. 14-9. Temperature distribution for LTE and non-LTE models with $T_{\text{eff}} =$ 15,000°K and log $g = 4$, including the effects of $H\alpha$. Note that since this calculation was somewhat simplified, the results are not identical to those of Figure 14-10.

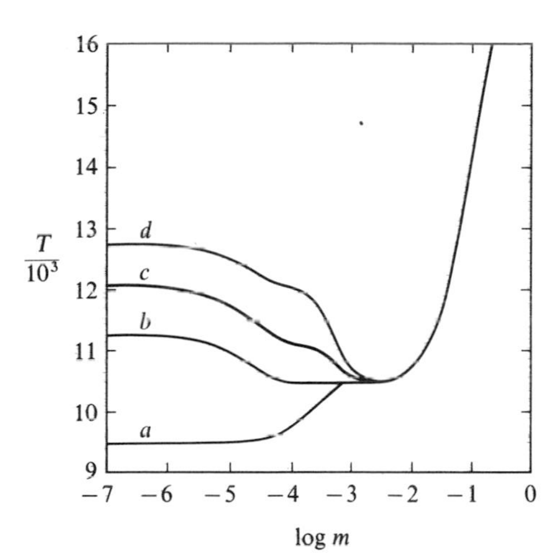


FIG. 14-10. Temperature distributions showing effects of lines: (*a*) LTE model, no lines; (*b*) non-LTE model, no lines, showing heating in Lyman continuum; (*c*) non-LTE model including $H\alpha$; and (*d*) non-LTE model including $H\alpha$, $H\beta$, and $P\alpha$. The addition of $H\gamma$, $P\alpha$, and $B\alpha$ causes only very small changes relative to curve *d*.

Work upon the intriguing problem of non-LTE line blanketing has clearly just begun, and a great deal remains to be done. The few results obtained so far have been so interesting that one might easily hope that much is yet to be learned from further efforts in this area.

# References


1. Abramowitz, M., and J. A. Stegun. *Handbook of Mathematical Functions.* Washington, D.C., U.S. Department of Commerce, 1964.
2. Abt, H. A., A. B. Meinel, W. W. Morgan, and J. W. Tapscott. *An Atlas of Low-Dispersion Grating Stellar Spectra.* Kitt Peak National Observatory, Steward Observatory, and Yerkes Observatory, 1969.
3. Alder, B., S. Fernbach, and M. Rotenberg, eds. *Methods in Computational Physics,* Vol. 7. New York, Academic Press, 1967.
4. Aller, L. H., *The Atmospheres of the Sun and Stars.* New York, Ronald Press, 1963.
5. Aller, L. H., and D. B. McLaughlin, eds. *Stellar Structure.* Chicago, University of Chicago Press, 1965.
6. Bates, D. R. *Atomic and Molecular Processes.* New York, Academic Press, 1962.
7. Bode, G. *Die Kontinuierliche Absorption von Sternatmosphären,* Kiel, Institut für Theoretische Physik, 1965.
8. Breene, R. G. *The Shift and Shape of Spectral Lines.* Oxford, Pergamon Press, 1961.
9. Chandrasekhar, S. *Radiative Transfer.* New York, Dover, 1960.
10. Cooper, J., ed. *Proceedings of Workshop Conference on the Lowering of the Ionization Potential,* J.I.L.A. Report No. 79. Boulder, Joint Institute for Laboratory Astrophysics, 1966.
11. Dirac, P. A. M. *Principles of Quantum Mechanics,* 4th ed. Oxford, Clarendon Press, 1958.
12. Glennon, B. M., and W. L. Weise. *Bibliography on Atomic Transition Probabilities.* Washington, D.C., U.S. Department of Commerce, 1966; supplement, 1968.

13. Greenstein, J. L., ed. *Stellar Atmospheres.* Chicago, University of Chicago Press, 1960.
14. Griem, H. R. *Plasma Spectroscopy.* New York, McGraw-Hill, 1964.
15. Harvard-Smithsonian Conference on Stellar Atmospheres. *Proceedings of the First Conference*, S.A.O. Special Report No. 167. Cambridge, Smithsonian Astrophysical Observatory, 1964.
16. Harvard-Smithsonian Conference on Stellar Atmospheres. *Proceedings of the Second Conference*, S.A.O. Special Report No. 174. Cambridge, Smithsonian Astrophysical Observatory, 1965.
17. Heitler, W. *Quantum Theory of Radiation.* Oxford, Clarendon Press, 1964.
18. Hiltner, W. A., ed. *Astronomical Techniques.* Chicago, University of Chicago Press, 1962.
19. Hynek, J. A., ed. *Astrophysics: A Topical Symposium.* New York, McGraw-Hill, 1951.
20. Jefferies, J. T. *Spectral Line Formation.* Waltham, Blaisdell, 1968.
21. Kepple, P. C. *Improved Stark Profile Calculations for the First Four Members of the Hydrogen Lyman and Balmer Series*, University of Maryland Report No. 831. College Park, University of Maryland, 1968.
22. Kieffer, L. J. *Bibliography of Low Energy Electron Collision Cross-Section Data*, N.B.S. Misc. Pub. No. 289. Washington, D.C., U.S. Department of Commerce, 1967.
23. Kourganoff, V. *Basic Methods in Transfer Problems.* New York, Dover, 1963.
24. Linsky, J. L. *Calcium Line Formation in the Solar Chromosphere*, S.A.O. Special Report No. 274. Cambridge, Smithsonian Astrophysical Observatory, 1968.
25. Menzel, D. H., ed. *Selected Papers on Physical Processes in Ionized Plasmas.* New York, Dover, 1962.
26. ———. *Selected Papers on the Transfer of Radiation.* New York, Dover, 1966.
27. Merzbacher, E. *Quantum Mechanics.* New York, Wiley, 1961.
28. Morgan, W. W., P. C. Keenan, and E. Kellman. *An Atlas of Stellar Spectra.* Chicago, University of Chicago Press, 1943.
29. National Center for Atmospheric Research. *Resonance Lines in Astrophysics.* Boulder, The Center, 1968.
30. Panofsky, W. K. H., and M. Phillips. *Classical Electricity and Magnetism*, 2nd ed. Reading, Addison-Wesley, 1962.
31. Slater, J. C. *Quantum Theory of Atomic Structure.* New York, McGraw-Hill, 1960.
32. Spitzer, L., Jr. *Physics of Fully Ionized Gases.* New York, Wiley, 1956.
33. Thomas, R. N. *Some Aspects of Nonequilibrium Thermodynamics in the Presence of a Radiation Field.* Boulder, University of Colorado Press, 1965.
34. Traving, G. *Über die Theorie der Druckverbreiterung von Spektrallinien.* Karlsruhe, Verlag G. Braun, 1960.
35. Underhill, A. B., and J. H. Waddell. *Stark Broadening Functions for the Hydrogen Lines*, N.B.S. Circular No. 603. Washington, D.C., U.S. Department of Commerce, 1959.
36. Unsöld, A. *Physik der Sternatmosphären.* Berlin, Springer-Verlag, 1955.

37. Weise, W. L., M. W. Smith, and B. M. Glennon. *Atomic Transition Probabilities*, Vol. I (NSRDS-NBS4). Washington, D.C., U.S. Department of Commerce, 1966.
38. Woolley, R. v. d. R., and D. W. N. Stibbs, *The Outer Layers of a Star*. Oxford, Clarendon Press, 1953.
39. Zirin, H. *The Solar Atmosphere*. Waltham, Blaisdell, 1966.

# Glossary of Physical Symbols

Below we list each symbol with its physical significance and the page on which it first appears. Standard mathematical symbols, dummy variables and indices, and notations used only on a single page are not included.

|  |  |
|---|---|
| $a$ | Ratio of damping width to Doppler width, $\Gamma/4\pi\,\Delta\nu_D$, 251 |
| $a_0$ | Bohr radius, 69 |
| $a_i$ | Quadrature weight, 47 |
| $a_\nu$ | Macroscopic absorption coefficient (per gram of stellar material), uncorrected for stimulated emission, 82 |
| $a_\nu$ | Coefficient in linear expansion of Planck function, 153 |
| $a_\nu^*$ | LTE value of absorption coefficient (per gram of stellar material), uncorrected for stimulated emission, 217 |
| $a_\nu(\mu)$ | Line absorption intensity at angle $\mu$ on disk, 240 |
| $A, A'$ | Surface area, 1 |
| $A$ | Ratio of hydrogen abundance to abundance of heavy elements ("metals"), 73 |
| $\mathbf{A}$ | Vector potential, 86 |
| $A_i$ | Atomic weight of element $i$, 72 |
| $\mathbf{A}_i$ | Difference matrix coupling depth points $i$ and $i-1$ in difference-equation method, 166 |
| $A_0$ | Central absorption depth in line, 333 |

|  |  |
|---|---|
| $A_\nu$ | Line-flux absorption depth, 240 |
| $A_{ji}$ | Einstein spontaneous emission probability for transition $j \rightarrow i$, 82 |
| $A(\rho)$ | Fourier transform of Holtsmark distribution, 266 |
| $b_\nu$ | Coefficient in linear expansion of Planck function in an atmosphere, 153 |
| $b_i$ | Non-LTE departure coefficient, 79 |
| $b_\nu(\tau)$ | Planck function normalized to center-of-disk intensity, $B_\nu(\tau)/I_\nu(0, 1)$, 208 |
| $\mathbf{B}$ | Magnetic induction, 86 |
| $B^*$ | Continuum source term, 378 |
| $\mathbf{B}_i$ | Coupling matrix at depth point $i$ in difference-equation method, 166 |
| $B_{ij}$ | Einstein absorption probability for transition $i \rightarrow j$, 81 |
| $B_{ji}$ | Einstein-induced emission probability for transition $j \rightarrow i$, 82 |
| $B(T)$ | Integrated Planck function, $\int_0^\infty B_\nu(T)d\nu$, 35 |
| $B_\nu(T)$ | Planck function, 10 |
| $B_\nu^*(\tau)$ | Planck function distribution in radiative equilibrium, 182 |
| $c$ | Velocity of light, 5 |
| $C_p$ | Specific heat at constant pressure, 200 |
| $C_v$ | Specific heat at constant volume, 200 |
| $\mathbf{C}_i$ | Difference matrix coupling depth points $i$ and $i+1$ in difference-equation method, 166 |
| $C_k$ | Interaction coefficient of $k$th Stark component, 269 |
| $C_o$ | Numerical factor for collision-rate calculations, 142 |
| $C_p$ | Coefficient of perturbation function of order $p$, 255 |
| $C_{ij}$ | Collisional transition rate from state $i$ to state $j$, 136 |
| $\mathscr{C}_{ij}$ | Net collisional transition rate from state $i$ to state $j$, 136 |
| $\mathscr{C}_{i\kappa}$ | Net collisional transition rate from state $i$ to continuum, 137 |
| $d_i$ | $b_i - 1$, 141 |
| $\mathbf{d}$ | Electric dipole moment, 93 |
| $\mathbf{d}_{mn}$ | $\langle\phi_m^*\|\mathbf{d}\|\phi_n\rangle$, matrix element of dipole moment, 95 |

|  |  |
|---|---|
| $D$ | Distance from star to observer, 3 |
| $D$ | Debye length, 130 |
| $\mathbf{D}$ | Electric displacement vector, 85 |
| $D_B$ | Balmer jump (magnitudes), 187 |
| $\mathbf{D}_i$ | Auxilliary matrix in difference-equation method, 166 |
| $D_P$ | Paschen jump (magnitudes), 187 |
| e | Electron charge, 88 |
| $\mathbf{E}$ | Electric field vector, 86 |
| $E, E_\nu$ | Energy emission, energy transported across a surface, 1 |
| $E_i$ | Energy of state $i$ relative to ground, 82 |
| $E_0$ | Magnitude of electric field, 86 |
| $E(\omega)$ | Energy spectrum, $F^*(\omega)F(\omega)/2\pi$, 245 |
| $E_n(x)$ | Exponential integral of order $n$, 21 |
| $E(\omega,T)$ | Energy spectrum of wavetrain of duration $T$, 255 |
| $f_c$ | Continuum oscillator strength, 114 |
| $f_i$ | Fraction of all atoms of a given species in ionization state $i$, 71 |
| $f_\nu$ | Flux measured by an observer, 4 |
| $\mathbf{f}^1$ | Outgoing radiation field representation in Riccati transformation method, 161 |
| $\mathbf{f}^2$ | Incoming radiation field representation in Riccati transformation method, 161 |
| $f_{i,j}$ | Fraction of all atoms of element $i$ in ionization state $j$, 72 |
| $f_{ij}$ | Oscillator strength of transition $i \to j$, 92 |
| $f(t)$ | Time-varying oscillation amplitude, 244 |
| $f(v)$ | Velocity distribution function, 133 |
| $f_\nu(\mu)$ | Incoming radiation field at upper boundary of atmosphere, 17 |
| $f(\xi')$ | Absorption profile in atom's rest-frame, 301 |
| $f(n', n)$ | Total oscillator strength for transition $n' \to n$, 98 |
| $f_K(n', n)$ | Kramers' formula oscillator strength for transition $n' \to n$, 105 |
| $f(n', l'; n, l)$ | Oscillator strength for transition from sublevel $l'$ of level $n'$ to sublevel $l$ of level $n$, 98 |
| $F$ | Frequency-integrated flux, $\int_0^\infty F_\nu d\nu$, 26 |
| $F$ | Stark field strength, 264 |
| $F_c$ | Flux in continuum near line, 240 |
| $F_{conv}$ | Flux transported by convection, 202 |

|  |  |
|---|---|
| $F_o$ | Normal field strength, 264 |
| $F_{\text{rad}}$ | Flux transported by radiation, 205 |
| $\Gamma_{\text{rad}}$ | Radiative damping force on an oscillator, 90 |
| $F_\nu$ | $\frac{1}{\pi}\mathscr{F}_\nu$, "astrophysical" flux, 4 |
| $\mathscr{F}_\nu$ | Energy flux through horizontal plane surface, 3 |
| $\mathbf{F}_\nu$ | Vector energy flux, 3 |
| $F(v)$ | Spontaneous recapture probability for electrons of $v$ velocity, 109 |
| $F(v)$ | Dawson's function, 252 |
| $F(\omega)$ | Fourier transform, 244 |
| $F(\omega, T)$ | Fourier transform of wavetrain of duration $T$, 254 |
| $F(a, b, c, x)$ | Hypergeometric function, 104 |
| $g$ | Acceleration of gravity at the surface of the star, 148 |
| $g_i$ | Statistical weight factor of excitation state $i$, 68 |
| $g_\nu(\mu)$ | Incident radiation field at lower boundary of atmosphere, 17 |
| $g(\mathbf{n}', \mathbf{n}), g(\mu', \mu), g(\mu', \mu, \phi)$ | Angular phase functions in scattering processes, 12 |
| $g_A(\mathbf{n}', \mathbf{n})$ | Isotropic angular phase function, 302 |
| $g_B(\mathbf{n}', \mathbf{n})$ | Dipole angular phase function, 302 |
| $g_{\text{I}}(n', n)$ | Bound-bound Gaunt factor for transition $n' \to n$, 105 |
| $g_{\text{II}}(n', k), g_{\text{II}}(n', \nu)$ | Bound-free Gaunt factor, 113 |
| $g_{\text{III}}(k, l), g_{\text{III}}(\nu, v), g_{\text{III}}(\nu, T)$ | Free-free Gaunt factor for hydrogen, 116 |
| $g_1(\tau, \mu)$ | Intensity weighting function, 344 |
| $G(v)$ | Induced recapture probability for electrons of velocity $v$, 109 |
| $G_1(\tau)$ | Flux-weighting function, 344 |
| $h$ | Planck's constant, 6 |
| $\hbar$ | $h/2\pi$, 92 |
| $\mathbf{h}^i$ | Source terms in Riccati transformation method, 161 |
| $H$ | Frequency-integrated Eddington flux, $\int_0^\infty H_\nu d\nu$, 35 |
| $H$ | Pressure scale height, 202 |
| $\mathscr{H}$ | Nominal Eddington flux, $\sigma T_{\text{eff}}^4/4\pi$, 178 |
| $\mathbf{H}$ | Magnetic field, 86 |
| $H_A$ | Atomic Hamiltonian, 93 |

| | |
|---|---|
| $H_0$ | Magnitude of magnetic field, 87 |
| $H_P$ | Perturber Hamiltonian, 278 |
| $H_\nu$ | Eddington flux $\frac{1}{4}F_\nu = \frac{1}{4\pi}\mathscr{F}_\nu$, 5 |
| $H_\nu^\circ, H^\circ$ | Current value of Eddington flux in Avrett-Krook method, 161 |
| $H_\nu^1, H^1$ | Perturbed value of Eddington flux in Avrett-Krook method, 161 |
| $H_n(v)$ | Expansion function in power series expression for Voigt profile, 252 |
| $H(\mu)$ | Limb-darkening function, 54 |
| $H(a, v)$ | Voigt profile, 251 |
| $H(p_i, q_i)$ | Hamiltonian operator, 93 |
| $\hat{\mathbf{i}}$ | Unit vector in $x$ direction, 87 |
| $I$ | Frequency-integrated specific intensity, $\int_0^\infty I_\nu d\nu$, 34 |
| $I_k$ | Fractional intensity of $k$th Stark component, 269 |
| $I^\circ$ | Current value of specific intensity in Avrett-Krook method, 176 |
| $I_\nu^1$ | Perturbed value of specific intensity in Avrett-Krook method, 176 |
| $I_+$ | Specific intensity moving in $+\mu$ direction, 159 |
| $I_-$ | Specific intensity moving in $-\mu$ direction, 159 |
| $I_\nu(\mathbf{r}, \mathbf{n}, t), I_\nu(\mu), I_\nu$ | Specific intensity of radiation, 1 |
| $I(\omega)$ | Power spectrum, 245 |
| $j$ | Current density, 86 |
| $\hat{\mathbf{j}}$ | Unit vector in $y$ direction, 87 |
| $j_\nu$ | Emission coefficient (per gram of stellar material), 9 |
| $j_\nu^s$ | Scattering emission coefficient, 13 |
| $j_\nu^t$ | Thermal emission, 10 |
| $J$ | Frequency-integrated mean intensity, $\int_0^\infty J_\nu d\nu$, 35 |
| $\bar{J}$ | Mean intensity averaged over line profile, $\int_0^\infty J_\nu \phi_\nu d\nu$, 440 |
| $J_\nu^\circ, J^\circ$ | Current value of mean intensity in Avrett-Krook method, 177 |
| $J_\nu^1, J^1$ | Perturbed value of mean intensity in Avrett-Krook method, 177 |
| $J_\nu(z), J_\nu(\tau), J_\nu$ | Mean intensity of radiation, 3 |

| | |
|---|---|
| $k$ | Boltzmann's constant, 26 |
| $k_c$ | Continuum opacity, 337 |
| $k_\alpha$ | Root of characteristic equation, 48 |
| $k_\nu$ | Extinction (or total opacity) coefficient, 9 |
| $\hat{\mathbf{k}}$ | Unit vector normal to reference surface, or in $z$ direction, 3 |
| $K$ | Frequency-integrated $K$-integral, $\int_0^\infty K_\nu d\nu$, 36 |
| $\mathscr{K}$ | Numerical coefficient in hydrogen cross-sections, 114 |
| $K_\nu$ | Second moment of radiation field over angle, 5 |
| $K_{1,r}$; $K_{2,r}$ | Generalized line formation kernel functions, 370 |
| $K_1(\tau)$, $K_2(\tau)$ | Line formation kernel functions, 257 |
| $K(AB)$ | Molecular dissociation coefficient, 74 |
| $l$ | Path length, 6 |
| $l$ | Azimuthal quantum number, 98 |
| $l$ | Convective mixing length, 202 |
| $l$ | Mean photon scattering length, 351 |
| $l_o$ | Line opacity assuming Voigt profile, $l_\nu = l_o H(a, v)$, $l_o = l_{ij}/\pi^{1/2}\Delta\nu_D$, 333 |
| $l_\nu$ | Line opacity coefficient, 84 |
| $l_{ij}$ | Line opacity in $i$ to $j$ transition, 333 |
| $L$ | Characteristic photon destruction length by collisions, 351 |
| $L_\alpha$ | Integration constant in discrete-ordinate solution, 49 |
| $L_{1,r}$ | Continuum kernel function, 370 |
| $L_n^l(r)$ | Associated Laguerre polynomial, 100 |
| $m$ | Electron mass, 70 |
| $m$ | Magnetic quantum number, 98 |
| $m_H$ | Mass of hydrogen atom, 77 |
| $M_\nu(z, n)$ | $n$th moment of radiation field over angle, 4 |
| $M_\alpha$ | Integration constant in discrete-ordinate solution, 50 |
| $n$ | Principal quantum number, 69 |
| $\mathbf{n}$, $\mathbf{n}'$ | Directions of light propagation, 1 |
| $N_e$ | Electron density, 10 |
| $N_i^*$ | LTE number density in state $i$, 79 |
| $N_{i,j}$ | Number density of atoms in excitation state $i$ and ionization state $j$, 68 |
| $N_j$ | Number density of atoms in ionization state $j$ summed over all excitation states, 68 |

| | |
|---|---|
| $N_i(\nu)$ | Number of atoms in state $i$ capable of absorbing radiation at frequency $\nu$, 81 |
| $p$ | Pressure, 32 |
| $p_e$ | Electron pressure, 69 |
| $p_g$ | Total gas pressure, 72 |
| $p_i$ | General momentum coordinate, 70 |
| $p_N$ | Pressure from atoms and ions, 72 |
| $p_R$ | Radiation pressure, 6 |
| $p_\nu$ | Probability of photoionization at frequency $\nu$, 109 |
| $p_\nu$ | Expansion coefficient for Planck function on $\tau_\nu$ scale, 325 |
| $p(\xi', \xi)$ | Redistribution probability in atom's rest-frame, 301 |
| $P, P'$ | Labels for test points, 5 |
| $P$ | Characteristic photon destruction length by photoionization, 352 |
| $\mathbf{P}_i$ | Radiation field representation at $i$th depth-point in difference-equation method, 166 |
| $P_{ij}$ | Rate of transitions (of all types) from state $i$ to state $j$, 137 |
| $P_{nl}(r)$ | Radial charge density, 100 |
| $P^{\|m\|}(\mu)$ | Associated Legendre polynomials, 99 |
| $P_\nu(\tau, \mu)$ | $\frac{1}{2}[I_\nu(\tau, \mu)+I_\nu(\tau, -\mu)]$, 164 |
| $\overline{P(\omega)}$ | Average radiated power at frequency $\omega$, 89 |
| $q_i$ | General space coordinate, 70 |
| $q(\tau)$ | Hopf function, 35 |
| $Q$ | Integration constant in discrete-ordinate solution, 50 |
| $Q$ | Correction factor allowing for ionization effects, 202 |
| $Q_i$ | Source term representation at $i$th depth point in difference-equation method, 166 |
| $Q_{ij}$ | Collisional cross-section in units of Bohr cross-section, $\sigma_{ij}/\pi a_0^2$, 143 |
| $r$ | Position in the stellar atmosphere, 1 |
| $r$ | Radial distance between two test points, 5 |
| $r$ | Ratio of continuous to line opacity, 369 |
| $r_0$ | Mean interparticle distance, 69 |
| $r_{ij}$ | $\langle i\|\mathbf{r}\|j\rangle$, matrix element of radius vector, 97 |
| $r_\nu(\mu)$ | Residual intensity in line at angle $\mu$ on disk, 240 |
| $R$ | Radius of star, 4 |

| | |
|---|---|
| $\mathscr{R}$ | Rydberg constant, 99 |
| $\mathbf{R}$ | Auxiliary ("reflection") matrix in Riccati transformation method, 161 |
| $R_0$ | Residual flux at line center, 388 |
| $R_\nu$ | Residual line flux, $1-A_\nu$, 240 |
| $R_{ij}$ | Radiative transition rate from state $i$ to state $j$, 81 |
| $R_{i\kappa}$ | Direct photoionization rate from state $i$ to continuum, 133 |
| $R_{\kappa i}$ | Radiative recombination rate to level $i$, 141 |
| $\mathscr{R}_{ij}$ | Net radiative transition rate from state $i$ to state $j$, 136 |
| $\mathscr{R}_{i\kappa}$ | Net radiative transition rate from state $i$ to continuum, 137 |
| $R(r)$ | Radial wave function, 98 |
| $R(x', x)$ | Angle-averaged redistribution function in dimensionless frequency units, 313 |
| $R(\nu', \nu)$ | Angle-averaged redistribution functions for scattering process, 12 |
| $R_\nu(\tau,\mu)$ | $\frac{1}{2}[I(\tau,\mu)-I(\tau, -\mu)]$, 164 |
| $R(x', \mathbf{n}'; x, \mathbf{n})$ | Redistribution functions in dimensionless frequency units, 309 |
| $R(\nu', \mathbf{n}'; \nu, \mathbf{n})$ | Redistribution functions for scattering processes, 11 |
| s | Path length, 9 |
| $\hat{\mathbf{s}}$ | Normal vector of a reference surface, 1 |
| $S$ | Frequency-integrated source function, 34 |
| $\mathbf{S}$ | Poynting's vector, 87 |
| $S$ | Line source function, 354 |
| $S_\nu$ | Source function; ratio of emissivity to absorptivity, 14 |
| $S_{ij}$ | Source function in line transition $i$ to $j$, 140 |
| $S(\alpha)$ | Normalized Stark profile, 269 |
| $S(-\mu)$ | Angular distribution of emergent intensity, 53 |
| $\overline{S(\omega)}$ | Average energy flux at frequency $\omega$, 87 |
| $S(\tau)$ | Line source function for $\epsilon=0$, 368 |
| $S_\epsilon(\tau)$ | Line source function for thermalization parameter $\epsilon$, 368 |
| $S(i,j)$ | Line strength in transition $i\rightarrow j$, 97 |
| $t$ | Time, 1 |
| $t$ | Current optical depth scale in Avrett-Krook method, 176 |

| | |
|---|---|
| $t_c$ | Self-relaxation time for electrons in a plasma, 130 |
| $t_r$ | Average recombination time, 130 |
| $T$ | Absolute thermodynamic temperature, 10 |
| $T_\nu$ | Total optical thickness of a finite slab, 17 |
| $T_e$ | Kinetic temperature of electrons, 132 |
| $T_{exc}$ | Excitation temperature, 338 |
| $T_k$ | Kinetic temperature of atoms and ions, 132 |
| $T_{eff}$ | Effective temperature, 27 |
| $T_0$ | Boundary temperature of atmosphere, 43 |
| $T_1$ | Temperature perturbation in Avrett-Krook method, 176 |
| $T(k^2)$ | Characteristic function, 49 |
| $T_0(t)$ | Current temperature distribution in Avrett-Krook method, 176 |
| $T(t, 0)$ | Time-development operator, 279 |
| $T_A(t, 0)$ | Time-development operator of atom, 281 |
| $T_P(t, 0)$ | Time-development operator of perturber, 281 |
| $\mathbf{u}$ | Dimensionless velocity, $\mathbf{v}/(2kT/m)^{1/2}$, 307 |
| $u_0$ | $h\nu_0/kT$, 327 |
| $U_\nu$ | Energy density, 5 |
| $U_j(T)$ | Partition function of ionization state $j$, 68 |
| $U(a, v)$ | Normalized Voigt function, 251 |
| $U(t, 0)$ | Time-development operator in interaction representation, 283 |
| $v$ | Velocity, 70 |
| $\bar{v}$ | Mean velocity of convective elements, 202 |
| $v$ | Frequency shift from line center measured in Doppler widths, $(\nu - \nu_0)/\Delta\nu_D$, 251 |
| $V$ | Volume, 5 |
| $V$ | Perturbation potential, 94 |
| $V_{mn}$ | $\langle\phi_m^*\|V\|\phi_n\rangle$, matrix element of perturbing potential, 94 |
| $V_{cl}(t)$ | Classical perturbation potential, 281 |
| $V'_{cl}(t)$ | Canonical transformation of classical interaction potential, 283 |
| $w$ | Doppler width for thermal velocity, $\nu_0(2kT/\nu m)^{1/2}/c$, 307 |
| $w_1, w_2$ | Relative probabilities of line and continuum bands in picket fence model, 423 |
| $W_\nu, W_\lambda$ | Equivalent width (Å or $m$ Å), 240 |
| $W(r)$ | Nearest-neighbor probability distribution, 236 |

| | |
|---|---|
| $W_\lambda(\mu)$ | Equivalent width at angle $\mu$ on disk (Å or $m$ Å), 240 |
| $W(\xi)$ | Line-of-sight velocity distribution function, 250 |
| $W(\mathbf{\Phi})$, $W(\Phi)$, $W(\beta)$ | Holtsmark distribution function, 265 |
| $W(\beta, \delta)$ | Field distribution including shielding effects, 273 |
| $x_\alpha$ | $1/k_\alpha$, where $k_\alpha$ is a characteristic root, 49 |
| $\bar{x}$ | Maximum (absolute) of incident and scattered frequencies in dimensionless units, 318 |
| $\underline{x}$ | Minimum of incident and scattered frequencies in dimensionless units, 485 |
| $X_\alpha$ | $x_\alpha^2 = 1/k_\alpha^2$, where $k_\alpha$ is a characteristic root, 49 |
| $X_H$ | Fraction of hydrogen that is ionized, 73 |
| $X_m$ | Fraction of heavy elements ("metals") that are ionized, 73 |
| $X_0$ | $u_0/(1 - e^{-u_0})$, 327 |
| $X_\tau[f(t)]$ | $K$-integral operator, 22 |
| $Y$ | Ratio of helium abundance to hydrogen abundance (by *number*), 122 |
| $Y_{ij}$ | Net collisional bracket, 391 |
| $Y_l^m(\theta, \phi)$ | Spherical harmonic, 99 |
| $z$ | Depth in atmosphere relative to some reference level, 3 |
| $Z$ | Total geometrical thickness of a finite slab, 17 |
| $Z$ | Nuclear charge; ionic charge, 69 |
| $Z_i$ | Charge on ion, 271 |
| $Z_{ji}$ | Net radiative bracket between levels $j$ and $i$, 140 |
| $\alpha$ | Stark shift in Å per unit normal field strength, $\Delta\lambda/F_0$, 269 |
| $\alpha_i$ | Abundance of element $i$ relative to hydrogen, 72 |
| $\alpha_\nu$ | Energy absorption cross-section per atom, 112 |
| $\alpha(t)$ | Atomic wave function, 281 |
| $\beta$ | Field strength in units of normal field strength, $F/F_0$, 264 |
| $\beta_\nu$ | Parameter measuring departure of $\kappa_\nu$ from mean value, 64 |
| $\gamma$ | Classical damping constant, 91 |
| $\gamma$ | Ratio of specific heats for a perfect gas, 199 |
| $\gamma$ | Convective efficiency parameter, 204 |
| $\gamma_\nu$ | Generalized noncoherent scattering coefficient in non-LTE source functions, 223 |
| $\Gamma$ | Reciprocal lifetime of excited state, 247 |

| | |
|---|---|
| $\Gamma_2$ | Ratio of specific heats for a nonperfect gas, 200 |
| $\Gamma_c$ | Collisional damping constant, 255 |
| $\Gamma_L$, $\Gamma_U$ | Reciprocal mean lifetime for lower (upper) states of a transition, 248 |
| $\Gamma_{\text{res}}$ | Resonance damping parameter, 294 |
| $\Gamma_R$ | Reciprocal mean radiative lifetime; radiative damping constant, 255 |
| $\Gamma_w$ | Collisional damping constant in Weisskopf theory, 257 |
| $\Gamma_6$ | Van der Waals damping constant, 298 |
| $\mathbf{\Gamma}^{ij}$ | Coupling matrices in Riccati transformation method, 161 |
| $\Gamma_{ij}(T)$ | Secondary temperature-dependent factor for collision rates, 143 |
| $\delta$ | Number of perturbers in Debye sphere, 273 |
| $\delta$, $\delta_C$, $\delta_R$, $\delta_L$, $\delta_U$ | Reduced damping widths, $\Gamma/2$ for circular frequency units, $\Gamma/2\pi$ for ordinary frequency units, 248 |
| $\delta$, $\delta_\nu$ | Radiation field parameter in schematic Lyman continuum problem, 226 |
| $\delta a_\nu$ | Change in absorption coefficient due to non-LTE effects, 217 |
| $\delta\kappa_\nu$ | Change in absorption coefficient due to non-LTE effects, ($\equiv \delta a_\nu$), 218 |
| $\boldsymbol{\delta\psi}$ | Perturbation of solution vector, 410 |
| $\delta(\nu - \nu')$ | Dirac delta function, 12 |
| $\Delta T(\tau)$ | Temperature perturbation at optical depth $\tau$, 170 |
| $\Delta\theta$ | Difference in $\theta_{\text{exc}}$ between sun and star, 342 |
| $\Delta\lambda$ | Wavelength shift, 269 |
| $\Delta\lambda_C$ | Classical damping width in wavelength units, 247 |
| $\Delta\nu$ | Frequency shift because of Doppler effect, 250 |
| $\Delta\nu_D$ | Doppler width in frequency units, 250 |
| $\Delta\tau_{d+\frac{1}{2}}$ | Optical depth increment between depth points $d$ and $d+1$, 410 |
| $\Delta\chi$ | Lowering of ionization potential, 69 |
| $\Delta\omega_0$ | Line shift in Lindholm theory, 261 |
| $\Delta\omega_0$ | Normal phase shift, 263 |
| $\Delta\omega_w$ | Frequency shift caused by a perturber at Weisskopf radius, 276 |
| $\Delta\omega_g$ | Frequency shift (from line center) between impact and statistical regimes, 276 |

|  |  |
|---|---|
| $\Delta\omega(t)$ | Frequency shift of a line due to a perturber, 255 |
| $\Delta(n', n)$ | Function appearing in Gaunt factors, 105 |
| $\epsilon$ | Fraction of emission that is thermal (classical theory), 324 |
| $\epsilon, \epsilon'$ | Collisional thermalization parameter, 355 |
| $\epsilon_0$ | Electric permittivity of vacuum, 86 |
| $\epsilon_\nu$ | Generalized thermal emission parameter in non-LTE source functions, 223 |
| $\zeta_\nu$ | $\sigma_\nu/(\kappa^* + \delta\kappa_\nu + \sigma_\nu)$, 218 |
| $\eta'$ | Phase change in time $ds$, 259 |
| $\eta$ | Photoionization coupling parameter, 378 |
| $\eta_0$ | Critical phase shift in Weisskopf approximation, 256 |
| $\eta_0$ | $l_0/(\kappa + \sigma)$, 333 |
| $\eta_\nu$ | Ratio of line opacity to continuum opacity, $l_\nu/(\kappa + \sigma)$, 324 |
| $\eta, \eta(\infty)$ | Total phase shift in an impact, 257 |
| $\eta(t)$ | Instantaneous phase shift induced by perturbations, 257 |
| $\eta(t, s)$ | Change in phase occurring in interval $s$ from time $t$, 259 |
| $\theta$ | Polar angle between radiation direction and normal to atmospheric layers, 1 |
| $\theta_{\text{eff}}$ | $5040/T_{\text{eff}}$, 191 |
| $\theta_{\text{exc}}$ | Reciprocal excitation temperature, $5040/T_{\text{exc}}$, 338 |
| $\Theta$ | Angle between incident and scattered pencils of radiation, 13 |
| $\Theta(\theta)$ | Polar-angle wave function, 98 |
| $\kappa_\nu, \kappa$ | Absorption coefficient (per gram of stellar material), corrected for stimulated emissions, 8 |
| $\kappa_\nu^*$ | LTE value of absorption coefficient (corrected for stimulated emissions), 218 |
| $\bar{\kappa}_F$ | Flux-weighted mean opacity, 37 |
| $\bar{\kappa}_J$ | Absorption-mean opacity, 40 |
| $\bar{\kappa}_P$ | Planck-mean opacity, 40 |
| $\bar{\kappa}_R$ | Rosseland-mean opacity, 39 |
| $\lambda$ | Wavelength (Å), 136 |
| $\lambda$ | Separation parameter in Avrett-Krook method, 276 |
| $\lambda\!\!\!^-$ | De Broglie wavelength, 276 |
| $\lambda_\nu$ | $\kappa_\nu/(\kappa_\nu + \sigma_\nu) = (1 - \rho_\nu)$, 153 |

| | |
|---|---|
| $\lambda_\nu$ | $[(1-\rho)+\epsilon\eta_\nu]/(1+\eta_\nu)$, parameter in classical line formation theory, 324 |
| $\Lambda$ | thermalization length, 361 |
| $\Lambda_\tau[f(t)]$ | Lambda operator; mean intensity operator, 21 |
| $\mu$ | $\cos\theta$ (cosine of polar angle of radiation flow), 3 |
| $\mu$ | Mean molecular weight, 77 |
| $\mu'$ | Number of grams of stellar material per gram of hydrogen, 78 |
| $\mu_i$ | Angle point in discrete-ordinate approximation, 47 |
| $\mu_0$ | Magnetic permeability of vacuum, 86 |
| $\mu_H$ | Reduced mass of hydrogen atom, 98 |
| $\nu, \nu'$ | Frequency, 1 |
| $\bar{\nu}$ | Maximum of incident and scattered frequencies, 315 |
| $\underline{\nu}$ | Mimimum of incident and scattered frequencies, 315 |
| $\nu_n$ | Threshold frequency for ionization from *n*th state of hydrogen, 114 |
| $\nu_0$ | Line center frequency, 250 |
| $\nu_{ij}$ | Frequency of transition $i \rightarrow j$, 82 |
| $\xi$ | Line-of-sight velocity, 250 |
| $\xi, \xi'$ | Photon frequencies in atom's rest-frame, 301 |
| $\xi$ | Mean coupling constant in line formation, 371 |
| $\xi_0$ | Total random line-of-sight velocity, 337 |
| $\xi_{\text{thermal}}$ | Most probable thermal velocity along line of sight, 338 |
| $\xi_{\text{turb}}$ | "Turbulent" velocity along line of sight, 338 |
| $\xi_\nu$ | $\kappa_\nu^*/(\kappa_\nu^* + \delta\kappa_\nu + \sigma_\nu)$, 218 |
| $\xi_\nu$ | Non-LTE source function parameter, 231 |
| $\pi(t)$ | Perturber wave function, 281 |
| $\rho$ | Mass density, 9 |
| $\rho$ | Charge density, 86 |
| $\rho_D$ | Debye shielding distance as an impact cutoff, 291 |
| $\rho_J$ | Density matrix element, 278 |
| $\rho_L$ | Lewis cutoff, 291 |
| $\rho_0$ | Effective impact parameter for collisions, 255 |
| $\rho_\omega$ | Weisskopf radius, 257 |
| $\rho_\nu$ | $\sigma_\nu/(\kappa_\nu + \sigma_\nu) = (1-\lambda_\nu)$, 151 |
| $\sigma$ | Stefan-Boltzmann constant, 26 |
| $\sigma_I$ | Imaginary part of collision integral, 260 |

| | |
|---|---|
| $\sigma_R$ | Real part of collision integral, 260 |
| $\sigma_T$ | Thomson scattering cross-section for free electrons, 125 |
| $\sigma_\nu$, $\sigma$ | Scattering coefficient, 9 |
| $\sigma_{ij}(v)$ | Cross-section for transitions from state *i* to state *j* induced by collisions with electrons of velocity *v*, 141 |
| $\tau'$ | Volume element, 87 |
| $\tau$ | Mean time between collisions, 255 |
| $\bar{\tau}$ | Mean optical depth scale, 37 |
| $\tau_1$ | Optical depth perturbation in Avrett-Krook method, 176 |
| $\tau_c$ | Line center optical depth, 345 |
| $\tau_c$ | Optical thickness of convective element, 205 |
| $\tau_l$ | Optical depth in line, 343 |
| $\tau_s$ | Effective collision time, 257 |
| $\tau_\nu$, $\tau$ | Optical depth, 15 |
| $\tau(\mathbf{r}_j)$ | Probability that position of particle *j* lies on range $(\mathbf{r}_j, \mathbf{r}_j + \mathbf{dr}_j)$, 264 |
| $\phi$ | Azimuthal angle of radiation propagation around normal to atmospheric layers, 3 |
| $\phi$ | Scalar potential, 86 |
| $\phi(s)$ | Reduced autocorrelation function, 258 |
| $\phi(\nu)$, $\phi(\nu')$ | Absorption or scattering profiles, 11 |
| $\phi_j(\mathbf{r})$ | Time-independent wave functions, 93 |
| $\phi_j(\mathbf{r}_j)$ | Field because of perturber *j*, 264 |
| $\phi_i(T)$ | Saha-Boltzmann factor, 79 |
| $\phi_\nu(\mu)$ | Limb-darkening function, 208 |
| $\Phi$ | Total perturbing field, 264 |
| $\mathbf{\Phi}_0$ | Nominal field, 265 |
| $\Phi_\nu$ | Absorption rate cross-section, $4\pi\alpha_\nu/h\nu$, 223 |
| $\Phi(s)$ | Autocorrelation function, 245 |
| $\Phi(\phi)$ | Azimuthal-angle wave function, 98 |
| $\Phi_\tau[f(t)]$ | Phi operator; flux operator, 21 |
| $\chi_i$ | Excitation potential of state *i* relative to ground level, 68 |
| $\chi_{\text{ion}}$, $\chi_I$ | Ionization potential, 68 |
| $\chi_\nu$, $\chi$ | Opacity ratio at frequency $\nu$ to standard opacity, 163 |

| | |
|---|---|
| $\boldsymbol{\psi}$ | Auxiliary radiation field vector in Riccati transformation method, 161 |
| $\boldsymbol{\psi}_d$ | Solution vector at depth point $d$ in complete linearization method, 409 |
| $\boldsymbol{\psi}_i$ | Auxiliary vector in difference-equation method, 166 |
| $\psi_p$ | Numerical factor in calculation of $\eta(\infty)$, 257 |
| $\psi_\nu$ | Non-LTE source function parameter, 230 |
| $\psi(\nu)$ | Emission profile, 11 |
| $\psi(\mathbf{r}_1, \ldots, \mathbf{r}_N)$ | Wave function of $N$ electron atom, 92 |
| $\Psi(\tau, a)$ | Saturation function, 345 |
| $\omega$ | Solid angle, 1 |
| $\omega$ | Circular frequency, 86 |
| $\omega$ | $\chi/\rho$, 410 |
| $\omega_0$ | Resonant frequency for an oscillator, 90 |
| $\omega_{mn}$ | Circular frequency of transition $m \to n$, 95 |
| $\Omega_{ij}(T)$ | Collisional rate between states $i$ and $j$, per atom, per electron, averaged over Maxwellian velocity distribution at temperature $T$, 142 |
| $\nabla, \nabla_R, \nabla_A, \nabla_E$ | Logarithmic temperature-pressure gradient, 200 |
| $\odot$ | Sun symbol, denotes value appropriate in solar atmosphere, 340 |
| $\oint$ | Integral over all solid angles, 3 |

# Index